Conversion Factors

Quantity	To convert from:	To:	Multiply by:
Moment of force	lbf-ft	N·m	1.355818
	kgf·cm	N·m	0.0980665
Power	hp	kW	0.7456999
	ft-lb/sec	W	1.356
	kcal/h	W	1.162
	Btu/h	W	0.2931
Pressure	atm	kPa	101.325
	lbf/in^2	kPa	6.894757
	bar	kPa	100
	torr	kPa	0.133322
	inHg (60°F)	kPa	0.37685
	inH$_2$O (60°F)	kPa	0.24884
	mmHg	kPa	0.133322
	mmH$_2$O (60°F)	kPa	0.009797
Specific fuel consumption	lb/hp-h	g/MJ	168.9659
Surface tension	dyne/cm	mN/m	1
Velocity	mph	km/h	1.609344
Viscosity, dynamic	centipoise	mPa·s	1
Viscosity, kinematic	centistoke	mm^2/s	1
Volume	in^3	L	0.01638706
	gal (US)	L	3.785412
	gal (UK)	L	4.5459012
	gal (UK)	gal (US)	1.2009
	m^3	L	1000
	bbl	L	158.9873
	bbl	gal (US)	42 (nominal)
	bbl	gal (UK)	35 (nominal)

Temperature
$$°C \times 9/5 + 32 = °F$$
$$(°F - 32) \times 5/9 = °C$$

Automotive Fuels Reference Book

Second Edition

Keith Owen and Trevor Coley

with a chapter contributed by Christopher S. Weaver

Published by:
Society of Automotive Engineers, Inc.
400 Commonwealth Drive
Warrendale, PA 15096-0001
U.S.A.
Phone: (412) 776-4841
Fax: (412) 776-5760

Publisher's Note:
The *Automotive Fuels Reference Book* is an extensively revised edition of the *Automotive Fuels Handbook* published by SAE in 1990. The change of title has been made to avoid confusion with the annually produced *SAE Handbook.*

Library of Congress
Cataloging-in-Publication Data

Owen, K. (Keith)
    Automotive fuels reference book / Keith Owen and Trevor
Coley ; with a chapter contributed by Christopher S. Weaver. --
2nd ed.
        p. cm.
    Rev. ed. of: Automotive fuels handbook. c1990.
    Includes bibliographical references (p. -) and index.
    ISBN 1-56091-589-7
    1. Motor fuels. I. Coley, Trevor, 1925- . II. Weaver,
Christopher S. III. Owen, K. (Keith). Automotive fuels
handbook. IV. Title.
TP343.089 1995
629.25'38--dc20   95-3199
                                                                    CIP

TP
343
084
1995

First edition published in 1990.

SAE Order No. R-151

Foreword

In this book we have tried to include those aspects of automotive fuels that are most likely to be needed by automobile engineers, oil technologists, commercial and academic research establishments, fuel additive manufacturers, and personnel from governmental agencies dealing with the control and specification of these fuels. Because such a wide range of people of varying disciplines are likely to be interested, we have included, as Appendix 1, a brief outline of fuel chemistry for those readers who are unfamiliar with the chemical symbols and equations used, even though these have been kept to a minimum. A glossary of terms has also been included, although most terms will have been explained at some point in the text.

Automotive fuels are of interest worldwide and for this reason this book has been written to cover, as far as possible, the global situation rather than the position in any single country. Thus, although leaded fuels are now of limited interest in many countries, their use will continue in some parts of the world well into the next century and so we thought it necessary to include a full discussion on such fuels.

By far the most common of the currently used automotive fuels are gasoline and diesel, so it is inevitable that most of the book is taken up by these fuels. Alternative fuels such as LPG or ethanol are, however, very important in some countries and there are indications that other alternative fuels could become important in the future. In particular, methanol looks as if it eventually could become the fuel of the future and is discussed in some detail, along with ethanol, in Chapter 19. A chapter on racing fuels has also been added because this is a subject of some interest on which comparatively little has been written.

Without government intervention, the use of alternative fuels in diesel engines appears to be a fairly remote possibility. Environmental pressures are leading to ever-tighter legislation against diesel exhaust emissions, but the major variations in levels of particulates and gases are associated with the engine rather than the fuel. However, although the costs of modifications to engines, vehicles and distribution systems are substantial, methanol is being evaluated as a "clean" alternative diesel fuel by several bus fleet operators in the U.S., Japan and Canada.

Because vehicle emission regulations are indirectly related to automotive fuels, we have incorporated as an Appendix a summary of current emissions legislation as compiled by CONCAWE (the oil companies' European Organization for Environmental and Health Protection) and included with their kind permission. Legislation is changing at a rapid rate and so we have generally restricted our comments in the text to the general principles governing the formation of these emissions and particularly how they are influenced by fuel quality.

Chapter 15, which covers diesel fuel low-temperature characteristics, includes a section on the measurement of low-temperature performance. Direct comparison of results obtained by different organizations is often difficult because of variations in test techniques. This measurement procedure, which includes testing in the field as well as in climatic chambers, provides guidance on vehicle instrumentation, cooling phases, driving patterns, etc. We hope that it will be accepted as the basis for an industry standard so that future test data will be more readily comparable with each other.

Regarding units, we have tried to stick to the SI system although there are some instances where alternative units are used. In these cases the units are widely accepted and commonly used.

Finally, we would like to acknowledge with grateful thanks the assistance of many friends and colleagues who read various parts of the text and who improved the book with their helpful comments. In addition, we would like to thank Lucas Diesel Systems for their general assistance with the chapters concerning diesel fuels. We are extremely grateful to the Esso/Exxon Chemicals Research Centre at Abingdon in the U.K., who made available to us the use of their Information Centre, and we thank particularly Miss Andrea Strafford and her colleagues who carried out for us many computer-based literature searches and who provided copies of papers, and without whose help we would never have been able to complete this task.

Preface to Second Edition

The pace at which changes are taking place in the domain of automotive fuels has necessitated a number of revisions to bring the Handbook up to date.

Although the first edition was published as recently as October 1990, there has been a significant change of emphasis with the wider introduction of reformulated fuels for both spark-ignition and compression-ignition engines, and in the types of additive for primary and secondary treatment of diesel fuels.

The other major area of change—and growth—is in environmental legislation against noxious and undesirable exhaust emissions from road vehicles. We are grateful to CONCAWE, the oil companies' European organization for environment, health and safety, for permission to incorporate their latest update on the situation in the appendices.

A new chapter has been contributed by Mr. C. S. Weaver to give more attention to the growing interest in the use of liquefied and compressed natural gas (LNG and CNG) and liquefied petroleum gas (LPG) as more environmentally friendly automotive fuels.

In view of concerns about automotive fuels and emissions, a section giving the health and environmental effects of gasolines and diesel fuels has been added.

The opportunity has been taken to increase the usefulness of the Handbook (and to resolve some uncertainties about units commonly used in the petroleum industry) by including Tables of Conversion and Heating Values. Inevitably, it has also been necessary to extend the list of abbreviations and acronyms which proliferate, often without explanation, in many technical publications.

Restructuring in Europe is also bringing in new names and abbreviations. At the end of 1993, the Single Europe Act came into effect, replacing the European Community (EC) with the European Union (EU). In May 1994, approval was given by the European Parliament to applications by four member countries of the European Free Trade Association (EFTA) to join the EU. The application by Norway was withdrawn after a referendum rejected joining but the other three countries, Austria, Finland, and Sweden, formally became members on January 1, 1995. Further enlargement and designation changes may occur if Eastern European countries are admitted but efforts will be made to minimize confusion in a situation which is still evolving.

The authors are appreciative of the advice and guidance provided by Don Goodsell, Ryozo Kato, former Esso/Exxon colleagues and other contributors during the preparation of this revision of the Handbook.

Keith Owen and *Trevor Coley*
January 1995

Contents

Chapter 1

Automotive Fuels and their Specification

Automotive fuels are probably the most important products manufactured and marketed by oil companies because such a large proportion (between 30 and 70%) of the crude oil run in a refinery is converted into gasoline and diesel fuel. Figure 1.1 shows how the sale of gasoline has progressed over a twenty year period in a number of countries. The depressive impact of increased crude oil prices in the

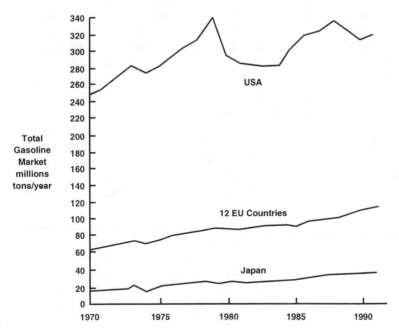

Fig. 1.1. Total gasoline sales over period 1970 - 1991.
(source: Associated Octel)

1

1970s can be readily seen. The low rate of growth that occurred in the early 1980s, followed by an upturn in consumption as crude oil prices stabilized, is also clearly illustrated.

Because of the high volumes involved and the competitive marketing situation in most countries, these products determine, to a very large extent, the profit or loss situation for an oil company. The high fixed costs involved in manufacture and distribution mean that changes in market share can have a dramatic influence on profitability. For this reason it is important for personnel involved with the production and marketing of these fuels to understand the quality aspects that can influence market share. Of course there are many other factors that affect the profitability of these fuels that are outside the control of oil companies, such as the price of crude oil as mentioned above and interest rates, but here we are concerned mainly with quality aspects that can dictate to some extent the performance of the vehicle and can be controlled by the fuel manufacturer.

The main aims of an automotive fuel technologist are to ensure that the product being manufactured meets required legal specifications, and that any "in-house" specifications imposed by the manufacturing or marketing company are realistic and worthwhile. National or other legally enforceable specifications represent the minimum quality that must be supplied and it is implicit that engine designers should ensure that their vehicles will run satisfactorily on such a quality. Appendix 8 summarizes national specifications (as of 1994) for gasoline and diesel fuel for a number of countries.[1]

Not all the fuel parameters that influence vehicle performance are fully defined in national specifications. In order for vehicle and engine manufacturers to ensure that their products meet legislative requirements in terms of emissions, power and fuel economy, and perform well on the range of fuels marketed, special legislative and reference fuels are available.[2] Legislative fuels are used by the motor industry for certification or homologation and generally represent the average quality to be found in the marketplace. Reference fuels can be used for engine development and these represent qualities close to the worst case that may be marketed for particular properties such as volatility or octane. Appendix 9 summarizes the specifications for both legislative and reference fuels used in Europe for testing with spark ignition and compression ignition engines.

Internal or in-house specifications are more severe in many respects than national specifications and usually include additional controls. They are used to define the quality a particular oil company or blender is producing. Differences from the national or standard specification exist in order to meet a competitive situation, capture a greater market share, meet a seasonal need, or overcome an existing field problem. Such specifications will include controls on the type and concentrations of additives to be used, and dates and levels at which seasonal specifications are to be introduced.

Close cooperation is important between fuel producers and motor manufacturers in the face of changes and regulatory pressures occurring in both industries. Joint committees have been set up in most countries to ensure that new developments, whether in legislation or engines or fuels, can be accommodated. In this respect, fuel quality and engine design is a chicken and egg situation. Designers of engines need to know what fuel will be available in a particular market, but equally, oil companies need to know the fuel requirements of vehicles on the road in order to maximize the number of vehicles, and hence customers, that they satisfy. Governmental bodies also need to be aware of the limited scope for modifying fuel properties and the influence such changes can have on vehicle performance, since they are involved in setting national fuel specifications.

One other group with a close interest in fuel quality comprises the fuel additive manufacturers. Additives can reduce manufacturing costs, enhance quality and provide benefits to the consumer that are recognizable — although sometimes a little help is needed from advertisements — and so they can significantly influence market share and profitability.

Gasoline and diesel fuel qualities have never been static, as is clear from the history of these fuels. There have been many different pressures causing the changes. Sometimes it has been the cost of crude oil or the need to protect our environment. At other times improvements in refinery processes have given rise to the potential to improve, and from time to time changes in vehicle design have led to the need to modify fuel composition. Occasionally the changes are for purely commercial reasons in that they reduce manufacturing costs or increase sales potential.

Although gasoline and diesel fuel are by far the most important automotive fuels when considering the world as a whole, there are areas where other fuels are used extensively. Examples of alternative fuels are the use of ethanol in Brazil, LPG in many countries such as Italy and The Netherlands, and CNG (compressed natural gas) in Canada and elsewhere. In Sweden, development of a low-emission CNG engine primarily for city buses was undertaken by Volvo Bus Corporation for the city of Gothenburg.[3] The favoring of such fuels almost always requires governments to allow beneficial taxation rates to encourage their use.

Methanol could possibly become an important fuel of the future because of its clean-burning characteristics, although its lower heat content means that larger fuel tanks and/or a lower mileage range will be necessary, as is discussed in detail in Chapter 21.

There have been moves in some countries towards the greater use of the diesel engine in cars to take advantage of its improved fuel economy over that of gasoline-powered vehicles. This advantage is mainly achieved at low throttle openings so that vehicles such as taxis, with high mileage under city driving conditions, reap the biggest benefit. Part of the fuel economy benefit is due to the higher density of diesel fuel and, hence, its greater energy content. Other factors giving the diesel engine this advantage are that it has lower throttling losses, and the fuel is burned at a much leaner air-fuel ratio than in the gasoline engine. The disadvantage of higher initial cost for a diesel-engined passenger car over the gasoline version has been virtually eliminated in Europe, enhancing the existing benefits of lower maintenance costs and longer life expectancy. Emissions of carbon monoxide and hydrocarbons are generally lower for diesel engines, although particulate emissions are less satisfactory.

References

1. CONCAWE, "Motor vehicle emission regulations and fuel specifications — 1994 update," Report 4/94, CONCAWE, 1994.

2. J.K. Pearson and M.J. Hawkins, "Fuels for the automotive industry," Paper C304/86, International Conference on Petroleum Based Fuels and Automotive Applications, Institution of Mechanical Engineers, London, November 1986.

3. H. Larsson and L. Karlsson, "The Volvo THG103 low-emission CNG engine," Volvo Technical Report No. 1, 1992.

Chapter 2

A History of Gasoline and Diesel Fuel Development

2.1 Gasoline

2.1.1 The Evolution of the Gasoline Engine

The first practical working internal combustion engines appeared in the middle of the 19th Century. A British engineer, James Robson, built a double-acting, atmospheric, spark-ignited gas engine in 1857, and three were in service before 1860, the year in which a Frenchman, Jean Etienne Lenoir, patented his own design of double-acting, spark-ignited gas engine, probably also operating at atmospheric pressure. There is no record of a patent for Robson's 1857 invention but he later patented a two-stroke, compression gas engine with spark ignition.

Although earlier engines built or designed around the beginning of that century had not been successful, a number of significant and innovative concepts were introduced.

An English patent granted in 1791 to John Barber had all the elements of a modern gas turbine and Barber was the first to consider the use of manufactured coal-gas as fuel.[1] Robert Street's English patent of 1794 is the first record of a mixture of a gaseous fuel (vaporized turpentine) and air being burned in a reciprocating internal-combustion engine (ICE).

The Swiss engineer Isaac de Rivaz (1752-1828)[1] is credited with building the first vehicle powered by an internal-combustion engine, using his 1813 design of spark-ignited atmospheric gas engine. Its limited range of travel prompted Rivaz to put forward the need for a network of gas generators to enable the vehicle's leather gas storage bag to be refueled at regular intervals.

The earliest carburetor for liquid fuel (alcohol or turpentine) for an internal combustion engine was the "preparation vessel" of a portable version of the gas engine patented in 1826 by the American engineer Samuel Morey (1762-1843).[1] Morey also visualized the application of the internal-combustion engine to road and rail transport.

The limitation of virtually all early engines was the absence of compression of the fuel-air mixture, the need for which was indicated by Carnot[1, 2] in 1824 but which was only applied effectively 40 years later in the Otto and Diesel engines.

In 1861, Dr. Nikolaus August Otto (1832-1891) conceived the idea of building a compression engine with controlled ignition, working on a four-stroke cycle: a suction stroke as the piston moves down to draw in an explosive charge; an upward stroke to compress the charge; a power stroke as the piston is forced down by the burning gases; and an upward exhaust stroke to expel the burnt gases. Ignition was controlled by introducing a flame, a heated tube or an electric spark into the compression space near the end of the compression stroke. The revolutionary concept of the "Otto cycle" of operation was successfully applied in 1876. The majority of the world's reciprocating internal combustion engines work on the same four-stroke principle and the term "Otto engine" is still used to describe the spark ignition engine.

Together with Eugen Langen, Otto founded the world's first engine factory in 1864 at Deutz, Germany, near Cologne, which is now the home of the engine builder Klockner-Humboldt-Deutz. Their museum includes the first atmospheric gas engine developed by Otto and Langen and one of the first four-stroke gas engines built in 1878, both still operable!

Exploitation of the early stationary Otto engines was limited by availability of a gas supply, but the use of the more convenient liquid petroleum spirit as fuel made the engines applicable for transport purposes. Early pioneers in Germany were Gottlieb Daimler and Karl Benz who built their first single-cylinder engined motor car in 1885 (and whose names are still eminent in the world of motoring), but interest soon spread to other European countries and to the U.S.

Since that time, more than a century ago, development of the gasoline engine has progressed steadily, from the modest single-cylinder unit of the 1880s with its rudimentary carburetor and ignition device, to the powerful multicylindered, turbocharged models of today with their sophisticated management systems for fuel injection, ignition timing and emissions control. The next section describes how the fuel itself has evolved to meet the progressively changing needs of the automobile engine and, more recently, to help preserve the environment in which we live.

2.1.2 Gasoline Development

From the time that the first horseless carriages appeared on the road at the end of the 19th Century, fuels and vehicles have evolved together. In the beginning, the only fuels available were the lighter fractions from the distillation of crude oils and shale oils. These boiled within the range of about 50 to 200°C, much the same as today's gasolines, but with a very poor octane quality. Vehicles had to have low compression ratios at that time to help them run without detonation, which could cause severe damage to the engine and restrict the power output obtainable.

The 1914/18 war changed both the demand and the quality requirements of gasoline, or motor spirit, as it was generally called at that time. The relatively high power-weight ratio of the internal combustion engine was quickly appreciated in terms of its potential for warfare, particularly as the Wright Brothers had already made their first manned flight in 1903. However, the restriction of having to use low compression ratios had become very apparent since the only way of achieving improved fuel quality at that time was to use crude oils rich in aromatics such as those from Borneo or the Dutch East Indies rather than the more paraffinic crudes from Pennsylvania or Oklahoma.

The importance of the composition of gasoline relative to its performance had become very clear by the end of the war when it became desirable to improve mechanical efficiency by increasing compression ratio. Work then started both in the U.S. with Midgely and Boyd at the General Motors Research laboratory and in Great Britain by Ricardo to establish the factors that prevented fuels from burning smoothly without detonating in an engine. It was soon established by Ricardo that the hydrocarbons with most resistance to knock were

*Fig. 2.1. The first gasoline pumps appeared soon after the First World War
(source: Esso Petroleum Company Ltd.).*

aromatics and that those with the least resistance were the normal
paraffins. The use of unrefined benzole (a mixture of benzene, toluene and xylenes produced by the distillation of coal tar) in motor spirit
was found to greatly improve performance at higher compression
ratios. The benefits of alcohols as fuel components that would prevent detonation were also discovered at that time.

Midgely, in parallel work, had developed a similar test engine to the
one used by Ricardo, but with the important addition of a device
known as the "bouncing pin" for detecting the onset of detonation.
This engine was ultimately adopted internationally as the standard for
measuring the antiknock quality of fuel, and was called the CFR engine after the Cooperative Fuel Research Committee formed in 1921.

One of Midgely's findings using this engine was that branch chain
paraffins, unlike straight chain paraffins, had excellent resistance to
knock. This encouraged his team to take on a large-scale systematic
investigation into chemical additives that would suppress knock.

They screened thousands of compounds and, at the end of 1921, selected tetraethyl lead (TEL) as having the greatest effectiveness with the best potential for commercial development. Problems concerning lead-oxide deposits in the combustion chamber were overcome by combining with the lead compound some ethyl halide which acts as a scavenger by volatilizing the lead in the combustion chamber.

Since the late 1920s, lead compounds have been the most important method of achieving required octane levels, and it is only since the 1970s and '80s that the use of lead has diminished due to environmental considerations. It is interesting that concerns regarding the toxicity of lead were widespread when it was first introduced, and an investigation was started by the Surgeon General of the U.S. Public Health Service in May 1925 to find out if there was any public health danger by the use of gasoline containing TEL. The sales of gasoline containing the additive were suspended until the results of this investigation were known. In January 1926, the Surgeon General decided that there were no good grounds for prohibiting the use of lead antiknock compounds, provided certain safety regulations were observed, and the sale of leaded gasolines restarted in the summer of 1926. Even so, the method of incorporating the lead into gasoline was rather questionable. Dunstan and Card,[3] writing about the work of Midgely and Boyd in 1924, gave the following description:

> "The antiknock is sold through the medium of specially fitted petrol pumps. These pumps are normal except that a small apparatus, called the ethylizer, is attached. If ethylized gasoline is required, the apparatus can be put in connection with the pump by operating a small lever, and every gallon of gasoline delivered will automatically have added to it the requisite quantity of lead tetraethyl and ethylene dibromide. Fresh supplies of the ethyl are supplied in sealed metal bottles containing about a quart, which are inverted and dropped in the top of the ethylizer, a pointed tube puncturing the seal and allowing the antiknock to run out as required."

In Great Britain the sale of gasoline containing the lead antiknock compound started in 1928, but its use was severely hampered by adverse press comment. Eventually a Government inquiry was held which cleared the product from a public health viewpoint. These concerns delayed the general acceptance of lead, not only in the U.K. but in many other countries, for several years.

In 1929, the octane scale was established in which two hydrocarbons were selected as references: one that tended to knock in an engine under almost all conditions (n-heptane) and the other having a much higher knock resistance than any known gasoline at that time (isooctane). The antiknock performance of a gasoline is compared with a blend of these two compounds which had been given arbitrary values of 0 and 100, respectively. This scale is still the standard method for defining the antiknock quality of all marketed gasolines throughout the world.

Considerable advances were also made in refining processes during this period between the wars. Although it had been found as early as 1861 that heating the heavier fractions of crude oil caused so-called "cracking reactions" to take place, leading to the formation of lighter hydrocarbons, the first commercial cracking process was not developed until the end of the 1920s. Cracking gives rise to the formation of olefinic compounds which, although they have relatively good antiknock properties, are also quite easily oxidized to form gums. These gums form both during storage, causing gels and sludges in the bottom of tanks, and also during engine operation, giving rise to deposits and causing piston ring and valve sticking. Additives or additional processing were found to be necessary to overcome these difficulties. Another problem was that these cracked streams often had an extremely unpleasant odor.

Cracking processes had become necessary in order to increase the yield of gasoline from crude oil beyond that obtainable from simple distillation. A number of variants of the first cracking process soon became available, all of which involved thermally cracking the larger molecules in the crude oil to smaller molecules. Continuous fractionation was also developed during this period both for atmospheric and vacuum distillation. During the 1930s, processes to convert the gases formed during cracking into gasoline components were introduced and included polymerization and alkylation.

These process improvements not only improved the yield of gasoline from crude oil, which was very necessary since demand had risen dramatically, but also the quality. By the mid-1930s octane values were around 70, enough to allow compression ratios to increase to about 5.5:1.

*Fig. 2.2. Winner of the King's Cup air race in 1926 being fueled with cans
and an early bowser (source: Esso Petroleum Company Ltd.).*

The Second World War created an enormous demand for gasoline.
Fighter planes needed 100 octane fuel in order to achieve the required
power output and this was effected by using such processes as alkylation and by adding high levels of lead. There was consequently a
severe shortage of high-octane components, and octane values for
gasoline for home use had to be reduced. However, during this period
rapid progress was made in refining technology, and continuous catalytic cracking and reforming processes were developed as well as
catalytic desulfurization.

After the war the demand for high-octane components for aviation use
subsided and they were released back to the domestic gasoline markets, allowing a steady improvement in octane quality. There was a
period of rapid growth in demand for gasoline during which time
brand names reappeared on the European market. From about 1950,
octane values started to increase very rapidly and had reached about
95 RON by 1955. This was higher than necessary for the vehicles of
that time but allowed engine designers to increase compression ratios
in their new designs.

Brand images and advertising were features of the 1950s and '60s as was the introduction of additives to overcome such problems as ice formation in the carburetor, fouling of spark plugs, and deposits in carburetors and on valves.[4] National specifications for gasoline were introduced towards the end of this period.

Another factor which had already become apparent in the late 1960s was concern about the environment. Air quality deficiencies had been recognized for some time in the Los Angeles basin and the blame for this was placed largely upon exhaust emissions from vehicles. Similar problems had been recognized in other major cities where there

*Fig. 2.3. Tiger advertisement of the early 1960s
(source: Esso Petroleum Company Ltd.).*

was a high concentration of automobiles and where long periods of sunshine were the norm. Exhaust emission regulations were introduced in California in the 1960s, and by the 1970s, controls had been put in place for the entire U.S. and in Japan.

The first emission controls on vehicles involved engine modifications which resulted in the sacrifice of performance. However, the introduction of catalytic converters in these countries gave better control of emissions such as carbon monoxide, unburnt hydrocarbons and nitrogen oxides, and overcame any performance debits. Such vehicles required unleaded gasoline because lead poisons the precious metal catalyst used in these converters. This led to considerable changes in refining operations in order to achieve adequate octane levels without the use of lead. Later, there were concerns about the toxicity of lead itself and the effect it might be having on the health of people having to breathe in the exhaust fumes from vehicles for several hours a day. These worries caused lead levels to be reduced in stages in many of those countries, particularly Europe, where catalytic converters had not yet been considered necessary as a method of improving exhaust emissions.

The introduction of unleaded and low-lead gasolines caused dramatic changes in gasoline composition and initially in Europe a real shortage of high-octane blend components. Oxygenated compounds such as alcohols and ethers were used to make up this deficiency and these, in turn, gave rise to a whole series of new problems (see Chapter 11).

In 1973 the Arab/Israeli (Yom Kippur) war occurred and the Arab nations nationalized their oil reserves. These were the first of a series of Middle East oil crises that drastically increased the price of crude oil and led to a strong emphasis on fuel economy and to enormous changes in both the refining and automotive industries. They included the Iranian revolution in 1977 and the Iran/Iraq war in 1980, after which the price of crude gradually reduced apart from a brief period following the invasion of Kuwait by Iraq in 1990. However, the high gasoline prices had already depressed sales and the consequent spare capacity in most refineries led to a rationalization of the oil refining industry with the closure of many smaller and less efficient refineries.

The high price of crude oil had also caused balance of payments problems in many countries without domestic sources of crude oil and this intensified the search for new oil fields and made some countries such as Brazil look for alternatives to petroleum-based fuels. Ethanol production from the fermentation of sugar cane became a viable alternative to gasoline in that country and necessitated the production of engines modified for satisfactory operation on this fuel.

In addition there was a dramatic effect on product demand and on the relative volumes of different products needed from a refinery. The price of fuel oil had increased so much that alternative fuels such as coal became more economic for large energy users such as power stations and steel works. This gave rise to an imbalance in product demand with an excess of fuel oil relative to gasoline, particularly in Europe. In order to redress the balance, a massive increase in cracking processes took place. The changes in chemical composition of the resulting gasolines had an influence on the performance of the gasolines, as discussed in later chapters, and the high prices put a new emphasis on fuel economy. The pressure on fuel economy relaxed somewhat in the late 1980s and early 1990s since oil prices had stabilized to some extent by then, apart from the short period of the Gulf War.

By 1989, because air quality, particularly in California, was still unsatisfactory in spite of very severe legislation to restrict noxious emissions from vehicles, a "new" type of gasoline was marketed by one oil company—Atlantic Richfield. It was called Reformulated Gasoline and was marketed on the basis that it helped reduce emissions from vehicles, particularly older vehicles not fitted with emission control devices. This gasoline was significantly different from the other gasolines of that time because it had a low RVP (to minimize evaporative emissions), low benzene content (to reduce its toxicity), low aromatics (to reduce benzene exhaust emissions), a fixed level of oxygenated components (to reduce HC and CO emissions), and zero lead. Other oil companies followed suit, and by the early 1990s, reformulated gasolines were mandatory in many parts of the U.S. and were beginning to appear in Europe and other countries.

The long-term future for gasoline, as we know it, is uncertain because of the finite nature of crude reserves and because of the continuing pressures to improve the environment. One factor that should be pointed out is the impact that developing countries will have on crude reserves. In 1991, the number of cars per thousand of the population[5] was 588 in the U.S., between about 300 and 500 in most European countries and 300 in Japan—but in India it was only 2.8 and in China 1.5! Even a modest increase in the developing countries will have a large impact on the demand for crude oil.

Clearly lead will eventually disappear completely in every country when the percentage of vehicles needing this type of fuel drops to a low enough level. Reformulated gasolines will probably become more widespread and will delay the introduction of alternative fuels such as methanol.

2.2 Diesel Fuel

2.2.1 The Evolution of the Diesel Engine

The compression ignition (CI) engine was conceived by Dr. Rudolf Diesel to improve on the relatively poor thermal efficiency of the early, spark-ignited gasoline engines by employing a higher compression ratio. The modern CI engine has evolved out of a combination of Diesel's original concept and that of a contemporary English engineer, Herbert (or Hubert) Akroyd-Stuart. Diesel had visualized a compression-ignition engine which could be run on a variety of fuels, injected by air blast and ignited by the hot compressed air. Possible fuels included a low grade oil derived from bituminous coal, a light Russian crude oil, lamp oil (kerosene), gasoline (benzin), and pulverized coal. Akroyd-Stuart's independent research work was based on "airless" injection of a liquid fuel, but relied on an external heat source rather than the heat of compression to start the engine, a fore-runner of the hot-bulb engines still used in some fishing boats.

Diesel (1858-1913) was born in Paris of German parents and studied engineering at Augsburg and Munich Technical College. After completing his doctorate, he went to Sulzer Bros., in Switzerland, where he studied thermodynamic principles, particularly as expounded by Sadi Carnot,[2] and developed his theories on improving the efficiency of the internal combustion engine. Various patents were filed around 1890, and a British patent in 1892 refers to the gradual introduction of fuel into air which has been highly compressed so that its temperature becomes high enough to ignite the fuel spontaneously when it comes into contact with the compressed air.

A paper on Diesel's theories, published in 1893, received wide acceptance among German engineers, and experimental work was undertaken by both Krupp and M.A.N. The first prototype engine, made in the same year, was tested after a period of motoring. The purpose of running without power was to smooth high spots off the bearings but it showed the need to reduce the piston diameter to avoid seizure. Changes were also made to the sealing devices to minimize leakage and ensure adequate compression.

The initial test, which was made with benzin (gasoline), very clearly demonstrated that automatic combustion had been achieved. The ignition was so violent that the pressure indicating mechanism was blown off the cylinder and only narrowly missed hitting Dr. Diesel! The indicator diagram of this first firing on August 10, 1893, carrying the brief handwritten comment: "1st explosion (benzin)," is held in the M.A.N. archives.[6]

Akroyd-Stuart was born in 1864, and after completing his technical training in London, worked in his father's tinplate factory at Bletchley, England. In 1885, his interest in fuel combustion was aroused by a small explosion which occurred when some paraffin oil vaporized and ignited after accidentally spilling into a pot of hot metal. This incident lead to Akroyd-Stuart's development of the vaporizer engine, which was exhibited in 1891 at the Royal Agricultural Show in Doncaster, England. This was the first oil engine to run without a spark for ignition, relying instead on an externally heated vaporizer in the cylinder head to start the engine. External heating was only needed for starting, as fuel combustion maintained the bulb of the vaporizer at a high temperature while the engine was running.

Akroyd-Stuart's most relevant patent, registered in 1890 and predating that of Diesel by two years, described drawing pure air, rather than a mixture of hydrocarbon vapor and air, into the engine so there was no risk of premature ignition of an explosive mixture. Near the end of the compression stroke, liquid fuel was sprayed onto the heated vaporizer where it vaporized and mixed with the hot air, igniting automatically and propelling the piston on its power stroke. It is believed that the compression ratio was about 2:1.

Since that time around the turn of the century, the diesel engine, its fuel injection equipment (developed predominantly by Robert Bosch) and the fuel itself have been evolving in parallel with the spark ignition engine and fuels, but at a different rate.

Progress was slow through the First World War but the pace increased afterwards. Early use of the heavy, slow-revving diesel engine was mainly in industrial, marine and railway installations, with only limited activity in the areas of road and air transport. However, shortages in the availability of gasoline in Germany after 1918 stimulated development of diesel engines in that country, notably by Daimler-Benz and M.A.N., with the application to commercial road vehicles particularly in mind. Interest spread quickly to other European countries, stimulating development of diesel engines specifically for road transport. By the mid-1930s there were significant numbers of diesel-powered trucks and buses in service, although, despite successful efforts by racing and rally enthusiasts, there was only limited acceptance of the diesel as a car engine.

By the time war broke out in 1939, the automotive diesel was well established in Europe but, except for the German army, it had only a limited role in the military sphere. The German air force also had a few bomber aircraft powered by Junkers Jumo diesel engines. However, due to the logistics problems of dealing with more than one type of fuel, most of the military vehicles used by the Allied Forces had gasoline engines. Although diesels were widely used in rail and sea transport, progress on automotive diesel engines virtually stopped until after the 1939-45 war. It was then that the importance of its better fuel economy was more generally recognized, bringing rapid growth in diesel-powered vehicles in the European commercial road transport sector.

Most of the diesels in commercial vehicles are direct injection (DI) engines, generally larger than 2.5 liters, in which the fuel is injected directly into the cylinder and burned in a combustion chamber defined by the cylinder head and the piston. Smaller-capacity engines of a size suitable for passenger cars were developed around the indirect injection (IDI) system, where the fuel is injected and ignited in a prechamber connected to the working cylinder by a narrow passageway through which the burning gases expand to force down the piston (see Chapter 13). The aim of designers of prechamber engines was to achieve steady and progressive burning of the fuel by means of a fairly high temperature in the prechamber and turbulent air movement to give good mixing with the fuel spray. Early prechamber designs include the Acro system and the Lanova air cell, both developed by the German engineer Franz Lang. A number of diesel engine manufacturers in Europe and the U.S. built engines with Lanova combustion chambers but did not continue to use the system. The prechamber type of engine was adopted by the newly formed Daimler-Benz company, which was supplying engines for buses and trucks. An important advance in prechamber design was made in Britain with the development of the Ricardo Comet combustion chamber design by H.R. (Sir Harry) Ricardo. The Ricardo Comet head, patented in 1931, gave greater flexibility of operation and freedom from smoke than previous designs and was adopted by a number of British diesel engine builders.

Over the last 30 years Europe and Japan have seen a significant growth in the number of passenger cars fitted with diesel engines, most of which have indirect injection systems. Prechamber engines have a lower efficiency than DI engines because of higher thermal and pumping losses, but they are not as critical of fuel quality. While initially used mainly for taxis, where high annual mileages and lower running costs justified the higher initial outlay, the economy and performance of the modern small IDI diesel engine have now established it firmly in the private car sector.

In the U.S. the availability of low-priced gasoline reduced the incentive to adopt the more economical diesel engine for road transportation of heavy goods and passengers. Although acceptance of the automotive diesel was slower than in Europe, production of diesels for off-highway use in marine, railroad and earthmoving equipment was started as early as the 1920s by U.S. companies such as Cummins, General Motors and Caterpillar. As these developments were taking place without the availability of the Bosch-type unit pump, American engine makers designed their own injection systems, usually a combined pump and injector assembly for each cylinder, operated by a pushrod and rocker from a camshaft in the engine.

Since the end of the 1939-45 war, the automotive diesel has found its place in the U.S. powering buses and long-distance freight transport, the leading engine suppliers being GMC/Detroit Diesel, Cummins and Mack. Cummins are also well established as suppliers of truck engines in several countries around the world, where manufacturing plants have been set up. For passenger cars in the U.S., however, the gasoline engine is still supreme, with barely 2% of the market taken by diesels which are mainly imported from Europe or Japan.

2.2.2 Diesel Fuel Development

In the years immediately following the exploitation of petroleum resources by drilling (initiated by Colonel Drake at Titusville, Pennsylvania in 1859), kerosene lamp oil was the most valuable petroleum fraction. Surplus gasoline was disposed of by burning, surplus heavy residue was dumped into pits and the "middle distillate" was used to enrich town's gas, which explains why it is often still referred to as gas oil. Only with the invention of the diesel engine was a specific role found for the middle distillate fraction.

Fuel quality requirements of the early, heavy and slow-revving diesel engines were not very stringent but, with engine design improvements to increase power-weight ratio, more refinement of the fuel became necessary.

Elimination of fuels having high viscosities and high levels of hard combustion residues resulted in benefits in engine behavior. The tightening up of fuel quality also enabled engine speeds to be increased, with consequential improvements in output, efficiency and reliability. These engines were the forerunners of the modern, high-speed diesel engine.

As a result of these improvements, the next fuel characteristic to come under scrutiny was ignition quality. Fuels which were satisfactory in slow-speed engines operating at a few hundred revolutions per minute had problems of poor startability and combustion noise under certain load conditions in engines designed to run at speeds of 2000 rpm and higher.

The ignition quality of a diesel fuel was originally expressed by the calculated Diesel Index, which is a function of the fuel's API gravity and aniline point:

$$\text{Diesel Index} = \frac{(\text{API Gravity})(\text{Aniline Point})}{100}$$

Crude oils and their distillates are mixtures of hydrocarbon types: paraffins, naphthenes and aromatics, in various proportions. Of the three, the paraffinic types of fuel have the best engine startability characteristics but the poorest cold properties because of the waxes which come out of solution as the fuel temperature decreases. The opposite is true for aromatic fuels. The aniline point of a fuel is the lowest temperature at which equal quantities of fuel and aniline will go into solution. Aromatics have a higher density and better solvent power than paraffins, so a fuel which is high in aromatics will have a lower aniline point than a predominantly paraffinic fuel. Consequently, a highly paraffinic fuel will have a high aniline point and a high API gravity (low density), giving a high calculated Diesel Index to indicate good startability. This empirical approach, however, was not sufficiently discriminating and, before long, the CFR Cetane Engine test (ASTM D 613) was adopted as the standard procedure to define ignition quality.

Unfortunately, the CFR engine test method has poor precision and so another calculated value, the Cetane Index (ASTM D 976), was introduced as a means of estimating the cetane number of a diesel fuel from its API gravity and mid-boiling point. It is now widely used for monitoring and controlling diesel ignition quality. The formula has been revised from time to time, as fuels have evolved, to maintain its predictive validity. This is illustrated by Figure 2.4,[7] which compares the cetane numbers of a large sampling of typical production fuels with calculated values derived from the 1980 formula.

In the near future, a new cetane index method, ASTM D 4737, is expected to replace ASTM D 976 in many diesel specifications, and the older method will be removed from the ASTM Book of Standards. Details of the new equation are given in Chapter 16, Section 16.2.

As crude oils differ in their hydrocarbon composition, the ignition quality of diesel fuels depends heavily on the type of crude oil from which they are distilled. During the 1930s, when European interest in diesel-powered vehicles was starting to develop, most of the fuel production in Europe came from Middle East crudes which were paraffinic and of good ignition quality. In other parts of the world, where different crude types were more common, some selectivity was necessary to ensure production of good quality diesel fuel.

Demand for automotive diesel fuel in Europe rose steadily through the 1950s and 1960s as highway networks were extended to cope with increased travel by commercial operators and also by private motorists who were beginning to move to diesel-engined passenger cars because of their better fuel economy.

Measured CFR
Cetane Number

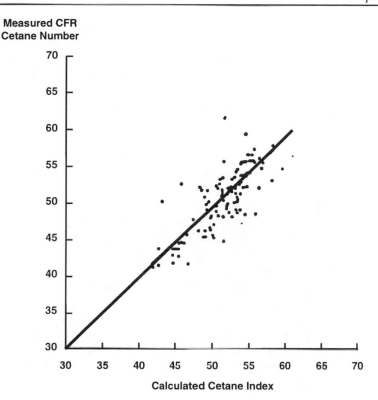

Calculated Cetane Index

Fig. 2.4. Correlation between calculated Cetane Index and measured CFR
Cetane Number.[7]

1980 CI equation:

$$CI = 454.74 - 1641.416\,D + 774.74\,D^2 - 0.554\,B + 97.803\,(\log B)^2$$

Where: D = density at 15°C, g/mL, by ASTM D 1298
B = mid-boiling temperature, °C, by ASTM D 86

Ignition quality during that period was generally satisfactory and, although additives to improve cetane number had been developed, there were no pressing technical or economic incentives for their use. Specifications were tightened to limit viscosity range, sulfur and contaminants (water, sediment and ash), while volatility had been established more as a means of product classification than of quality. The dominant fuel characteristic from both the refining and end-use viewpoint had now become its low-temperature performance, specified by cloud point or pour point. Cloud point and pour point are laboratory tests giving the temperature at which waxes start to separate and produce a cloudy appearance as the fuel is cooled and the lowest temperature at which the fuel can be poured. These tests are described in Chapter 15, Section 15.1. Seasonal grades were marketed because lowering the cloud point during winter months reduced the amount of diesel produced.

Very severe weather conditions over northern Europe during the 1962-63 winter had created havoc with the operations of diesel-powered trucks and buses. Exposure to long periods of unusually low temperatures had resulted in wax plugging of fuel lines and filters, causing problems of starting and running diesel vehicles. Stalled vehicles were abandoned at the roadside when drivers were unable to clear the fuel systems and restart their engines. Similar problems had been experienced in previous years from time to time but never to such an extent, largely because in those earlier days the diesel vehicle population had been much smaller.

A committee comprising members of the European Motor and Petroleum Industries was set up to investigate the difficulties that had been encountered and to recommend preventive measures to minimize future occurrences.

The committee identified design features of some vehicles which made them more sensitive to waxing problems. Recommendations were made for vehicle manufacturers to review their fuel system designs and to carry out corrective modifications where necessary. Fuel suppliers were also asked to ensure that the quality of diesel fuel marketed would be adequate for winter conditions. Meteorological records giving long-term average values were used to help establish realistic low-temperature specification levels for winter diesel fuel.

Since that time, diesel fuel quality in Europe and elsewhere has changed in a number of ways. A major influencing factor was the nationalization of oil reserves by the Arab nations referred to earlier and the ensuing changes in crude oil price and governmental policies relating to the use of non-renewable petroleum resources.

An important change soon after the severe winter problems of 1962-63 was the introduction of cold flow improver additives as a routine treatment of winter-grade diesel fuels. These additives enabled refiners to meet winter quality standards with little or no reduction in the quantity of diesel produced.

The current trend is for greater use of additives in automotive diesel. Cold flow additives are often used in summer fuels as well as in winter grades, while ignition improvers are sometimes needed to adjust the cetane value of fuels prepared from certain crude types or in blends containing cracked gas oils. These additives are now regarded as standard diesel fuel blend constituents, available to help refiners produce fuels of the required quality in the quantities needed by the market. They are normally added at the refinery to ensure that the fuel conforms to the appropriate specification before release for sale.

Additionally, multifunctional treatments may be introduced into on-grade product to provide extra benefits which the fuel marketer can exploit. In Europe, the biggest diesel market, advertising campaigns are used to promote the concept of a premium quality product. Claims, which will depend on the additive combination used, can include: better low-temperature performance; easier starting; protection against rust, corrosion and injector fouling; reduced exhaust smoke and noise; lower foaming tendency, enabling faster refueling with less likelihood of spillage; and a more acceptable odor.

Diesel fuels today are defined by specifications which have to take into account several factors: legal definitions; sulfur content; permitted levels of water, sediment and acidity; ash and carbon residues from combustion; safety standards; climatic factors and environmental considerations. All of these controls on quality are additional to the basic demands of the engine builder for a fuel which performs well in service. The relevance and importance of the various aspects of diesel fuel quality will be covered in the following chapter.

References

1. Lyle Cummins, <u>Internal Fire</u>, Society of Automotive Engineers, Inc., Warrendale, PA, 1989.

2. Carnot, Nicholas Leonard Sadi, <u>Reflections on Heat Engines</u> <u>(Reflexions sur la puissance motrice du feu)</u>, Ed. E. Mendoza, New York, 1960.

3. Dunstan, A.E. and Card, S., "The work of Midgely and Boyd." Report of the Empire Motor Fuels Committee, embodying other allied researches. Institution of Automobile Engineers, 1924, Volume XVIII, Part 1, pp. 3-12.

4. L.M. Gibbs, "Gasoline Additives—When and Why," SAE Paper No. 902104, 1990.

5. American Automobile Manufacturers Association, *World Motor Vehicle Data*, ISSN 0085-8307, 1993 Edition.

6. Lyle Cummins, Diesel's Engine, Carnot Press, 1993.

7. T.R. Coley, F. Rossi, M.G. Taylor and J.E. Chandler, "Diesel Fuel Quality and Performance Additives," SAE Paper No. 861524, 1986.

Further Reading

I.D.G. Berwick, "Key developments in a century of road transport fuels." Paper C318/86, Conference on petroleum based fuels and automotive applications, November 1986, Institution of Mechanical Engineers, London.

F.R. Banks, "The influence of tetraethyl lead on engine design and performance." *J. Inst. Pet. 35*, 1949, pp. 264-292.

L. Raymond, "Today's fuels and lubricants and how they got that way." *Automotive Engineering*, October 1980, pp. 27-32.

F.A. Jackman, "The history of gasoline." Technology of Gasoline, edited by E.G. Hancock, pp. 2-19, Society of Chemical Industry, 1985.

Earl Bartholomew, "Four decades of engine fuel technology forecast future advances." SAE National Fuels and Lubricants Meeting, November 1966.

Report of the Empire Motor Fuels Committee, The Institution of Automobile Engineers, Volume XVIII, Part 1 (1924).

Diesel Engine Reference Book, Ed. L.C.R. Lilley, Butterworths, 1984.

J.H. Boddy, "Diesel Fuels," Modern Petroleum Technology, Ed. G.D. Hobson, 5th Edition, John Wiley & Sons, 1984.

A.W. Judge, Maintenance of High Speed Diesel Engines, Chapman and Hall, 1956.

W.A. Gruse, Motor Fuels, Reinhold Publishing Company, 1967.

Historical Notes, International Training School Manual, Cummins Engine Company.

H.O. Hardenberg, Samuel Morey and His Atmospheric Engine, SAE SP-922, Society of Automotive Engineers, Inc., Warrendale, PA, 1992.

Daniel Yergin, The Prize, Simon and Schuster Ltd., 1991.

Chapter 3

Manufacture of Gasoline and Diesel Fuel

3.1 Introduction

As discussed in the previous chapter, the refining industry, particularly in the U.S. and Europe, has been undergoing a transition due to changes in product demand and in the quality of the gasoline and diesel blend components required. This has arisen because of increases in the price of crude oil and the requirement for vehicles to meet legislated targets for emissions and fuel economy. This chapter discusses the effect that these changes have had on refining operations; it describes the most important processes for manufacturing fuels and the characteristics of each component in terms of how it blends into a gasoline or diesel fuel.

It must be appreciated that gasolines, and to a lesser extent diesel fuels, can differ quite widely one from another in composition, even if they are of the same grade. This is because refineries are rarely the same in their general configuration and in their processes, and also because the crude oils that are processed can be very different and can even vary from day to day. In addition, the operating conditions on the plants themselves have to be modified according to seasonal and other factors influencing product demand and product quality and this changes the chemical composition of the finished fuel.

3.2 Crude Oil

Crude oil is the liquid part of the naturally occurring organic material composed mostly of hydrocarbons that is trapped geologically in underground reservoirs. It is by no means uniform and varies in density, chemical composition, boiling range, etc., from oil field to oil field and also with time from any given oil field. Table 3.1[1] shows the yields of a number of fractions when four different crude oils are split into different boiling ranges by a distillation process.

Table 3.1. Yield (wt %) of main products from crude oil by distillation.

Crude Type	Arabian Light	Nigerian	Brent	Maya
LPG	0.7	0.6	2.1	1.0
Naphtha	17.8	12.9	17.8	11.7
Gas oil/Kerosene	33.1	47.2	35.5	23.1
Residue	48.4	39.3	44.6	64.2

In this table, "naphtha" represents the percentage of hydrocarbons boiling in the gasoline range and "gas oil/kerosene" the hydrocarbon boiling roughly in the diesel fuel range (but including also jet fuel and kerosene). It can be seen that there are large differences in the percentages of the various streams that can be obtained from each crude by simple distillation.

3.3 Influence of Product Demand Pattern on Processing

If there were a free choice of which crude oils to purchase and if the product demand pattern could be met by selecting appropriate crudes, then a relatively simple refinery is all that would be necessary. Such refineries were commonplace in Europe in the 1960s and '70s, that is, up to the time that the crude oil price increases of 1973 and later had taken effect. Until then there had been a large demand for fuel oil (shown as residue in Table 3.1) but the availability of alternative cheaper fuels such as coal changed this, and the simple refineries then had no way to process the excess residue.

These refineries consisted basically of distillation units with processes for upgrading the octane quality of the naphtha and for removing malodorous sulfur compounds. They are called "hydroskimming refineries" and the layout of a typical one is shown in Figure 3.1.

It can be seen that this type of refinery produces a gasoline consisting of just a few components and that the diesel fuel is even simpler, with

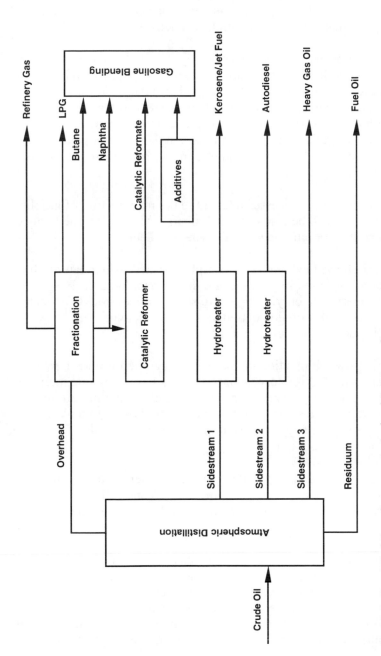

Fig. 3.1. Simplified flow diagram of a hydroskimming refinery.

only one component. Fuel for the refinery is normally any stream that is in excess of market requirements and which cannot command a higher price than the cheapest purchased petroleum fuel.

Nowadays, however, refineries generally need to be much more complex in order to be able to run any crude oil that happens to be available and to meet any reasonable change in product demand. In addition, the range of products which are made is often much greater than with a hydroskimming refinery, particularly if petrochemical processing to produce such materials as ethylene (for polyethylene manufacture), butadiene (for synthetic rubber manufacture), etc., is also part of the refinery operations.

In this chapter only processes relevant to the manufacture of gasoline and diesel fuel will be discussed and others will be mentioned only to show how they fit into the general refinery scheme.

Figure 3.2 shows the layout of a complex refinery which has all the major processes used in gasoline and diesel manufacture. It is a theoretical refinery plan only, designed to show the feedstocks for each type of process that could be used and to indicate where in the general processing scheme each process fits.

It is perhaps worth mentioning that heat integration is an extremely important aspect of refinery layout. It refers to making maximum use of all the heat energy available by, for example, using hot product streams which need to be cooled down before they go to tankage to partially heat other streams prior to putting them through a furnace on a distillation or other unit.

Processes such as cracking which convert heavier streams to lighter streams are known as "conversion processes," and refineries able to carry out such operations are sometimes called "conversion refineries."

Refineries can normally cope with the changes in product demand that occur as a result of seasonal or other short-term, but predictable, influences. However, when massive changes occur, as was the case when fuel oil demand in Europe fell away dramatically due to oil price rise, the ability of individual units to produce the required product pattern may not be adequate. In such instances there will often be a shortfall

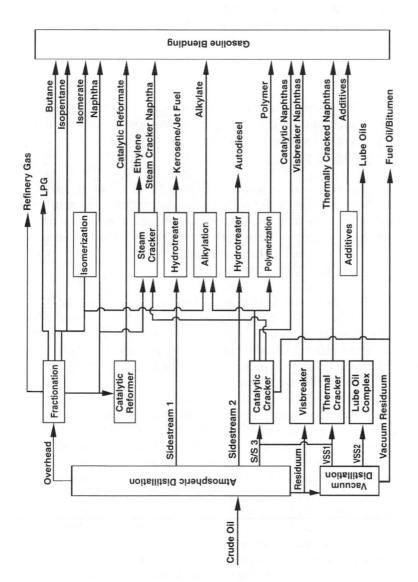

Fig. 3.2. Simplified flow diagram of a complex refinery.

while existing units are expanded or new units installed, and this has to be made up in some way. Purchases of suitable streams from other refineries with spare capacity is one way out, and another, which is still in use to meet the demand for unleaded gasoline in many countries, is to use oxygenated blendstocks such as alcohols or ethers (see Section 3.9 and Chapter 11).

3.4 Distillation

This is the initial process used in all refineries and its aim is to separate the crude oil into different boiling range fractions, each of which may be a product in its own right, a blend component or feed for a further processing step.

Crude oil contains many thousands of different hydrocarbons, each of which has its own boiling point. The lightest are usually gases at normal ambient temperature but can remain dissolved in the heavier liquid hydrocarbons unless the temperature is raised. The heaviest hydrocarbons are solids at ambient temperature but are able to stay in solution except at low temperatures.

Distillation does not chemically change the crude oil in any way but splits it up into fractions, each of which has its own boiling range. In practice, the distillation process is not able to effect perfect separation, as illustrated in Figure 3.3, and there is always a small amount of lighter and heavier materials present in any fraction, depending on the efficiency of the distillation tower.

Figure 3.4 illustrates the primary distillation process which is carried out in a unit called a "pipestill." The crude oil, which may have had to undergo a preliminary treatment to remove undesirable materials such as salt, is preheated by one or more of the refinery streams through heat exchangers and is then further heated in a furnace to about 400°C. The resulting mixture of liquid and vapor is then put into a fractionating tower at atmospheric pressure, the internal construction of which is designed to effect the maximum separation possible between compounds of different boiling points. Gases such as ethane, propane and butane are part of the overhead stream, and then successive sidestreams will contain progressively higher boiling range groups of hydrocarbons. The residue (fuel oil) stays liquid and flows out of the bottom of the tower.

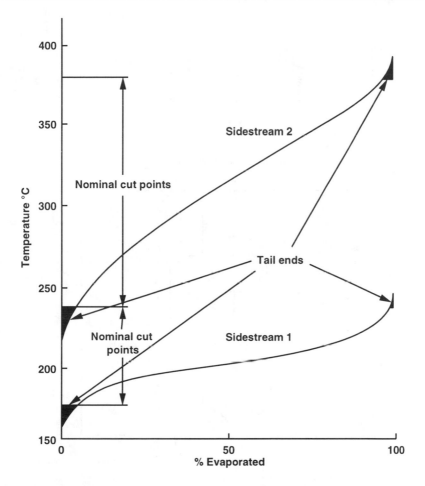

Fig. 3.3. Distillation curve for Sidestreams 1 and 2 showing imperfect separation.

In a complex refinery, the residue from the atmospheric tower would be further fractionated under vacuum to produce other streams which might then be used for lubricating oil manufacture, feed for a cracking unit, bitumen manufacture or heavy fuel oil.

A typical pipestill is shown in Fig. 3.5.

35

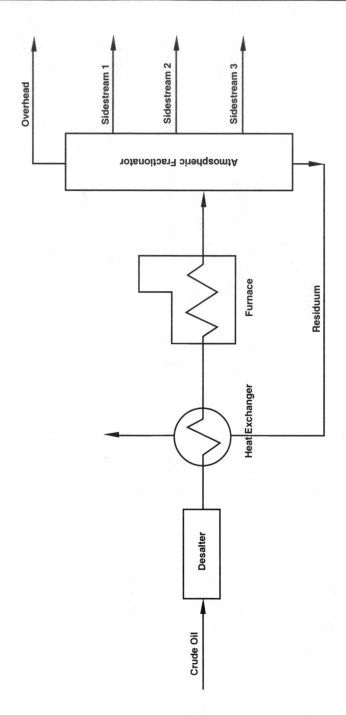

Fig. 3.4. Flow through a pipestill.

Fig. 3.5. A pipestill (source: Esso Petroleum Company Ltd.).

3.5 Cracking Processes

There are two general types of cracking process — thermal and catalytic — and each proceeds by quite different mechanisms.[2] *Thermal cracking* is believed to take place through the creation of hydrocarbon free radicals by carbon-carbon bond scission, for example:

$$CH_3(CH_2)_{14}CH_3 \rightarrow 2CH_3(CH_2)_6CH_2 \bullet$$
free radical

37

These free radicals are extremely reactive and undergo various reactions to form lower molecular weight hydrocarbons and free radicals, as for example:

$$CH_3(CH_2)_4CH_2.CH_2.CH_2 \bullet \rightarrow CH_3(CH_2)_4CH_2 \bullet + CH_2=CH_2$$
ethene

or $\quad CH_3(CH_2)_4CH \bullet \rightarrow CH_3(CH_2)_2CH_2 \bullet + CH_2=CH.CH_2.CH_3$
$$\mid \qquad\qquad\qquad\qquad\qquad\qquad \text{n-butene}$$
$$CH_3CH_2$$

Eventually the chain reactions can be terminated by two free radicals reacting with each other to form a new paraffinic molecule.

Dehydrogenation of olefins can also occur under severe thermal cracking conditions to form dienes:

$$CH_2=CHCH_2CH_3 \rightarrow CH_2=CH-CH=CH_2 + H_2$$

Naphthenes may undergo ring cleavage or may dehydrogenate to form aromatic, which are the most stable class of hydrocarbons to further cracking.

Catalytic cracking, unlike thermal cracking, proceeds by the formation of carbonium ions, for example:

$$CH_3(CH_2)_4CH_2-CH_2-H \rightarrow CH_3(CH_2)_4CH_2-\overset{+}{C}H_2$$
carbonium ion

Once formed, carbonium ions rapidly undergo a number of subsequent reactions such as isomerization to give secondary or tertiary carbonium ions:

$$R-\overset{+}{C}H-CH_2-CH_3 \rightarrow R-\overset{+}{C}-CH_3$$
$$\mid$$
$$CH_3$$

Cracking at the C-C bond in the β position relative to the positively charged carbon atom also occurs to form olefins:

$$\overset{+}{CH_3}\text{-}CH\text{-}CH_2\text{-}CH_2\text{-}R \rightarrow CH_3\text{-}CH=CH_2 + \overset{+}{R\text{-}CH_2}$$

The olefins themselves then undergo further reactions to form branch chain and aromatic compounds. The reactions are terminated when the carbonium ion is destroyed by donating a proton to an acid site on the catalyst or by abstracting a hydride ion from the catalyst.

The yields and the characteristics of the finished product can therefore be quite different for thermally cracked naphthas (and this includes naphthas from visbreakers and cokers as well as thermal crackers) as compared to catalytic crackers. Thermally cracked naphthas tend to contain lower aromatics levels and fewer branch chain compounds than catalytically cracked naphthas but more lighter olefins and dienes.

It is the olefin and aromatics content that are important in cracked streams because they have a large effect in determining the behavior of the gasoline or diesel fuel into which they are blended. Olefins are very reactive compounds that are easily oxidized and have better octane characteristics than the paraffinic streams from which they are derived, although their cetane values are somewhat worse. Aromatic compounds have excellent octane properties but have very poor cetane numbers.

The yields obtained depend on, among many other factors, the hydrocarbon makeup of the feed. Paraffins are the easiest hydrocarbons to crack and aromatics the most difficult, with naphthenic compounds being intermediate.

Individual cracking processes are described below.

3.5.1 Thermal Cracking

Conventional thermal cracking is now virtually obsolete for the production of motor gasoline since there are many other superior cracking processes available. The process requires the feed to be heated to around 500°C at a pressure of up to 70 bars when it is passed into a soaking chamber where the cracking reactions take place. The cracked products are fractionated and the yield of cracked naphtha obtained can be as much as 60% by volume.

The product is relatively unstable and requires the use of antioxidants and/or other treatments to prevent gum formation in storage or during use. It has a relatively poor Motor Octane quality.

3.5.2 Visbreaking

Visbreaking is a low-severity form of thermal cracking originally developed to reduce the viscosity of residual fuel oils, thereby minimizing the amount of higher-value distillate needed to achieve the required viscosity. Nowadays it is most often used as a relatively cheap process for converting some of the excess fuel oil into the more profitable distillate. The process is similar to conventional thermal cracking using temperatures up to about 500°C and pressures up to about 20 bars. Yields of visbreaker naphtha for gasoline blending vary from about 4 to 8%. For diesel fuel manufacture the yield of gas oil varies from about 12 to 25% using atmospheric residuum as feed.

3.5.3 Coking

This is a severe form of thermal cracking designed to convert residual products such as fuel oils into gas, naphtha, heating oil, gas oil and coke. It is particularly valuable as a process when the market for residual fuel oils is restricted. The coke produced can be used as a fuel but is also marketed for other uses such as electrode manufacture. Two major forms of coking process are in use — delayed coking and fluid coking.

Delayed coking is a semi-continuous process in which the heated charge is subjected to a long residence time in a soaking drum to allow all the cracking reactions to go to completion. Coke is deposited in the drum—there are two of these so that while one is on-stream the other can be cleaned. The cracked products are fractionated and heavy distillates are often recycled with feed.

Fluid coking is a continuous process which converts vacuum residuum to gaseous and liquid products using a fluidized solids technique. Fluidization involves subjecting small particles, in this case of coke, to a stream of gas of an appropriate velocity so that the suspended particles behave as if they were a liquid. Coking of the residuum is achieved by spraying it into a fluidized bed of hot, fine coke particles. The use of a fluidized bed allows the coking reactions to be carried out at higher temperatures and with shorter residence times than can be employed in delayed coking. Two vessels are used in fluid coking — a reactor and a burner. A portion of the coke is burned to provide heat and this heat is transferred to the reactor by the circulation of hot particles of coke.

"Flexicoking" is an extension of the fluid coking process which is able to process very heavy, high-sulfur residues. The coke formed, which contains most of the sulfur and heavy metals from the feed, is converted to a low-calorie gas suitable for use in process burners.

3.5.4 Catalytic Cracking

Catalytic cracking is the most important and widely used process for converting heavy refinery streams to lighter products, and it is the most popular method of increasing the ratio of light to heavy products from crude oil. It has superseded conventional thermal cracking because of its higher yields, better product quality (particularly for gasoline) and superior economics.

Most processes use a fluidized bed of catalyst into which feed is introduced. Fluidization has the advantage that the catalyst can be made to flow through pipes and from one vessel to another. The gas velocity is quite critical; at low velocities the catalyst will flow down pipes while at higher velocities it can be made to flow upward. This allows the catalyst to be moved readily between the reactor and the regenerator. Fluidized beds also allow high mass and heat transfer rates between the catalyst surface and the gas phase to give fast reaction rates.

The catalyst itself is a form of aluminum silicate known as zeolite which has a high activity and which suppresses the formation of light olefins.

Figure 3.6 shows diagrammatically one type of catalytic cracker, of which there are many different designs. The catalyst is moved continuously from the reactor, where carbon is deposited on it during the reactions, to the regenerator, where the carbon is burned off in a stream of hot air. As the catalyst is recycled back to the reactor it carries with it the heat needed to vaporize and crack the feedstock, which is usually a heavy gas oil from the atmospheric or vacuum dis-

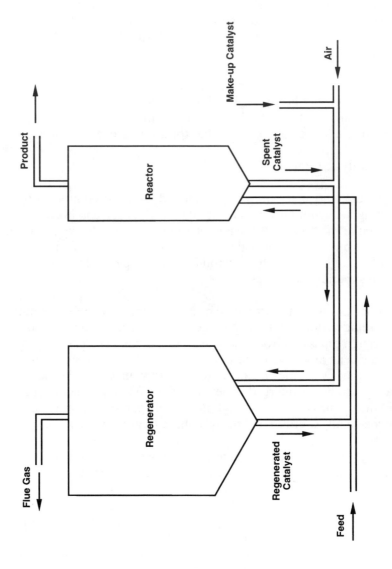

Fig. 3.6. Fluid catalytic cracker.

tillation towers. The vapor and the catalyst flow together into the reactor where the cracking takes place, although with today's high-activity catalysts, much of the cracking can take place in the transfer lines.

The cracked oil vapors pass to fractionating towers in which the new, smaller molecules are separated from the heavier products. The products from such a unit would be gas, catalytic naphthas, cycle oils and residue.

New developments in catalyst technology and in regeneration techniques allow the use of a mixture of residue and heavy distillate as feed. Without these developments the catalyst would become deactivated very rapidly when residuum is present.

3.5.5 Hydrocracking

The mechanism of hydrocracking is similar to that of catalytic cracking, but with hydrogenation superimposed. Carbonium ions are formed which crack, and the fragments are rapidly hydrogenated under the high hydrogen partial pressures used in the process. The rapid hydrogenation prevents coke deposition on the catalyst as occurs in catalytic cracking so that long runs without catalyst regeneration are achieved.

A dual-function catalyst is needed which performs satisfactorily for both cracking and hydrogenation reactions. This is achieved by the use of a silica alumina base to promote the cracking reactions, on which is dispersed metals such as molybdenum, tungsten, cobalt, nickel, etc., to promote hydrogenation reactions.

Hydrocracking is used mainly to produce low boiling fractions from feedstocks such as heavy gas oils, waxy distillates and even residues. The heavy feedstocks contain polycyclic aromatic compounds and these are quickly partially hydrogenated followed by splitting of the rings to form substituted monocyclic hydrocarbons. The side chains are then removed to form isoparaffins. The degree of cracking depends greatly on the feedstock and, in general, the heavier the feed the more middle distillate is produced.

The hydrogen requirement of hydrocrackers is extremely high so that unless there is plenty available from such units as catalytic reformers,

it is necessary to have as part of the hydrocracker complex, a hydrogen production plant.

The yield and product quality achieved depend on the feed to the unit
and the severity of operation. The product boiling in the gasoline
range is usually further processed by passing it through a catalytic
reforming process since it can have a relatively low octane number.
The units are often run to maximize production of middle distillates.

3.5.5.1 Catalytic Distillate Dewaxing

Distillates with a high wax content can make it difficult for the refinery planner to meet low-temperature quality specifications. While
refrigerative dewaxing is a normal procedure for lowering the pour
point of lubricant base stocks, it would be impractical and uneconom-
ical for diesel fuel components, because of the large amounts of product (and wax) involved.

Catalytic dewaxing or Middle Distillate Dewaxing (MDDW) is a process developed for waxy middle distillate streams, to improve their
low-temperature properties. It is a type of mild hydrocracking in
which a very low pour point gas oil can be produced from a waxy
straight-run feedstock.

Figure 3.7[3] shows a flow plan of the process. A shape-selective catalyst preferentially hydrocracks the normal and near-normal paraffins
in the waxy feed, producing a dewaxed gas oil and small proportion
of lighter components. The effectiveness of the process is evident
from results obtained with two feed streams of different wax contents.

A Light Arabian heavy atmospheric gas oil with a pour point of +18°C
cracked in the MDDW unit yielded 85 volume percent of dewaxed
product with a –18°C pour point. From a more waxy Libyan HGO
with a +33°C pour point, a 65% yield of 0°C pour point distillate was
obtained.

The process uses a lot of hydrogen and can be very costly, particularly
if hydrogen availability is limited and a new plant has to be installed.
Although distillate dewaxing is justified in some refining situations, only
a few units are in operation around the world. In 1994 twelve catalytic
distillate dewaxing plants were operating under license from Mobil

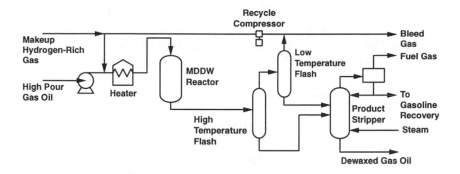

Figure 3.7. Flow plan of catalytic distillate dewaxing process.[3]

and there were two other licensees using similar technology developed by BP.

3.5.6 Steam Cracking

Steam cracking produces ethylene and other low boiling olefins used as chemical intermediates, particularly for the production of polythene. A by-product from this process used in gasoline blending is steam-cracked naphtha or, as it is often called, pyrolysis gasoline or pygas.

The process does not employ a catalyst and involves cracking naphtha or gas oil at temperatures in the range of 750 to 900°C in the presence of steam. The steam reduces the partial pressure of the feedstock, thereby increasing the yield of gaseous rather than liquid products. It also minimizes the formation of coke. Other feeds such as ethane or LPG can be used, but these produce very little steam-cracked naphtha for gasoline blending.

Steam-cracked naphtha must be pretreated before it is used in gasoline because it contains dienes and other very reactive olefins that reduce storage stability and can lead to excessive deposit formation in engines. The pretreatment can be a hydrogenation step or some other process such as heat soaking that removes the gum-forming materials.

Many steam-cracked naphthas contain high levels of benzene and other aromatics, depending on the feed and the operating conditions.

These aromatics are sometimes removed for use as chemical intermediates, but if they are not, care must be taken that the maximum
allowable level of benzene in gasoline is not exceeded when this component is used.

3.5.7 Middle Distillate Synthesis

Synthetic hydrocarbons are manufactured on a commercial scale in
South Africa by reacting coal, steam and oxygen together to produce a
synthesis gas which then undergoes Fischer-Tropsch reactions. The
efficiency of the process is low and it is generally used only where
supplies of crude oil are limited and cheap coal is available.

A more recent application of the Fischer-Tropsch approach is the
Shell Middle Distillate Synthesis (SMDS)[4,5] process, for which a
simplified flow scheme is shown in Figure 3.8.

The natural gas is first partially oxidized to hydrogen and carbon
monoxide by the Shell Gasification Process. The heart of the process
is the second step, Heavy Paraffin Synthesis (HPS), an updated version of the Fischer-Tropsch technique, using a highly selective catalyst. In the HPS, the hydrogen and carbon monoxide components of
the synthesis gas react to form predominantly long-chain paraffins
which extend well into the wax range.

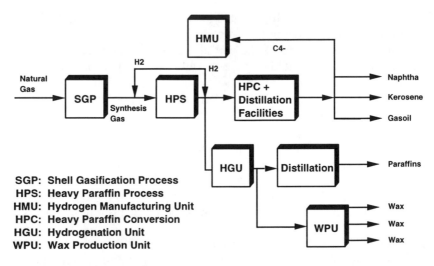

SGP: **Shell Gasification Process**
HPS: **Heavy Paraffin Process**
HMU: **Hydrogen Manufacturing Unit**
HPC: **Heavy Paraffin Conversion**
HGU: **Hydrogenation Unit**
WPU: **Wax Production Unit**

Figure 3.8. Shell Middle Distillate Synthesis (SMDS).[4] (Courtesy SIPM)

The subsequent steps comprise Heavy Paraffin Conversion and distillation sections to produce naphtha, kerosene, gas oil, paraffins and various grades of wax. Lighter grades are fed back for conversion into hydrogen while the long-chain waxy paraffins are either left intact or broken into smaller molecules, depending on market needs.

This process, which depends on easy availability of cheap natural gas, produces fuels which are completely free from aromatic and sulfur components, offering obvious benefits from an environmental viewpoint. Until full-scale plants are in operation, modest production capabilities will limit use of the synthesized gas oil to research applications rather than as a commercial blend component.

3.6 Catalytic Reforming

Catalytic reforming is an extremely important, widely used process for increasing the antiknock properties of naphtha. There are many variations developed by different companies such as Platforming (Universal Oil Products, UOP), Powerforming (Exxon) and Ultraforming (Amoco). They all utilize a catalyst in which platinum, often together with rhenium or other precious metal, is deposited on a support of alumina. The catalyst is readily poisoned by such materials as arsenic, copper and lead and its activity is reduced by the presence of sulfur and nitrogen compounds. For this reason the feed to a catalytic reformer is hydrodesulfurized (see Section 3.8.3) to remove these undesirables.

The main route for improving the antiknock properties of naphtha in catalytic reforming is the formation of aromatics by the dehydrogenation of naphthenes:

methylcyclohexane   toluene   hydrogen

It can be seen that the process is a net producer of hydrogen which can be used for hydrotreating and for hydrocracking.

Other reactions which take place are: isomerization, in which straight chain paraffins are rearranged into the higher octane quality branched hydrocarbons; dehydrocyclization, in which paraffins are cyclized to naphthenes and then dehydrogenated to aromatics; hydrocracking to form lower, smaller paraffinic molecules from larger ones; and de-alkylation, in which the side chains from higher aromatics are removed to form lower aromatics and paraffins. Most of these reactions simply involve rearranging the carbon atoms in the molecule so that the boiling range of the product is not vastly different from that of the feed.

Small amounts of polycyclic aromatics are also formed, particularly when the units are operated at high severities, and these are undesirable in a gasoline because of their tendency to lay down combustion chamber deposits and also because they tend to be carcinogenic. Catalytic reformates are sometimes rerun by passing them through a flash tower to remove these very high boiling materials.

Fig. 3.9. A catalytic reformer (source: Esso Petroleum Company Ltd.).

The process itself can be semicontinuous or continuous. In the semi-continuous form, long on-stream runs are achieved followed by a period when the catalyst, which is contained in several reactors arranged in series, is regenerated by carefully burning off the coke. There are two forms of continuous reforming: one involves the use of a swing reactor that is used in place of each of the other reactors in turn so that they can be regenerated without disrupting the run; the other has the reactors stacked one on top of the other and the catalyst flows through them by gravity and is then lifted to a regenerator and recycled.

3.7 Alkylation, Isomerization and Polymerization

3.7.1 Alkylation

Alkylation is the name given to a process for producing a high-octane gasoline component (alkylate) by combining light olefins with isobutane in the presence of a strongly acidic catalyst (sulfuric or hydrofluoric acid). It was an extremely important process during the Second World War when it was used to manufacture high-octane aviation gasoline for military use.

The olefins used in the feed are usually derived from catalytic cracking units and are normally a mixture of propenes and butenes. The reactions that take place are complex, although basically, the isoparaffins formed have the same number of carbon atoms as the sum of those in the isobutane and olefin reactants. An example of this type of reaction can be represented simplistically as:

$$CH_3-CH(CH_3)-CH_3 \; + \; CH_2=C(CH_3)-CH_3 \; \rightarrow \; CH_3-C(CH_3)(CH_3)-CH_2-CH(CH_3)-CH_3$$

isobutane   butene-1   isooctane (2,2,4 trimethylpentane)

In the example above, the product is isooctane (2,2,4 trimethylpentane), which has, by definition, a RON and a MON of 100. Alkylation produces a mixture of high-octane branch chain paraffins with a low sensitivity (the difference between RON and MON, see Chapter 6) and can be a valuable component when MON is a limiting specification point.

3.7.2 Isomerization

This is a process for converting straight chain paraffins to branch chain and may be used to provide isobutane feed for the alkylation process or simply to convert the relatively low-octane quality of straight paraffins to the more valuable branch chain molecules. Thus, n-pentane with a RON of 62 can be converted to isopentane having a RON of 92.

In principle, the process involves contacting the hydrocarbons with the catalyst (platinum on a zeolite base) and separating any unchanged straight paraffins for recycle through the unit. The product is clean burning and has good RON and MON qualities.

3.7.3 Polymerization

In this process light olefins such as propene and butenes are reacted together to give heavier olefins which have a good octane quality and do not increase unduly the vapor pressure of the gasoline. An example of the type of reaction that takes place is as follows:

$$CH_3CH_2CH=CH_2 + CH_3CH=CH_2 \rightarrow CH_3(CH_2)_4CH=CH_2$$

butene $\qquad$ propene $\qquad$ heptene

The catalyst most commonly used is phosphoric acid on keiselguhr.

The product is almost 100% olefinic and has a relatively poor MON compared with RON.

3.8 Finishing Processes

Gasoline and diesel fuel component streams produced by the above processes are often unsuitable for immediate use for a number of reasons including objectionable odor or instability. For this reason they are subjected to secondary treatments, the most important of which from a gasoline and diesel fuel viewpoint are caustic washing, UOP Merox treating and hydrodesulfurization. Other processes in this category are Copper Chloride Sweetening, Inhibitor Sweetening, Acid

Treatment, etc., but these are generally no longer used and will not be discussed here.

3.8.1 Caustic Washing

Washing with caustic soda or caustic potash will remove a number of undesirable contaminants such as hydrogen sulfide, mercaptans (thiols), cresylic acids and naphthenic acids, as well as acidic materials carried over from processes such as alkylation which use an acid catalyst. Caustic washing is mainly used nowadays as a final cleanup operation to remove the last traces of contaminant. It is sometimes followed by a water wash to ensure that there is no carry-over of caustic soda into the finished product.

Caustic washing is carried out by mixing together the hydrocarbon stream and aqueous caustic soda, and then allowing the mixture to settle into two layers in a horizontal drum so that each layer can be drawn off separately. Because of the environmental problems caused by the disposal of spent caustic soda, processes in which the caustic is regenerated, such as the Merox Process, are generally preferred.

3.8.2 Merox Treating

This process, developed by Universal Oil Products (UOP), catalytically oxidizes the evil-smelling mercaptans to nonodorous disulfides, which can either be left in solution (Merox Sweetening) or removed altogether (Merox Extraction). The basic reactions are:

1. Extraction of mercaptans from the oil phase by means of aqueous caustic soda:

$$RSH + NaOH \rightarrow NaSR + H_2O$$

2. Regeneration of the caustic and oxidation of the mercaptides by air blowing in the presence of a catalyst:

$$2NaSR + \tfrac{1}{2}O_2 + H_2O \rightarrow RSSR + 2NaOH$$

The disulfides formed are insoluble in caustic soda and so are allowed to redissolve in the hydrocarbon in the sweetening process; but in the extraction process the sodium mercaptide solution is separated from the hydrocarbon phase before air blowing so that the separated disulfides can be removed prior to the return of the regenerated caustic for reuse. The catalyst consists of a metal chelate and can be supplied either in a form that is soluble in caustic soda or as a solid supported on a carrier.

3.8.3 Hydrodesulfurization

Naphtha, kerosene and gas oils, which contain high molecular weight sulfur compounds in addition to hydrogen sulfide and light mercaptans, can be desulfurized by hydrogen treatment. The process also improves stability by saturating olefinic compounds, particularly the more reactive ones, and by removing nitrogen- and oxygen-containing compounds.

Examples of the reactions which take place are:

$$RSH + H_2 \rightarrow RH + H_2S$$

$$RSSR' + 3H_2 \rightarrow RH + R'H + 2H_2S$$

The catalyst used contains cobalt and molybdenum on an alumina base and can be regenerated in situ or by removing it for regeneration offsite. The hydrogen sulfide formed is normally converted into elemental sulfur in a sulfur recovery plant using the Claus process.

There are many variations on the hydrodesulfurization process developed by different companies such as Hydrofining (Exxon Research and Engineering) and Ultrafining (Standard Oil Co. (Indiana)), but the basic principles of all of them are the same and are as outlined above.

3.9 Oxygenated Gasoline Components

As a result of the phasedown of lead, there have been difficulties in many parts of the world in meeting both octane quality and the required volume of gasoline. These have been overcome to some extent by the use of oxygenated blend components such as methanol, ethanol, tertiary butanol, mixtures of alcohols, methyl tertiary butyl ether (MTBE), tertiary amyl methyl ether (TAME), or mixtures of ethers. The behavior of these components in gasoline is discussed in Chapter 11; this section covers the most important processes that are used in the manufacture of these materials.

3.9.1 Alcohols

Methanol (CH_3OH) has been widely used, together with one or more higher alcohols, within certain restrictions whenever the price has made it economical to do so. Virtually all methanol is manufactured from natural gas although it can be produced from any carbonaceous raw material such as coal or wood that can be converted to synthesis gas (carbon monoxide and hydrogen). The reactions are:

$$CH_4 + \tfrac{1}{2}O_2 \rightarrow CO + 2H_2$$

$$CO + 2H_2 \rightarrow CH_3OH$$

Highly selective catalysts are used to maximize the yield of methanol, although less selective catalysts will give higher alcohols such as ethanol (C_2H_5OH), butanol (C_4H_9OH), etc:

$$2CO + 4H_2 \rightarrow C_2H_5OH + H_2O$$

$$4CO + 8H_2 \rightarrow C_4H_9OH + 3H_2O$$

Mixtures of alcohols can also be obtained from synthesis gas using the above route and have been used as gasoline components because the higher alcohols present reduce the water sensitivity of the methanol. The process developed by Dow/UCC uses one of a family of active and stable molybdenum-sulfide catalysts, which have a good selectivity to the production of alcohols. The water produced is removed in this process by the use of molecular sieves.

Ethanol is most commonly produced by hydration of ethylene or by fermentation of biomass. Most of the ethanol used industrially is made synthetically by mixing ethylene with steam at 60-70 atmospheres and about 300°C over a phosphoric-acid catalyst supported on diatomaceous earth. The reaction is:

$$CH_2{=}CH_2 + H_2O \rightarrow CH_3CH_2OH$$

The fermentation route to ethanol requires the action of certain enzymes derived from yeast cells upon carbohydrates such as glucose. Carbon dioxide is given off during the process and the resulting aqueous mixture contains about 10% ethanol. This must be concentrated to remove virtually all of the water by means of distillation and other processing steps.

Tertiary butyl alcohol (TBA, t-butanol, $(CH_3)_3COH$), which has been widely used as a cosolvent for methanol, is most often made by the controlled oxidation of isobutane to tertiary butyl alcohol and tertiary butyl hydroperoxide (TBHP). The hydroperoxide is then reacted with propene to give propylene oxide and additional butyl alcohol. The reactions are:

$$4(CH_3)_2CHCH_3 + 3O_2 \rightarrow 2(CH_3)_3COH + 2(CH_3)_3COOH$$
isobutane   oxygen   t-butanol   t-butylhydroperoxide

$$(CH_3)_3COOH + CH_3CHCH_2 \rightarrow (CH_3)_3COH + CH_3CHOCH_2$$
TBHP   propene   TBA   propylene oxide

Although the TBA produced by this route tends to be somewhat impure, it is satisfactory for use in gasoline without further refining.

3.9.2 Ethers

The most important ether used in gasoline blending is MTBE (methyl tertiary butyl ether), although others such as TAME (tertiary amyl methyl ether) and ETBE (ethyl tertiary butyl ether) are receiving more and more attention.

MTBE is manufactured by reacting methanol with isobutylene, i.e:

$$(CH_3)_2C=CH_2 + CH_3OH \rightarrow (CH_3)_3COCH_3$$
isobutylene   methanol   MTBE

Different processes for manufacturing MTBE vary only in the route that is used to make the isobutylene. Thus, the Arco process dehydrates TBA whereas the Houdry process dehydrogenates isobutane.

TAME is produced commercially in a similar way to MTBE in that methanol is reacted with an isoamylene such as 2-methyl-2-butene or 2-methyl-1-butane:

$$(CH_3)_2C=CHCH_3 + CH_3OH \rightarrow (CH_3)_2C_2H_5COCH_3$$
2-methyl-2-butene   methanol   TAME

ETBE is again similar to MTBE except that ethanol is used instead of methanol.

Mixed ethers have also been produced for use as a gasoline blend component and here, instead of using relatively pure olefins such as tertiary butylene or isoamylene, a mixture of C_4 to C_7 olefins is reacted with methanol.

3.10 Gasoline Blending

Gasoline blending, when carried out on a routine basis with a relatively small number of blend components, provides few difficulties once the blending data for the components in question have been determined. However, in a complex refinery or when untried components are purchased, it can require a number of trial blends to establish blending behavior. One reason is that the blend does not always behave as expected in terms of how the calculated quality compares with actual measured quality, particularly if there are plant changes or new components being used. The main causes of these discrepancies are:

- Many of the important specification parameters such as octane and vapor pressure do not blend linearly.

- The octane behavior of a given component is modified by the nature of other components in the blend, i.e., other factors than RON and MON, such as hydrocarbon composition, influence the way in which a component blends.

- The blender is trying to meet several specification points at once and any move to meet one specification point may well result in another being put out of grade.

There are other factors that conspire to make the life of a gasoline blender difficult, such as:

- Restrictions in the maximum level allowed for some components or additives, such as oxygenates, anti-knock additives, sulfur, etc.

- Tankage limitations — a refiner has to find a home for all of the product streams made, and if all of a product stream cannot be used it may necessitate cutting back in plant throughput to avoid overfilling tanks. This, in turn, may mean a deficiency in other streams and an increase in costs.

In addition, many specification points are legally enforced so that the penalties in terms of adverse publicity and fines if out-of-grade product is put on the market can be very expensive. Because test procedures often have a poor precision, this has to be taken into account when setting "internal" specification standards so as to be sure that any sample picked up by a controlling body has a very high probability (usually at least 95%) of meeting the required specifications.

The cost of "giving away" quality, i.e., the cost of making a higher quality than is required by the specification, can be very high indeed, particularly for important points such as octane and vapor pressure. This is not to say that all refiners manufacture as close to the legal specifications as they can; many will have their own "internal" specifications that can be significantly better than the minimum legal limits. They do this in order to make their product more attractive to customers so that they can increase, or at least maintain, their share of the gasoline market.

Fungibility, i.e., the interchangeability of a gasoline between suppliers and the compatibility of gasolines from different sources, is another problem and is important for two reasons: First, a refiner may wish to sell to or purchase from another manufacturer either to make up a deficiency in manufacturing capacity or to minimize distribution costs. In these cases the material must meet the specification of the purchasing company, which may be different from the internal specification of the supplying company. Second, if a customer purchases gasoline from two different sources and they are not compatible, then the resulting mixture in his tank may not perform satisfactorily even though both batches of gasoline meet the required specifications. An example of this latter effect is when two gasolines, one containing methanol and the other not, are mixed together, the vapor pressure of the mixture may be higher than that of either of the two constituents. If both were on the limit of the vapor pressure specification, then it is possible that the mixture would be out of specification.

3.10.1 Blending Operations

Blending can be either continuous or batchwise, although it is always necessary to check each finished tank of gasoline prior to shipment or sale to ensure that it is satisfactory and meets the required specification limits. Batches of gasoline are normally blended by line mixing through a manifold to ensure that the blend in the tank is homogeneous and not layered, as can happen when components are added to the tank one by one. It is important to check for layering by taking samples from the top, middle and bottom of a tank and, if they are identical in terms of density (or some other simple-to-measure characteristic), to then combine these samples for final testing of the tank. If layering is found, some method of mixing the tank contents must be used such as recirculation through a pump. Great care has to be taken in the handling of samples used for measuring vapor pressure, because it is very easy to lose light ends from them.

Continuous analyzers are frequently used for measuring octane quality so that the overall quality going into a rundown tank from a processing plant or into a blend tank can be accurately assessed by integrating the individual results. Such instruments have, as an internal standard, a gasoline or stream having an accurately known octane level close to that of the product. The standard is run through the instrument automatically at regular intervals and compared with the test result, so that the blender has confidence in the data being produced. These instruments can allow a refiner to operate closer to the specification limit than if he had to rely on individual CFR ratings, and this can represent a very real cost saving.

The blender has to identify which specification points are critical, i.e., which points are most likely to be difficult to meet in terms of cost and the available component streams. For many refiners this will be octane quality or Reid Vapor Pressure, and it will be necessary to concentrate particularly on these points. Many specification limits will always be satisfactory whatever the blending method, and these are only monitored on an occasional basis.

3.10.2 Blending Calculations

Blend calculations are not often made by hand but are computed using linear programming in order to optimize the gasoline pool within the restraints imposed by the specification, the volumes required, the need to produce other products, tankage limitations, and so on. Sophisticated software packages are now available which control and optimize blends by manipulating the set points of a digital blender, using on-line analyzer measurements.[6]

The main difficulty in achieving an accurate prediction of the properties of a gasoline blend is that most of the important parameters, such as RON, MON and RVP, do not blend linearly and the way in which they blend will vary according to the other components in the pool. A number of different ways of overcoming this problem have been described,[7] as summarized below.

3.10.3 Octane Blending

When blending leaded gasolines, the octane values at the lead level of the final blend must be used. Lead response curves, in which octane quality (RON or MON) is plotted against lead concentration, are used (see Figure 3.10), although, since lead levels are usually very restricted, most operators will operate at the maximum level allowed in order to minimize manufacturing costs. The response to lead is non-linear, with the greatest octane benefit occurring with the first increment of lead and with progressively smaller benefits as the lead concentration increases.

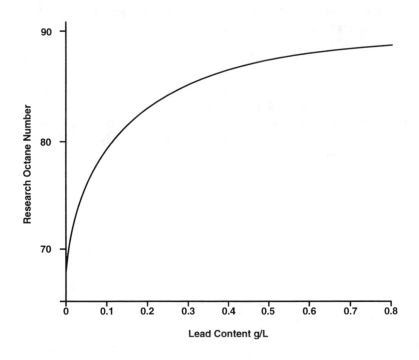

Fig. 3.10. Lead response curve for straight run naphtha.

Unleaded gasolines are blended in the same way except, of course, that clear (i.e., without lead) octane qualities are used.

The most widely used procedures for octane blending are:

1. The Blending Bonus Approach. The properties of a blend can be calculated simply by making a weighted average using measured values for each of the components, as shown below for a three-component blend:

$$P_B = V_1 P_1 + V_2 P_2 + V_3 P_3$$

where P_B is the property (such as RON, MON, etc.) of the blend; P_1, P_2, and P_3 are the properties of the components, and V_1, V_2, and V_3 are the volume fractions.

However, this approach will usually give a significant error, in that it will overestimate the MON of unleaded and low-lead gasolines and the RON of low-lead gasolines. It will also underestimate the RON of unleaded gasolines. Nevertheless, if all the blends are fairly similar in composition, then the error is usually quite constant and an equation can be used in which this under- or overestimation is corrected:

$$P_B = V_1 P_1 + V_2 P_2 + V_3 P_3 + BONUS$$

The value of the bonus, which is the expected deviation from linear blending, is determined by measuring the RON of a series of trial blends. For more universal application, it can be calculated from a knowledge of the olefin and aromatic contents of each component. One such equation for determining the value of the bonus in a specific gasoline pool is given below:

$$BONUS = 0.43(\bar{R}) - 0.004(\bar{O}^2) + 0.004(\bar{A}^2) - 0.01(\bar{R}^2) - 0.005(\bar{O}\bar{A}) - 55.18$$

where $\bar{R}$ is the volumetric average RON, and $\bar{O}$ and $\bar{A}$ are the average olefin and aromatic contents. Because this method uses compositional data on each component, it takes into account changes in blending bonus when larger than normal changes in blend composition occur.

A variation of this method[8] uses an "Excess Octane Number" which can be positive or negative and represents the deviations from the ideal blending behavior. It is claimed to require little data, to produce simpler and more accurate correlations, and is highly suited to use in a computer-based LP-optimized blending model.

2. The Component Blending Value Approach. Here blending numbers that do blend linearly are used for each component, so that the blending equation becomes:

$$P_B = V_1 B_1 + V_2 B_2 + V_3 B_3$$

where B_1, B_2, and B_3 are the linear blending numbers for each component. These blending values are usually obtained by determining the octane quality of a carefully designed set of fuels blended from all the components available and on which the octane levels have been determined.

3. Blending Interaction Coefficient Approach. In this procedure, the effect of different components on blending behavior is taken into account by assuming that each blend component will interact on the other components, as shown by the following equation:

$$P_B = V_1 P_1 + V_2 P_2 + V_3 P_3 + V_1 V_2 IC_{12} + V_1 V_3 IC_{13} + V_2 V_3 IC_{23}$$

where IC_{12} is the interaction coefficient between components 1 and 2, IC_{13} is the interaction coefficient between components 1 and 3, and so on.

3.10.4 Reid Vapor Pressure Blending

Deviations from linear blending mainly occur when light hydrocarbon components such as butane or isopentane are used and also when alcohols such as methanol or ethanol are blend components.

Two methods are used for hydrocarbon blends. One widely used procedure is to blend by molar proportion and the other is to use determined linear blending values, as for octane. For example, although the determined RVP of butane is 4.5 bars, it blends as if it is actually about 5.5 bars, depending, as always, on what other components are present.

The effect of alcohols is more difficult to take into account although the interactive coefficient approach has been found to give good results.

3.10.5 ASTM Distillation Blending

It is better to use the "percent evaporated at a given temperature" rather than the "temperature at which a given amount evaporates" for blending calculations. The "levels of percent evaporated at given temperatures" (e.g., % evaporated at 100°C) blend linearly, unless alcohols are present or unless the component boiling ranges are very different. With alcohols, interaction models give good prediction.

There are, however, no simple equations that enable initial boiling points, final boiling points, or the temperatures for any distillation level (e.g., temperature °C for 50% evaporated) of blends to be predicted.

3.11 Diesel Fuel Blending

Diesel fuels are normally blended from at least two refinery streams. The operation becomes more complicated as the number of components increases because, as with gasoline, some of the most important fuel characteristics do not blend in a linear manner. In a simple hydroskimming refinery, blending two or possibly three components is a relatively straightforward procedure once the blending parameters of the individual components are known. It is a more complex operation with a conversion refinery having as many as seven or eight possible blend components that can differ significantly in key characteristics. Processing a variety of crude oil types may also contribute to the complexity.

The planner responsible for blending will be restricted by the scheduled, long-term pattern of refinery operation as regards crude types, throughput and cut points of the various streams. However, within these limitations, the planner has scope to respond to short-term market-driven requirements. In most countries, the dominating factor is the demand for high value gasoline, followed by the jet fuel/kerosene market, with diesel and distillate heating oil next in line. A refinery with a big market for jet fuel will try to minimize the amount of kerosene in the diesel blend and, if possible, use an alternative stream such as light cracked gas oil to lower the diesel cloud point.

The constraints on blend composition are the specifications to be met, the volume of product needed by the market, and the quality and quantity of available component streams. The relative values of those streams are also pertinent because the aim is to make the blend that will be most profitable for the refinery.

Refinery planners cannot formulate a diesel fuel blend in isolation. For logistic and economic reasons, other products the refinery is making also need to be taken into consideration because the same stream may have to be shared among several different products. The refiner has to find an outlet for all the product streams, and tankage capacity provides only limited possibilities for stockpiling if all of a particular stream cannot be used. Cutting back on crude throughput would be a last resort, as it results in shortages of other products and increased operating costs.

As with gasoline, many specification points are legally enforceable, and failure to comply can be very expensive for a company if it results in adverse publicity as well as heavy fines for marketing off-grade product. As test procedures used to measure fuel properties often have poor precision, this has to be taken into account to ensure that any sample picked up by a controlling body has at least a 95% probability of meeting the required specification points. For this reason, refiners usually work to limits that are slightly tighter than those required by the specifications.

Many other aspects of diesel blending are similar to those of gasoline blending, such as compatibility of diesel fuels from different sources and the need for fuels to be fungible (or interchangeable) for deals between suppliers. These are important for the same reasons as given in Section 3.10 on gasoline blending.

3.11.1 Diesel Blending Operations

As with gasoline, blending can be either continuous or in batches, but it is always necessary to check each finished tank of fuel prior to shipment to ensure that it is satisfactory and meets all the required specification limits. Blends of diesel fuel are normally prepared by simultaneous transfer of streams from the various component tanks into the finished product tank by pumping them through a line containing a mixing section. Where multiple stream blending is not possible, component streams are added one by one to the tank and the contents are blended by recirculation around an external system which transfers fuel from the bottom of the tank to the top. Components will settle into different density layers if mixing has been inadequate, so it is important to check for layering by sampling at various levels in the tank. Pumping round is continued until top, middle and bottom samples show the product to be homogeneous.

Some refineries are equipped with line analyzers; this enables continuous monitoring of a specification property by drawing samples either from the rundown line to intermediate storage or from the blending line to the finished product tank. Line analyzers are available to measure the cloud point and cold filter plugging point of diesel fuels. The cold filter plugging point (CFPP) is described in Chapter 15, Section 15.1.5. An optional capability which may be incorporated with a stream analyzer for cloud point or CFPP is a closed-loop feedback system to adjust the blend proportions or the treat rate of flow improver additive to maintain a consistent low-temperature quality.

3.11.2 Diesel Blend Calculations

Before the blender can start to formulate a blend to meet the quality criteria in the quantity needed for the market, the following information is needed:

- the identity of the most critical/difficult specification points to be met.

- the quantity of product to be made.

- the qualities and quantities of the available component streams.

For a diesel fuel, the cold properties tend to be the principal constraint, but other characteristics are also important for a variety of reasons and most specifications also include the following parameters:

- Flash point — for safe handling and storage; also a legal requirement.

- Volatility — for complete combustion, but also a legal requirement.

- Cetane number — for ease of ignition and to control legislated emissions.

- Sulfur — to minimize engine wear and meet legislation.

- Viscosity — for consistency and good atomization.

- Density — for consistency and good fuel economy.

The relevance of these various fuel properties will be discussed in Chapters 14 and 16.

Instead of imposing a minimum or maximum limit for viscosity and density, some specifications define a tolerance band for one or both of these properties to avoid extreme values causing wide variations in engine performance and emissions.

Blend calculations to optimize the diesel formulation within the specification restraints are usually made by linear programming rather than by hand. However, as some of the important fuel parameters such as cloud point and viscosity do not blend linearly, they must be converted to linear blending numbers, or blending indices, before the program can be run.

For Cloud Point Blending, component cloud points are used in the formula below to calculate their cloud point blending indices. From these the index for the blend is obtained by summation of the product of volume and index for each component.[9] Back substitution of the blend index in the formula then gives the cloud point of the blend.

$$CBI = 1.618 \ e^{(0.0426 \ CL)} + 0.0433 \ CL$$

where   CBI = Cloud Point Blending Index
        CL = Cloud Point, °C

A certain amount of trial and error will be required to finalize the formulation and, for convenience, cloud blending indices can be taken from the graph in Figure 3.11[9] which has been derived from the blending index formula.

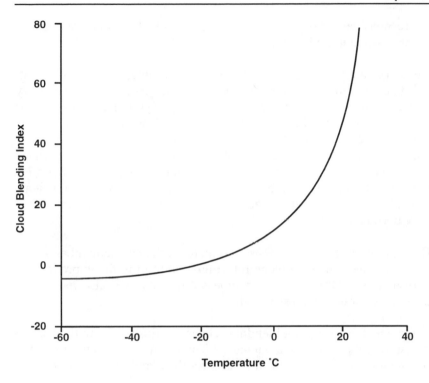

Fig. 3.11. Cloud Point blending curve.[9]

A similar procedure is followed for Viscosity Blending, with component viscosities converted into viscosity blending numbers by means
of the Refutas blending chart, which uses scales devised to express the
temperature-viscosity relationship for Newtonian fluids as a straight
line:

$$VBN_1 = 14.534 \log [\log (CS_1 + 0.8)] + 10.975$$

where VBN_1 = Viscosity Blending Number for component 1
        CS_1 = Viscosity, cSt, for component 1

Component VBNs are summated by weight and the resulting blend
VBN converted to blend viscosity by back substitution in the formula.

If additives such as flow improvers or cetane improvers are to be used
as a blend component to enable specification targets to be met, the
planner will need to take account of other relationships to ascertain

the responsiveness of the blend, because the result of additive treatment is also a nonlinear function.

Response to a flow improver additive, for example, is very dependent on the distillation characteristics of the blend, particularly at the heavy end, as well as the wax content.[10] One way of stimulating response is to incorporate a small amount of a heavier stream (see Chapter 15, Section 15.2.7). While this can often enable the CFPP target to be attained, it may have an unfavorable effect on cloud point and perhaps other quality criteria. This can be corrected by using more light component material to bring the blend back on specification, but if a catalytically cracked gas oil is used for this purpose the cetane number may become too low.

Diesel ignition additives are used to increase the cetane number and do not normally involve blend reformulation to limit the improvement permitted by additive treatment, unless cetane index is also specified, as in the current British standard.

A formula for flash point blending relates the flash point to the vapor pressure of the blend components.[11] There is also a formula for color in which light absorbance is the basis for the blending index table.[11]

References

1. C.W.C. van Paasen, "Changing Refining Practice to meet Gasoline and Diesel Demand and Specification Requirement," Paper No. C316/86, Institution of Mechanical Engineers Symposium on Petroleum Based Fuels, London, UK, 1986.

2. Modern Petroleum Technology, Edited by G.D. Hobson, 5th Edition, John Wiley and Sons, 1984.

3. S.P. Donnelly and J.R. Green, "Catalytic Dewaxing Process Improved," *Oil and Gas Journal*, pp. 77-82, October 1980.

4. D. Parkes, "Matching Supply and Demand for Transportation Fuels in the Pacific Rim Countries Post 1990," Fourth International Pacific Conference on Automotive Engineering, Melbourne, Australia, March 1988.

5. M. van der Burgt, J. van Klinken, Tjong Sie, "The Shell Middle Distillate Synthesis Process," Paper presented at the 5th Synfuels Worldwide Symposium, Washington D.C., 1989.

6. J. White and H.V.D. Kemp, "Optimized Gasoline Blending at Lindsey Oil Refinery," *Petroleum Review*, 46, No. 546: 317-19, July 1992.

7. G. Searle, "Gasoline Blending," College of Petroleum and Energy Studies Manual for *Gasoline Technology Course RF5*, 1993.

8. A. Muller, "New Method Produces Accurate Octane Blending Values," *Oil and Gas Journal*, March 23, 1992.

9. M.H.T. Auckland and D.J. Charnock, "The Development of Linear Blending Indices for Petroleum Properties," *Journal of the Institute of Petroleum*, Vol. 55, No. 545, pp. 322-329, Sept. 1969.

10. J. Zielinski, F. Rossi and A. Stevens, "Wax and Flow in Diesel Fuels," SAE Paper No. 841352, 1984.

11. R.M. Butler, G.M. Cooke, G.G. Lukk and B.G. Jameson, *Industrial Engineering Chemistry*, Vol. 48, No. 808, 1965.

Further Reading

C.Y. Tsang, V.S.F. Ker, R.D. Miranda and J.C. Wesch, *Technology, Oil and Gas Journal*, pp. 33-38, March 28, 1988.

"Index Predicts Cloud, Pour and Flash Points of Distillate Fuel Blends," *The Oil and Gas Journal*, pp. 66-69, November 9, 1970.

R.C. Hutcheson and C.W.C. van Paasen, "Diesel Fuel Quality into the Next Century," Institute of Petroleum Symposium on The European Auto Diesel Challenge, London, England, April 1990.

Chapter 4

Storage, Distribution, and Handling of Gasoline and Diesel Fuel

4.1 Introduction

A number of factors can influence the quality and performance of gasoline and diesel fuels after they leave the refinery and before they get to the customer's tank. They can be contaminated with water, held for long periods in contact with air, subjected to severe mixing, brought into contact with metals and elastomeric materials which they can attack, and they can be subjected to a range of temperature and humidity conditions. At an oil terminal they can be reblended, mixed with other batches of the same sort of material, treated with various additives, accidentally contaminated with other products, and subjected to electrostatic discharges during pumping operations.

Most of these can modify product quality, although great efforts are extended by oil companies to minimize the effect of these factors.

4.2 Influences on Product Quality During Distribution

4.2.1 Sea Transport

Contamination of the product by water and from inadequately cleaned tanks can occasionally occur during transport by sea tanker. The presence of water can give rise to haze formation, especially when it is mixed in with the fuel by the centrifugal pumps during discharge. When this occurs it can require the use of dehazer additives to avoid excessive holdup in storage tanks while the product clears.

Evaporation losses can also take place but are not usually large enough to have a significant effect on product quality. When light products such as gasoline are transported long distances, the cargo can undergo considerable changes in temperature that can accentuate evaporative losses.

4.2.2 Pipeline

This method of transporting fuel is growing because it is safe, cost-effective and energy conserving. It does have potential difficulties, however, because most pipelines carry several products, although black and white oils are usually segregated.

Contamination between one product and the next in a pipeline does occur to some extent since there is no physical barrier between them, although they are always put through in a predetermined sequence to minimize the effects of contamination. A typical sequence might be gasoline, followed by kerosene or aviation fuel, diesel fuel, kerosene and then gasoline again. It is essential to prevent contamination of unleaded gasoline by leaded material since the maximum lead content is mandated by law to avoid damage to the sensitive three-way catalysts fitted to vehicles. The rate of flow through the pipeline is kept above a certain minimum speed (depending on pipe diameter) to minimize mixing between products.

The interface between one product and the next is detected by the use of continuous-recording density meters or by other means. The interface material is segregated and blended back at a later stage into a product where the specification will not be put off-grade.

Apart from interface contamination, there are other minor possibilities for product quality problems during pipeline transportation. Some additives or components used in gasoline can loosen rust from the sides of the pipe and hold it in suspension, causing filter blockage. Anticorrosion additives (see Chapters 9 and 18) are often added to keep the inside of the pipeline free from rust since it can restrict the maximum flow rate through the pipeline. The anticorrosion additives themselves cause no problems provided they are used at appropriate levels.

Drag-reducing agents (see Chapters 9 and 18) have also been used in products flowing through pipelines. These additives reduce turbulent flow and enable higher pumping rates to be achieved. They are used only as a short-term measure when pipelines become limited in capacity.

Surfactant additives are often used in gasolines to keep the engine clean and free from deposits, but they can be adsorbed onto the sides of the pipe and desorbed into other products going through the pipeline. When this occurs, problems such as haze stabilization and additive loss can ensue.

4.2.3 Road and Rail

Rail is often used to transport fuel to terminals and road is almost invariably the final method of transportation used to reach the service station. Contamination problems are quite rare since dedicated tankers are usually used. The biggest potential problem is in ensuring that human error does not result in the wrong materials being put into the service station tanks.

4.3 Influences on Product Quality During Storage

Quality can improve in the tanks used to hold the product at various stages in its transportation from the refinery to the service station, if the fuel is held long enough for particles of dirt and rust that might have become entrained to separate out. Water, which can originate from processing steps or be picked up during transportation, may also separate out at this stage and settle to the bottom of the tank.

In gasoline, the presence of water and sediment is identified using ASTM Method D 4176 in which a sample is swirled in a clean glass jar and examined for visual sediment or water droplets. For diesel fuel, the centrifuge method ASTM D 1796 is used, which is included in the ASTM D 975 specification for diesel fuel oils. A 50 mL sample of the fuel is mixed with an equal volume of water-saturated solvent and centrifuged to concentrate the water and sediment at the bottom of conical centrifuge tubes. Some diesel specifications impose separate limits for water and sediment, while others limit the combined amount of water and sediment. Typical maximum levels are 0.05% for water and sediment together, 0.05% for water alone and 0.01% for sediment alone.

Lower levels have been set in the new European specification for automotive diesel fuel, EN 590 (see Appendix 8), which calls for maximum limit values of 200 mg/kg (0.02%) for water and 24 mg/kg for particulate matter. However, a maximum limit value of 500 mg/kg may be specified until December 31, 1995, by those European countries where regulations and/or codes of practice result in inherently wet distribution systems.

Oxidation stability problems (see Chapter 8) may occur with excessive storage times, as is sometimes required for strategic fuel stocks. For storage at high ambient temperatures, inert gas blanketing (often nitrogen) is sometimes used to minimize oxidation of the fuel.

4.3.1 Water Contamination in Tankage

Water can be introduced into the gasoline while it is in tankage because many tanks have floating roofs and rainwater can find its way past the seals into the fuel. Water is also sometimes used to push fuel out of the tank; such a fuel is, therefore, always saturated with water. A drop in temperature of the product may also cause dissolved water to separate out as a haze which will eventually clear unless certain types of surfactant additives are present which can stabilize such hazes.

Most storage tanks have a layer of water on the bottom which should be kept to a minimum by frequently drawing it off. The product draw-off point is usually well above any water layer so that there should be little chance of contamination from tank water bottoms. However, if the water layer is allowed to build up excessively, or if a tank is not given enough time to allow the water to settle out, then free water can be present in the product being discharged. In cases where the amount of tankage is restricted, gasolines or diesel fuels may be pumped out of a tank at the same time as fresh product is being pumped in, and in these cases, particularly if the tank level is low, it is possible for the water bottoms to be stirred up and contaminate the product.

In cases where alcohols are present in the gasoline, the presence of water bottoms causes special problems which are discussed in Chapters 11 and 21.

4.3.2 Microbiological Contamination

Bacterial and fungal growths can sometimes lead to operational problems such as cloudy fuel and filter blocking if storage tanks are not cleaned regularly. Gasoline is generally less susceptible to this problem than diesel fuel, particularly leaded gasoline because lead antiknock additives have some biocidal properties. The organisms, which can be aerobic or anaerobic, live in water bottoms of tanks and feed on the fuel at the interface. Anaerobic, or sulfate-reducing bacteria, derive their oxygen from sulfates and this gives rise to a slimy sulfide deposit and hydrogen-sulfide gas. The hydrogen sulfide can cause corrosion problems as well as give an unpleasant odor. Both the water and oil phases become darker when this occurs. Biocides[1] can be used to kill these organisms (see Chapters 9 and 18), but cleaning and good housekeeping are also necessary to remove the by-products of their activity and to minimize the likelihood of further problems.

4.3.3 Sludge in Tankage

Other materials that may separate out in tankage are gums and lead compound degradation products from gasoline, and wax from diesel fuel. These can contribute to a layer of sludge at the bottom of the tank.

Wax settling from diesel fuels can occur when the fuel has cooled to below its cloud point as a result of unexpectedly low temperatures or long storage in small exposed tanks. Although settled wax usually redissolves when the tank is refilled or as the temperature rises, its removal is advisable if the amount of wax remaining in a fuel could put a new batch of fuel out of specification. Additives to overcome wax separation problems are discussed in Chapter 16.

The sludge layer in gasoline tanks can also build up to a considerable level and it is necessary from time to time to take tanks out of service to remove this material before it becomes a potential contaminant. The gums present in the sludge can cause severe deposit problems in engines.

4.3.4 Evaporative Losses

A significant amount of evaporation is possible with very volatile gasolines. This can change the performance of a gasoline and increase its costs, as well as have an adverse environmental effect. The amount of evaporative loss from a tank will depend on whether it has a fixed or a floating roof. In fixed roof tanks, which operate with a space above the liquid, vapor loss can occur during tank filling and in tank breathing as the temperature changes. The losses are reduced to some extent by the provision of pressure/vacuum (P/V) valves that allow some positive or negative pressure to develop before vapor expulsion or air induction can occur. Better control of evaporative emissions is possible with the use of internal floating roofs. With floating roof tanks, losses can occur from the vapor seals between the roof and the tank shell, from various fittings in the roof for taking samples, etc., and from the wet sides as the tank level drops during discharge. Improved tank seals can minimize these losses. Control of evaporative emissions during manufacture and distribution are known as "Stage 1 recovery" and include the installation of vapor recovery facilities at terminals and depots.

Evaporative emissions are also of concern during the refueling of vehicles, and vapor recovery equipment in service stations to minimize these losses can be installed—the so-called "Stage 2" controls—when there is a requirement to do so.

4.3.5 Oxidation

Both gasoline and diesel fuel can oxidize during storage, giving rise to the formation of gums and gum precursors that can cause deposit formation in engines and seriously influence their performance. Fuels containing olefinic components, arising mainly from cracking operations, are the most susceptible to gum formation and, therefore, may need some special processing or the use of antioxidants (see Chapter 9). When gums form, they initially remain in solution but, as the amount increases, they begin to separate out to give cloudy fuels and, in extreme cases, blocked lines and filters and high sludge levels in tankage.

The mechanism by which the oxidation of hydrocarbons progresses occurs in several stages, as follows:

1. Chain initiation involving the generation of free radicals:

$$R\text{-}H \rightarrow R\bullet + (H\bullet)$$

2. Chain propagation:

Once a hydrocarbon free radical ($R\bullet$) has been formed, it can combine with oxygen to form a peroxide radical ($R\text{-}O\text{-}O\bullet$)

which, in turn, can react with another hydrocarbon molecule thereby generating other hydrocarbon free radicals and a hydropereroxide (R-O-OH):

$$R\bullet + O_2 \rightarrow R\text{-}O\text{-}O\bullet$$

$$R\text{-}O\text{-}O\bullet + R'H \rightarrow R\text{-}O\text{-}OH + R'\bullet$$

The oxidation process is therefore self-perpetuating. The free radicals can also give rise to polymerization as well as oxidation reactions to form high-molecular-weight materials. These can deposit in the fuel system.

3. Chain termination:

The chain reaction is only terminated, in the absence of an antioxidant, by reactions which lead to non free radical products:

$$R\bullet + R\bullet \rightarrow R\text{-}R$$

$$R\text{-}O\text{-}O\bullet + R\bullet \rightarrow R\text{-}O\text{-}O\text{-}R$$

The chain breaking or terminating function of antioxidants is thought to proceed by the donation of a hydrogen atom from the reactive center of the antioxidant to the peroxy radical. The activity is then sufficiently stabilized by resonance to discontinue chain propogation.[2,3]

The presence of copper and certain other metals dissolved in the gasoline will actively catalyze the oxidation process. The use of metal deactivating additives (see Chapter 9, Section 9.1) can overcome this problem.

With leaded gasolines, tetraethyl lead can be oxidized and degrade on storage (unless antioxidant is present) to form insoluble lead compounds which can manifest themselves as a cloudiness in the gasoline and a light-colored heavy deposit in the bottom of the tank. This deposit can block filters and jets.

4.4 Safety Considerations for Storage and Handling

4.4.1 Flash Point

To minimize the likelihood of an accidental fire while petroleum fuels are being handled and transferred into and out of storage containers, it is important that the appropriate codes of practice are carried out. Typical regulations include restrictions on smoking and define the type of electrical equipment that may be used. Flammable liquids are usually classified according to their flash points and, for example, in the U.K., gasoline with a flash point below -21°C (actually it is normally below -40°C and too low to measure by standard methods) is given an Institute of Petroleum classification of I, and diesel fuel with a minimum flash point of 56°C is in Class III. The flash point is the temperature to which the fuel must be heated under specified conditions to produce a vapor-air mixture that will ignite when a test flame is applied.

A minimum flash point of around 56°C is typical of diesel fuel in many countries, although other flash point levels may apply where climatic or other considerations prevail. In the U.S., for example, the permitted minimum flash point for No. 1D grade diesel fuel is 38°C, and in Canada it is a minimum of 40°C.

4.4.2 Electrical Conductivity

Although conductivity is not a specification requirement for either gasoline or diesel fuel, it is occasionally measured because there is a potential safety risk due to the buildup of a static electricity charge during bulk handling at high pumping rates. Grounding leads are normally used to conduct away any charge when large quantities of fuel are pumped into or out of storage tanks, but aviation kerosene is also treated routinely with a static dissipator additive.

Gasoline tends to be more conductive than diesel fuel which, in turn, is more conductive than kerosene, and there is little risk of an excessive charge building up when refueling vehicles at a service station. However, faster pumping rates are used when loading and offloading road and rail fuel tankers and occasionally an antistatic additive (see Chapters 9 and 18) is used in these cases, even though the equipment is almost always grounded.

4.5 Health and Environmental Effects of Gasoline

The currently available data on the health, safety and environmental properties of gasoline has been collated by CONCAWE[4] (the oil companies' European organization for environmental and health protection), and the following is a very brief resume of their findings. It is primarily concerned with hydrocarbon-only gasolines although some information on oxygenates has been added (see also Chapter 11). The references given by CONCAWE in their report have not been included in this summary.

4.5.1 Health Aspects

The International Agency for Research on Cancer (IARC) has reviewed the carcinogenic risks from gasoline by looking at occupations where gasoline exposure may have occurred, such as service station attendants and car mechanics. IARC allocated gasoline an overall classification of Group 2B, i.e., possibly carcinogenic to humans, based on what was accepted as inadequate and limited evidence in experimental animals plus other evidence including the presence of benzene and 1,3 butadiene in gasoline.

4.5.1.1 Inhalation

High concentrations of gasoline vapor can accumulate in confined spaces since it is heavier than air, and can present both a safety and a health hazard. Short and infrequent exposures are unlikely to result in a health risk. Exposure to gasoline vapor concentrations in the range of 500 to 1000 ppm can cause irritation of the upper respiratory tract and, if continued, will give rise to a narcotic effect with symptoms such as headache, dizziness, nausea and mental confusion with eventual loss of consciousness. These central nervous system effects can occur rapidly with sudden loss of consciousness even after a brief exposure to very high concentrations. Cardiac irregularities have also been reported after exposure to high vapor concentrations.

4.5.1.2 Ingestion

The taste and smell of gasoline will usually limit ingestion to a small amount—in adults ingestion is usually the result of siphoning attempts and with children from drinking from unlabeled or wrongly labeled containers. Gasoline is of moderate to low oral toxicity for adults, but for children, ingestion of even small quantities may prove dangerous or even fatal.

Spontaneous vomiting is common after ingestion with the likely consequence of aspiration of liquid gasoline into the lungs—this is the principal hazard and no attempt should be made to induce vomiting.

4.5.1.3 Aspiration

Aspiration of even small amounts of liquid gasoline into the lungs, either directly or as a consequence of vomiting, can have very serious results. Irritation of the respiratory tract may lead rapidly to difficulty in breathing and development of a potentially fatal chemical pneumonitis.

4.5.1.4 Skin Contact

Repeated or prolonged skin contact can result in drying, cracking and possible dermatitis. In rare cases, an individual may become sensitized to the dyes used in some gasolines. Repeated contact may also make the skin more susceptible to irritation and penetration by other materials.

4.5.1.5 Eye Contact

Contact of liquid gasoline with the eye may cause moderate to severe irritation and conjunctivitis. This effect is transient and permanent injury is unlikely to result.

4.5.2 Exposure Limits

Because of the complex and variable composition of gasoline there is no widely accepted exposure limit which is generally applicable. Limits which have been set are:

1. In the U.S., the American Conference of Governmental Industrial Hygienists has set an 8-hour time-weighted average threshold limit value (TLA-TWA) of 890 mg/m^3 (300 ppm) and a short-term exposure limit (TLV-STEL) of 1480 mg/m^3 (500 ppm) over a period of 15 minutes. These limits are based on the typical vapor composi-

tion of gasoline in the U.S., where the aromatic content is generally lower than in Europe.

2. In Sweden an 8-hour TLV-TWA of 220 mg/m^3 (about 70 ppm) and a STEL of 300 mg/m^3 (about 100 ppm) have been set. These are based on a typical liquid composition with an assumed aromatic content of 46%.

4.5.3 Ecotoxicity

The environmental effect of gasoline spills is mainly due to the water solubility of its components. Each of the several hundred hydrocarbon components in gasoline will have its own very low solubility, and of these, aromatic hydrocarbons have the greatest solubility and therefore represent the major part of the water soluble fraction. The monoaromatic hydrocarbons are of most concern, and of these, benzene, toluene, ethylbenzene and xylene are the most important with naphthalene and methylnaphthalene as the most important diaromatics. When oxygenated compounds are also present, since these are very much more soluble in water than hydrocarbons (with the lower alcohols being completely miscible), such spills can represent a much more severe environmental problem.

Following a spillage, the more volatile components are rapidly lost by evaporation and, to a lesser extent, by dissolution into water. Local environmental conditions such as temperature, wind, wave action, soil type, etc., together with photo-oxidation, biodegradation and adsorption onto suspended material, all contribute to the weathering of the remains of the spilled gasoline.

Microorganisms present in sediments and in the water are capable of degrading gasoline components and the time to reduce the fuel concentration to 50% of the initial amount has been reported as between 1.2 and 2.7 days in sand, loam or clay soils. Nutrient addition and inoculation with bacterial isolates enhances biodegradable losses.

Gasoline exhibits some short-term toxicity to freshwater and marine organisms. The components which are most prominent in the water-soluble fraction and cause aquatic toxicity are also highly volatile and can be readily biodegraded by microorganisms. Because of these factors, spilled gasoline, unless it contains alcohols, is unlikely to remain in water in sufficient quantities to cause aquatic effects, and as a result presents a minimal overall risk to the environment.

4.5.4 Disposal

It is seldom necessary to dispose of large quantities of gasoline, but when it is, such as with residues from tank cleaning, it should be done by incineration. Materials that have been highly contaminated with gasoline should also be incinerated, but less contaminated materials may be acceptable for authorized landfill sites. Contaminated soil may be treated by land farming.

4.6 Health and Environmental Effects of Diesel Fuel

The dossier on diesel fuel being prepared by CONCAWE to complement that on gasoline was not available in time to be included, so the following comments have been abstracted from another source.[5]

4.6.1 Health Aspects

Diesel fuels are complex mixtures of hydrocarbons produced by distillation of crude oil. Cracked components which contain polycyclic aromatic hydrocarbons (PCAs) may also be present and some PCAs have been shown by experimental studies to induce skin cancer. Performance enhancing additives may also be included but at concentrations well below 0.5% they do not constitute a risk to health.

4.6.1.1 Inhalation

The higher boiling range of diesel fuel makes it less volatile than gasoline, but if any vapors, mists or fumes are generated, they will be slow to disperse and in confined spaces can present a health hazard. Exposure to vapor, mists or fumes may cause irritation to eyes, nose and throat, but short and infrequent exposures are unlikely to result in a health risk.

4.6.1.2 Ingestion

The taste and smell of diesel fuel will usually limit ingestion to a small amount. In adults, ingestion is usually the result of siphoning attempts and, with children, from drinking from unlabeled or wrongly labeled containers. Diesel fuel is unlikely to cause harm if accidentally swallowed in small doses, though larger quantities may cause nausea and diarrhea.

4.6.1.3 Aspiration

Aspiration of even small amounts of diesel fuel, either directly or as a consequence of vomiting, will injure the lungs.

4.6.1.4 Skin Contact

Brief or occasional contact with diesel fuel is unlikely to cause harm to the skin but prolonged or repeated exposure may lead to dermatitis. This material contains significant quantities of polycyclic aromatic hydrocarbons (PCAs), some of which have been shown by experimental studies to induce skin cancer.

Injections through the skin, resulting from contact with the product at high pressure (for example, when testing fuel injectors) constitute a major medical emergency and surgical exploration should be undertaken without delay. Injuries may not appear serious at first but within a few hours tissue becomes swollen, discolored and extremely painful with extensive subcutaneous necrosis.

4.6.1.5 Eye Contact

Diesel fuel is unlikely to cause more than transient stinging or redness if accidental eye contact occurs.

4.6.2 Exposure Limits

There is no appropriate occupational limit for diesel fuel. If vapor, mists or fumes are generated, their concentration in the workplace air should be controlled to the lowest reasonably practicable level.

4.6.3 Ecotoxicity

Spillages of diesel fuel may penetrate the soil, causing ground water contamination and it may be harmful to aquatic organisms. Films of fuel on water surfaces may cause physical damage to organisms and also impair oxygen transfer. The product is inherently biodegradable and there is no evidence to show that bioaccumulation will occur.

4.6.4 Disposal

If disposal of large quantities of diesel fuel is necessary, it should be done by incineration, under conditions approved by the local authority or by a licensed waste disposal contractor. Materials that have been highly contaminated with diesel fuel should also be incinerated but less contaminated materials may be acceptable for authorized landfill sites. Hazard warning labels are a guide to the safe handling of empty packaging and should not be removed.

References

1. T. Coley, "Diesel fuel additives influencing flow and storage properties." Gasoline and Diesel Fuel Additives, John Wiley and Sons, p. 130, Chichester, 1989.

2. P. Polss, "What additives do for gasoline," *Hydrocarbon Processing* V.52, N.2, pp. 61-68, 1973.

3. H.N. Giles, J.N. Bowden and L.L. Stavinoha, "Overview on assessment of crude oil and refined product quality during long term storage." U.S. Department of Energy and U.S. Army Fuels and Lubricants Laboratory, June 1985.

4. CONCAWE Product Dossier No. 92/103, "Gasolines," CONCAWE, Brussels, July 1992.

5. "Automotive Diesel Fuel," Material Safety Data Sheet No. STC2111, BP Oil Technology Centre, Sunbury-on-Thames, England, 1993.

Chapter 5

The Spark Ignition Engine and the Effect of Fuel Quality on Performance

5.1 Introduction

The effect of fuel quality on vehicle performance is relatively small compared with the influence of vehicle design, although fuel quality can have dramatic effects under certain conditions. As engine designs have progressed and changed, so has the need for improved fuels, particularly in terms of antiknock quality because knocking or detonation can limit the amount of power available and give rise to catastrophic engine damage.

Pressures to improve air quality have led to a whole range of design modifications and innovations by the motor industry which include the use of catalysts to minimize the main pollutants from vehicle tailpipes and the use of vapor adsorption systems which prevent the loss of light hydrocarbons to the atmosphere by evaporation. In addition, complex electronic control devices have been developed which can ensure that the car is operating under optimum conditions at all times. The oil industry, in turn, has made some important changes including removing lead, introducing additives to ensure that critical equipment is kept clean, and changing gasoline specifications to avoid excessive evaporative losses and to minimize undesirable exhaust emissions.

The price rises in 1973 and later brought the finite nature of natural fuel resources sharply into focus and made it clear that fuel economy was an important factor that had to be taken into account when designing vehicles. The importance of fuel economy was reemphasized when it was realized that CO_2 is a greenhouse gas and that emissions of it should be minimized. The steps taken reduced average fuel consumption levels by over 25% in a period of ten years. Although there is relatively little that can be done to fuel composition to help in this problem, the optimization of gasoline volatility and the use of additives play some part in achieving lower fuel consumption.

There is, however, a real need to improve the energy conversion from the fuel in the tank to work in turning the road wheels. At present no more than 15% of the energy in the gasoline is available to provide power as shown in Table 5.1.[1]

Table 5.1 Fuel Energy Balance in a Gasoline Engined Vehicle[1]

Energy Losses from:	% Lost
Exhaust	20 - 45
Radiation	1 - 5
Incomplete Combustion	2 - 5
Coolant	15 - 30
Transmission	1 - 5
Friction	7 - 38
Available as Road Power	0 - 15

Taxation also influences the fuel composition and type of fuel used in any particular area, as, for example, in the relative tax on unleaded versus leaded fuel and the favoring of diesel fuel and/or LPG as compared with gasoline in some countries.

5.2 The Spark Ignition Engine

This is only a brief guide to those aspects of the spark ignition internal combustion engine that are important with respect to gasoline performance.

In this section we will consider the most common spark ignition engine using a reciprocating piston to transmit power. The other engine type is the rotary or Wankel engine in which the combustion chambers rotate. This engine design is much less common and from an automotive fuel's viewpoint does not pose any significantly different problems to that of the conventional reciprocating engine.

The general principle of operation of the four-stroke internal combustion engine (also known as the Otto engine, after its German inventor in the late 1880s) is well known and will only be described briefly. The fuel system provides a controlled mixture of fuel vapor and air to the combustion chamber. This mixture is compressed by the piston being pushed upward by the action of the crankshaft and is then ignited by means of a spark plug. The resulting combustion gives rise to an increase in pressure in the combustion chamber which pushes the piston down and transmits this power to the crankshaft and, thence, to the wheels. The piston pushes the combustion gases out through the exhaust valve on its return upward stroke. The exhaust valve then closes and the inlet valve opens so that, as the piston travels downward, it draws in a fresh charge of air and fuel mixture. The next upward stroke of the engine compresses this mixture since both inlet and exhaust valves are closed, and then the mixture is again ignited and the four cycles of compression, ignition, exhaust and induction continue.

Two-stroke engines have been widely used in the past because they have the advantage of light weight, compactness, lower manufacturing costs than four-strokes and high specific power. In this design the air-fuel mixture is pumped into the combustion chamber through transfer ports during the piston's down stroke and, at the same time, the exhaust gases (and some of the incoming charge) escape through the exhaust port on the other side of the cylinder. Two-strokes were virtually abandoned by about 1970 because of a number of factors including the difficulty in controlling exhaust emissions, fouling of spark plugs, noise and the need to mix oil and gasoline. However, new designs are now being progressed[2,3,4] which overcome many of the problems and may lead to the return of this type of engine. The octane requirements of these engines, at least in terms of Research octane number, can be low and this could be an advantage for the future.

5.3 Vehicle Fuel Systems

The fuel system is of prime importance to the gasoline technologist. It starts with the fuel tank which is usually situated at the rear of the vehicle and should be manufactured from material that cannot be attacked by the gasoline. Plastic fuel tanks are generally satisfactory although problems of permeability have been encountered with some of them when fuels containing alcohols, and particularly methanol, have been used. Terne plate, which is a steel sheet coated with a lead-tin alloy, is also generally satisfactory provided that there are no pin-holes for corrosion to start by attacking the underlying steel. A small amount of water is not uncommon in a fuel tank since it can separate from the fuel during a decrease in temperature. Alcohol-containing fuels can also attack terne plates, the extent of the attack depending on its exact composition. Therefore, it is important to ensure that the correct terne plate grade is used to minimize the risk of attack. Other coatings on steel are used by different manufacturers and it is impor-tant that they are fully evaluated using as wide a range of fuels as possible before being introduced.

Most car tanks have a vent system to allow air to escape during fill-ing, and in countries where there are restrictions on evaporative emis-sions, the vent pipe (and other parts of the fuel system where it is possible for vapors to escape) is connected to a charcoal canister to prevent vapors being released to the atmosphere. Clearly the vapor pressure of the gasoline will be a factor in determining the size of canister required. The canister is regenerated by sucking intake air through it when the engine is running so that the vapors are burned. The filler cap often incorporates a two-way relief valve which allows a slight pressure or vacuum to develop, and there is frequently a rollover valve which prevents fuel from discharging if the vehicle rolls over in an accident. In some countries it is also required that gasoline vapor escaping from the tank during filling be controlled. Special filling nozzles are used in these cases which provide a seal at the filler pipe and a vapor return line which takes the vapor back to the storage tank.

The car tank will also incorporate a strainer/filter in the bottom of the tank near the outlet line. Particles of dirt and rust which may be intro-duced with the fuel or which may appear due to corrosion of the tank are held here to prevent problems later in the system.

The fuel pump delivers fuel from the tank to the carburetor or injec-tors and its siting and design are critical to the performance of the vehicle in hot weather. Mechanical fuel pumps are usually mounted on the side of the cylinder block and operate by raising and lowering a diaphragm which draws fuel in and then forces it out through a series of valves. Many vehicles are fitted with a vapor return line from the fuel pump to the tank in order to return any vapors and excess fuel to the tank. Pumping stops if vapor forms in the pump causing "vapor lock" which can halt a car completely. Heat can be transmitted to the pump body from the cylinder block and this, together with the fact that the pump operates on the principle of using a partial vacuum to draw in the fuel, means that this type of pump is somewhat prone to fuel vaporization problems (see Chapter 7), resulting in poor hot weather driveability. The fuel lines from the tank to the pump are also subject to vapor formation unless they are carefully sited because they are on the suction side of the pump and may be subjected to heat from the engine.

Electric fuel pumps do not generally suffer from these problems because they are increasingly sited in the fuel tank itself so they are kept cool and fuel from them is under pressure and less likely to vaporize.

The pump supplies fuel directly to the carburetor or fuel injectors. Carburetors provide a mixture of atomized fuel and air in the required ratio, depending on the engine operating conditions. A diagram of a simplified carburetor is shown in Fig. 5.1. As air flows through the venturi it produces a partial vacuum which causes the fuel nozzle to deliver a fine spray of gasoline into the air stream. The throttle or butterfly valve controls the air passage and, hence, the amount of gasoline entering the air stream. Fuel is delivered into the carburetor body via a float bowl that maintains a constant head of fuel. The float bowl has a vent to atmosphere or to a carbon canister. Air-fuel ratio variations are required as driving conditions change; for example, when starting on cold days it must be rich, whereas for normal cruising it can be quite lean. These variations are achieved by means of a choke or some other system. Electronic control of air-fuel ratio by means of an oxygen sensor in the exhaust gases is used in many car models, particularly when they are fitted with certain types of exhaust catalysts which operate most effectively at stoichiometric air-fuel ratio.

During idle operation the throttle is virtually closed so that very little air can get by the butterfly valve, and it is necessary to have a special system to allow the required amount of mixture at the correct air-fuel ratio to get to the combustion chamber. This involves a fairly complex series of passages in the carburetor through which the air-fuel mixture passes until it emerges just upstream of the throttle valve, where it is mixed with the small amount of air getting past to give the air-fuel ratio required for smooth idle performance. Deposits from the

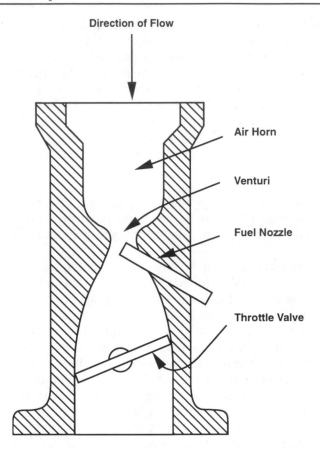

Fig. 5.1. A simple carburetor. (Reproduced with permission from <u>Automotive Engines</u>, Crouse and Anglin, 5th Edition, 1975, McGraw-Hill)

fuel and from gases and particulates pulsing back from the combustion chamber can build up on the throttle plate and the throttle body and restrict the air going past. This is only of concern at low throttle openings, as occurs during idle, when the mixture going forward may be too rich for smooth running and give rise to uneven firing and stalling. Some fuel compositions accentuate this problem but additives are available which can minimize deposit formation (see Chapter 9).

There are many different designs of carburetor which are now extremely sophisticated. They vary in the extent to which they are subject to deposit formation and in their sensitivity to such deposits in terms of performance.

The position of the carburetor is also very important. If it is positioned so that excessive heat soak-back can occur from the engine after it is switched off, then a number of hot-fuel handling problems can occur, as described in Chapter 7.

Fuel injection is an alternative to carburetion that is now dominating new car design because it is capable of superior control, which is very important in view of the tight restrictions on exhaust emissions. Injection is normally into the inlet manifold via a throttle body injector (TBI), similar to a carburetor system, or into the intake ports by means of port fuel injectors (PFIs). The advantages over carburetion are increased power output relative to engine size, lower specific fuel consumption, higher torque at low engine speeds, lower exhaust gas pollutants and improved hot and cold weather driveability performance. Although it is more expensive, fuel injection is capable of much better control of air-fuel ratio than carburetion because injection time can readily be varied to cover all engine operating conditions. It is also suitable for use with an electronic engine management system.

There are two types of injector — mechanical and electronic. In the mechanical type, fuel is supplied to each injector from a fuel distributor and the amount of fuel delivered is proportional to the quantity of air inducted by the engine. From a fuel viewpoint, this type of injector is relatively free of problems.

The electronic system (see Fig. 5.2) uses electromagnetically actuated injectors which spray fuel into the inlet ports of the engine either on each revolution of the crankshaft or on every other revolution. The action is intermittent and is controlled by a number of engine factors. Good atomization in the spray is vital to ensure low hydrocarbon emissions and good driveability.[5] Because injection is not continuous, there can be a tendency for deposits to occur on the pintle due to any residual gasoline present being subjected to heat from the engine. These deposits can affect the spray pattern and upset vehicle driveability. The problem is most acute in stop/start driving conditions which maximizes opportunities for heat soak-back into the injectors. It has been largely overcome by the use of appropriate additives in the fuel (see Chapter 9) and, in new models, by improved injector design.

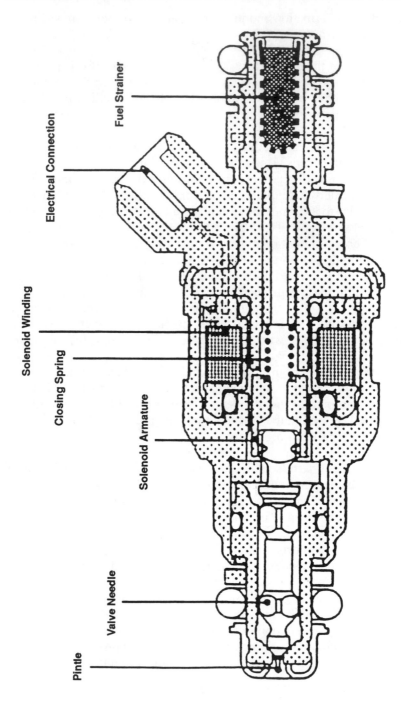

Fig. 5.2. A typical port fuel injector.[6]

Three other areas of the fuel system can be susceptible to fuel quality: the intake manifold, the intake ports, and the intake valves. The likelihood of different gasoline qualities forming deposits in these areas is discussed in Chapter 8, but such deposits can cause driveability problems. The intake manifold is probably less sensitive except that, if a "hot spot" is used to warm up the charge and help vaporization, deposits are more likely to form on it and, if so, will increase warm-up time.

Deposits on the intake valves and in the ports will interfere with the breathing of the engine and adversely affect cold starting and acceleration performance. In extreme cases, engine damage can occur as a result of the valve being held off the seat. Such damage is sometimes caused by hot gases escaping past the open valve which can burn away part of the seat, or, in some designs, by the open valve being struck by the rising piston.

5.4 Ignition Systems

For the air-fuel mixture to ignite in the combustion chamber and provide the maximum amount of useable work requires a spark of adequate energy at the right time in the cycle, the air-fuel ratio to be in the burnable range, and for the plug to be positioned so that the spark occurs within a combustible mixture of air and fuel. Most of these are outside the influence of the fuel but the fuel factors that can impinge upon the correct ignition and combustion of the charge are the anti-knock quality, the lead level and, to some extent, the gasoline composition. The combustion process itself is discussed in some detail in Chapter 6 so that at this stage it need only be mentioned that the minimum electrical energy to ignite a mixture of gasoline and air is when the mixture is in stoichiometric proportions. Going on either side of stoichiometric requires more and more energy until the limits of burnability are reached. The fuel can influence ignition because some components are able to burn over a wider range of air-fuel ratio than others. Additives are available that help the spark and overcome to some degree the problems of spark plug fouling (see Chapter 9). Improved spark ignition systems[7] can also help when the air/fuel ratio is lean so that toxic exhaust emissions are minimized (see Section 5.5).

The timing of the spark is vital because if it occurs too soon, i.e., if the timing is over-advanced, knock can occur causing damage to the engine and loss of power. This can be avoided by the use of higher octane quality fuels, but suitable fuels are not always readily available. If the spark is later than the optimum point, i.e., if it is retarded, then loss of power can result. In some engines there is a knock sensor fitted[8] which detects knock and puts into effect some action to prevent knock and so avoid engine damage. Usually this means retarding the ignition, but other actions are possible. Because knock is influenced by driving conditions, it will often only occur for the time that the engine is being driven under knock-critical modes. Of course, there will be no knock at all if the fuel has a higher antiknock value than required by the engine. The knock sensor system is designed so that after a few cycles without knock, the ignition is gradually advanced back to the optimum setting until the next knocking cycle. Quite complex systems for retarding and re-advancing the ignition timing have been developed to minimize the time that the engine is running in a nonoptimized state. From a fuel quality viewpoint this can mean that the engine will function on a wide range of octane quality fuels without damage, although lower quality fuels will give poorer acceleration performance.

5.5 Combustion and Exhaust Emission Control Systems

The combustion of gasoline is discussed in detail in Chapter 6; in this section we discuss engine design factors that influence the fuel quality required. Many of the design features introduced to control exhaust emissions also impact gasoline quality, and these are also discussed.

The most important aspects of combustion are that the fuel should burn smoothly and efficiently, without knocking, and without giving rise to unacceptable levels of toxic exhaust gases. The exhaust gases that are of most concern, and for which there is control legislation, are carbon monoxide, unburnt hydrocarbons and oxides of nitrogen. Another exhaust-gas pollutant that is still important in many countries is, of course, lead. There are other pollutants such as benzene and polyaromatic hydrocarbons (PAHs) for which there is also concern.

Considering first the combustion chamber itself, there are two major factors that are important, and these are the compression ratio and the shape. Compression ratio is the ratio of the volume of the chamber when the piston is at bottom dead center (BDC) to when it is at top dead center (TDC), and is a measure of how much the incoming charge is compressed in the cylinders. It is important because the higher the compression ratio, the better the thermal efficiency and, hence, the better the fuel economy and power for a given-sized engine. Unfortunately, the tendency for the fuel to knock also increases with increasing compression ratio. The use of turbochargers or superchargers to increase power from an engine also effectively increases the compression ratio, and so increases the detonation tendency.

Knock occurs when the unburnt gases ahead of the flame front (the "end gases") spontaneously ignite causing a sudden rise in pressure accompanied by the characteristic pinging or pinking sound. This results in a loss of power and can lead to damage to the engine. It should not be confused with preignition which occurs when the charge is ignited, perhaps by glowing deposits or a hot spark plug, before the spark appears.

There are two common shapes for combustion chambers — hemispherical and wedge (Fig. 5.3). The shape influences three factors that are important for both detonation tendency and exhaust gas quality: turbulence, squish and quench.

Turbulence ensures good mixing of the air and fuel and reduces the time for a flame front to sweep through the mixture, thereby allowing less time for detonation to occur. Squish is the term used for the squeezing of the mixture away from the end gas region in a wedge-shaped combustion chamber, which causes further turbulence. Quench is the cooling effect of the sides of the combustion chamber which can make the mixture in the vicinity of the walls too cool to detonate. The squish region of a wedge-shaped combustion chamber is also the quench area since heat is readily conducted away, thereby reducing the tendency for the end gas to explode. Quenching also has the effect of preventing the thin layer of air-fuel mixture next to the walls of the combustion chamber from completely burning so that some unburnt hydrocarbons are present in the exhaust gases.

For low hydrocarbon emissions a compact combustion chamber is required with a small surface-to-volume ratio to minimize quench effects. The spark plug should be located centrally so that the flame front has a relatively short distance to travel and, hence, allows complete combustion without detonation. There also needs to be turbulence. The hemispherical combustion chamber meets a number of these requirements in that it has a low surface-to-volume ratio and the spark plug is located near the center of the dome. However, there are no squish or quench areas as in the wedge type of chamber and the

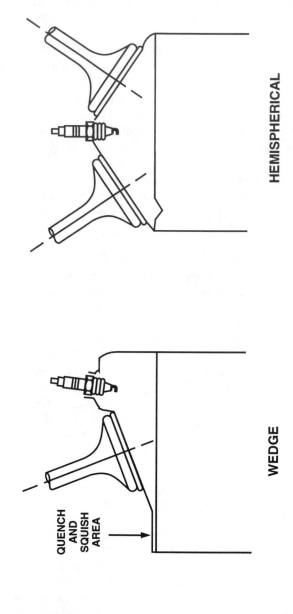

Fig. 5.3. Wedge and hemispherical combustion chambers. (Reproduced with permission from Automotive Engines, Crouse and Anglin, 5th Edition, 1975, McGraw-Hill)

lack of turbulence prevents good mixing so that exhaust gas quality is not particularly good.

The use of a very small squish zone clearance, i.e., below about 1.0 mm, helps to minimize hydrocarbon emissions. However, it has been found that if high levels of deposits build up on the top of the piston, contact can be made with the cylinder head in the squish zone[9] when the piston is at top dead center (see also Chapter 9, Section 9.3.6). This mechanical contact will give rise to a knocking noise which has been called "carbon knock" or "carbon rap," among other names, and is totally unrelated to combustion knock. It mostly disappears after the engine has warmed up. Excessive carbon build-up causing this problem may result from high treats of certain types of additive[10] or from the presence of too great a level of deposit-forming high-boiling point compounds in the gasoline.

Mixture strength has an important influence on exhaust emissions as shown in Fig. 5.4. Here the symbol λ represents the air-fuel ratio, and a value of 1.0 represents the stoichiometric ratio; lower values refer to mixtures richer than stoichiometric and higher values to leaner than stoichiometric.

It can be seen that emissions of hydrocarbons (HC) and carbon monoxide (CO) are high when λ levels are low, although nitrogen-oxide emissions are low. Maximum torque is obtained when the mixture strength is about 0.9λ and air-fuel ratios are usually adjusted to this value when maximum power is required, i.e., when running at wide open throttle. Specific fuel consumption is high under this condition.

At stoichiometric the CO and HC levels drop to a low level as does the specific fuel consumption, and as λ values increase these parameters continue to fall and then the hydrocarbon emissions start to increase as the misfire limit is approached. Nitrogen oxides reach a maximum just lean of stoichiometric and are reduced by the use of exhaust gas recirculation (EGR) which lowers the peak temperature of combustion. EGR is achieved either by means of valve overlap or by feeding exhaust gas back through valves into the inlet manifold or elsewhere in the intake system. EGR has been blamed for contributing to deposits in the intake system and its use has increased the need for additives to keep the intake system clean.

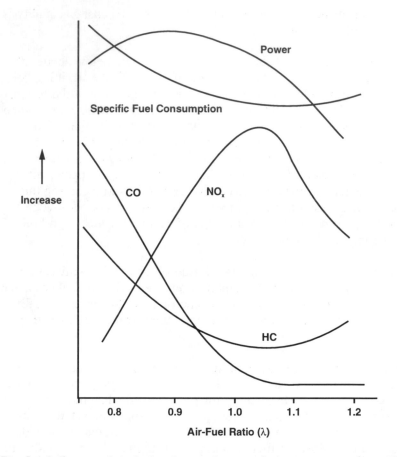

*Fig. 5.4. Influence of air-fuel ratio on exhaust emissions, power and specific
    fuel consumption.*

It is clear that by using EGR and a lean mixture, exhaust emissions
can be reduced to a fairly low level without the use of a catalytic system. However, as the λ value approaches the misfire region, vehicle
driveability begins to worsen. One way of avoiding this is to use a
stratified charge system in which the gas in the region of the spark
plug is much richer than the remaining gas so that it ignites readily
and forms a good flame kernel that completely combusts the remainder of the charge. This system allows a much leaner mixture to be
used and the more complete combustion gives improved fuel
economy without driveability problems. Stratified charging can be
achieved in a number of ways such as by giving the air-fuel mixture a

swirling motion as it enters the chamber, or by having a precombustion chamber containing the spark plug that receives a much richer mixture than the main part of the combustion chamber.

Gasoline volatility and composition, especially the presence of oxygenates, can have an effect on emissions and on driveability, as is discussed in Chapter 7, and these become particularly important when vehicles are designed to operate near the lean limit of combustion.

Three-way catalytic converters, which use a catalyst containing platinum, palladium and/or rhodium, have been mandatory for all new cars in the U.S. and Japan since 1975. They convert the carbon monoxide, unburned hydrocarbons, and nitrogen oxides in the exhaust to carbon dioxide, water and nitrogen, which are much less harmful. The use of these catalyst systems is increasing in Europe and other parts of the world and is, of course, a primary reason for the change to unleaded gasoline. The efficiency of catalytic systems to reduce exhaust emissions is influenced by fuel sulfur levels since sulfur does poison catalysts, but to a much lesser extent than lead. Because of this, as discussed in Chapters 10 and 12, there is pressure to reduce the maximum allowable levels of sulfur in gasoline.

Approaches to meet the improved fuel efficiency and exhaust emission levels required by Low Emission Vehicles (LEVs) and Ultra Low Emission Vehicles (ULEVs) after 1995 in California have been reviewed[11] and these designs may well be equally sensitive to fuel quality—as may other novel engine concepts aimed at improving emissions and efficiency.[12,13]

References

1. F.H. Palmer, "Environmental Challenges Facing Road Transport Fuels and Vehicles Outside the U.S.A.," SAE Paper No. 932682, 1993.

2. L. Brooke and P.J. Mullins, "To Stroke...or Not Two Stroke?" *Automotive Industries*, May, 1988.

3. P. Duret, A. Ecomard and M. Audinet, "A New Two-Stroke Engine with Compressed Air Assisted Fuel Injection for High

Efficiency Low Emissions Applications," SAE Paper No. 880176, 1988.

4. "New Developments in Two-Stroke Engines and Their Emissions," SAE SP-835, 1990; SP-833, 1991; and SP-849, 1991.

5. J. Senda, T. Tsukamoto, *et al.*, "Atomization of Spray Under Low Pressure Field from Pintle Type Gasoline Injector," SAE Paper No. 920382, 1992.

6. R.C. Tupa and C.J. Dorer, "Gasoline and Diesel Fuel Additives for Performance/Distribution Quality II," SAE Paper No. 861179, 1986.

7. M.A.V. Ward, "A New Spark Ignition System for Lean Mixtures Based on a New Approach to Spark Ignition," SAE Paper No. 890475, 1989.

8. S.M. Dues, J.M. Adams and G.A. Shinkle, "Combustion Knock Sensing: Sensor Selection and Application Issues," SAE Paper No. 900488, 1990.

9. Proceedings of the CRC Workshop on Combustion Chamber Deposits, November 15-17, 1993. Published by Coordinating Research Council, Inc.

10. P. Schreyer, K.W. Starke, *et al.*, "Effect of Multifunctional Fuel Additives on Octane Number Requirement of Internal Combustion Engines," SAE Paper No. 932813, 1993.

11. G.K. Fraid, F. Quisseck and E. Winkelhoffer, "Improvement of LEV/ULEV Potential of Fuel-Efficient High Performance Engines," SAE Paper No. 920416, 1992.

12. K. Craven, N. Clark and J.E. Smith, "Initial Investigations of a Novel Engine Concept for Use with a Wide Range of Fuel Types," SAE Paper No. 920057, 1992.

13. "Il Motori Verde," *La Rivista Dei Combustibili*, 45, No. 32:83-88, February 1991.

Chapter 6

Gasoline Combustion

Combustion in the reciprocating internal combustion engine is quite different from the relatively simple continuous combustion that takes place in engines such as the gas turbine. It is intermittent and occurs under complex and continually varying conditions. Combustion efficiency in such engines is very sensitive to fuel quality and the fuel quality requirement of the engine is strongly dependent on operating conditions.

6.1 Normal Combustion

The amount of air required to combust a hydrocarbon fuel completely to carbon dioxide and water can readily be calculated. For example, heptane, a hydrocarbon boiling in about the middle of the gasoline range, under perfect combustion conditions undergoes oxidation as follows:

$$C_7H_{16} + 11O_2 \rightarrow 7CO_2 + 8H_2O$$

In this example, one volume of heptane requires 11 volumes of oxygen or 52.6 volumes of air. On a weight basis this is equivalent to an air-fuel ratio of 15:1. One part of gasoline, because it differs from heptane in its carbon-hydrogen ratio, requires about 14.5 parts by weight of air for complete combustion, although the exact stoichiometric amount will depend on the composition of the fuel. Generally speaking, for a purely hydrocarbon gasoline, if the air-fuel ratio is less than about 7:1 it will be too rich to ignite and if it is more than about 20:1 in a conventional engine it will be too weak. Lean burn engines are now manufactured that will operate satisfactorily on air-fuel ratios leaner than 20:1.[1] These engines have the advantages of improved engine efficiency and low levels of carbon monoxide and oxides of nitrogen in their exhaust gases, but not as low as can be achieved using a catalytic system.

When a fuel is mixed with oxygen, so-called "preflame" reactions will commence even before the mixture has reached the combustion chamber and will continue after ignition until all the fuel has been consumed by the advancing flame front.[2,3] The extent of these reactions will depend on a number of factors such as the chemical composition of the fuel and the temperature and pressure of the mixture. The nature of the reactions will determine whether the fuel will burn "normally", i.e., whether it will combust smoothly and efficiently or whether it will give rise to some form of "abnormal" combustion such as, for example, knock or preignition.

Normal combustion occurs when a flame front moves smoothly, but in a somewhat irregular fashion, across the combustion chamber after being initiated by the spark, until combustion is complete. The irregular movement is because of turbulence and incomplete mixing. The pressure changes in the chamber during this process are shown in Figure 6.1. It will be seen that the pressure begins to rise as the mixture is compressed and then rises rapidly after ignition due to the temperature increase and the formation of the combustion gases, mainly carbon dioxide and water vapor. Maximum pressure occurs soon after the piston has reached the top of its travel (TDC).

Even with normal combustion, all spark ignition engines show variations in the maximum cylinder pressure and rate of pressure rise from cycle to cycle (this is known as cyclic dispersion) in spite of strict control of the running conditions.[4] It is believed to be due to variations in turbulence between cycles which vary flame speeds across the combustion chamber. If reductions in cyclic dispersion could be achieved, significant benefits in terms of improved fuel consumption and lower octane requirements would be possible.[5]

6.2 Spark Knock

Spark knock, so called because it is influenced by the timing of the spark, is one of the most important forms of abnormal combustion; it determines, to some extent, the thermal efficiency that can be achieved in an engine. The higher the compression ratio, the better the thermal efficiency, but the greater the tendency for spark knock to

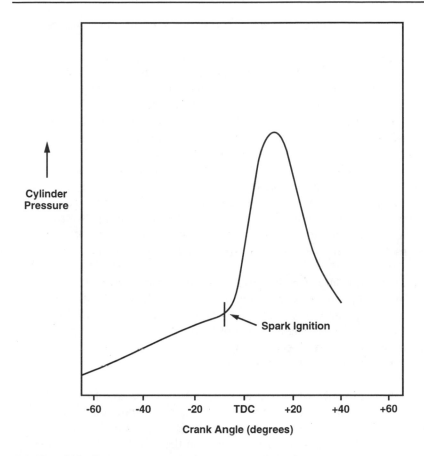

Fig. 6.1. Cylinder pressure trace during normal combustion.

occur and so the higher the fuel octane quality that is required. By retarding the ignition timing, the tendency for knock will decrease (and vice versa), but going beyond a certain limit can adversely affect power output. This response to ignition timing distinguishes it from other forms of abnormal combustion such as preignition or run-on.

Even vehicles operating on a fuel for which they have been designed will sometimes knock, and this may be due to a number of factors such as excessive deposit formation in the combustion chamber, over-advancement of the ignition timing, particularly severe driving conditions, or a combination of several factors during manufacture in which the production tolerances all conspire to increase octane requirement.

Of course, it can also be due to the fuel being of a poorer quality than specified.

The sequence of events when knock occurs is shown diagrammatically in Fig. 6.2. As the flame propagates from the spark plug, the temperature and pressure of the unburnt gas ahead of the flame front are raised due to heat from the flame front itself and the increase in pressure from the expanding burning gases. Because of this, preflame reactions will take place at an increasing rate and eventually may reach the point when the mixture will self-ignite. In normal combustion, this stage is never reached because there is insufficient time for the preflame reactions to progress to the point of autoignition. As engine speed is increased, the time for preflame reactions to take place is reduced and so the tendency to knock decreases, although normally the ignition timing is advanced as engine speed increases.

The autoignition of the end gases causes a rapid increase in pressure, setting up a pressure wave which resonates in the combustion chamber at a frequency of between 5000 and 8000 Hz, depending on the geometry of the chamber. This produces the characteristic pinking sound associated with knock. Fig. 6.3 shows a typical cylinder pressure trace during a knocking cycle.

Knock during acceleration at wide open throttle from a low engine speed is of such short duration that it does not usually cause damage and, unless it is very severe, will not cause discernible loss of power.[6] High constant-speed knock, on the other hand, can cause loss of power and severe damage usually to the cylinder head gasket, the spark plug electrodes and the pistons.[7,8] In extreme cases, knock can lead to preignition (see Section 6.8.1) and/or so-called "runaway knock" where the knock intensity gets progressively higher until catastrophic engine damage occurs. This is because the temperature in the chamber is raised during heavy knock and this, in turn, increases the tendency for more knock to occur, and so on.

Knock continues to be one of the most important aspects of engine design and of gasoline technology, and many aspects of it are still incompletely understood and under investigation.[9,10,11,12,13]

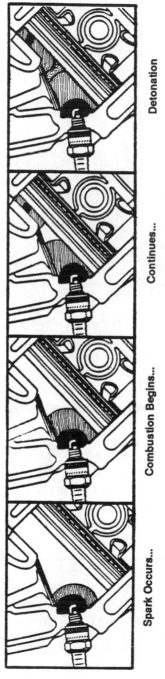

Fig. 6.2. Sequence of events when knock occurs (source: Champion Spark Plug Company).

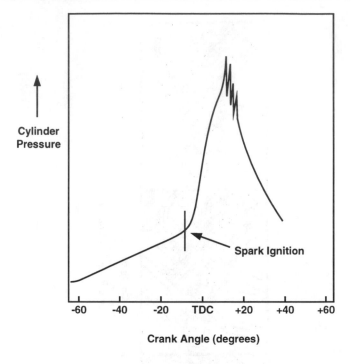

Fig. 6.3. Cylinder pressure trace during knock.

6.3 Measurement of Gasoline Antiknock Quality

Prior to 1929, fuels were rated using an engine in which the compression ratio could be varied between 2.7:1 and 8:1. Each fuel was run in this engine at various air-fuel ratios and ignition timings to obtain conditions for maximum power, and the highest compression ratio was then established beyond which knock and power loss occurred. Fuels were assigned values in terms of Highest Useful Compression Ratio (HUCR), a scale which was later standardized against toluene.

The octane scale proposed by Graham Edgar in 1926 was established in 1929 and has been used ever since. In this scale, two pure paraffinic hydrocarbons of similar physical characteristics have been selected as standards. One, isooctane (2,2,4 trimethylpentane), having a very high resistance to knock, was arbitrarily assigned a value of 100, and the other, n-heptane, with an extremely low knock resistance, was assigned a value of zero. The octane number of a fuel (Research Octane or Motor Octane) is the volume percentage of isooctane in a blend with n-heptane that shows the same antiknock performance as the test fuel when tested in a standard engine under standard conditions. For fuels having octane values above 100, mixtures of isooctane and tetraethyl lead are used and the relationship between octane number and the lead content of isooctane is published in the standard test method and reproduced in Appendix 10.[14]

The test engine used for determining octane values of fuels was developed in the U.S. by the Waukesha Company under the direction of the Cooperative Fuel Research Committee (CFR), and is a single-cylinder, variable compression ratio engine known as a CFR engine. It was found in a series of tests in the early 1930s that it was not possible to correlate the performance of cars on the road with just one type of octane number because engine designs and driving conditions were continually changing, and this modified the response to octane quality. Nowadays, it is usual to define octane quality using at least two octane parameters and sometimes three. These are Research Octane Number (RON), Motor Octane Number (MON), and a number concerned with the distribution of Research Octane quality through the boiling range of the gasoline, which is discussed later.

The Research Octane Number test correlates best with low-speed, relatively mild driving conditions whereas the Motor Octane Number relates to high-speed, high-severity conditions. Clearly, vehicles on the road are operating most of the time at a severity somewhere between these two levels, and in some countries, notably North America, gasoline is specified by an Antiknock Index which is the average of the RON and MON of the fuel, i.e.:

$$\text{Antiknock index} = \frac{1}{2} \, (\text{RON} + \text{MON})$$

In Europe, RON and MON are usually specified separately and gasoline is often designated for identification or marketing purposes only by the Research Octane rating.

Most gasolines have a higher RON than MON, and the difference between these two ratings is called the "sensitivity." For fuels of the same RON, a high-sensitivity gasoline will have a lower MON than a low-sensitivity gasoline. It represents the sensitivity of the fuel to changes in the severity of engine operating conditions in terms of antiknock performance. This is illustrated in Fig. 6.4[15] in which octane number is plotted against engine severity. The location of the Research and Motor test procedures in terms of their relative severities is shown and most engines will operate at a severity somewhere between these two positions for most of the time. However, there are conditions when some engine designs may be more severe than the Motor method (as with some two-stroke engines) and others which may be less severe than the Research method (as with some turbocharged or supercharged engines). Three fuels are shown: Fuel A having a high RON (97) and a low sensitivity (S=3), Fuel B having a somewhat lower RON (96) and a high sensitivity (S=12), and Fuel C having a much lower RON (84) but a very low sensitivity (S=1). It can be seen that if the vehicle happens to be operating at a severity corresponding to X, then Fuel B performs better than Fuel A even though it has a lower RON and MON. Similarly, if the vehicle is operating at a severity corresponding to Y, Fuel C is better than Fuel B even though, again, it has a lower RON and MON than Fuel B. Of course, at the severity levels that occur with most vehicles for most of the time, Fuel A is better than Fuel B which is better than Fuel C.

The test procedures[14] for RON and MON are carried out under the conditions summarized in Table 6.1 below:

Table 6.1 Test Conditions for the Research and Motor Test Procedures

Test	Research	Motor
CRC Designation	F-1	F-2
ASTM Method	D2699	D2700
Engine speed, rpm	600	900
Intake air temperature, °C	depends on barometric pressure	38
Mixture temperature, °C	not specified	149
Coolant temperature, °C	100	100
Ignition advance, deg	13· btdc	linked to compression ratio

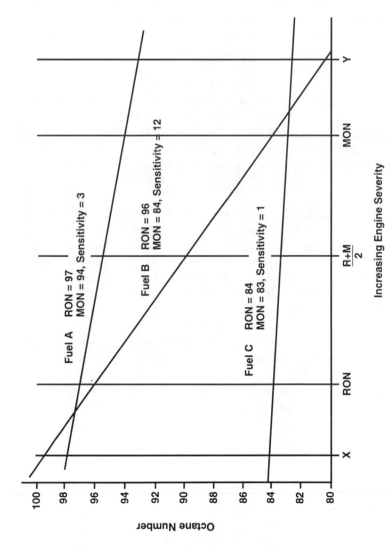

Fig. 6.4. Octane quality and engine severity.[15]

Both procedures are very similar and involve adjusting the compression ratio to obtain a standard knock intensity as indicated by a knock meter, using the fuel under test. The air-fuel ratio is then adjusted to give maximum knock by raising or lowering the carburetor bowl and the compression ratio is then readjusted to give a mid-scale reading on the knock meter. A blend of isooctane and n-heptane (known as Primary Reference Fuels, PRFs) of approximately the same octane number as the test fuel is placed in a second carburetor bowl which is then switched through to the engine and the height again adjusted to give maximum knock. The reading on the knock meter is noted provided it is within a certain range, and a second PRF differing by two octane numbers from the first is tested. This is continued until the reading for the test fuel is bracketed by the readings for two PRFs which differ by no more than two octane numbers. The octane quality of the test fuel is then obtained by interpolation and is usually reported to the nearest 0.1.

A third octane parameter which gives a measure of the distribution of RON quality through the boiling range of the gasoline has been widely used in Europe, although its importance is declining as engine designs improve and become more sophisticated. It is possible to make two fuels which have the same RON, but the way that the octane quality is distributed throughout the boiling range can be quite different as shown in Fig. 6.5.

The RON levels of the lighter parts of Fuel A are much lower than the corresponding portions of Fuel B, and the RON levels of the heavier fractions are higher than those of Fuel B. In Fuel B all the components have similar octane qualities, whereas Fuel A is composed of light components of relatively poor octane quality and heavier components of high octane quality. In practice, most gasolines have a lower octane quality in the front, or low boiling end, than in the back, or high boiling end.

The importance of octane distribution in the fuel is that during full throttle accelerations in some vehicles, large amounts of liquid fuel are forced into the inlet manifold and initially only the lighter parts are swept on into the combustion chamber by the air stream. If these have a low octane quality, transient knock may occur until the heavier ends catch up as the acceleration continues. Although this type of transient knock does no damage to the engine because it does not last long enough for combustion chamber temperatures to rise very much,

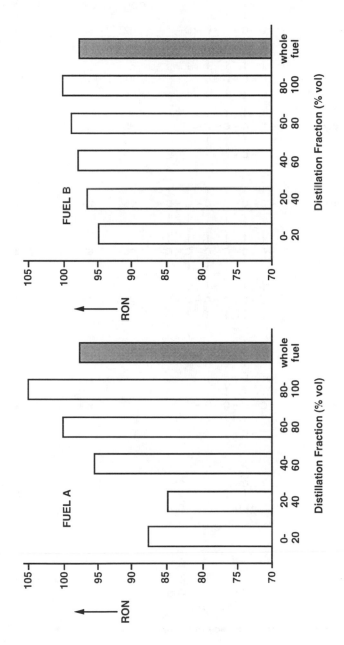

Fig. 6.5. Distribution of octane quality through the boiling range of two gasolines.

		RON	MON	Sensitivity (RON-MON)
n-Pentane	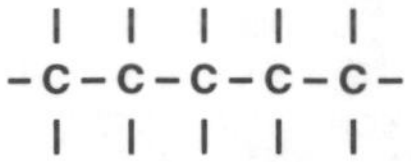 $-C-C-C-C-C-$	62	62	0
Pentene 1	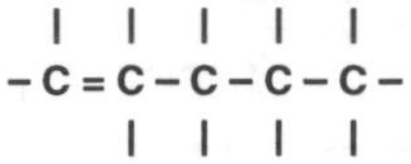 $-C=C-C-C-C-$	91	77	14
Iso pentane	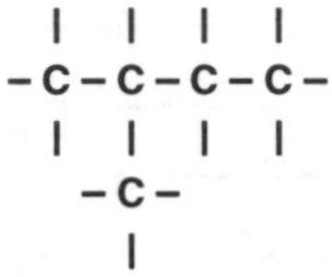 $-C-C-C-C-$ $-C-$	92	90	2
n-Hexene	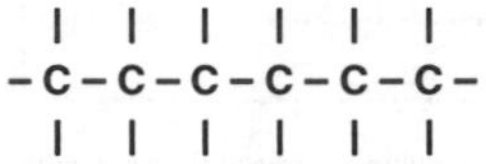 $-C-C-C-C-C-C-$	25	26	1
Hexane 1	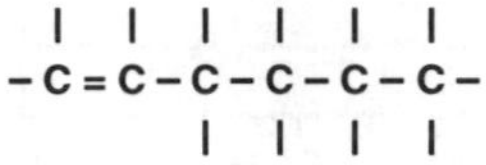 $-C=C-C-C-C-C-$	76	63	13
2 Methyl pentane	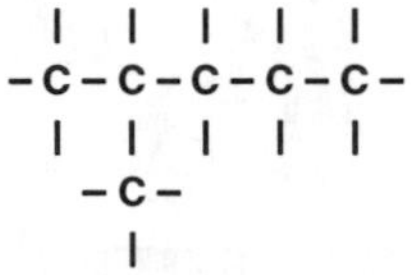 $-C-C-C-C-C-$ $-C-$	73	73	0
2, 2 Dimethyl butane	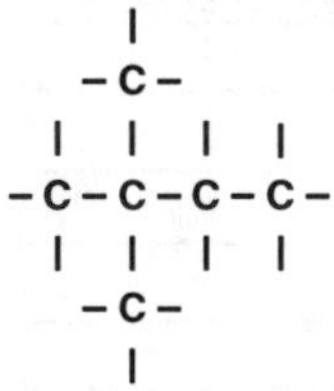 $-C-$ $-C-C-C-C-$ $-C-$	92	93	-1
Cyclohexane	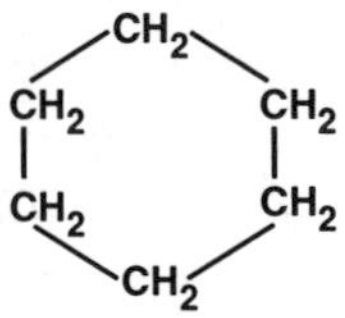 (cyclohexane ring: CH_2)	83	77	6
Benzene	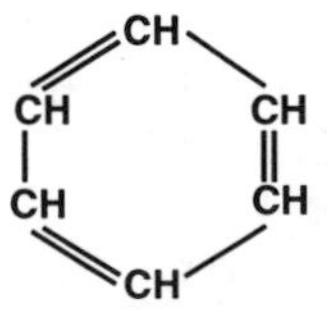 (benzene ring: CH)	>100	>100	Approx 12

Note: H atoms have been omitted from some formulae above for simplicity.

Fig. 6.6. Chemical structure influences octane quality.

the noise can be disturbing to the driver. The increasing use of fuel injection and more sophisticated fuel intake systems have reduced this problem of fuel segregation in the inlet manifold.

Octane distribution in the fuel can be measured by a number of different tests, but probably the most well-known involves simply distilling off the portion of fuel boiling up to 100°C and measuring its RON. This value is usually designated as the R100°C of the fuel. The difference between the RON of the whole fuel and the R100°C is abbreviated to ΔR100°C. Other tests that have been used to define front-end octane quality include measuring the RON of the first 75% to distill off (R75%) and the use of a modified CFR engine with a manifold which segregates some of the heavier ends before they reach the cylinder (DON, Distribution Octane Number, measured using ASTM D 2886).

Other methods of measuring octane quality by correlating laboratory test data with the CFR engine results have been devised, mainly because the engine test is slow and expensive. They are often based on composition[16] or on infrared spectra[17] and there are numerous variations of these and other methods for predicting octane quality without going through the expensive and time-consuming engine test procedure. Many of the procedures can be operated continuously and have been developed purely for process control.

6.3.1 Influence of Chemical Structure on Octane Quality

The chemical structure of hydrocarbons has a great influence on the octane quality, as illustrated in Figure 6.6. This shows that the introduction of a double bond in a straight chain hydrocarbon to make an olefin (see Appendix 1 for an explanation of chemical structure) has a large effect on increasing RON although MON is increased to a smaller extent (compare n-pentane with pentene 1). Cracking processes give rise to olefinic compounds and so cracked streams tend to have high sensitivities.

Similarly, branch chain compounds have better octane qualities than straight chain, as shown in Figure 6.6 by a comparison of n-hexane with 2 methyl pentane and with 2,2 dimethyl butane. Each of these three compounds has the same number of carbon and hydrogen atoms, but both RON and MON increase as the degree of branching increases, and the sensitivity of all of them is low. Branch chain paraffinic hydrocarbons are present in the products from isomerization and alkylation processes, i.e., in isomerate and alkylate.

Saturated cyclic compounds (i.e., naphthenes) such as cyclohexane can be seen to have better octane levels than the corresponding straight chain compound, in this case hexane, but with an intermediate sensitivity. Aromatic compounds such as benzene have still higher octane values but with a relatively high sensitivity. Aromatics are formed in the catalytic reforming process and, to a lesser extent, in cracking processes.

6.4 Antiknock Additives

6.4.1 Lead Alkyls

Lead alkyls are still the most important antiknock additives, even though they are being phased out in many countries for environmental reasons. They usually represent the most economical way to achieve required octane levels. With unleaded gasoline the required octane levels are generally obtained by more severe refining but this reduces yield and increases operating costs. However, the need to reduce noxious emissions from exhaust gases is well documented and, because lead poisons the catalysts used to achieve this and is itself toxic, it is both inevitable and desirable that lead compounds in gasoline should be phased out. Nevertheless, they are still used in many countries, and their use is likely to continue for some time since many current vehicles need lead in the gasoline to prevent valve seat wear and because the use of lead alkyls reduces gasoline manufacturing costs and crude usage.

They function, as do all organometallic antiknocks, by decomposing at the appropriate temperature in the combustion cycle to form a cloud of catalytically active metal-oxide particles.[18,19,20] These particles interrupt the chain branching reactions which lead to the rapid combustion known as knock.

The two most common lead antiknock compounds used are tetraethyl lead (TEL) and tetramethyl lead (TML), and their compositions and characteristics are shown in Table 6.2. Commercial lead fluids contain 1,2 dibromoethane and 1,2 dichloroethane which act as lead scavengers and prevent a buildup of lead compounds in the combustion chamber. Volatile lead halides are formed which are exhausted from the engine. Without scavengers, hard deposits can build up in the combustion chamber. These deposits can flake off and cause valve burning by holding valves off their seats, thus allowing the hot combustion gases to escape past the valves.

Table 6.2 Composition and Properties of Lead Alkyls

	TEL	**TML**
Composition, wt%		
Lead alkyl	61.5	50.8
1,2 dibromoethane	17.9	17.9
1,2 dichloroethane	18.8	18.8
Dye, diluent, inhibitor, etc.	1.8	12.5
Lead content, wt%	39.39	39.39
Properties		
Specific gravity, 20°/4°C	1.60	1.58
Vapor pressure @ 20°C, mbar	67	87
Boiling point of lead alkyl, °C	200 (decomposes)	110

Other lead antiknock compounds are available that contain mixtures of the various ethyl and methyl lead alkyls such as, for example, the compound designated as CR-50 which has the composition shown in Table 6.3.

Table 6.3 Composition of the Lead Compound CR-50

	wt% in compound
Tetraethyl lead	2.7
Triethylmethyl lead	14.2
Diethyldimethyl lead	24.0
Ethyltrimethyl lead	13.0
Tetramethyl lead	2.2
Dibromoethane	17.9
Dichloroethane	18.8
Dye, toluene, etc.	7.2
Lead content	39.39

It will be seen that all the lead compounds contain the same fixed amount of elemental lead and all have the same amount of scavenger. For aviation gasoline, however, only dibromoethane is used as a scavenger. The relative effectiveness of different lead alkyls depends on their volatility, when they decompose in the cycle, and the type of base gasoline being used. There can be significant benefits to refiners in selecting the most appropriate lead alkyl for their own gasolines. In some cases this may mean using mixtures of TEL and TML, and in others a chemical mix such as CR-50 is the most effective.

The response of gasoline components to lead alkyls is illustrated in Figure 6.7. The curves show that the greatest octane benefits are when they are blended into low-octane quality components, and that the octane improvements reduce as lead concentration increases. Paraffinic components tend to have a greater response than aromatic components.

In finished gasolines, selection of the optimum type and concentration of lead alkyl is extremely important, particularly in view of the limitations on the maximum amount of lead that can be used. Figs. 6.8 to 6.10[21] show lead response curves for gasolines produced by hydroskimming and conversion refineries (see Chapter 3, section 3.3). Gasolines from hydroskimming refineries contain no olefins, unlike those from conversion refineries which have cracking processes.

The following guidelines[21] may help in the selection of the most appropriate lead alkyl to use:

- For Regular gasolines, i.e., below about 93 RON, TEL is usually the best.

- For Premium gasolines where RON is the most critical parameter, TEL is normally preferred for lead concentrations below 0.40 g/L. Above this level, physical mixtures of TEL and TML or CR compounds may be beneficial.

- For Premium gasolines where MON is critical, again, mixtures of TEL and TML or CR compounds are likely to be best.

- For Premium gasolines where R100°C is critical, TML or mixtures of TML with TEL are likely to be most beneficial.

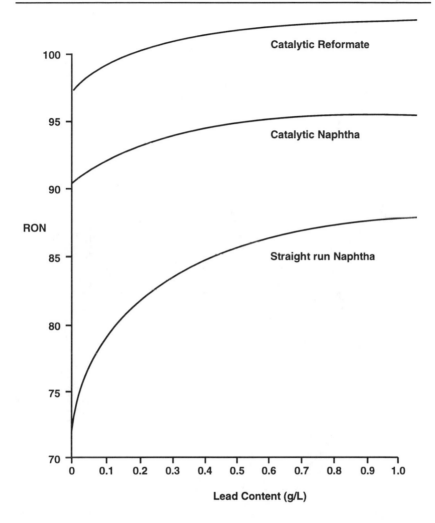

Fig. 6.7. Lead response curves for different gasoline components.

As lead is removed from the gasoline pool it requires more and more energy to make the same volume of gasoline at the same octane quality. Although maintaining a high octane level enables vehicle manufacturers to use high compression ratios and, therefore, achieve a better efficiency, this is pointless if the benefits are more than offset by efficiency losses at the refinery. A number of studies have been carried out to find the balance between these two conflicting factors. One of these, called RUFIT (Rational Utilization of Fuels In Trans-

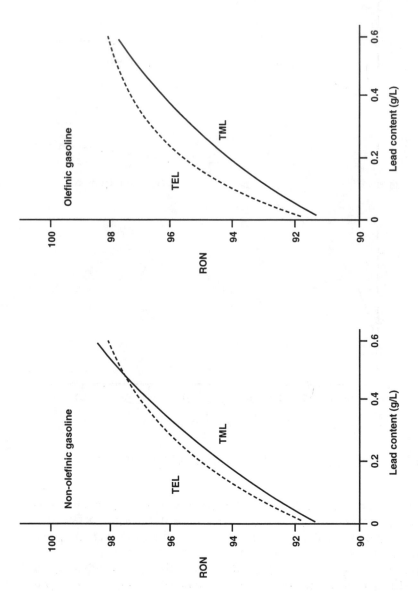

Fig. 6.8. Lead response of gasolines in terms of RON.[21]

port), was a joint study by the oil and motor manufacturers with the objective of identifying the optimum octane quality to use at different lead levels. Input data used included the improvement in vehicle fuel economy with increasing compression ratio, and the quantity of crude oil consumed to make a given volume of gasoline at different octane qualities and lead levels.

The combination of this vehicle and refinery fuel consumption data[22] is given in Figure 6.11, which shows that at 0.4 g/L the optimum RON level is 96, at 0.15 g/L it is 95.5, and at zero lead it is about 92. This was re-evaluated in 1983[23] in the light of the availability in the future of improved refinery equipment, and the optimum octane quality for unleaded moved to 94.5.

The presence of sulfur in the gasoline reduces the effectiveness of lead alkyls by promoting lead alkyl decomposition and by deactivating the lead-oxide species formed.[24,25,26] The order of antagonism of different sulfur compounds to lead alkyls, which is the same for all gasolines, is as follows:

Polysulfides > thiols > alkyl disulfides > alkyl sulfides > elemental sulfur > aryl disulfides > aryl sulfides > thiophenes

Two other factors are important when considering the benefits and disadvantages of lead. One is that it prevents wear of exhaust valve seats by its lubricating action (see Chapter 9), and the other is that it may give lower levels of octane requirement increase (ORI) (see Section 6.7).

6.4.2 MMT — Methyl Cyclopentadienyl Manganese Tricarbonyl

Although over the years there have been many investigations into possible organometallic antiknock compounds to replace lead alkyls,[27] the nearest to a successful alternative that has been found is MMT. This compound is more effective than lead on a metal concentration basis, as shown in Fig 6.11,[21] and was commercialized in 1958 by the Ethyl Corporation as a supplement to lead under the trade name AK33X.[28,29] It can only be used at relatively low concentrations (below about .0165 g Mn/L) because of problems with fuel instability, deposit buildup in engines, lack of response at higher concentrations and its adverse effect on hydrocarbon emissions from catalyst-controlled cars.[21,30,31,32,33] It is relatively expensive and it

119

Base Case: 1000 ton Gasoline, 96 RON, 0.4g Pb/L
Car Efficiency Parameter: 1.0

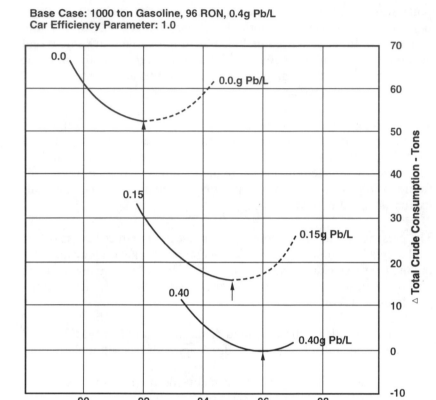

Fig. 6.11. The effect of lead content on crude consumption/optimum octane number.[22]

cannot be used in countries where there are restrictions on the use of organometallic compounds in gasoline.

MMT shows a synergistic octane effect with lead in many fuels, particularly those high in paraffins. It acts as a co-antiknock with low lead levels and it is for this reason that it has been used together with lead to boost octane levels. It was present, for example, in many of the very low-lead gasolines that were produced in the U.S. It has also been used on its own in unleaded gasoline in Canada. However, its future widespread use looks doubtful because of the limitations summarized above and the use of highly aromatic gasolines, where its response is poor.

6.4.3 Other Metallic Antiknocks

Figure 6.12 shows the relative effectiveness of a number of alternative metallic antiknocks in addition to lead and MMT. The only ones to have been used commercially contain iron and were used during the

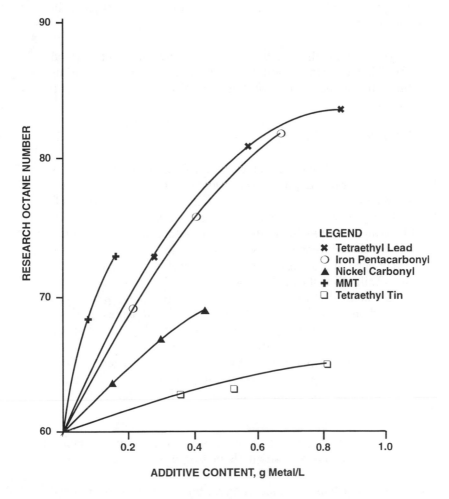

Fig. 6.12. Effectiveness of organometallic antiknock additives.[21]

121

1930s in both the U.S. and Europe either as pentacarbonyl or as fer-rocene (dicyclopentadienyl iron). Ferrocene has been reassessed as an antiknock more recently[34,35,36] at low concentrations (15 to 30 ppm) and, although it increases RON by up to 1.5 units and MON by up to 0.9 units at 30 ppm, it is not clear whether the drawbacks of iron com-pounds, namely excessive engine wear and spark plug fouling, have been overcome at the low concentrations suggested. In one set of tests[36] which indicated that ferrocene may also give some improve-ment to emissions and fuel economy, one pair of cars fitted with a 3-way catalyst ran for 85000 km without any reported problems and another pair completed their scheduled 20000 km also apparently without problems.

Although other metals have been shown to have antiknock effective-ness, they have generally failed for a number of reasons including combustion chamber and spark plug deposits, toxicity, wear, and cost-effectiveness.

6.4.4 Organic Antiknocks

Most of the undesirable characteristics of metallic antiknocks arise because they leave a deposit on combustion. Organic antiknocks, being ashless, have always been of considerable interest and, in fact, the first antiknocks to be identified were the aromatic amines. Of the readily available aromatic amines, N-methylaniline is one of the most effective, although many others have been investigated.[37] It requires about one percent by volume of N-methylaniline (NMA) to give an activity similar to that of 0.1 g lead/L, and so it is much less cost-effective than lead. More recently[38] a comprehensive investigation was carried out in which a wide range of potential ashless antiknocks were evaluated. Four groups of compounds showed up as having comparable activity to N-methylaniline: aromatic compounds con-taining nitrogen; aromatic compounds containing oxygen; iodine and aliphatic iodine compounds; and selenium compounds. The best of these are compared in Table 6.4 in terms of their relative effectiveness with N-methylaniline.

It was concluded from the study that none of the compounds tested were as cost-effective as lead alkyls or further processing.

Some types of ashless compound have, however, proved to be valu-able in terms of antiknock performance and have been used widely

Table 6.4 Relative Effectiveness of Organic Antiknocks

Compound	Relative Effectiveness (Weight basis, NMA=1)
Best amine	1.1
Best hydrazine	0.9
Best N-nitrosamine	1.5
Best phenol	0.5
Best formate	0.5
Best oxalate	0.4
Iodine	1.1
Best selenium compound	2.8

since lead began to be phased out. These are the oxygenated compounds such as alcohols and ethers. They are used at concentrations of several percent and so could be classed more as blend components than antiknock additives. They are, however, very important in helping refiners to meet both the volumes required and the octane quality of low-lead and unleaded gasolines. Their use is discussed separately in Chapter 11.

6.5 Octane Blending

The constraints imposed on gasoline blenders are many and diverse. First, they must meet all the octane specifications, usually RON, MON and R100°C. Because octane numbers, as measured, do not blend linearly and because their blending behavior is dependent on the other components that are present in a blend, it can be difficult to predict the octane quality of a finished gasoline. If blending is carried out at a refinery rather than using only purchased components, maximum use must be made of all the components produced, particularly if their alternative uses are limited and/or much less profitable. Whatever the source of the components, each blend must be made at the lowest possible cost. Octane quality is only one of several specification points that must be met with minimum quality giveaway, and often, correcting a blend for one parameter will put it out of specification for another. Tankage restrictions and the relatively poor precision of many of the tests also increase the difficulties, although the use of continuous analyzers, particularly for octane quality, helps in this respect.

Octane blending is discussed in more detail in Chapter 3 and covers the several methodologies that are available to help blenders in their task. Many of them require lengthy calculations and detailed input due to the interactions that occur between components. The blending characteristics of each component can change with time due to different crudes being run, changes in plant conditions, changes in blend composition, the use of oxygenated components, etc.

Appendix 2 summarizes the physical properties of some of the hydrocarbons that can be present in gasoline, and Appendix 11 summarizes the octane quality for a number of hydrocarbons that are in the boiling range of gasoline.

6.6 Octane Requirements of Vehicles and Engines

Vehicles differ widely in the way they respond to octane parameters and in the level of octane quality they require to be clear of knock. It is important to both the oil and the motor industry to know what the octane requirements of vehicles are under both normal and severe driving conditions so that fuels can be made available to satisfy essentially all cars in a given population regardless of driving conditions. It is also important that vehicles are not produced requiring higher octane levels than are available.

The measurement of octane requirement is difficult because the test is rather imprecise and the results are influenced by weather conditions and by the way the engine/vehicle has been driven immediately prior to the test. The situation is further complicated because vehicle manufacturing tolerances can give rise to large variations between different cars of the same model. A spread of seven octane numbers is quite normal[15] and this means that in order to define with reasonable accuracy the octane requirement of a model, it is necessary to test at least ten examples. Manufacturing tolerances of compression ratio account for much of this spread of requirements and it has been suggested[39] that reducing the manufacturing tolerances on compression ratio alone would enable vehicles to make better use of available octane quality and give an overall benefit of about two percent in fuel consumption.

Clearly it would be extremely expensive for an individual company to test enough vehicles to be able to estimate the octane requirement distribution of the cars in any one country, and for this reason such testing is normally carried out cooperatively. In the U.S. such tests are carried out by the Coordinating Research Council (CRC), in Europe by the Cooperative Octane Requirement Committee (CORC), in Japan by the Japanese Petroleum Institute (JPI), and in Australia by the Australian Cooperative Octane Requirement Council (ACORC). Each of these groups prepares its own reference fuels and uses standard test procedures so that all the data obtained are compatible within a geographical area.

The reference fuels consist of a number of fuel series in which RON is increased in one-number increments from a level below that of the lowest commercial grade available up to as high as possible without deviating too far from commercial reality; typically, each series may vary from about 88 to a little over 100 RON. The number of fuel series necessary depends on how the vehicles in the area under consideration respond to octane parameters. In Europe, for many cars, R100°C is important as well as RON and MON and this means that at least three series in which these parameters are varied independently of each other are required. In the U.S., R100°C is not an important variable and so only two series are used. Where R100°C and sensitivity are important, it is preferable to use fuel series having two levels of sensitivity (i.e., RON - MON) and two levels of ΔR100°C (i.e., RON - R100°C). Vehicles are also tested using Primary Reference Fuels (PRFs) which are blends of n-heptane (RON and MON = 0) and isooctane (RON and MON = 100).

The test procedures used[40,41] vary somewhat from country to country but all involve accelerations at full and at part throttle from a low speed and finding at which octane level on each series trace knock occurs. Trace knock is also called borderline knock and is usually defined as the lowest level of knock that can be detected by ear, although some laboratories use instrumentation to detect knock and relate the readings back to the trace knock level. In addition, tests are also frequently carried out at constant speeds and wide open throttle to simulate travel on high-speed roads under various conditions of load. In these cases some instrumentation is allowed in order to assist in the detection of knock. These latter tests do not include fuels in which R100°C is varied, since this parameter is not considered important at constant speed. Constant-speed tests are almost invariably carried out using a chassis dynamometer (rolling road) to allow more accurate control.

6.6.1 Vehicles with Knock Sensors and/or Engine Management Systems

Knock sensors[42,57] are now fitted to many engines in order to protect them from damage and to enable higher compression ratios to be used so as to achieve an improvement in fuel economy. They are usually attached to the cylinder head and when they detect sound frequencies in the knock range, they activate a mechanism which reduces the octane requirement of the vehicle. This is most often done by retarding the ignition although other methods are available such as the use of the waste gate in turbocharged vehicles. The ignition timing (or other mechanism) is then gradually relaxed back to the optimum for power and economy until driving conditions put the engine back into incipient knock again.

Figure 6.13 is a simplified illustration of what happens to knock intensity in a vehicle fitted with a knock sensor when the octane level of the fuel is gradually lowered. With high octane numbers the vehicle is clear of knock. As the octane quality is gradually reduced, knock will eventually start and will gradually increase in intensity until it reaches a level at which it is detected by the knock sensor. At this stage the ignition timing begins to be retarded—or some other method of reducing the octane requirement of the engine is activated—and this will hold the knock intensity at the level set by the vehicle manufacturer. If the octane of the fuel is reduced still further, the mechanism for reducing knock intensity continues to function until it reaches the limit of its operation. At this stage knock level will then increase rapidly from the level set by the engine manufacturer up to medium or heavy knock.

Unfortunately, as the timing of a vehicle is retarded, so is its performance and Fig. 6.13 also shows what happens to acceleration time while the knock sensor is functioning. This worsening of acceleration performance can be very noticeable[43] and can be used as a criterion for determining octane requirement, if desired, in this type of vehicle. The conventionally defined octane requirement would be at the end of the plateau where the knock sensor is operating, just before it starts increasing rapidly, although the vehicle acceleration performance is likely to be severely curtailed at this point.

Some vehicles with complex engine management systems have a learning curve which adapts to the way the vehicle is normally driven

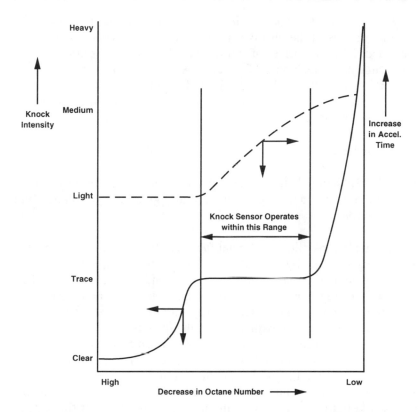

Fig. 6.13. Effect of a knock sensor on knock intensity as octane quality is lowered.

so that engine conditions can be optimized. This means that if the vehicle is driven in a completely different manner to the way the management system has adopted, such as when the octane requirement of the engine is being determined, a value will be obtained that will depend on the type of driving on which the learning curve has optimized. Thus it is necessary to "precondition" such a vehicle in a standard manner before measuring its octane requirement, in order to get consistent results.

6.6.2 Data Analysis

After tests have been carried out cooperatively on a large number of vehicles, the results are pooled so that each participating company can carry out its own data analysis if it wishes, and, in addition, some general analysis of the results is usually carried out which may be published.[44,45] A number of studies have been carried out on methods of analyzing such data[15,46,47] and these can involve preparing satisfaction curves as in Figure 6.14 and developing equations which relate how a car population responds to the various octane parameters. Such equations might be of the form:

$$\text{Road Octane Number} = a.\text{RON} + b.\text{MON} + c.\text{R100°C} + d$$

where Road Octane Number is the octane requirement in terms of PRFs and a, b, c and d are constants for any given car population and driving mode.

These correlations allow fuel manufacturers to predict the percentages of vehicles satisfied by a gasoline in any given car population and to design a gasoline to maximize the satisfaction level. However, what they really need to know is how many *customers* are satisfied rather than how many *vehicles* are technically satisfied, i.e., are free of knock when driven by technical raters using standard driving procedures. A considerable amount of work has been carried out on this problem[48,49,50,51,52,53] and all agree that customers do not hear or recognize knock as readily as technical raters and, in addition, do not drive in as severe a manner as required by the test procedure and so are less likely to provoke knock in the engine. In some of the tests it has been suggested that customers are less sensitive than technical raters by as much as five octane numbers on average, although there are large differences between customers. The methods of analysis of the data are complex[54] but the procedures available do enable reasonable estimates of customer satisfaction levels to be made.

6.6.3 Octane Rating of Fuels Using Vehicles or Engines

Octane requirements are valuable in establishing the fuel quality required to satisfy a vehicle or population of vehicles. It is sometimes useful to be able to compare the performance of a number of different fuels in a vehicle on the road in much the same way as is done using a CFR engine. The octane numbers thus obtained are also called Road Octane numbers because they are related to the performance of Primary Reference Fuels (PRFs) and are carried out on the road. To

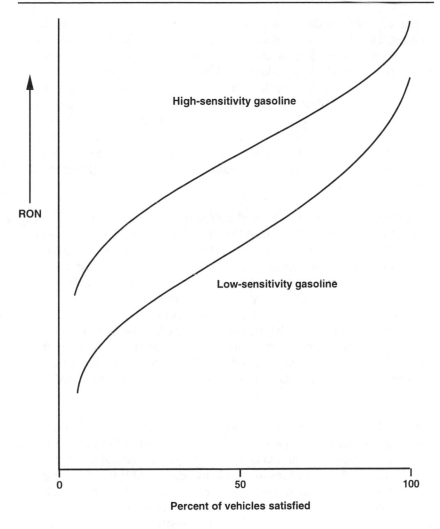

RON

High-sensitivity gasoline

Low-sensitivity gasoline

0   50   100

Percent of vehicles satisfied

Fig. 6.14. Examples of vehicle satisfaction curves.

carry out such fuel ratings, the ignition timing of the vehicle is adjusted to find a setting which gives trace knock for a particular fuel and driving mode, whereas when measuring octane requirements, the fuel quality is varied to find an octane level at which trace knock occurs.

There are two standard procedures commonly used to rate fuels in vehicles: the Modified Uniontown procedure and the Modified Borderline procedure. The Modified Uniontown procedure was developed from an earlier procedure, the Uniontown method, which involved assessing the knock intensity of different fuels at a fixed ignition setting and comparing the results with those from a series of standard reference fuels. This procedure is no longer used because knock above the trace level can remove deposits and thereby lower octane requirements. It is also more difficult to compare different knock severities than a series of readings all at trace knock but at different ignition settings.

The Modified Uniontown procedure requires the basic ignition timing of the vehicle to be modified so that it can be readily adjusted and its position measured. Full throttle accelerations are made from a low speed using primary reference fuels, and the timing adjusted so that a trace level of knock is detected during some part of the acceleration. These results are plotted to give a curve relating ignition timing with PRF octane value at the trace knock level as in Figure 6.15. The test fuel is then run and the PRF octane number (the Road Octane Number) corresponding to the trace knock ignition setting is read off this curve. The procedure is rapid and simple but does not give much information about how the octane rating changes over the speed range.

The Modified Borderline knock procedure is the second method used and here the automatic spark advance mechanism of the vehicle is put out of action and arrangements made to adjust it manually during an acceleration. Accelerations are made as before but this time the spark advance is adjusted continually in order to maintain a trace knock level throughout the acceleration. The data obtained can be handled in a number of ways. One method is to construct a series of curves for different PRFs and then to read off these values against the test fuel trace, as shown in Figure 6.16. Here it can be seen that Fuel A has a higher Road Octane Number than Fuel B at low speeds and a lower value at high speeds.

With many modern vehicles it is not possible to change the ignition timing manually and complex electronic equipment tailored to a specific car is needed for this purpose.

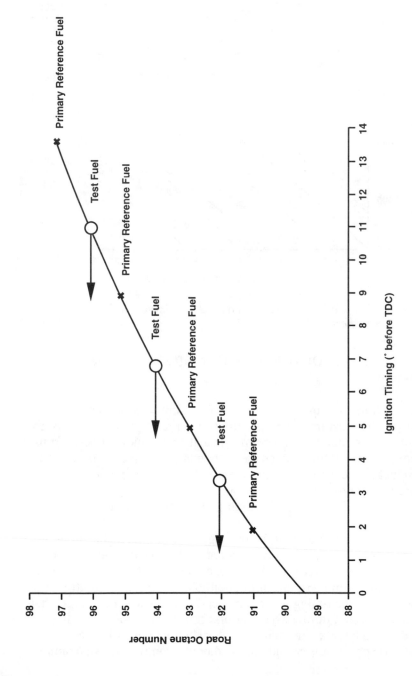

Fig. 6.15. Modified Uniontown method for determining Road Octane Number.[90]

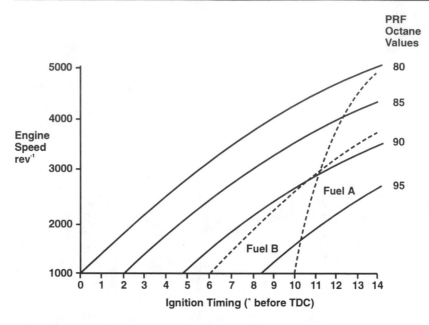

Fig. 6.16. Modified Borderline method for determining Road Octane Number.[90]

6.6.4 Engine and Other Factors That Influence Octane Requirements

Of the factors that influence the octane requirement of an engine, probably the most important is compression ratio. However, it is not possible to predict the octane requirement of a vehicle directly from its compression ratio because there are a large number of other variables that influence it.[55] Nevertheless, for a given engine, a unit increase in compression ratio will increase octane requirement by anything from 3 to 6 units depending on other design features.

The second most important factor is ignition timing; a change in ignition timing by advancing it two degrees will typically increase octane requirement by between a half and one unit. The selection of the optimum ignition timing to use throughout the speed range is extremely important. If the timing is too advanced in the low-speed range, then knock may occur during acceleration. If it is too advanced at higher speeds, then knock, although present, may not be heard because of wind and engine noises, and engine damage can result. An example of the data required to help an engine designer make the best choice of ignition characteristics is shown in Figure 6.17.[56]

As engine speed increases, knocking tendency at a fixed ignition timing will normally decrease because less time is available for preflame reactions to take place. However, as indicated in Fig. 6.17, ignition timing is advanced as engine speed increases in order to ensure that maximum power is obtained over the whole speed range. This means that high-speed operation can be as critical as low-speed operation. On some models, as discussed in Section 6.6.1, knock sensors are fitted that activate a mechanism to reduce octane requirement immediately when the engine begins to knock. Then, if no further knock occurs, the engine is allowed to gradually revert back to normal.[57] In many such vehicles, there can also be a benefit in that improved acceleration performance can be obtained with fuels having octane levels above the nominal octane requirement of the vehicle. The accelera-

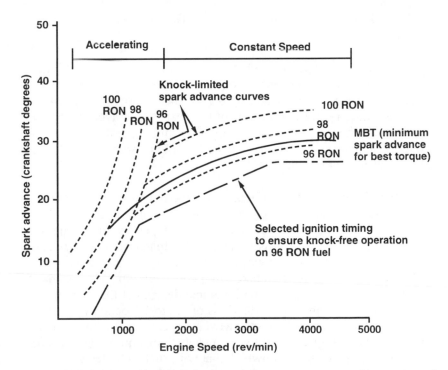

Fig. 6.17. Selection of ignition characteristic to ensure acceptable power and antiknock performance.[56]

tion performance of vehicles without knock sensors is largely unaffected by fuel octane quality.

Other vehicle design or operating factors that influence knock are load, mixture strength and charge temperature. Increasing the load increases the engine temperature and the end gas pressure and, hence, increases the tendency to knock. With regard to mixture strength, knocking tendency is at a maximum at an air-fuel ratio of about 13.5:1, but this varies with the characteristics of the fuel. Charge temperature is also important because the higher the charge temperature, the higher the tendency for knock, as would be expected from the consideration of knock as an end gas explosion.

The effect of engine variables on knock means that because of manufacturing tolerances, apparently identical cars of the same model can differ significantly in octane requirement. When this is coupled with variations in deposit laydown in the combustion chamber and errors in measuring octane requirements, it can be seen that it is difficult to establish the octane requirement characteristics of a car population with any precision.

External factors such as the ambient temperature, pressure and humidity can also influence the octane requirements of cars, and these can be taken into account when setting octane specification for specific locations and for different seasons of the year.[58] On average, octane requirements increase with increasing ambient temperature by 0.097 MON/°C and decrease with increasing specific humidity by 0.25 MON/g of water/kg of dry air.[59]

Atmospheric pressure also influences octane requirements; a reduction lowers the octane requirement because the mixture density is reduced. Atmospheric pressure changes are not taken into account in seasonal octane specifications because they are not predictably dependent on the time of year. However, atmospheric pressure does reduce with altitude and this is often taken into consideration. The extent of the effect varies considerably with vehicle design but the current ASTM specifications allow reductions of 1.0 to 1.5 octanes per 300 meters.[59] Many new vehicles, however, are equipped with sophisticated electronic engine control systems which can minimize this altitude effect and tests have shown[60] that with such vehicles the reduction in octane requirement is only about 0.2 octane numbers per 300 m.

6.7 Octane Requirement Increase (ORI)

Combustion chamber deposits significantly increase octane requirements and it is normal to measure the octane requirement of vehicles
only after they have accumulated several thousands of miles when, on
leaded fuels at least, most of the increase in octane requirement will
have taken place, as shown in Figure 6.18.[15] The increase in octane
requirement is usually between 2 and 10 numbers from a new engine[61] and this needs to be taken into account by engine designers to
ensure that vehicles operate satisfactorily on the fuels that are available.

There are conflicting views as to whether leaded fuels show different
ORIs and stabilization times than unleaded. Several investigators
have found[62,63] that there is no significant difference between leaded
and unleaded ORIs whereas other data[64,65,66] have suggested that
unleaded fuel gives higher ORIs than leaded. Some averaged results

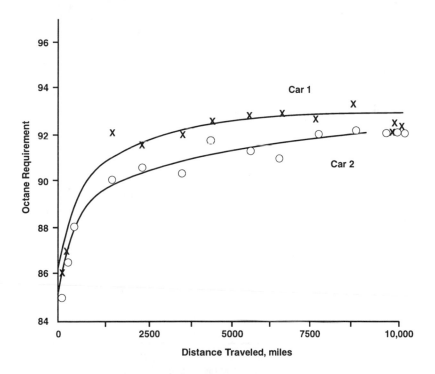

*Fig. 6.18. Typical curves of the increase in octane requirement with distance
traveled under normal everyday driving conditions.[15]*

using unleaded gasolines obtained by CRC[66,67] are shown in Figure 6.19 and suggest that the rise is quite slow and may still be increasing even after 15,000 miles. The vehicles in this survey showed average ORI levels of about 5 RON units and 3 MON units and there was an indication that the higher the initial requirement of the vehicle, the lower the ORI.

The measurement of ORI is difficult and the test is rather imprecise. This is not surprising since there are a number of variables that are either difficult or impossible to control. These arise because the octane requirement measurements are taken at intervals of 1000 or more miles, and during this time the weather conditions may have changed and the rater's appraisal of knock intensity may have varied. In addition, changes to the engine may have occurred and there may have been variations in the mileage accumulation driving cycle because it is difficult to maintain a constant test cycle on the road.

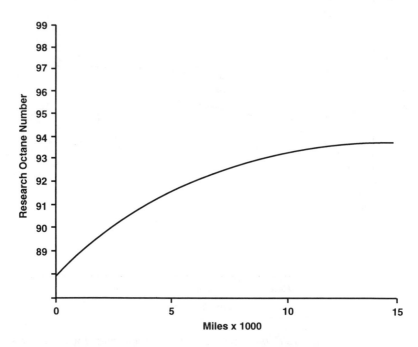

Fig. 6.19. ORI data on unleaded fuels — best-fit analysis.[66,67]

City-type driving involving a great deal of stopping and starting tends to lay down more deposits than high-speed driving. If, during the octane requirement test, knock is allowed to increase above the borderline level, deposits may be removed and the octane requirement reduced. Similarly, if a vehicle is subjected to a high-speed run immediately prior to carrying out the test, the requirement may be lower since some of the deposits may have burned off.

Deposits outside the combustion chamber, particularly ridge-type deposits in the inlet ports, have also been claimed to increase octane requirements,[68,69] although other work has suggested that combustion chamber deposits account for virtually all of the ORI in cars.[62] It seems likely that engine design plays an important part in determining whether ridge deposits in the inlet ports will influence octane requirements.

The nature of the combustion chamber deposits has been investigated[70] and it has been shown that they are partly carbonaceous and partly inorganic. They are derived from the combustion of the fuel and from any lubricant that finds its way into the chamber via the piston rings or valve guides. Thus, any metals present will have come from either or both of these sources. The organic part arises from incomplete combustion and some components or additives are known to contribute more than others to ORI.[71] Aromatic compounds, and particularly those with condensed ring structures, are prone to give rise to deposits.[72] Heavy lubricants and especially brightstocks also contribute disproportionately to ORI,[62,73] as can some multifunctional fuel additives.[74]

The most important factor in causing ORI is the thermal insulation effect of the deposits[70] which increases the temperature in the combustion chamber. It has been estimated[62] that this accounts for 90% of the octane requirement increase. The remaining 10% is due to the increase in compression ratio caused by the presence of the deposits and the chemical nature of the deposits which themselves may catalyze those reactions known to favor detonation.

Additives are available which can reduce ORI and these are discussed in Chapter 9.

6.8 Other Abnormal Combustion Phenomena

6.8.1 Surface Ignition and Preignition

Surface ignition is the ignition of the charge by hot deposits or surfaces and can occur before or after the spark from the plug. It can be distinguished from spark knock in that it is not controllable by the spark advance mechanism. It may not be recurrent or repeatable since it can die out as deposits are burned off, or it can increase in severity. This form of abnormal combustion should not be confused with knock which is the autoignition of the end gas after it has been heated and compressed by the advancing flame front. Mild surface ignition gives rise to rough operation of the engine and can be accompanied by a rumbling noise, and so this form is known as *rumble*.

Preignition is a form of surface ignition in which the charge is ignited before the occurrence of the spark. Clearly, if the ignition occurs only fractionally before the spark, the effect will be the same as advancing the ignition. If, however, surface ignition occurs some time before the spark, then negative work is obtained resulting in very high temperatures. In extreme cases the high temperatures can promote yet more preignition, and when this occurs it is often called *runaway surface ignition* (RSI). This is a violently unstable condition in which so much heat is produced that damage such as a hole burned in the piston crown can occur in a very short time. Other types of preignition have been recorded[75] such as *wild ping* and *thudding,* but are rarely experienced with today's vehicles.

The ignition source is quite often the spark plug which can get very hot if it has the wrong heat range. The temperature that a plug reaches depends on how far the heat has to travel from the center electrode to the cooler outer shell and, thence, to the cylinder head. A long path will result in a hotter plug that is more prone to preignition but is less likely to be fouled by carbonaceous deposits. The exhaust valve can also be the ignition source, so that engine design factors are extremely important in determining whether or not an engine is likely to be prone to preignition problems.

Engine operating conditions also influence preignition since they affect combustion chamber temperatures. High-speed and high-load driving obviously result in high temperatures, although low-speed stop/start operation tends to maximize deposits. Advancing the ignition timing will increase temperatures, but changes in air-fuel ratio have little effect. If an engine is knocking, this can increase combustion chamber temperatures and lead to preignition. For this reason it can sometimes be difficult to establish whether the primary cause of an engine failure is knock or preignition.

Deposit induced preignition (DIPI) has a somewhat random nature. Once it has started it may die away as the carbonaceous deposits are burned off, or it may increase in severity. Test work has shown that the combustion chamber surface temperature is the dominant factor in deposit formation,[76] and hence driving mode is correspondingly important. Lead deposits are less likely to cause preignition than those from other metals[77] such as those derived from the lubricant. Deposits on the exhaust valves are particularly likely to cause preignition problems. Phosphorus and boron additives have been used to reduce the glowing tendency of deposits (see Chapter 9) but are rarely used nowadays because they adversely affect exhaust gas catalysts.

Engines can be checked for their susceptibility to preignition. This is done by first running them under a low-speed, low-load cycle to lay down deposits, and then changing to a high-speed, high-load operation to raise the temperature of the deposits. The onset of preignition can be detected in a variety of ways such as fall in power output, increase in peak pressure, increase in temperature, presence of a flame front prior to the spark by using an ionization detector,[77] displacement of peak pressure curves,[78] etc. A fast response is required so that steps can be taken to avoid damaging the engine. The test can be made more severe by incorporating a small amount of a critical lubricant in the gasoline.

The fuel is usually only a minor variable in influencing preignition although several different aspects of the fuel are important. These are its deposit-forming tendency, its resistance to ignition by a hot source and its ability to heat up a potential ignition source.

Turning first to the deposit-forming tendency of fuels, this can be due to all or any of the following:

- The presence of heavy non-volatile hydrocarbons in the gasoline. These may arise during manufacture from poor fractionation and/or from high-severity catalytic reformer operation. In the latter case, polynuclear aromatics are formed which are particularly difficult to combust completely and so readily form carbonaceous deposits. It is best to remove these heavy compounds at the refining stage by rerunning the reformate through a distillation tower.

- Deposits derived from metals. These can glow, although lead deposits are not particularly bad in this respect[77] and usually only have an effect on preignition when high levels of lead are used, i.e., above about 0.6 g/L.

- Additives used in gasoline can give rise to carbonaceous deposits. Mineral oils are sometimes added for a number of reasons (see Chapter 9) and these can be deposit-forming. Brightstocks (heavy lube oils derived from residuum) are particularly prone to deposit formation.[62,77] Other additives that have the potential to cause problems are any metal containing compounds and some polymeric surfactant additives[75] (see Chapter 9).

It has already been mentioned that some lubricant can find its way into the combustion chamber and cause deposits which can be especially bad from a preignition point of view.[77,79] The metals which are often present are one factor, but so also are some of the polymeric additives such as VI improvers.

The second important factor is the tendency for the fuel to be ignited by a hot source. This preignition resistance has been evaluated for a wide range of compounds using an electrically heated hot spot to find the relative ease with which they can be ignited.[77,78,80,81,82] A rating scale has been used[80] in which isooctane (having a high resistance to preignition) is given an arbitrary value of 100 and cyclohexane (low resistance to preignition) is given a value of zero. Ratings outside this range are determined by extrapolation.

Ratings obtained using this scale show that hydrocarbons, even of the same chemical type, vary enormously in their preignition resistance, as, for example:

Aromatics: Benzene (26) is poor but toluene (93) is good.

Naphthenes: Cyclopentane (70) is good but cyclohexane (0) is poor.

Olefins: Di-isobutylene (64) is good but hexene-2 (-26) is poor.

Paraffins: All are good and mainly in the range 50 to 100.

Alcohols: Methyl alcohol (<0) is poor but isopropyl alcohol (62) is good.

Similar rankings are obtained using hot spot temperature to rate the fuel as shown in Figure 6.20.[77]

Lead, apart from any deposit-forming tendency, improves the resistance of fuels to preignition, having the greatest effect on fuels of low preignition resistance.

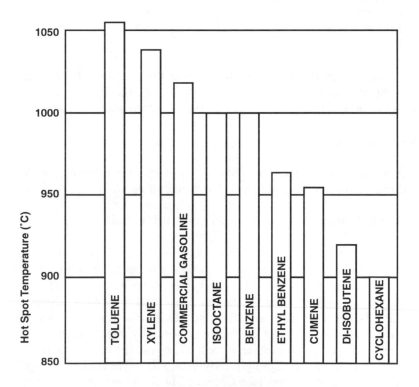

Fig. 6.20. Hot spot ignition temperature data obtained using a single-cylinder engine at 4000 rpm and equivalence ratio of 1.10.[77]

141

The final fuel factor which influences preignition is the ability of the fuel to heat up a potential hot spot. This was found to be partially correlated with flame speed[77] — the higher the flame speed, the higher the hot spot temperature. The temperature of a metal filament in the combustion chamber of an engine running at 4000 rpm and full throttle was measured using different pure hydrocarbon fuels; the data obtained are shown in Table 6.5:

Table 6.5. Deposit Heating Tendency of Different Fuels

Hydrocarbon	Temperature Attained by Hot Spot
Isooctane	705°C
Xylene	750
Toluene	770
Cyclohexane	790
Cumene	800
Ethylbenzene	800
Benzene	835

It has been shown[77] that a combination of the resistance to hot spot ignition and the heating ability of the fuel gives excellent agreement with the preignition tendency for a given ignition source.

6.8.2 Misfire

Misfire, unless it is due to mechanical or electrical problems, can be influenced by gasoline quality. The main causes are:

1. Air-fuel ratio is outside the combustible limit.

With most vehicles this type of failure is likely to be due to the mixture being too weak to burn, particularly since many vehicles operate under very lean conditions. This can often happen in cold weather during accelerations when the excess liquid fuel forced into the intake manifold does not volatilize sufficiently to give an ignitable mixture. Increasing the volatility of the fuel will normally overcome this malfunction. Driveability problems such as hesitations and stumbles can result until the mixture strength moves into the right range. Oxygenated gasoline components (see Chapter 11) can aggravate this situation because they effectively lean the mixture still further, depending on the oxygen content of the compound being used.

Additives are available which can improve the spark, thereby minimizing such misfire and reducing, to some extent, any driveability malfunctions (see Chapter 9, Section 9.4).

Too rich mixtures can also give rise to misfiring and can be due to incorrect carburetor settings, certain types of hot weather driveability problems aggravated by high fuel volatility (see Chapter 7), or extended use of the choke. During idle, deposit or ice formation in the carburetor can lead to rough idle and even stalling due to misfire. The deposits prevent enough air from flowing past the closed throttle plate to give an ignitable mixture. Additives are available that can prevent deposit and ice formation.

2. Deposit formation on the plug electrodes.

Deposits may simply bridge the gap so that there is no spark at all or they may provide an insulation layer that makes it more difficult to achieve a spark of the required energy to ignite the mixture. The deposits may be largely carbonaceous due to the mixture being too rich or because of heavy non-volatile compounds in the gasoline, or they may be inorganic arising from the antiknock additive (lead or manganese) or from any lubricant that has found its way into the combustion chamber.

Lead fouling is usually a problem only at relatively high lead levels, i.e., greater than about 0.6 g/L. In countries where higher levels than this are used, a problem known as plug glazing can occur.[83] Here the lead forms a glassy coating on the ceramic of the plug which is electrically conducting at high temperatures and so allows the charge to leak away. This problem can be overcome to some extent by the use of phosphorus additives (see Chapter 9) that combine with the lead coating and reduce its electrical conductivity.

A problem sometimes experienced by vehicle manufacturers is that vehicles coming off the production line frequently need to be moved for short distances before they are delivered to the dealer. Almost every time the vehicle is moved it will involve starting from cold and will require the use of the choke. This can cause so much cold fouling that misfire can occur and the manufacturer may need to fit new plugs prior to delivery to the customer. The problem can be overcome or alleviated by the use of a special factory-fill gasoline[84] containing, somewhat surprisingly, a lube oil which can be anything from a light spindle oil to a brightstock. The amount used will depend on the viscosity of the oil, but 5% of a light mineral oil or 0.5% of a brightstock have been found to be effective in eliminating misfire. Other materials have also been found to be effective[84] such as a polybutene or 5% methanol; the latter presumably works because it leans the mixture. It is important to appreciate that these additives should only be used in factory-fill gasolines and are not necessarily suitable for normal operation requiring much less frequent use of the choke.

3. Erosion of the plug gap or damage to the plug electrodes.

This occurs as a result of a combination of electrical erosion and corrosion. The fuel does not normally have any significant effect on electrical erosion but the presence of lead compounds and, in particular, the halogen compounds used as scavengers have been blamed for corrosive wear of the electrodes.[85] The elimination of lead is claimed to reduce spark plug electrode erosion although it has also been suggested[85] that unleaded fuels are more prone to predelivery cold fouling. This may be due to the higher aromatics levels in some unleaded fuels.

Preignition can cause mechanical damage to the electrodes and thus give rise to misfire.

6.8.3 Run-on

This is also sometimes called after-running and refers to the tendency for an engine to continue to fire after the ignition has been switched off. It is most frequently due to the autoignition of the fuel-air mixture that is drawn into the combustion chamber before the engine stops rotating. Surface ignition can also cause run-on but this is much less common and is characterized by the engine firing on every cycle and maintaining its speed rather than the very erratic firing typical of run-on caused by autoignition.

A detailed investigation of this phenomenon[86] has shown that during run-on, preflame reactions take place in the combustion chamber after the ignition has been switched off which result in spontaneous ignition of the charge. For these reactions to proceed to the stage of autoignition, the charge has to be hot, it has to be under pressure and it has to have sufficient time. The combustion chamber will be hot after switching off and, because engines continue to rotate to some extent, at least one of the cylinders will be under compression. If the engine speed is too high, there will be insufficient time for the preflame reactions to reach the autoignition point, and if the temperature is not high enough, they will not progress fast enough. If, however, the temperature and speed fall within the critical region for spontaneous ignition, the engine will refire. This may result in the temperature and speed increasing out of the critical range, so that the engine may not fire again for one or two cycles, that is, until conditions are once again in the critical region. This gives rise to the very uneven type of running that is characteristic of run-on. Its duration is limited by the overall cooling of the engine.

Run-on is less of a problem nowadays because fuel cut-off valves are frequently fitted. Too high an idle speed setting and inadequate cooling of the combustion chamber are mechanical factors that worsen run-on.

The fuel factor which has the greatest influence on run-on is its resistance to spontaneous ignition and this is its Research Octane quality.[87] There are many examples of vehicles requiring a higher RON to overcome run-on than is needed to avoid knock.

6.9 Exhaust Emissions and Fuel Economy

The need to control exhaust emissions and to achieve acceptable levels of fuel economy have been major factors in causing design changes in automobiles and compositional changes in gasolines. The control of combustion in an engine is important in both cases, although there are other aspects of vehicle design than that of the engine which have a large influence. Knowledge of the vehicle design changes puts into perspective the corresponding modifications that have followed in gasoline composition.

6.9.1 Exhaust Emissions

The first engine design feature to reduce emissions to the atmosphere was probably the PCV (positive crankcase ventilation) system, which allowed blowby and other gases from the crankcase to pass back to the intake manifold instead of being discharged directly to the atmosphere. The first attempt to control and regulate exhaust emissions based on vehicle testing was introduced in California in 1964. From then on, increasingly severe regulations have been imposed which have spread to most countries throughout the world and these have required more and more complex and sophisticated equipment to be fitted to vehicles. A summary of the current (1994) regulations for vehicle emissions and the testing requirements are included in Appendices 3, 4, and 5.[88]

The emissions of greatest concern are carbon monoxide (CO), unburnt hydrocarbons (HC) and nitrogen oxides (NO_x). If a fuel could be completely combusted, there would be no emissions of CO and HC from the exhaust, only CO_2 and water vapor; NO_x would still be present because it is formed from atmospheric nitrogen, the amount depending on the peak temperature in the combustion chamber. The use of lead alkyls has been reduced in most countries and stopped completely in some. They cannot be used in gasolines designated for use in vehicles fitted with catalyst systems or in countries where there are particular concerns regarding the toxic effects of lead.

There are other toxic pollutants that are not yet regulated, but which are of increasing concern. These include aldehydes, benzene and 1,3 butadiene. In addition, the need for global anthropogenic carbon dioxide emissions to be reduced has been emphasized at various World Climate Conferences and steps to reduce them have been taken or are being considered in some countries. These generally involve a combined energy and carbon tax and/or a requirement to continually improve fuel consumption in new models.

The major vehicle factors that influence these exhaust emissions have been mentioned in Chapter 5 and are:

- **Control of the air-fuel ratio.** Figure 5.4 shows how the air-fuel ratio influences these emissions and indicates that HC and CO can be minimized by operating on the lean side of stoichiometric provided that the lean limit of combustion, when misfire can occur, is not reached. NO_x can also be controlled by this means but requires a special design that allows the mixture to burn under very lean conditions. For starting and during warm-up there is a need to have a rich mixture even though it worsens CO and HC emissions, and design features ensure that the warm-up period is minimized. The use of fuel injection systems rather than carburetion enables the air-fuel ratio to be closely controlled under all engine operating conditions and, hence, they have a positive effect in controlling exhaust emissions.

Vehicles fitted with some types of catalyst system operate efficiently only when the air-fuel ratio is close to stoichiometric, so that, again, control of air-fuel ratio is extremely important and is often achieved by a feedback system using an oxygen sensor in the exhaust gases.

- **Design of combustion chamber.** The layers of air-fuel mixture next to the relatively cool cylinder head and piston crown do not readily burn and so contribute to the HC in the exhaust gases. Carbon buildup in the combustion chamber can also absorb fuel which is released on the exhaust stroke when the pressure is comparatively low. Combustion chamber design, including the use of a stratified charge system, can minimize these effects.

- **Exhaust gas recirculation (EGR).** If a small part of the exhaust gas is recirculated back through the engine, the combustion temperature is reduced which lowers the formation of NO_x. Increased valve overlap has much the same effect since on the intake stroke some exhaust gas can be sucked into the cylinder.

- **Use of catalyst systems.** Two-way catalyst systems use an oxidation catalyst which converts the CO and HC in the exhaust gases to CO_2 and water, but relies on EGR to minimize oxides of nitrogen. The three-way catalyst systems handle all three pollutants by converting the nitrogen oxides to nitrogen and oxygen as well as the CO and HC to CO_2 and water.

These systems are necessary to meet the most stringent exhaust emission regulations and are often used with fuel injection to give better control of the air-fuel mixture. Two-way systems are normally open loop, which means that there is no feedback control to compensate for fuel differences. Three-way catalyst systems have closed-loop control so that the mixture can be maintained at stoichiometric under most driving conditions and regardless of the fuel composition. Such systems are typically deactivated during cold start, initial warm-up and maximum power operation so that at these times the engine is operating under open-loop mode.

To minimize the adverse effect of the catalyst being non-functional when cold, various methods of preheating it have been considered so that exhaust emissions can be controlled even during the cold start and warm-up period.

6.9.2 Fuel Effects on Exhaust Emissions

Although engine design features must always be the prime factor in reducing undesirable exhaust emissions, the fuel can play a significant role. The fuel quality factors that are important are covered in some detail in other chapters and particularly in Chapter 12. Emissions Regulations are summarized in Appendix 3 and Emissions Test Cycles and Procedures in Appendix 4. The most important gasoline characteristics that can influence emissions and air quality are:

- Unleaded gasoline to allow the use of catalyst systems.

- Presence of oxygenated blend components to reduce CO and HC emissions.

- Gasoline volatility control since this influences warm-up time and evaporative emissions.

- Control on maximum aromatics content to minimize benzene emissions from exhaust.

- Use of additives to keep fuel system clean and free from deposits.

- Control on light olefins since they are very photochemically active if allowed to evaporate into the atmosphere.

- Control on maximum benzene content to minimize level of this carcinogen in the atmosphere by evaporation.

- Control on sulfur content, since this can deactivate the catalyst and so increase emissions.

6.9.3 Fuel Economy

The interest in fuel economy was accentuated when the price of crude oil increased sharply during the 1970s and '80s, and now, in the U.S. and Japan, continuing improvements in fuel economy have been enforced. In some European countries, governmental control is limited to mandatory publication of vehicle fuel consumption data (see Appendix 6).

Vehicle design features, as with emissions, are the most important factors in determining the fuel consumption of a vehicle. Testing is carried out using standard procedures (see Appendix 4) that simulate both city and constant-speed driving modes.

Compression ratio is important because the higher the compression ratio (within limits), the more efficient the combustion. Unfortunately, as explained in Chapter 5, Section 5.5, octane requirement also increases with increasing compression ratio.

Many of the features that reduce emissions also improve fuel consumption such as lean-burn operation, absence of deposits in the fuel system, and fast warm-up. Other design features which are not directly fuel related are reduced drag coefficients, lower vehicle weight, and better control of air-fuel ratio such as by the use of fuel injectors.

Gasoline effects on fuel economy are covered elsewhere in this volume and may be summarized as follows:

- **Heating value and relative density.** There is an empirical correlation between heating value and relative density (see Chapter 10, Sections 10.5 and 10.8) for hydrocarbon fuels so that on a volumetric basis, as the density increases so does the heating value. Thus, high density components give a positive benefit in volumetric fuel economy, and it has been estimated that an increase in relative density of 0.1 will give a long-trip economy advantage (i.e., where warm-up effects are small) of between 7 and 10%.[89]

- **Oxygenated components** clearly do not fit into this correlation because these materials have lower heating values and relatively high densities. There is a gain in combustion efficiency due mainly to the leaning of the air-fuel ratio (see Chapter 11), which at least partly offsets the penalties due to the low heating value, provided the oxygen content of the fuel is within the allowable limits.

- **Fuel volatility.** Chapter 7, Section 7.9 summarizes the effect of fuel volatility on economy and indicates that for good short-trip

149

economy, a fast warm-up is required which is provided by a volatile gasoline. However, for good long-trip economy, heavy components with high volumetric heating values are required which are provided by relatively non-volatile gasolines.

- **Gasoline additives.** It has already been mentioned that deposits in parts of the fuel system such as the carburetor can upset the air-fuel ratio by making the mixture richer and, hence, worsening emissions of HC and CO. Since fuel consumption is worsened by rich mixtures, the use of additives to keep the fuel system free of deposits can show a real fuel economy benefit in open-loop control vehicles when taken over a distance of several thousand miles (see Chapter 9).

- **Octane quality** does not directly influence fuel economy. However, for most fuels, the higher the octane quality, the higher the relative density and so the better the fuel economy. This is because aromatics, which have high densities, are often used to improve octane quality.

It is also true that the higher the octane, the higher the compression ratio that can be used and, hence, the better the fuel economy. However, this effect cannot be utilized by motorists because it is a vehicle design feature.

References

1. G.J. Germane, C.G. Wood and C.C. Hess, "Lean Combustion in Spark-ignited Internal Combustion Engines — a Review." SAE Paper 831694, 1983.

2. W. Haskell and J. Bame, "Engine Knock — an End Gas Explosion." SAE Paper 650506, 1965.

3. C.F. Taylor, The Internal Combustion Engine in Theory and Practice, Volume 2: Combustion, Fuels, Materials, Design. Cambridge, Mass., The M.I.T. Press, 1968.

4. D.J. Patterson, "Cylinder Pressure Variations, a Fundamental Combustion Problem." SAE Paper 660129, 1966.

5. D. Lyon, "Knock and Cyclic Dispersion in a Spark Ignition Engine." Paper C307/86, Institution of Mechanical Engineers Conference on Petroleum Based Fuels and Automotive Applications, London, November 1986.

6. J.L. Addicott, "Some Considerations on Detecting and Preventing Spark Knock in the Car Engine." Institution of Mechanical Engineers Proceedings 1967-68, Vol. 182, Part 2A.

7. F. Sezzi, G. Cornetti, V. Arrigoni, F. Vicenzetto and S. Biancucci, "Possible mechanisms of piston failure due to detonation and preignition," Paper 3, *Giornale Ed Atti Della Assoziazione Tecnica Dell'Automobile*, July 1969.

8. W.E. Betts, "Knock and engine damage." Paper EF/2/2 CEC International Symposium on the Performance and Evaluation of Automotive Fuels and Lubricants, Rome, June 1981.

9. W.R. Leppard, "The Chemical Origin of Fuel Octane Sensitivity," SAE Paper No. 902137, 1990.

10. C.K. Westbrook, W.J. Pitz, and W.R. Leppard, "The Autoignition Chemistry of Paraffinic Fuels and Pro-Knock Additives: A Detailed Chemical Kinetic Study," SAE Paper No. 912314, 1991.

11. G. Knonig and C.G.W. Sheppard, "End Gas Autoignition and Knock in a Spark Ignition Engine," SAE Paper No. 902135, 1990.

12. W.R. Leppard, "The Autoignition Chemistries of Primary Reference Fuels, Olefin/Paraffin Binary Mixtures, and Non-Linear Octane Blending," SAE Paper No. 922325, 1992.

13. G. Xiaofeng, R. Stone, *et al.*, "The Detection and Quantification of Knock in Spark Ignition Engines," SAE Paper No. 932759, 1993.

14. 1988 Annual Book of ASTM Standards, Volume 05.04, ASTM, 1988.

15. A.G. Bell, "The Relationship between Octane Quality and Octane Requirement." SAE Paper 750935, 1975.

16. "Real-Time Measurement of Octane Number and Other Critical Refinery Process Variables," *Hydrocarbon Technology International*, 167-69, 1993.

17. "Schnelle Octanzahlmessung," *Erdoel Erdgas Kohle*, 107, No. 7-8:339, 1991.

18. A.D. Walsh, "The Mode of Action of Tetraethyl Lead as an Antiknock." Lectures on the basic combustion process, Ethyl Corp., Detroit, 1954.

19. I.C.H. Robinson, "Knock in the Gasoline Engine." Critical Reports on Applied Chemistry, Vol 10 — Gasoline Technology, p. 81, Society of Chemical Industry, Blackwell Scientific Publications, 1983.

20. E.B. Rifkin, "The Role of Physical Factors in Knock." A.P.I Division of Refining, Vol. 38 (III), 1958.

21. T.J. Russell, "Motor Gasoline Antiknock Additives." Associated Octel Company Report 87/10. 1987.

22. CONCAWE. Ad Hoc Group "RUFIT". "The Rational Utilization of Fuels in Private Transport (RUFIT): Extrapolation to the Unleaded Gasoline Case." Report 8/80, Den Haag, Concawe, 1980.

23. CONCAWE. Ad Hoc Group Automotive Emissions Fuel Characteristics. "Assessment of the Energy Balances and Economic Consequences of the Reduction and Elimination of Lead in Gasoline." R. Kahsnitz *et.al.*, Report 11/83, Den Haag, Concawe 1983.

24. H.K. Livingstone, "Sulfur-Tetraethyl-Lead Interaction in Motor Fuels." *Ind. and Eng. Chem.*, Vol. 41, No. 5, p. 888, 1949.

25. H.K. Livingstone, "The Effect of Sulfur Compounds on the Octane Number of Leaded Fuels." *Oil and Gas Journal*, Vol. 46, No. 45, p. 81, 1948.

26. R.L. Mieville and G.H. Megeurian, "Mechanism of Sulfur-Alkyl-Lead antagonism." *Industrial and Engineering Chemistry*, Vol. 6, No. 4, 1967.

27. R.M. Whitcombe, "Non-lead Antiknock Agents for Motor Fuels." Noyes Data Corporation, Park Ridge, New Jersey, 1975.

28. J.D. Bailie, G.W. Michalski and G.H. Unzelman, "The MMT Outlook, 1977." API Refining Department 42nd Midyear meeting, Product Quality and Conservation, May 1977, Chicago, Illinois.

29. "Ethyl MMT, Manganese Octane Improver for Leaded and Unleaded Gasolines." Ethyl International, Petroleum Chemicals Division Publication, September 1983.

30. D.L. Lenane, "Effect of a Fuel Additive on Emission Control Systems," SAE Paper No. 902097, 1990.

31. K. Otto and R.L. Sulak, "Effects of Mn Deposits from MMT on Automobile Catalysts in the Absence and Presence of Other Fuel Additives," *Environmental Science and Technology 12*, pp. 181-184, 1978.

32. R.G. Hurley, L.A. Hansen, *et al.*, "The Effect on Emissions and Emission Component Durability by the Fuel Additive Methylcyclopentadienyl Manganese Tricarbonyl (MMT)," SAE Paper No. 912437, 1991.

33. R.G. Hurley, L.A. Hansen, *et al.*, "Particulate Emissions from Current Model Vehicles Using Gasoline with Methylcyclopentadienyl Manganese Tricarbonyl," SAE Paper No. 912436, 1991.

34. S. Toma, P. Elecko, M. Salisova and V. Vesely, "Ferrocene Derivatives as Gasoline Additives." Acta Fac. Rerum Nat. Univ. Comenianae, Form. Prot. Nat. 7, 18792, 1981.

35. G. Wilke, "Veba Develops Octane Improver," *Eur. Chem. News,* Vol. 46, No. 1229, 32, 1986.

36. K.P. Schug, H.-J. Guttmann, A.W. Preussand, K. Schadlich, "Effects of Ferrocene as a Gasoline Additive on Exhaust Emissions and Fuel Consumption of Catalyst Equipped Vehicles," SAE Paper No. 900154, 1990.

37. J.E. Brown, F.X. Markley and H. Shapiro, "Mechanism of Aromatic Amine Antiknock Action." *Ind. and Eng. Chem.*, Vol. 47, No. 10, p. 2140, 1965.

38. R. Mackinven, "A Search for an Ashless Replacement for Lead in Gasoline." Jahrestagung, DGMK, West Germany, 1974.

39. W.E. Betts, "Improved Fuel Economy by Better Utilization of Available Octane Quality." SAE Paper 790940, 1979.

40. CRC Report No.539, "1983 CRC Octane Number Requirement Survey." Attachment 2, 1983.

41. Cooperative Octane Requirement Committee, "Summary of the Technique for Determining Road Octane Requirements under Accelerating Conditions." 1977.

42. S. Dues, *et al.*, "Combustion Knock Sensing: Sensor Selection and Application Issues," SAE Paper No. 900488, 1990.

43. M.J. McNally, J.C. Callison, *et al.*, "The Effects of Gasoline Octane Quality of Vehicle Acceleration Performance—A CRC Study," SAE Paper No. 912394, 1991.

44. CRC Report, "A 1986 CRC Octane Number Requirement Survey." 1987.

45. The Subcommittee of Octane Number Requirement Survey, Gasoline Section, Products Division of the Japanese Petroleum Institute, "Octane Number Requirement Survey of 1985 Japanese Passenger Cars." *Sekiyu Gakkaishi,* Vol. 29, No. 6, 1986.

46. C.S. Brinegar and R.R. Miller, "Statistical Estimation of the Gasoline Octane Number Requirement of New Model Automobiles." *Technometrics*, Vol. 2, No. 1, 1960.

47. J.C. Ingamells and E.R. Jones, "Developing Road Octane Correlations from Octane Requirement Surveys." SAE Paper 810492, 1981.

48. E.S. Corner, A.M. Hochhauser and H.F. Shannon, "Technical versus Customer Knock Satisfaction — Two Decades." SAE Paper No. 780322, 1978.

49. E.R. Jones and J.C. Ingamells, "Predicting Customer Octane Satisfaction." SAE Paper No. 810493, 1981.

50. W.E. Bettoney, J.D. Benson, B.D. Keller, G.W. Stanke and J.D. Rogers, "Knock Perception a 1975 Customer/Rater Study by CRC." SAE Paper No. 780321, 1978.

51. R.F. Becker and M.A. Taylor, "The Results of the European Interindustry Study on Driver Reaction to Knock." International Symposium on Knocking of Internal Combustion Engines, Wolfsburg, 1981.

52. W.E. Bettoney, J.D. Rogers, G.W. Stanke and B.Y. Taniguchi, "Customer versus Rater Octane Number Requirements — a 1978 Survey." SAE Paper No. 801355, 1980.

53. J.P. Vihlein, W.F. Biller, *et al.*, "CRC Customer Versus Rater Octane Requirement Program," SAE Paper No. 932673, 1993.

54. R.N. Rodriguez and B.Y. Taniguchi, "A New Statistical Model for Predicting Customer Octane Satisfaction using Trained Rater Observations." SAE Paper No. 801356, 1980.

55. W.E. Betts, R. Gozzelino, B. Poullot and D. Williams, "Knock and Engine Trends." Paper EF2, CEC Second International Symposium on The Performance of Automotive Fuels and Lubricants, Wolfsburg, June 1985.

56. F.H. Palmer and A.M. Smith, "The Performance and Specification of Gasoline." Critical Reports on Applied Chemistry, Vol. 10 — Gasoline Technology, p. 114, Society of Chemical Industry, Blackwell Scientific Publications, 1985

57. H. Decker and H.U. Gruber, "Knock Control of Gasoline Engines — A Comparison of Solutions and Tendencies, with Special Reference to Future European Emission Legislation," SAE Paper No. 850298, 1985.

58. B.D. Keller, J.H. Steury and T.O. Wagner, "Seasonal Octane Specifications." SAE Paper No. 780668, 1978.

59. 1988 Annual Book of ASTM Standards, Volume 05.01, ASTM D 439, 1988.

60. J.C. Callison, "Octane Number Requirements of Vehicles at High Altitude." SAE Paper No. 872160, 1987.

61. J.C. Callison, T. Wusz, and W.F. Biller, "Coordinating Research Council Trends in Octane Number Requirement Increase," SAE Paper No. 892036, 1989.

62. J.D. Benson, "Some Factors which affect Octane Requirement Increase." SAE Paper No. 750933, 1975.

63. Cooperative Octane Requirement Committee, "The Effect of Removing Lead Alkyl Antiknock from European Motor gasolines." October 1972.

64. E.J. Forster and L.E. Stinson, "Effect of Leaded versus Unleaded Fuels on Stabilized Octane Requirements." National Fuel and Lubricants Meeting of the National Petroleum Refiners Association, New York, 1970.

65. H.T. Niles, R.J. McConnell, M.A. Roberts and R. Saillant, "Establishment of ORI Characteristics as a Function of Selected Fuels and Engine Families." SAE Paper No. 750451, 1975.

66. CRC Report, "Octane Requirement Increase of 1982 Model Cars." Report No. 540, September 1984.

67. CRC Report, "Octane Requirement Increase of 1984 Model Cars." Report No. 549, October 1986.

68. H.E. Alquist, G.E. Holman and D.B. Wimmer, "Some Observations of Factors Affecting ORI." SAE Paper No. 750932, 1975.

69. L.B. Graiff, "Some New Aspects of Deposit Effects on Engine Octane Requirement Increase and Fuel Economy." SAE Paper No. 790938, 1979.

70. L.B. Ebert (Editor), <u>Chemistry of Engine Combustion Chamber Deposits.</u> Plenum Press, New York, 1985.

71. M.K. Megnin and J.B. Furman, "Gasoline Effects on Octane Requirement Increase and Combustion Chamber Deposits," SAE Paper No. 922258, 1992.

72. P.J. Choate and J.C. Edwards, "Relationship Between Combustion Chamber Deposits, Fuel Composition, and Combustion Chamber Deposit Structure," SAE Paper No. 932812, 1993

73. P.A. Barber, T.F. Lonstrup and N. Tunkel, "The Role of Lubricant Additives in Controlling Abnormal Combustion (ORI)." SAE Paper 750449, 1975.

74. P. Schreyer, K.W. Starke, *et al.*, "Effect of Multifunctional Fuel Additives on Octane Number Requirement of Internal Combustion Engines," SAE Paper No. 932813, 1993.

75. J.R. Sabina, J.J. Mikita and M.H. Cambell, "Preignition — does it provide roadblock to greater efficiency in fuel utilization." *The Oil and Gas Journal*, June 29, 1953.

76. S.S. Cheng and C. Kim, "Effect of Engine Operating Parameters on Combustion Chamber Deposits," SAE Paper No. 902108, 1990.

77. J.C. Guibet and A. Duval, "New Aspects of Preignition in European Automotive Engines." SAE Paper No. 720114, 1972.

78. V. Arrigoni, G. Cornetti, B. Gaetani, F. Vicenzetto and F. Sezzi, "A quantitative system for measuring preignition." Paper No. 2, *Giornale Ed Atti Della Associazione Tecnica Dell'Automobile*, July 1969.

79. A. Marciante and P. Chiampo, "The Influence of Lubricating Oil Ash on Surface Ignition Phenomena." SAE Paper 700458, 1970.

80. D. Downs and J.H. Pigneguy, "An Experimental Investigation into Preignition in the Spark Ignition Engine." Institution of Mechanical Engineers, Automobile Division Proceedings, 1950-51.

81. D. Downs and F.B. Theobald, "The Effect of Fuel Characteristics and Engine Operating Conditions on Preignition." Institution of Mechanical Engineers, Automobile Division Proceedings, 1963-64.

82. V.F. Massa, "A Study of the Normal and Abnormal Combustion Behavior of Gasolines." SAE Paper No. 293C, Detroit 1961.

83. G.B. Toft, "Effect of insulation deposits on spark plug misfire in the gasoline engine." Paper EF/4/3, CEC International Symposium on Performance Evaluation of Automotive Fuels and Lubricants, Rome, June 1981.

84. K. Owen, Comments at Champion Ignition and Engine Performance Conference, Paragraphs 111 and 121, Brussels, 1984.

85. R.C. Teasel, Comments at Champion Ignition and Engine Performance Conference, Paragraph 421-424, Brussels, 1984.

86. W.S. Affleck, P.E. Bright and R.J. Ellison, "Run-on in Gasoline Engines." Institution of Mechanical Engineers, Automobile Division Proceedings, Vol. 183, Part 2A, 1968-69.

87. J.C. Ingamells, "Effect of Gasoline Octane Quality and Hydrocarbon Composition on After-run." SAE Paper No. 790939, 1979.

88. CONCAWE, "Motor vehicle emission regulations and fuel specifications — 1994 update," Report 4/94, CONCAWE, 1994.

89. D.R. Blackmore and A. Thomas, <u>Fuel economy of the gasoline engine,</u> The MacMillan Press Ltd, 1978.

90. G.D. Hobson, <u>Modern Petroleum Technology,</u> Chapter 20, "Fuels for Spark Ignition Engines" by K. Owen, John Wiley, 1984.

Chapter 7

Gasoline Volatility

Gasoline is a mixture of many different compounds, each having its own boiling point and vapor-forming characteristics. Thus, when distilled using a simple still with little fractionation, gasolines show a boiling range covering a temperature spread of around 170°C from the initial boiling point (IBP) to the final boiling point (FBP). The temperature range over which the gasoline distills will depend on the composition of the gasoline and the fractionation efficiency of the distillation column. A very efficient column may be able to separate individual compounds if they have moderate differences in boiling point and if each is present in a reasonable amount. Such a column would give a distillation curve looking something like Fig. 7.1 for a mixture containing equal amounts of five different components each having a different boiling point. Gasolines usually contain so many readily identifiable compounds (up to about 400) that the distillation curve is quite smooth, even when a high degree of fractionation is used. Figure 7.2 shows gasoline distillation curves obtained at two different levels of fractionation efficiency; curve A is the sort of curve that one might obtain with good fractionation, whereas curve B is the same gasoline but with very little fractionation using the standard ASTM D86 test.

It will be seen that the temperature at which the mixture begins to distill is much lower with a high degree of fractionation, and the final temperature is much higher, showing that individual compounds or groups of similar boiling compounds are separated much more readily in this case. Of course, if one wanted an even better separation, then techniques such as gas/liquid chromatography could be used.

However, it is not necessary to be concerned with high-efficiency fractionation when considering the behavior of a gasoline in an engine, because the type of evaporation that occurs represents only a

161

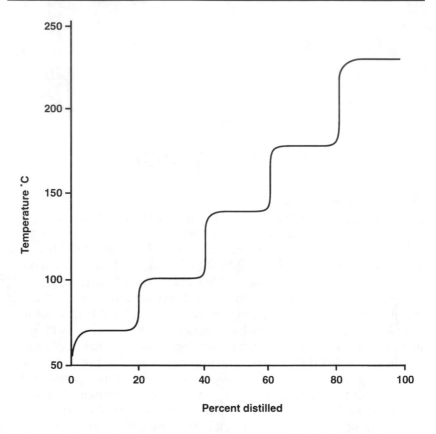

Fig. 7.1. Good fractionation can separate compounds differing in boiling point.

coarse separation of different boiling range materials. This is not to say that the distillation curve is unimportant — it is very important indeed and correlates with a number of performance aspects of a vehicle, including its driveability. Driveability can be defined as the response of the vehicle to the throttle; a vehicle with good driveability will accelerate smoothly without stumbling or hesitating, will idle evenly and will cruise without surging.

Figure 7.3 shows a gasoline distillation curve obtained using a low fractionation-efficiency still and indicates the aspects of vehicle performance that different parts of it influence.

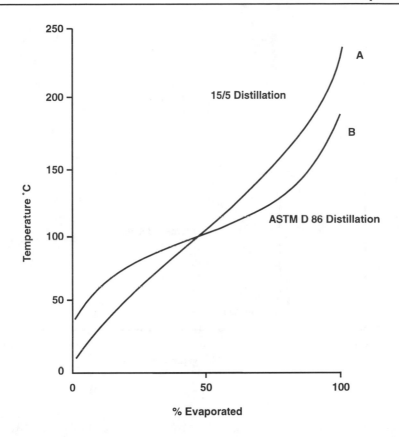

Fig. 7.2. Gasoline distillation curves.

It can be seen that if the distillation curve is displaced downwards the gasoline becomes more volatile, and vice versa. The front end, that is, the compounds in the gasoline having boiling points up to about 70°C, is the first to be distilled over, and this controls ease of starting and the likelihood of hot weather problems such as vapor lock occurring. The mid-range largely controls the way that the vehicle drives in cold weather, and particularly the time for the engine to warm up. It also influences to some extent the tendency for ice to form in the carburetor during cool, humid weather. The back end contains all the heavier, high boiling point compounds and these have a high heat content and so are important in improving fuel economy when the engine is fully warmed up. However, some of these heavier com-

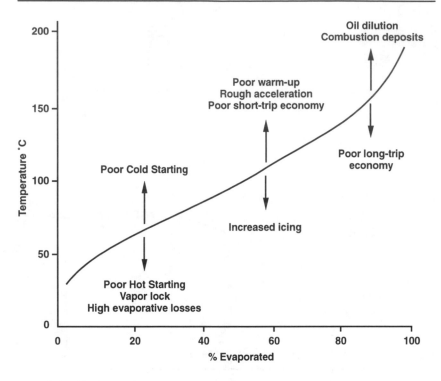

Fig. 7.3. Gasoline volatility is a compromise.

pounds may find their way past the pistons into the crankcase and dilute the crankcase oil. They are also not as readily combusted as the lighter components and give rise to combustion chamber deposits.

It can be seen from Figure 7.3 that if one goes too far in the direction of making a gasoline more volatile, one set of problems can occur, and if one goes too far in the other direction, another set of difficulties is possible. Weather conditions, particularly ambient temperature, influence the choice of volatility required for satisfactory operation. Altitude also has a small effect because atmospheric pressure affects the rate of evaporation of gasoline. Vehicles themselves vary enormously in the way that they respond to gasoline volatility. Some vehicles are extremely tolerant to changes whereas others can give severe driveability problems if the gasoline volatility is not closely matched to the weather conditions prevailing. The vehicle design aspect which is most important in this respect is the proximity of the fuel system to hot engine parts. It is necessary to avoid excessive vaporization during hot weather and yet to make sure that there is enough heat present during cold weather to adequately vaporize the gasoline.

Thus, the setting of volatility specifications is a compromise that is influenced by weather conditions, geographical location and the characteristics of the car population.

7.1 Measurement of Gasoline Volatility

The tests most commonly used to define gasoline volatility are the Reid Vapor Pressure, the ASTM Distillation test and the Vapor-Liquid Ratio.

7.1.1 Reid Vapor Pressure

This is the vapor pressure obtained under standard conditions, using an air-to-liquid ratio of 4:1 and a temperature of 100°F. It is determined by ASTM Procedure D 323[1] which involves filling a metal chamber with chilled sample and connecting it to an air chamber which is, in turn, connected to a pressure gauge as shown in Figure 7.4. The apparatus is immersed in a water bath at 100°F (37.8°C) and is shaken periodically until a constant pressure is obtained, which is the Reid Vapor Pressure (RVP). It can also be determined by other procedures such as the Micro-method (ASTM D 2551) which is calibrated against results obtained by the above standard procedure and which requires only a small amount of sample; or by calculation from the gasoline composition as determined by Gas/Liquid chromatography.

7.1.2 ASTM D 86 Distillation[1]

In this test a 100 mL sample of gasoline is distilled in a standard apparatus under specified conditions of heat input and coolant temperature so that the distillation rate is strictly controlled. The basic equipment is shown in Figure 7.5, although automatic versions are available[2] in which the mercury-in-glass thermometer is replaced by a thermocouple and the distillation rate is controlled by varying the heat input based on signals from a photocell following the rising meniscus of the distillate in the receiver. The distillation curve is plotted or printed out automatically.

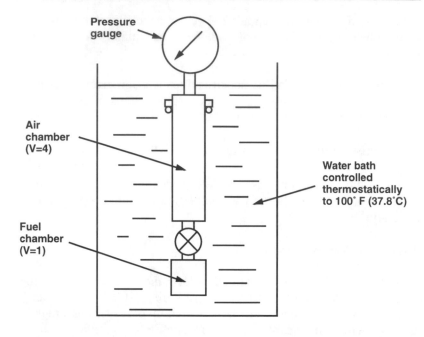

Fig. 7.4. Reid Vapor Pressure apparatus.[81]

The fractionation efficiency is equivalent to about one theoretical plate and a curve such as that shown in Figure 7.6 is obtained. Some of the various terms used in relation to distillation data are illustrated in this figure and others that are important when considering distillation data are:

- Percent distilled: the volume in milliliters of condensate in the receiver corresponding to a simultaneous temperature reading. Also called percent recovered.

- Percent residue: the volume of residue left in the flask when allowed to cool after the distillation is complete.

- Percent loss: 100 minus (maximum percent distilled plus percent residue). This represents mainly those very light hydrocarbons that are not condensed.

- Percent evaporated: the sum of the percent distilled and the percent loss.

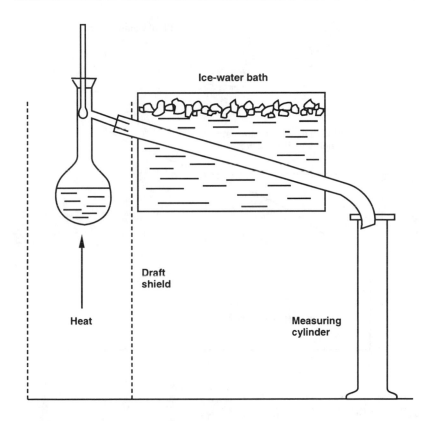

Fig. 7.5. ASTM D 86 distillation apparatus.[81]

Distillation data are often represented and specified by the temperature at which a given percentage of the gasoline is evaporated so that T10% (sometimes written T10 or $T_{10\%}$) is the temperature at which 10% of the gasoline is evaporated using ASTM D 86. Similarly, they can also be represented by the percentage evaporated at a given temperature so that E70°C (or $E_{70°C}$ or E70) represents the percentage evaporated at 70°C. It is considered preferable and more meaningful by many to use percentages evaporated rather than temperatures, particularly when carrying out blending calculations. It is important to remember that it is always percent *evaporated* rather than percent distilled or recovered when considering specifications or comparing gasolines.

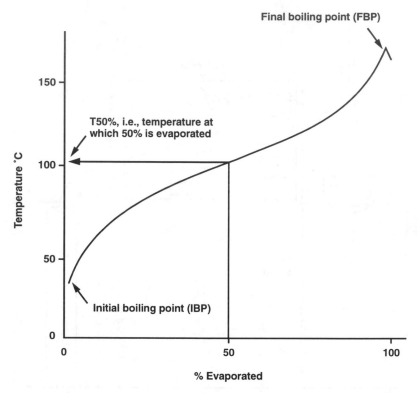

Fig. 7.6. ASTM D 86 distillation curve.

The distillation characteristics of a gasoline are not always similar to that shown in Figure 7.6 since the shape is dependent on the blend composition. Figure 7.7 illustrates two other types of distillation curve that can be found and compares them with a conventional curve. Gasolines containing only components that boil within the same narrow range are known as "narrow-cut" blends and show a fairly flat plateau for a large part of the curve. Blends consisting of light and heavy components with very little material boiling in the intermediate temperature range are known as "dumbbell" or "mid-cut" blends and these show a steep rise in temperature once the light materials have been distilled until the heavy components start distill-ing. Other types of distillation curve are possible, such as one with a low front-end volatility and a high back-end volatility, and this is sometimes known as a "front-cut" gasoline. The driveability perfor-

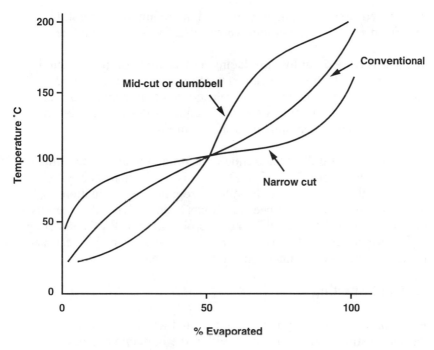

Fig. 7.7. Atypical gasoline distillation curves.

mance of fuels with these atypical distillation curves can be quite different from that of conventional gasolines.

7.1.3 Vapor-Liquid Ratio

The Reid Vapor Pressure test involves an arbitrary 4:1 ratio of vapor to liquid, but when a gasoline is subjected to the levels of temperature found in the fuel system of a vehicle, it can be important to know what ratio of vapor to liquid is likely to form. The amount of vapor produced can have a direct influence on the performance of the vehicle, particularly under conditions of high ambient temperature.

The Vapor-Liquid Ratio test (ASTM D 2533)[1] measures the volume of vapor formed at atmospheric pressure from a given volume of gasoline at a specified test temperature. Sometimes it is desirable to know the temperature corresponding to a given vapor-liquid ratio (V/L) for a particular gasoline and, in these cases, the V/L is

169

determined at several temperatures and the required temperature is read from a plot of temperature versus V/L for the test gasoline.

The test is carried out by introducing a measured volume of liquid fuel at about 0°C through a rubber septum into a glycerin-filled burette. The charged burette is then placed in a temperature-controlled water bath and the volume of vapor in equilibrium with liquid fuel is measured. The test equipment is shown in Figure 7.8.

The test is rather difficult and time-consuming to carry out in practice because quite often a V/L of 20 ($T_{V/L=20}$) is used when specifying gasolines; this means that at least three separate determinations at different temperatures are needed. It has been shown[3] that there is a good correlation between V/L and a combination of RVP and certain distillation points, and procedures have been established for estimating temperature-V/L values[4] from these parameters.

7.2 Cold Starting

For a spark ignition engine to start, the air-fuel ratio of the mixture in the combustion chamber must be within the ignitable range, i.e., in general it must be between about 7:1 and 20:1 by weight, although some engines may be able to start with air-fuel ratios outside these limits. When the engine is cold it may be difficult to achieve even the leanest ignitable air-fuel ratio because the fuel may not vaporize sufficiently, and under these conditions the mixture is richened to bring it within the ignitable range by increasing the injection time with fuel-injected vehicles or by the use of the choke with carburetted vehicles.

There are many other reasons for a vehicle to fail to start easily.[5] An adequate cranking speed is important because higher speeds favor good atomization of the fuel in carburetted vehicles and provide some warming of the engine due to the frequent compressions. Battery condition is also important because of its effect on cranking speed, and it is unfortunate that battery power output drops as the temperature reduces. The lubricant viscosity at low temperature also influences cranking speed and so affects cold starting. Finally, the ignition system and ignition timing are significant factors since it is important to have a spark of sufficient power at the optimum time to ignite the mixture. Thus, the fuel has a comparatively small effect on cold start-

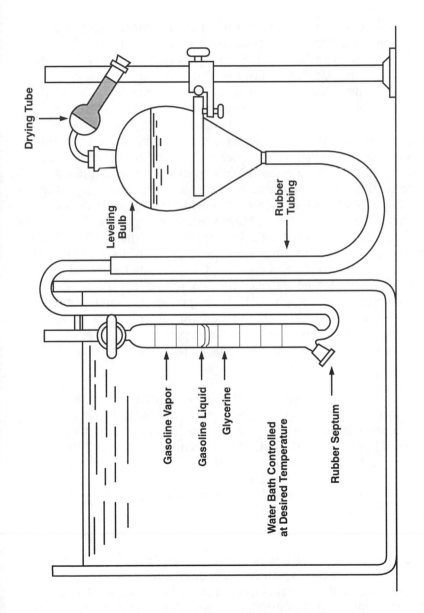

Fig. 7.8. Vapor-liquid ratio apparatus.[1]

ing compared to the mechanical and electrical factors mentioned above, except at low temperatures.

In spite of its minor role, fuel volatility becomes increasingly important the more the ambient temperature drops below about -10°C.[5] For fuel-injected vehicles and many modern carburetted cars, it is the low boiling components in a gasoline that are most relevant in cold starting, and these are characterized by such parameters as the RVP, the temperature at which ten percent evaporates (T10), the amount that evaporates at 70°C (E70), or a combination of some of these terms. In practice, these parameters are often highly intercorrelated so that a high E70 gasoline will have a low T10, and so it is of little consequence which of these is chosen in a specification to ensure adequate starting under low ambient temperature conditions. There are some carburetted vehicles, however, in which heat soak-back occurs from the hot engine into the carburetor after the engine has been switched off. This can cause the lighter portion of the gasoline in the float bowl to evaporate so that when the engine is being restarted after it has cooled down, the gasoline going forward initially will be depleted of volatile components and may not ignite. For this reason, in such vehicles, E100 may provide a better predictor of cold start performance than the terms relating to the front end only. It also explains why some highly volatile fuels containing perhaps excessive amounts of butane may show poor cold-starting characteristics in some vehicles.

Specification levels to ensure satisfactory cold-start performance will clearly depend on the climatic conditions in the geographical area under consideration, as well as the car population itself. Usually, minimum levels for RVP and E70 or a maximum level for T10 are specified. Levels such as 0.6 bar minimum for RVP and 10% minimum for E70[5,6] or 60°C maximum for T10[7] are satisfactory for most winter locations. Too high a front-end volatility, however, will increase the risk of driveability problems caused by excessive vaporization in the fuel system.

In many studies, cold starting performance is included in cold weather driveability, where the overall cold weather performance is related to a "Driveability Index," as is discussed in the next section. These driveability indices do not usually include RVP but rely primarily on the ASTM D86 distillation curve.

7.3 Cold Weather Driveability

The driveability of a vehicle has been defined as the degree to which a vehicle starts readily, idles evenly, drives smoothly when cruising and accelerating, and generally responds to the throttle. It is most critical when the vehicle is warming up. It is well known that vehicle driveability deteriorates as ambient temperature decreases and, generally speaking, drivers do not expect such a good performance during warm-up of their vehicles in very cold weather.

Driveability malfunctions are caused by variations in mixture strength giving an air-fuel mixture outside the ignitable range in one or more cylinders for a few cycles. When a single point injection or a carburetted engine is cold and the ambient temperature is low, a large proportion of the fuel going forward can be present in the inlet manifold as a liquid film[8,9] rather than as a vapor. It is this lack of vaporization that gives rise to a hesitation before a burnable mixture reaches the cylinders at the start of an acceleration. Under these circumstances, too rich a mixture can also occur during the acceleration as the excess liquid "catches up." An uneven idle and "surging" during cruise in carburetted or single point injection cars may be caused by maldistribution of the fuel between cylinders, and this can be another reason for a stumble during an acceleration. Poor vaporization in cold weather is overcome to some extent by the provision of a choke giving over-rich mixtures during warm-up.

It is clear from the above that the ease with which a gasoline vaporizes in the engine determines the cold weather driveability of the fuel/vehicle combination. Fuel volatility and ambient temperature are obviously both important factors, but even more so is fuel system design. The use of multipoint fuel injectors rather than carburetors has a positive effect on driveability, as also do heating systems for the inlet manifold and good air-fuel ratio control during warm-up. Unfortunately, the need to meet emissions limits can sometimes mean that driveability performance is not always satisfactory under all conditions.

There are a number of procedures that have been developed to quantitatively measure cold weather driveability such as the CRC procedure in the U.S.[10,11] and the CEC code of practice in Europe.[12,13] With many vehicles having engine management systems it is necessary to carry out preliminary tests to precondition the adaptive memory while the vehicle is warming-up.

One CRC procedure that has been used[14] is quite complicated and only a brief summary can be given here. The test involves evaluating the starting and idle performance and then, after 5 seconds in "drive," making a light throttle acceleration from 0 to 25 mph at a constant throttle opening and at a predetermined manifold vacuum. The vehicle is then cruised at 25 mph, and after 0.2 miles the throttle is opened to the detent position and the vehicle is accelerated to 35 mph in high gear at constant throttle. It is then decelerated to stop and a wide open throttle acceleration made from 0 to 35 mph. A deceleration is then made to 10 mph, and after a total of 0.4 miles, an acceleration at light throttle from 10 to 25 mph. During these maneuvers the severity of the various malfunctions are recorded. Finally, the idle quality is evaluated over a 30-second period in drive. The driving part of this cycle is repeated 3 times and then a further series of accelerations and decelerations carried out.

The CEC procedure uses a set drive cycle which involves a cold start, an idle period, and acceleration and cruise modes as shown in Figure 7.9. Tests can be carried out on the road or on a chassis dynamometer in a climate-controlled chamber.

Numerical ratings are assigned to each part of the test procedures so that an overall demerit rating can be calculated for each car/fuel/temperature combination. Also, when accelerating, the time to achieve a speed of 70% of that obtained with an appropriate standardization fuel and without choke is known as the warm-up time, as illustrated in Figure 7.10.

The fuel parameters that influence cold weather driveability are not simple and can vary widely from one vehicle to another. An overall Driveability Index (DI) has been developed in the U.S.[15] where:

$$DI = 0.5 \, T10 + T50 + 0.5 \, T90$$

This equation has been developed and refined, and a later version for use in the U.S. is as follows:[16]

$$DI = 1.5 \, T10 + 3 \, T50 + T90$$

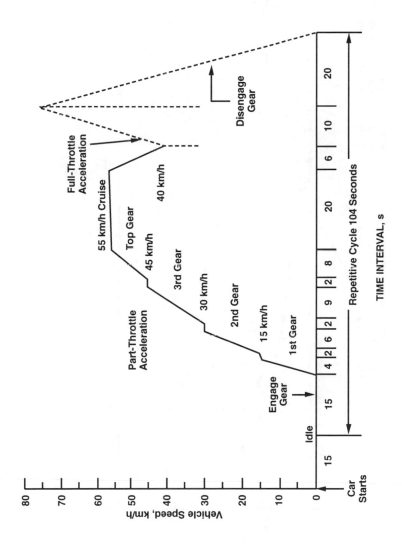

Fig. 7.9. CEC cold weather driveability test cycle. [12.13]

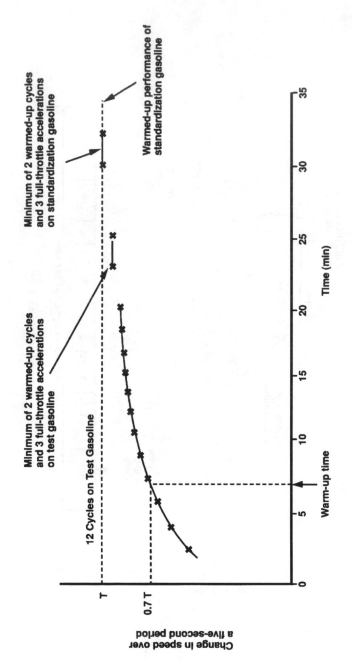

Fig. 7.10. Warm-up time curve.[13]

This version has been shown to correlate well with driver satisfaction both for hydrocarbon-only gasolines and also when MTBE is present—although at the same DI, the hydrocarbon fuels had a higher satisfaction level than gasolines containing MTBE.[17] For fuels representing typical commercial gasolines, RVP and DI are usually intercorrelated so that meeting low RVP levels in order to minimize evaporative emissions can lead to poor cold weather driveability.[18]

Another driveability equation developed for some U.S. vehicles[19] also uses the same parameters of T10, T50 and T90, together with a variable related to the shape of the distillation curve, to give a Driveability Number.

Often, and particularly in Europe, a single distillation parameter such as T50 or the E100°C point, is used as a Driveability Index, although this is not satisfactory when MTBE is blended into the gasoline.[20] The reason is that the MTBE vaporizes very readily and so gives a lean mixture under transient operation such as during an acceleration, resulting in a deterioration in driveability. To overcome this, a new Driveability Index has been suggested for fuels containing MTBE:[21]

$$DI = T50 + M/2 \quad \text{where M is the \%vol MTBE in the blend}$$

The distillation terms T50 and E100°C are highly intercorrelated for most gasolines and so are equally valid and both have been widely used. When such single predictors are used, it is assumed that the whole of the distillation curve has a conventional shape and that, for example, the T10 and the T90 points have the normal relationship to the T50 point.

A high level of involatiles such as heavy aromatics will increase the T90 point and this will cause poor driveability and also increase hydrocarbon exhaust emissions.[21,22,23]

Test work in France using some 25 vehicles and four fuel series[24] indicated that the best overall correlation with driveability demerits was obtained with an expression containing two variables: RVP + 22 E100 (where RVP is in mbars).

Two similar variables, i.e., RVP and T50%, were used in a study conducted in the U.S. by CRC at intermediate temperatures (30°F to

177

56°F).[25] Under these conditions, fuels containing MTBE behaved very similarly to hydrocarbon-only fuels, but fuels containing ethanol gave a somewhat poorer driveability performance at the same volatility level.

Tests have also been carried out using fuels with atypical distillation characteristics[26] and it has been shown that, in general, the further the fuel deviates from the conventional, the worse the cold weather driveability.

The relationship between these technical assessments of vehicle driveability and customer perception of driveability malfunctions has also been studied.[17,27] It was shown that there is a good correlation between the technical and customer assessments and that the test procedure is, therefore, a good indicator of consumer response to cold weather driveability.

7.4 Hot Weather Driveability

When gasoline vaporizes prematurely in the fuel system, i.e., upstream of the carburetor jets or injectors, driveability problems can also ensue. The likelihood of such problems occurring will depend, as with cold weather driveability, on engine design, ambient temperature and pressure, driving mode and fuel volatility.

Fuel system design is perhaps the most important factor because, although many vehicles appear to be virtually free of problems in hot weather even with relatively volatile fuels, others can exhibit very severe problems indeed. Thus vehicles with port fuel injectors are much less susceptible to driveability problems at high and intermediate temperatures than carburetted or throttle body injected vehicles.[25,28] The position of the fuel pump(s) is extremely important. If at least one of them is submerged in the fuel tank, then the gasoline being delivered to the carburetor or injectors will be under pressure and much less likely to vaporize prematurely. If the pump or the fuel lines are positioned so that they can be heated by the engine, this will increase the possibility of undesirable vaporization. Vehicles with low drag coefficients may sometimes suffer hot weather driveability problems because cooling air flowing into the engine compartment is often minimized. Equipment in the engine compartment that is likely to be hot, such as turbochargers, will also give rise to high under-hood temperatures and increase vaporization problems. Even the fuel in the tank can get hot due to recirculation and to the fact that the exhaust system is often positioned directly under it. As a result, a considerable amount of fuel weathering involving loss of light ends occurs as a car is being driven and can result in a very significant drop in RVP.[29]

Ambient temperature is obviously important in causing hot weather driveability malfunctions. High-altitude driving is also a problem not only because of the reduced pressure but also because it often means that the engine compartment is very hot due to the high-load, low-speed driving mode used when climbing, even though ambient temperature tends to drop with altitude. Both ambient temperature and altitude are taken into account when setting gasoline volatility specifications.[30]

Driving mode is clearly important because this determines how hot the engine compartment and the fuel system can get. The methods of test for Hot Weather Driveability[31,32] usually involve three phases: The first requires the vehicle to be driven at a relatively high speed to get it fully warmed up; the second allows time for the heat of the engine to soak back into the fuel system, and this is achieved by switching off the engine or by allowing it to idle for fifteen minutes or so; and the third measures the ease with which the engine restarts, the idle quality, the smoothness of the acceleration and the time to reach a given speed. These data are recorded and compared with results when a low-volatility reference fuel is used. Other test procedures have been used[28] and, as with testing for cold weather driveability, for vehicles fitted with engine management systems it can be necessary to carry out preliminary runs to precondition the adaptive memory of the system.

The type of hot weather driveability problem that occurs can be classified as "carburetor percolation," "vapor lock" or "foaming."

Carburetor percolation occurs when fuel in the float bowl starts to boil, either during or following a hot soak, and forces excess fuel into the inlet manifold via the carburetor vent (in vehicles with an internally vented system) or through the carburetor jet system. This gives rise to an over-rich mixture which can cause difficulty in restarting the engine and persistent stalling.

Vapor lock is probably the main field problem that occurs and this also causes poor restarting and other malfunctions such as hesitation and stumbles during accelerations. It can be so severe that it can cause the engine to cut out completely. These malfunctions are due to the mixture being too lean as a result of excessive vapor formation, preventing the fuel pump from operating satisfactorily so that the fuel in the float bowl is not replenished.[33]

Carburetor foaming[34,35] is caused by the rapid boiling of fuel as it enters a hot carburetor, thereby generating a foam which causes the float to sink. This allows more and more fuel to enter the bowl so that it fills with foam, thereby blocking the air vent and causing an increase in pressure inside the bowl. This pressure increase, in turn, forces excess fuel through the metering jet and vent so that the vehicle suffers similar malfunctions to those caused by vapor lock, even though they are caused by excessive enrichment of the mixture rather than the overleaning that occurs with vapor lock. In cases where the vent goes directly to atmosphere, fuel can be forced out into the engine compartment and cause a fire hazard.

It is generally accepted that it is the front-end volatility of a gasoline that determines the extent to which hot weather driveability problems are likely, and a number of volatility terms have been investigated to correlate with field and laboratory experience.[36,37] The gasoline temperature to obtain a V/L ratio of 20 is widely used to control front-end volatility[37,38,39] and is the basis of the front-end volatility specifications of ASTM D 439 (now superseded by ASTM D4814).[30,40] This U.S. specification defines minimum quality standards in different geographic regions, depending on their climatic conditions, to protect consumers against hot weather driveability and other problems.

The alternative to using V/L ratio is to use a combination of RVP and an ASTM D 86 distillation point such as E70°C. This is sometimes preferred because, as discussed earlier, such expressions are highly correlated with V/L ratio[3] and are easier to measure. The general expression:

$$RVP + n\ E70°C$$

where n is a constant, has been shown to give a good correlation with $T_{V/L=20}$ and, hence, with hot weather driveability performance.[41,42,43] In the U.S. this expression has been called the Vapor Lock Index

(VLI)[44] or Front-End Volatility Index (FEVI) when a value of n=9 has been used (this relates to the use of RVP in mbar and E70°C in %vol — if RVP is measured in psi, n is 0.13).[37] For European and Japanese vehicles the value of n has been shown[45] to give the best correlation with vehicle performance when it lies between about 4 and 7 (RVP in mbar and E70°C in %vol). This is consistent with values of 5[46] and 7.2[24] that have been obtained by other workers. In general, the expression RVP + 7 E70°C has been widely used for many years for European and Japanese vehicles and has been called the FlexiVolatility Index or simply the Vapor Lock Index. It still appears to be satisfactory in terms of its correlation with hot weather driveability malfunctions. RVP and T50% have also been used to correlate with hot driveability,[28] but since T50 is itself highly intercorrelated with E70 for most marketed hydrocarbon fuels, this does not necessarily mean that T50% is a better predictor of hot driveability than E70°C.

The front-end volatility indices discussed above have been correlated with consumer response and found to be satisfactory, although it has been shown that motorists are generally less critical than technical raters for most driveability malfunctions.[47]

The presence of oxygenates such as MTBE in the fuel can improve hot weather driveability[28] as compared with a hydrocarbon fuel of the same front-end volatility since such problems are usually caused by the air/fuel mixture being too rich.

Data on the driveability characteristics of different vehicle models are usually obtained cooperatively within the oil and other industries, both in the U.S. and Europe, using standard test procedures and fuels. The raw data obtained are then available to all participants in the test program. A volatility data bank can be built up and used to establish performance levels for individual car models or whole car populations for a range of fuel volatilities and ambient temperatures. The data can also be used to estimate levels of (RVP + 7 E70°C) to satisfy different car populations at various ambient temperatures[45] as shown in Figure 7.11. In this figure, the use of different units for RVP (kPa instead of mbar) means that the Vapor Lock Index equation becomes: (RVP + 0.7 E70°C).

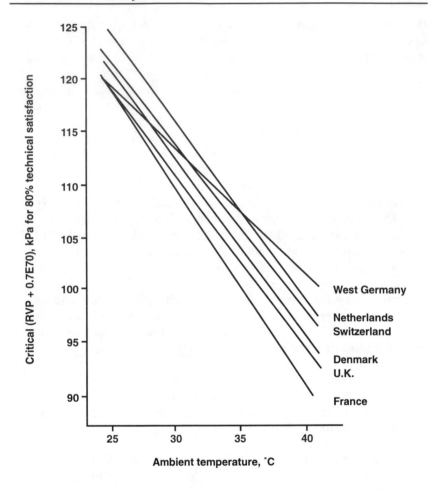

Fig. 7.11. Hot weather driveability performance of European car populations, 80% technical satisfaction.[45]

7.5 Evaporative Emissions from Vehicles

Atmospheric hydrocarbon emissions contribute to the formation of photochemical smog which is a major problem in certain cities such as Los Angeles and Tokyo, although in Europe the problem is much less severe. In the U.S., smog formation is still present in certain areas in spite of mandatory controls which severely limit sources of hydrocarbons from vehicles. The presence of volatile toxic compounds such as benzene in gasoline is another reason for wishing to minimize such vapors in the atmosphere. Evaporative loss control devices are capable of reducing evaporative emissions by over 90%. In Western Europe in 1986, it was estimated that about half of the volatile organic compounds in the atmosphere came from man-made sources and that of these, exhaust hydrocarbons contributed about 25%, evaporative losses from vehicles about 10% and evaporative losses during refueling about 2%.[48] Exhaust emission legislation has become progressively more severe in most countries and the use of, or intention to introduce, evaporative emission legislation is becoming more and more widespread.[49] Devices on the vehicle to control refueling evaporative emissions (onboard control) have also been the subject of much interest[50,51,52] and it has been shown that over 98% of these losses can be overcome by the use of suitable technology.

Evaporation losses from a vehicle can occur from several different sources on the vehicle unless control systems have been fitted. The most important of these are breathing losses from the fuel tank and carburetor bowl. Vehicle design features that have an important influence on evaporative losses are: the carburetor system, the presence of heat shields and insulation, the shape and liquid surface area of the fuel tank, and the fuel tank venting system.

It is usual to divide vehicle evaporative losses into three categories:

1. **Running Losses** — These are the losses that occur while a vehicle is being driven.

2. **Diurnal Losses** — These losses occur when the vehicle is stationary for an extended period with the engine switched off, and are due to the normal temperature changes that occur over a 24-hour period. Most of these emissions occur whenever there is an increase in temperature which causes an expansion of the vapor in the fuel tank.

3. **Hot Soak Losses** — These are defined as the losses that occur due to engine heat soaking back into the fuel system when a fully warmed-up vehicle is stationary. Most of these losses come from the carburetor bowl in vehicles without emission controls.

The measurement of evaporative losses from vehicles has been carried out by a variety of different methods, the two most important of which are: the adsorption of vapors from different parts of the fuel system in canisters containing activated charcoal; and the measurement of the hydrocarbon concentration within a sealed chamber in which the vehicle is contained. The most commonly used of these methods is the latter and is known as the SHED (Sealed Housing for Evaporative Determination) test procedure[53,54] (see Appendix 4). In the U.S., the diurnal and hot soak emissions are measured and the sum of these two is used for Federal certification of the vehicle. The running losses are of less concern because a purge system is fitted which draws vapors back into the engine when the vehicle is operating. The diurnal portion of the test originally simulated the temperature rise that occurs over 24 hours by raising the temperature of the fuel in the tank from 16 to 29°C in one hour while the vehicle is in the SHED. However, new regulations replace the one-hour diurnal simulation with a 72-hour test in which the entire vehicle will experience three successive 24-hour diurnal temperature cycles.[55] The hot soak part of the test takes place after the vehicle has undergone a standard warm-up procedure on a chassis dynamometer, when it is pushed into the SHED and the hydrocarbon concentration measured after a given time. From a knowledge of the volume of the SHED, the total hydrocarbons evaporated during both portions of the test can be calculated. The SHED procedure has a disadvantage in that it picks up all hydrocarbon emissions and not just those from the fuel system. With a new vehicle a surprisingly large amount of hydrocarbons can come from the tires, upholstery, undersealant, etc., and it is usual to steam-clean and weather vehicles before testing if fuel system hydrocarbon loss is the prime concern.

A comprehensive study of evaporative emissions was carried out by the U.S. Bureau of Mines in the late 1960s to early 1970s.[56,57,58] Emissions from a wide range of vehicles at several ambient temperatures were measured using fuels of varying volatility. The test procedure involved measurement of the losses from particular points in the fuel system by condensing the vapors and weighing, since the SHED procedure had then not yet been developed. RVP was used to categorize the fuels and it was found, for example, that at 21°C with a fuel of 0.7 bar, total evaporative emissions varied from 8 to 40 g/test depending on the vehicle. A study by API in 1978[59] showed that the relationship between evaporative losses and RVP is nonlinear, with losses increasing disproportionately as RVP is increased. Subsequent test work has confirmed the importance of RVP.[60,61,62]

Controls were introduced in 1970-71 in the U.S. that required only relatively simple modifications to enable vehicles to meet the initial emissions standards of 6 g/test. However, as the limits were progressively tightened, it became necessary to introduce more sophisticated methods of control. The technique that has been universally adopted to meet the current limits is the use of canisters containing activated carbon to which all fuel system vents are connected. Hydrocarbon vapors generated during diurnal and hot soak periods are adsorbed on the carbon and retained in the canister. The carbon is regenerated by drawing air back through the canister and into the engine when the engine is running, so that it is purged of hydrocarbons. This type of system is successful in reducing evaporative emissions by about 95% from uncontrolled levels[63] and allows vehicles to meet a 2 g/test limit.

Evaporative emission limits are also imposed in Japan and Australia. In Europe, the EC Commission has stated its intention to control evaporative emissions from motor vehicles by requiring all cars to be fitted with small carbon canisters and for emissions from the distribution system to be fitted with controls. Studies have been carried out in Europe by CONCAWE (the oil companies' European organization for environment, health and safety) to determine typical emission levels from a number of European and Asian cars, to find the effect of various fuel and vehicle parameters on them using a slightly modified SHED procedure and, in a later test, to determine the effects of temperature on them.[64,65,66] The work confirmed that vehicle and fuel system design have the greatest effect on evaporative emissions and that carbon canisters are very effective in controlling these emissions. It also showed that:

- The emissions from the 10 vehicles tested varied from 9 to 25 g/test on a winter-type fuel and from 4 to 16 g/test on a lower volatility summer-grade fuel.

- RVP is the only significant fuel parameter influencing these emissions. It has a relatively small effect — a reduction of 10 kPa reduced evaporative emissions by 23%.

- The cycle used to warm-up the vehicle is important in its influence on total evaporative losses.

- The presence of oxygenated compounds (see Chapter 11, Section 11.5.6) in a gasoline did not influence evaporative losses.

- Diurnal losses are significant, but are readily controlled by carbon canisters.

- Hot soak losses and running losses from uncontrolled vehicles increase progressively with ambient temperature and RVP. A 1°C change in ambient temperature was found to have the same effect on evaporative emissions as a 3.8 kPa change in RVP.

In a test using mainly U.S. vehicles, reducing RVP by 1 psi (0.7 kPa) reduced diurnal emissions in non-oxygenated fuels and ethanol-containing fuels by 46.4 and 53.7%, respectively, but there was an unexplained zero effect with MTBE-containing fuels.[67]

7.6 Influence of Fuel Volatility on Exhaust Emissions

A considerable amount of test work has been carried out to establish the effect of both fuel composition and fuel volatility on exhaust emissions. It is generally agreed that their influence is small relative to the effect of vehicle technology.[68]

Heavy hydrocarbons in gasoline contribute to exhaust emissions and lowering the 90% distillation temperature has been found to reduce exhaust hydrocarbon emissions[69] but to slightly increase NO_x. The effect appears to be due to the distillation characteristics of the fuel rather than its chemical composition.[70,71] Most of the above work has been carried out using American vehicles but a study using European cars showed a similar reduction in HC emissions on reducing T90 but at the expense of NO_x emissions.[68]

Although T90 was used as the volatility variable in the above work, it was not generally varied independently with other distillation points such as T50 and, in fact, it is to some extent correlated with it, so that some of the effect on HC emissions could be attributed to the T50 point. In support of this, a good correlation of HC emissions with Driveability Index (DI), defined as:

$$DI = 1.5 \ T10 + 3 \ T50 + T90$$

has been found[72] regardless of aromatics content, although there were no consistent effects on CO or NO_x.

An investigation into why only hydrocarbon emissions and not CO tailpipe emissions are significantly raised with increasing heavy hydrocarbon constituents (T90) showed that these hydrocarbon emissions are produced mainly in the first cycle of the Federal Test Procedure when the engine is warming up and the catalyst is only just becoming active.[73] CO emissions are influenced to a much lesser degree during this first cycle.

The effect of front end and mid-range volatility on exhaust emissions has also been investigated[74] in European cars at a range of medium to low temperatures and it has been shown that CO emissions increase to a small extent but that HC emissions marginally decrease with increasing volatility. NO_x emissions did not seem to be affected by gasoline volatility. However, ambient temperature has a dramatic effect with both CO and HC increasing as the temperature is reduced, but with a much lower effect on NO_x. In a U.S. test,[75] increased RVP produced slight increases in CO.

Other toxic air pollutants are influenced to some degree by gasoline volatility. In a study in which T90 was the only volatility variable studied, it was shown that when T90 was lowered, benzene, formaldehyde, acetaldehyde and 1,3 butadiene levels were reduced.[76] Aromatics and the presence of oxygenates also influenced these pollutants.

7.7 Intake System Icing

Ice formation in the intake system of a vehicle is associated mainly with carburetted vehicles although it can also occur in throttle-body-injected vehicles. It is much less common now than it has been in the past because improvements in fuel intake heating systems tend to minimize it. However, a surprising number of vehicles, particularly in areas where ambient conditions of temperature and humidity favor icing, still tend to suffer from this problem.

Ice deposition occurs when the water in cool, moist, intake air condenses in the carburetor or throttle body and then freezes because of the temperature drop caused by the evaporating gasoline. It occurs only in cool, humid weather, i.e., when the ambient temperature is roughly in the range -5 to +12°C and the Relative Humidity is above about 80%. At lower temperatures and at low humidity levels there is not enough water in the air to cause problems, and at higher ambient temperatures the temperature depression caused by the evaporating gasoline is not enough to freeze any condensed water. It is made more severe by increasing fuel volatility because this, in turn, increases the temperature drop that occurs as the gasoline evaporates. Figure 7.12 illustrates the ambient conditions that can cause intake system icing in a critical car. The areas enclosed by the curves depend on gasoline volatility. It is usually accepted that the mid-fill volatility as defined, for example, by the E100°C figure, is the best single fuel volatility parameter for controlling icing. It is, however, more usual and more economical to control intake system icing by the use of additives rather than by adjusting volatility (see Chapter 9).

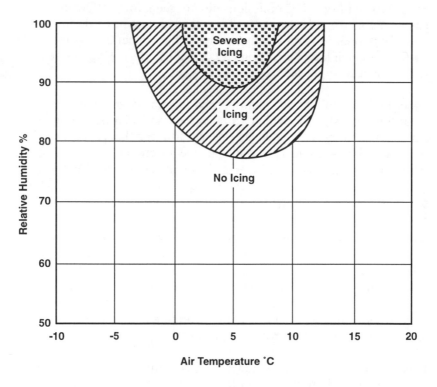

Fig. 7.12. Ambient temperature conditions causing icing.[77]

Intake system icing should not be confused with the freezing of any extraneous free water that may have found its way into the fuel system, causing a blockage of fuel lines. This occurs only when the ambient temperature is well below freezing and when severe contamination of the fuel with water has occurred. It is not influenced by fuel volatility, although it can be minimized by certain types of anti-icing additives.

The effect of ice deposition in a carburetor is to restrict the air or mixture flow. The area where this ice deposition occurs varies with different carburetors, and because of this it can have quite different effects on driveability. Where the ice forms on the throttle plate, as in Fig. 7.13, it will only have a significant effect on air-fuel ratio at very low throttle openings or when the throttle is closed completely such as

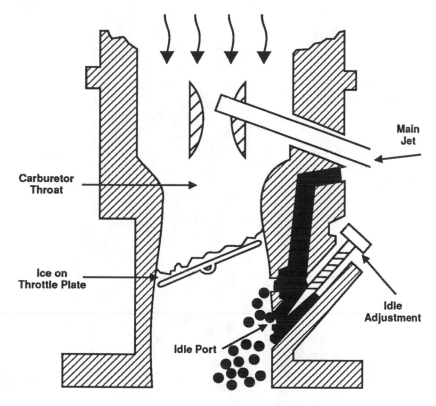

Fig. 7.13. Throttle plate icing causes stalling at idle.[51,53]

when the engine is idling. This type of icing, known as idle icing, is most common in city driving during the period that the vehicle is warming up, and tends to cause the engine to stall whenever it idles because the mixture is too rich. Fuel consumption is also increased during this period of idling and so are the CO and hydrocarbon exhaust emissions, particularly for non-catalyst cars.

Ice can also form in the venturi area, as in Figure 7.14,[77,78] and this tends to cause power loss during cruise conditions, again because of restrictions in the air flow. The drop in power can be so great that the vehicle can come to a complete halt. When this happens it can be quite puzzling because if the motorist waits for a little time, perhaps while looking for a fault, heat soak-back from the engine will cause the ice to melt and the vehicle will operate quite normally again on

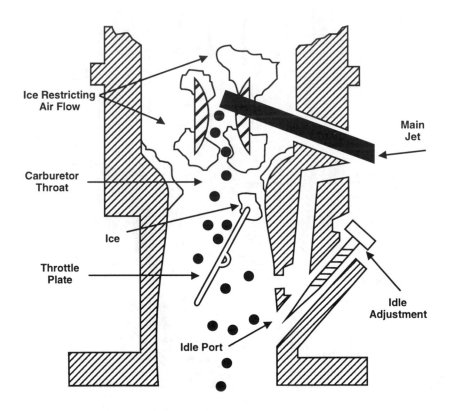

Fig. 7.14. Ice formation in venturi area causes loss of power.[77,78]

restarting. This type of icing is known as cruise icing and will also have an adverse effect on fuel economy and exhaust emissions.

Standard test procedures are available in Europe for evaluating the icing characteristics of a vehicle or of a gasoline or gasoline/additive combination.[79] The test procedures cover idle icing, cruise icing and an extended idle test meant to simulate the conditions that occur when cars are left idling unattended on cold mornings in order to allow them to warm up before the owner gets in and drives away. The tests can be carried out on the road or on a track, but it is more precise to use a chassis dynamometer in a temperature- and humidity-controlled climate chamber. In the idle icing test, a standard cycle is used which is repeated a dozen or so times and which consists of a full throttle acceleration to 50 km/h, a cruise at this speed for 30 sec, and then an idle period of 30 sec. The idle mode only is assessed and demerits applied which relate to the smoothness of the idle and the occurrence of stalls. The cruise icing test is usually carried out with a fully warmed-up vehicle, unlike the idle icing test, and involves running the car at a constant *throttle* position at a speed in the range 70 to 120 km/h. If ice forms, the speed will gradually drop and the vehicle may, in severe cases, completely stall. The extent of the speed loss is a measure of the icing that has occurred. When assessing gasolines or anti-icing additives it is necessary to select a test car that is known to be susceptible to icing problems and to use a relatively volatile base fuel free of oxygenates and anti-icing additives.

There are also many "in-house" tests for evaluating anti-icing additives that are used mainly by additive manufacturers.[78] They almost always involve an engine on a test bed, fed with inlet air having a controlled humidity and temperature. Additives are tested under idling conditions and the number of stalls, or the time for the first stall, is taken as a measure of the degree of icing that has occurred.

7.8 Oil Dilution and Combustion Chamber Deposits

The least volatile part of the gasoline will contain the molecules having the highest heat content and the highest density. This portion of the gasoline will be important, therefore, in contributing to fuel economy once the vehicle has warmed up. However, because this part of the gasoline is less volatile, it may not be fully vaporized by the time it gets to the combustion chamber, particularly when starting and driving away on a cold day with the choke in full use. Under these circumstances liquid fuel can flow down the bores past the piston rings and into the lubricant in the crankcase. This can have two effects: It can wash lubricant off the bores and increase bore wear, and it can dilute the lubricant itself and increase wear in general. It is not uncommon for as much as 10 percent of the lubricant to consist of gasoline heavy ends after starting and driving the first few miles, and in cold weather it can take several miles for the oil to heat up sufficiently to evaporate off these gasoline components.

Another potential problem caused by the heavy ends of the gasoline is that they frequently contain compounds that are difficult to combust. Some of the heavier aromatic compounds that arise from catalytic reforming fall into this category, and particularly any polycyclic aromatics. In the combustion chamber they may only partially burn and give rise to carbonaceous deposits which can gradually increase the octane requirement of the engine.[80] They can also cause spark plug fouling and misfire. In addition, these partially burnt hydrocarbons can find their way into the lubricant, where they increase sludge levels, and they can also appear in the fuel intake system due to valve overlap or exhaust gas recirculation, and thereby increase inlet system deposits.

The back end of the gasoline is usually controlled by the final boiling point (FBP) plus either the 90% distillation temperature or the percent evaporated at a temperature such as 180°C. FBPs in excess of about 220°C are usually not allowed by gasoline specifications, and values well below this are preferred in cold climates.

7.9 Fuel Economy and Gasoline Volatility

Gasoline volatility has a rather complex influence on fuel economy. Most of the elements have already been mentioned, and this section attempts only to try to pull all these effects together.

For short runs in cold weather, high front-end and mid-fill volatilities are important because a quick start and a fast warm-up (to minimize the use of the choke) can improve fuel economy in vehicles with manual chokes, as illustrated in Figure 7.15.[81] However, most cars nowadays have automatic chokes and the influence of volatility will then be less important, as shown by Figure 7.16.[81] The majority of journeys made are under 10 miles, so that vehicles in cold weather may never get fully warmed up.

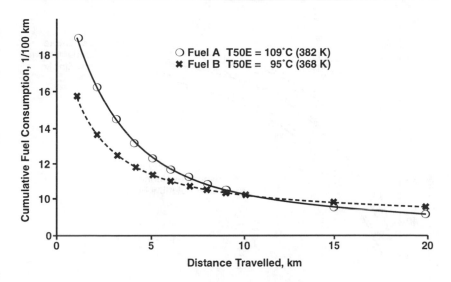

Fig. 7.15. Effect of volatility on fuel economy during warm-up with a manual choke vehicle.[81]

Cool, humid weather can, with high volatility gasolines containing no anti-icing additive, bring some vehicles into intake icing conditions, severely reducing their fuel economy. However, such problems only occur with a minority of cars and under restricted weather conditions. For most vehicles high volatility gasolines vaporize better and give a more complete mixing of air and fuel, and so are likely to combust more efficiently. Unfortunately, this benefit can be offset to some extent in vehicles without evaporative emission controls by reductions in economy due to evaporation losses.

Long-trip economy is provided by a high heat content in the gasoline; this means a high density and, therefore, a low volatility. As mentioned in Section 7.8, most of the energy in a gasoline is contained in the heavy ends and so setting back-end volatility specifications is a compromise between good long-trip fuel economy and increased crankcase lubricant dilution/combustion chamber deposits with all the attendant problems that these bring.

In an analysis of fuel economy data obtained using the Federal Test procedure in a study primarily designed to look at fuel effects on exhaust emissions, the following conclusions were drawn:[83]

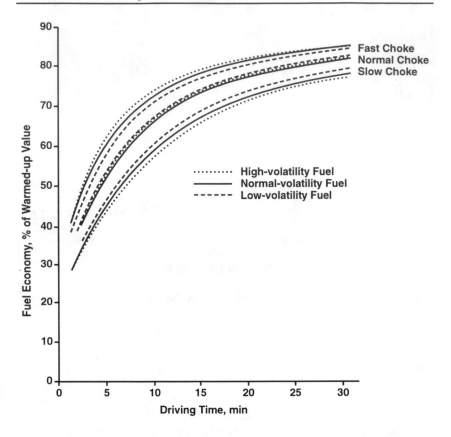

Fig. 7.16. Effect of volatility on fuel economy during warm-up with automatic choke vehicles.[81,82]

Volumetric fuel economy (VOLFE) is lowered when:

- Aromatics are reduced. A reduction from 45% to 20% lowered VOLFE by about 3%.

- Oxygenates are added. The addition of 2.7 wt% oxygen lowered VOLFE by 2.3% in current vehicles and by 1.6% in older vehicles.

- T90% is reduced from 360°F (182°C) to 280°F (138°C). This lowered VOLFE by about 1.5%.

- Olefins are reduced from 20% to 5%. This lowered VOLFE by 0.2 to 0.6%.

References

1. 1988 Annual Book of ASTM Standards. Volume 05.01, 1988.

2. H.C.V. Usherwood, "A New Automatic Standard Distillation Apparatus," *Petrol. Rev.*, March 1974, 1747.

3. G.I. Jenkins, "Control of the Front-End Volatility of Motor Gasolines. Calculation of Vapor/Liquid Ratios from the Reid Vapor Pressure and Distillation Test." *J. Inst Pet.*, 54, 8085.

4. 1992 Annual Book of ASTM Standards. Method D 4814 - 92C, Appendix X2, 1992.

5. Y. Nakajima, T. Saito, *et al.*, "The Influence of Fuel Characteristics on Vaporization in the S.I. Engine Cylinder During Cold Cranking," SAE Paper No. 780612, 1978.

6. European Standard for Unleaded Petrol, CEN, European Committee for Standardization, 1984.

7. F.H. Palmer and A.M. Smith, "The Performance and Specification of Gasoline," Critical Reports on Applied Chemistry, Vol. 10 — Gasoline Technology, p.107, Society of Chemical Industry, Blackwell Scientific Publications, 1983.

8. I.W. Kay, "Manifold Fuel Film Effects in an S.I. Engine," SAE Paper No. 780944, 1978.

9. H.B. Servati and W.W. Yuen, "Deposition of Fuel Droplets in Horizontal Intake Manifolds and the Behavior of Fuel Film Flow on its Walls," SAE Paper No. 840239, 1984.

10. Coordinating Research Council, Inc., "Driveability Evaluation in Cool Weather," New York, 1970.

11. J.D. Benson, H.A. Bigley and J.L. Keller, "Passenger Car Driveability in Cool Weather," SAE Paper No. 710138, 1971.

12. "CEC Cold Weather Driveability Code of Practice for Use on Road, Track and Vehicle Dynamometer for Vehicles with Spark Ignition Engines," CEC, London.

13. K. Falk, "The Development of a European Cold Weather Driveability Test Procedure for Motor Vehicles with Spark Ignition Engines," SAE Paper No. 831754, 1983.

14. Coordinating Research Council, Inc., "Driveability Performance of 1977 Passenger Cars at Intermediate Temperatures — Paso Robles," CRC Report No. 499, September 1978.

15. N.D. Brinkman, *et al.*, "Effect of Fuel Volatility on Driveability at Low and Intermediate Ambient Temperatures," SAE Paper No. 830593, 1983.

16. D.A. Barker, L.M. Gibbs and E.D. Steinke, "The Development and Proposed Implementation of the ASTM Driveability Index for Motor Gasoline," SAE Paper No. 881668, 1988.

17. A. Buczynsky, "Effects of Driveability Index and MTBE on Driver Satisfaction at Intermediate Temperatures," SAE Paper No. 932671, 1993.

18. L. Abramo, C.E. Baxter, *et al.*, "Effect of Volatility Changes on Vehicle Cold Start Driveability," SAE Paper No. 892088, 1989.

19. D.A. Barker and M.R. Dunn, "Driveability Number — a Gasoline Volatility Parameter Related to Cold Start Passenger Car Performance," SAE Paper No. 831756, 1983.

20. M. Tomita, *et al.*, "Effect of Gasoline Quality on Throttle Response of Engines During Warm-up," SAE Paper No. 900163, 1990.

21. K. Oda, *et al.*, "Effect of Gasoline Composition on Engine Performance," SAE Paper No. 930375, 1993.

22. J.A. Gething, "Distillation Adjustment: An Innovative Step to Gasoline Reformulation," SAE Paper No. 910382, 1991.

23. A.A. Quader, *et al.*, "Why Gasoline 90% Distillation Temperature Affects Emissions with Port Fuel Injection and Premixed Charge," SAE Paper No. 912430, 1991.

24. D. Le Breton, "Hot and Cold Fuel Volatility Indexes of French Cars: A Cooperative Study by the GFC Volatility Group," SAE Paper No. 841386, 1984.

25. J.P. Graham, *et al.*, "Effect of Volatility on Intermediate-Temperature Driveability with Hydrocarbon-only and Oxygenated Gasolines," SAE Paper No. 912432, 1991.

26. J.H. Baudino and L.C. Copeland, "Atypical Fuel Volatility Effects on Driveability, Emissions, and Fuel Economy of Stratified Charge and Conventionally Powered Vehicles," SAE Paper No. 780610, 1978.

27. J.K. Pearson, K.H. Reders and V.M. Tertois, "The Correlation of Consumer and Chassis Dynamometer Cold Weather Driveability," Paper EF4, CEC Symposium Wolfsburg, 1985.

28. S.W. Jorgensen and R.M. Reuter, "Hot-Start Driveability of Low T_{50} Fuels," SAE Paper No. 932672, 1993.

29. A.D. Brownlow, "Fuel Weathering," API Publication 4493, July 1989.

30. 1988 Annual Book of ASTM Standards, Vol 05.01, Method D 439, 1988.

31. Coordinating Research Council, "Driveability Performance of Late Model Passenger Cars at High Ambient Temperatures," CRC Report No. 490, November 1976.

32. "CEC Hot Weather Driveability Code of Practice for Use on Road, Track and Vehicle Dynamometer for Vehicles with Spark Ignition Engines," Ref. CEC M09T84, 1984.

33. J.K. Pearson, B.D. Caddock and P.L. Orman, "A Computer Model for the Prediction of Vapor Lock in the Fuel Pump of a Carburetted Engine," SAE Paper No. 821201, 1982.

34. V.M. Tertois and B.D. Caddock, "Carburetor Foaming and Its Influence on the Hot Weather Performance of Motor Vehicles," SAE Paper No. 821202, 1982.

35. L.J. Clark, "The Causes and Control of Carburetor Foaming," SAE Paper No. 841400, 1984.

36. E.R. Morrison, G.D. Ebersole and H.J. Elder, "Laboratory Expressions for Motor Fuel Volatility and Their Significance in Terms of Performance," SAE Paper No. 650859, 1965.

37. Coordinating Research Council, "Evaluation of Expressions for Fuel Volatility," Report No. 403, 1967.

38. R.F. Becker, U. Ciardiello, *et al.*, "Hot Weather Volatility Requirements of European Passenger Cars," SAE Paper No. 780651, 1978.

39. C.R. Morgan and C.N. Smith, "Fuel Volatility Effects on Driveability of Vehicles Equipped with Current and Advanced Fuel Management Systems," SAE Paper No. 780611, 1976.

40. R.K. Riley, "Vapor Lock in Late Model Cars," SAE Paper No. 831707, 1983.

41. P.J. Clarke, "Front-End Volatility Requirements of Late Model Cars at Intermediate Ambient Temperatures," SAE Paper No. 830595, 1983.

42. B.D. Caddock, *et al.*, "The Hot Fuel Handling Performance of European and Japanese Cars," SAE Paper No. 780653, 1978.

43. E. Yoshida, H. Nomura, *et al.*, "The Effects of Gasoline Volatility on the Driveability of Passenger Cars," 13th World Petroleum Congress (Buenos Aires), 1991.

44. P.J. Clarke, "The Effect of Gasoline Volatility on Emissions and Driveability," SAE Paper 710136, 1971.

45. J.M. Jones, J.K. Pearson and J.S. McArragher, "The Setting of European Gasoline Volatility Levels to Control Hot Weather Driveability," SAE Paper No. 852118, 1985.

46. F.H. Palmer, "The Development of a CEC Driveability Test Procedure for European Fuels and Vehicles," SAE Paper No. 811230, 1981.

47. F.H. Palmer, "Hot Weather Driveability — Does the CEC CF24 Test Method Reflect Motorists Requirements?" Paper EF3, CEC Symposium Wolfsburg, 1985.

48. CONCAWE, Position Paper on Gasoline Vehicle Evaporative Emissions, CONCAWE, The Hague, February 1987.

49. CONCAWE, "Trends in Motor Vehicle Emission and Fuel Regulations — 1994 Update," Report 4/94, CONCAWE, The Hague, April 1994.

50. W.J. Koehl, D.W. Lloyd and L.J. McCabe, "Vehicle Onboard Control of Refueling System Demonstration on a 1985 Vehicle," SAE Paper No. 861551, 1986.

51. J.N. Braddock, P.A. Gabele and T.J. Lemmons, "Factors Influencing the Composition and Quantity of Passenger Car Refueling Emissions — Part 1," SAE Paper No. 861558, 1986.

52. G.S. Musser and H.F. Shannon, "Onboard Control of Refueling Emissions," SAE Paper No. 861560, 1986.

53. U.S. Code of Federal Regulations (1977) Title 40: Protection of Environment; Chapter 1: Environmental Protection Agency; Subchapter C: Air Programs, Part 86: Control of Air Pollution from New Motor Vehicles and New Motor Vehicle Engines: Certification and Test Procedures; Subpart B: Emissions Regulations for 1977 and Later Model Year New Light-Duty Vehicles and Light-Duty Trucks, Test Procedures; Sections 86.10578 through 86.14478. Revised July 1, 1977.

54. "Measurement of Fuel Evaporative Emissions from Gasoline Powered Passenger Cars and Light Trucks Using the Enclosure Technique," SAE Recommended Practice J171, April 1991.

55. W.D. Dudek and D. Fisher, "Multiple Diurnal Evaporative Emissions Determinations with a Naturally Controlled Variable Volume Enclosure," SAE Paper No. 932674, 1993.

56. R.W. Hurn and B.H. Eccleston, "Evaporative Losses from Automobiles: Fuel and Fuel System Influences," Report C150/71, London: Institute of Mechanical Engineers, 1971.

57. B.H. Eccleston, B.F. Noble and R.W. Hurn, "Influence of Volatile Fuel Components on Vehicle Emissions," U.S. Bureau of Mines Report No. 7291, 1970.

58. B.H. Eccleston and R.W. Hurn, "Effect of Fuel Front-End and Mid-Range Volatility on Automobile Emissions," U.S. Bureau of Mines Report No. 7707, 1972.

59. American Petroleum Institute, "A Study of Factors Influencing the Evaporative Emissions from In-Use Automobiles," API Publication No. 4406, April 1985.

60. W.J. Koehl, *et al.*, "Effects of Gasoline Composition and Properties on Vehicle Emissions: A Review of Prior Studies—Auto/ Oil Air Quality Improvement Air Program," SAE Paper No. 912321, 1991.

61. "The Effects of Fuel RVP and Fuel Blends on Emissions at Non-FTP Temperatures," American Petroleum Institute, Preliminary Report, Washington, D.C., May 1991.

62. D. McClement, "Emissions and Vehicle Performance with Lower RVP Fuels," Final Report for the American Petroleum Institute, Automotive Testing Laboratories, Inc., January 1988.

63. C.M. Heinen, "We've Done the Job, What's It Worth?" SAE Paper No. 801357, 1980.

64. J.S. McArragher, *et al.*, "Evaporative Emissions from Modern European Vehicles and Their Control," SAE Paper No. 880315, 1988.

65. CONCAWE, "An Investigation Into Evaporative Hydrocarbon Emissions from European Vehicles," Report No. 87/60, 1987.

66. CONCAWE, "The Effects of Temperature and Fuel Volatility on Vehicle Evaporative Emissions," Report No. 90/51, 1990.

67. R.M. Reuter, J.D. Benson, *et al.*, "The Effects of Oxygenated Fuels and RVP on Automotive Emissions—Auto/Oil Quality Improvement Program," SAE Paper No. 920326, 1992.

68. J.G. Jeffrey and N.G. Elliott, "Gasoline Composition Effects in a Range of European Vehicle Technologies," SAE Paper No. 932680, 1993.

69. A.M. Hochhauser, *et al.*, "The Effect of Aromatics, MTBE, Olefins and T90 on Mass Exhaust Emissions from Current and Older Cars," SAE Paper No. 912322, 1991.

70. W.J. Koehl, J.D. Benson, *et al.*, "Effects of Heavy Hydrocarbons in Gasoline on Exhaust Mass Emissions, Air Toxics and Calculated Reactivity," SAE Paper No. 932723, 1993.

71. A.R. Guerrero and J.M. Lyons, "The Effect of Aromatics Content on NO_x Emissions," California Air Resources Board Report, March 1990.

72. J.A. Gething, "Distillation Adjustment: An Innovative Step to Gasoline Reformulation," SAE Paper No. 910382, 1991.

73. W.R. Leppard, J.D. Benson, *et al.*, "How Heavy Hydrocarbons in the Fuel Affect Exhaust Mass Emissions: Modal Analysis," SAE Paper No. 932724, 1993.

74. J.S. McArragher, *et al.*, "The Effect of Gasoline Volatility on Vehicle Exhaust Emissions at Low Ambient Temperatures," CONCAWE Report No. 93/51, 1993.

75. H.M. Doherty and W.H. Douthit, "Effect of Oxygenates and Fuel Volatility on Vehicle Emissions at Seasonal Temperatures," SAE Paper No. 902130, 1990.

76. R.A. Gorse, J.D. Benson, *et al.*, "Toxic Air Pollutant Vehicle Exhaust Emissions with Reformulated Gasolines," SAE Paper No. 912324, 1991.

77. K. Owen and R.G.M. Landells, "Precombustion Fuel Additives," <u>Critical Reports on Applied Chemistry, Vol 25 — Gasoline and Diesel Fuel Additives</u>, p. 20, J. Wiley and Sons, 1989.

78. R.C. Tupa and C.J. Dorer, "Gasoline and Diesel Fuel Additives for Performance/Distribution Quality II," SAE Paper No. 861179, 1986.

79. Coordinating European Council, Tentative Test Method No. CEC M10T87, "Intake System Icing Procedures for Use on Road, Track or Vehicle Dynamometer with Spark Ignition Vehicles," 1987.

80. P.J. Choate and J.C. Edwards, "Relationship Between Combustion Chamber Deposits, Fuel Composition, and Combustion Chamber Deposit Structure," SAE Paper No. 932812, 1993.

81. B.D. Caddock, "The Effect of the Physical Properties of Gasoline on Fuel Economy," Fuel Economy of the Gasoline Engine, Chapter 4, p. 73, edited by D.R. Blackmore and A. Thomas, Macmillan Press Ltd., 1978.

82. J.L. Keller and J. Byrne, American Petroleum Institute, 1966.

83. A.M. Hochauser, J.D. Benson, *et al.*, "Fuel Composition Effects on Automotive Fuel Economy," SAE Paper No. 930138, 1993.

Chapter 8

Influence of Gasoline Composition on Storage Stability and Engine Deposit Formation

8.1 The Influence of Gasoline Composition on Storage Stability

The type of hydrocarbon present in gasoline is very important in terms of the storage stability of that gasoline, as discussed in Chapter 4, where the mechanism of the free radical reaction involved in hydrocarbon oxidation is outlined. Olefinic compounds are much less stable to oxidation than aromatic or paraffinic compounds so that, when considering using gasoline components that are olefinic in nature, precautions must be taken to ensure that they do not give rise to undesirable oxidation products. These precautions normally consist of using an antioxidant (see Chapter 9) and/or using additional processing such as hydrogen treatment or heat soaking to improve the stability (see Chapter 3). Under these conditions, and provided that no metals are present that could catalyze hydrocarbon oxidation, long-term storage of gasoline can be successfully carried out.[1]

It is mainly the cracking processes such as catalytic cracking, thermal cracking, coking, steam cracking and visbreaking that give olefinic gasoline blend components. In particular, thermal cracking processes (visbreaking, coking, thermal cracking) can give rise to diolefinic compounds in addition to monoolefins, and these are extremely susceptible to oxidation and formation of gummy compounds. The presence of some sulfur and nitrogen compounds will also contribute to the autooxidation of gasoline,[2] as will the use of alcohols as blend components.[3]

The oxidation of hydrocarbons during storage proceeds via the formation of peroxide and hydroperoxide radicals to give eventually gums and acidic products.[4] The gums are only slightly soluble in gasoline and, if more and more of them are allowed to form, perhaps because of insufficient oxidation inhibitor being present, they will come out of solution and settle as a sludge on the bottom of the storage tank. If they are present to an excessive degree in the finished gasoline, they will deposit in the intake system of the engine, as discussed below. Certain metals, notably copper, catalyze the oxidation process and must be avoided in fuel systems or, if present in the fuel, must be deactivated by the use of appropriate additives.[5]

The hydroperoxides formed during the oxidation of hydrocarbons are proknocks and, if present, they will reduce the octane quality of the gasoline.[6] Antiknock performance can also be worsened if there is insufficient antioxidant present to prevent any tetraethyl lead that is present from degrading due to oxidation. If this occurs a white deposit of lead oxide and lead carbonate can form and block filters, etc., and the octane quality will reduce depending on the extent of loss of the tetraethyl lead. TML is much more stable to oxidation than TEL and is unlikely to give this problem.

8.2 Measurement of Storage Stability

There are a number of different laboratory tests that are used to establish the oxidation stability of a gasoline. Some of them are rather lengthy and so are unsuitable for routine control of gasoline stability, and none of them can be said to give a very precise idea of the resistance of a gasoline to oxidation during storage. The reason for this is that storage conditions are extremely variable. Some storage tanks have fixed roofs while others have floating roofs that reduce contact with air; ambient temperature conditions will vary according to geographical location; and the nature of the gasoline itself varies, with some blends containing high levels of olefins.

The most commonly used tests are as follows:

Oxidation Stability of Gasoline (Induction Period Method)
(ASTM D 525)[7]

In this test a sample of gasoline is oxidized in a bomb initially filled with oxygen at 100 psi (689 kPa). The bomb is held at a temperature of 100°C and the pressure is recorded until the break point is reached, that is, the point at which the pressure starts to drop at a rate of at least 2 psi (13.8 kPa) in a 15-minute period. The induction period (sometimes called breakdown time) is the elapsed time from placing the bomb in the bath at 100°C and the break point.

As a rule of thumb, it is generally accepted that an Induction Period of 240 minutes corresponds to a storage life of at least six months, although such a fuel can be satisfactory for up to two years depending on its chemical composition. This test has been shown to have a poor correlation with gum deposition under more realistic conditions.[8]

Existent Gum in Fuels by Jet Evaporation (ASTM D 381)

This test is used for aviation gasoline and aircraft turbine fuel as well as for motor gasoline. When testing gasoline, a measured volume is evaporated under controlled conditions of temperature in a stream of air. The residue is weighed, extracted with n-heptane and then re-weighed. The existent gum is the heptane insoluble part of the residue. The total residue, before extraction, is called the unwashed gum.

The existent gum test is intended only as a guide to the likelihood of a gasoline causing induction system deposits and valve sticking rather than as a measure of the fuel's storage stability. However, a high value will indicate that there has already been some oxidation of the gasoline and that more might be expected. The existent gum is really only a measure of the amount of gum already present in the fuel plus a small amount formed under the comparatively mild conditions of the test.

The unwashed gum gives an idea of the amount of non-volatile components present in the gasoline and, of course, includes the existent gum. These involatiles can include additives as well as contaminants that may be present as a result of poor handling practices in the refinery.

Oxidation Stability of Aviation Fuels (Potential Residue Method) (ASTM D 873)

Although this method is specified for aviation fuels, it can also be used for motor gasolines. A sample of the gasoline is oxidized in a bomb for a set period (often 16 hours) using the same conditions as in the Induction Period test. Any precipitate is filtered off and an Existent Gum test carried out on the filtered gasoline. The potential gum is the sum of the soluble and insoluble gum.

The potential gum represents the overall potential of the fuel to form gums and deposits and so is some indication of the storage stability of the fuel. However, the oxidation conditions used are much more severe than occur in practice and the correlation with storage stability is poor. It is probably true to say, however, that a low potential gum represents a stable fuel, although the reverse may not necessarily be true.

Long-Term Storage Stability Test[9]

In this test, fuels are stored in the dark for 12 weeks at 43°C and the existent gums then determined. Aging for one week by this method is generally accepted to be equivalent to approximately one month of storage under ambient conditions. Although this test procedure more closely represents actual storage conditions than the tests discussed earlier, it takes too long to carry out to be used for quality control.

8.3 Deposit Formation in Engines due to Gasoline Oxidation

Deposit formation in the engine due to lack of fuel stability can have a severe influence on vehicle driveability performance[10] and can cause ring and valve sticking. The nature and amount of the deposits laid down in different parts of the engine will depend very much on a variety of factors including temperature, driving mode, contact with air, and so on. In addition, foreign matter such as sand or rust can find its way into the fuel and may attach itself to the sticky gums formed by oxidation, thereby increasing the volume and adverse effects of the deposits.

8.3.1 Deposit Formation in the Fuel Tank and Fuel Lines

The antioxidants that are used to enhance storage stability (see Chapter 9) are most effective when the fuel is liquid rather than vaporized, since they are relatively non-volatile. There would normally be sufficient additive present to fully protect the fuel in its liquid state and so problems of oxidation in the tank or fuel lines are relatively rare. Too much of certain antioxidants can sometimes contribute to intake system deposits.[11]

The fuel in a vehicle's tank is normally no more than a few weeks old from the time it was blended, although it may sometimes have been stored for considerable periods. There have been cases during periods of economic recession when cars have been stockpiled for well over a year prior to being sold. These vehicles would have needed to have some gasoline in their tanks in order to move them from the production line to the storage area, and so in these cases there is a considerable potential for problems due to gasoline oxidation. When such long storage times are predicted, it would be normal for a vehicle manufacturer to use a special "factory fill" gasoline containing additional oxidation inhibitors and other additives.

The oxidation that occurs in the vehicle's tank is the same as in any storage tank, with gums being formed that will come out of solution, block filters and deposit in various parts of the fuel intake system. Even though the fuel is not normally held for long periods, in use it can be heated to temperatures as high as 50°C in the tank due to recirculation and heat transfer from the exhaust system. The tank will generally be under a slight positive pressure, and there may be metals present which can catalyze oxidation reactions. In addition, vehicle movement will agitate the tank contents and mix them with air, further increasing the likelihood of oxidation.

Gasolines which have undergone oxidation on storage and contain hydroperoxides are sometimes known as "sour" gasolines (although this term is also applied to petroleum streams containing mercaptans). These have been shown[12,13] to have an adverse effect on certain elastomeric materials used in parts of the fuel system such as the fuel hose. Sour gasolines have been shown to affect the tensile strength, hardness and volume swell of such materials as nitrile rubbers and epichlorohydrin copolymers. There has been concern that such attack could result in the hoses becoming permeable and contributing to increased hydrocarbon emissions.[14]

8.3.2 Deposit Formation in Carburetors and Fuel Injectors

The fuel is vaporized and mixed with air in the carburetor or at the fuel injector, so this is the time when oxidation really starts. Preflame oxidation reactions occur, as has been discussed in Chapter 6, and some already partially oxidized hydrocarbons may polymerize to form gums which will stick to surfaces so that deposits will build up. These deposits can occur in the carburetors or fuel injectors, intake manifolds and the intake valves and ports, depending on the temperature regime in each part of the system. Oxidation accelerated by heat soak-back from the engine when it is switched off can be a major cause of deposit formation. For this reason, driving which involves a great deal of stopping and starting is the most critical.[15,16] Vehicles with a large heat capacity in the engine compartment such as those making use of turbochargers, for example, can give rise to high underhood temperatures and may be more susceptible to this type of deposit formation. The use of electric cooling fans which continue to operate after the vehicle is switched off may often be beneficial in avoiding or minimizing these deposits, particularly if the air stream is directed at critical components.

Although the buildup of deposits probably starts with gums and non-volatiles sticking to the metal surfaces, it is increased by particulates and other materials from a number of possible sources. These include exhaust gases arising from other traffic, airborne materials which penetrate the air filter, combustion gases pulsed back into the fuel inlet system due to valve overlap, exhaust gas recirculation,[17] and blowby gases. Additives are available which can largely overcome these deposit problems (see Chapter 9).

In a carburetor, deposits can form on the walls of the throttle body, on the throttle plate, in the idle air circuit and in metering orifices and jets, as shown in Figure 8.1. These deposits cause the air-fuel ratio to move away from the optimum setting as selected by the vehicle manufacturer and so may cause driveability malfunctions, a decrease in fuel economy and, for non-catalyst vehicles, an increase in exhaust emissions.[18]

Fuel injectors can be particularly susceptible to deposit formation if they inject into the ports rather than into the inlet manifold, and if they are electronically controlled so that they inject only at the appropriate time rather than continuously.[16,19,20,21,22,23] Port Fuel Injection (PFI) is now widely used because of the benefits in terms of fuel economy, exhaust emissions and overall vehicle performance. The deposits form in the pintle area, as shown in Figure 8.2, and can reduce fuel flow rate and alter spray pattern. The fuel factors which increase injector deposit formation are diolefins, olefins and sulfur content.[24] Unwashed gum from heavy aromatics has no influence and there is no difference in deposit forming tendency between leaded and unleaded gasolines.

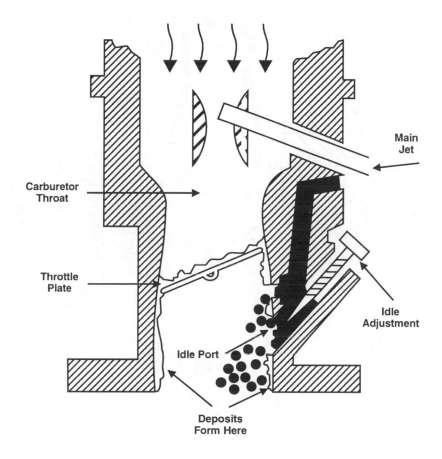

Fig. 8.1. Deposit formation in a carburetor.[67]

The deposits can build up very quickly, sometimes in under 1000 miles if a particularly severe stop/start driving mode is used, and can hinder cold starting, cause severe driving malfunctions and power loss, increase exhaust emissions in non-catalyst cars, and worsen fuel economy. Many catalyst vehicles have oxygen sensors and these will enable injector operation to be adjusted so as to maintain the correct air-fuel ratio as deposits build up. This delays the onset of problems but such devices are ultimately unable to compensate for distorted spray patterns.

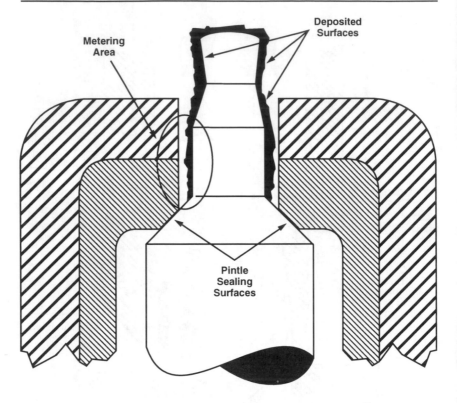

Fig. 8.2. Deposits form in the pintle area of port fuel injectors.[15]

8.3.3 Deposit Formation in the Inlet Manifold, Ports and on Valves

A "hot spot" is often incorporated in the inlet manifold to help fuel vaporization, and it is here that deposits are likely to concentrate and where they will have the greatest effect. They will increase warm-up time and so worsen fuel economy, and they may give driveability problems. Deposits in the ports and on the underside of the valves are particularly important. Port deposits have been claimed to contribute to ORI,[25,26] and the combination of port and valve deposits will influence the breathing of the engine and so will affect power, emissions and driveability.[27,28] The mechanism by which intake valve deposits adversely affect driveability and exhaust emissions is claimed to be due to two factors:[29]

1. C9 and C10 hydrocarbons are held up as liquids by the deposits, the amount increasing with increasing valve deposits, so that during accelerations the air/fuel ratio becomes momentarily lean, possibly to the extent of causing a misfire.

2. The heavy fuel absorbed on the valves is released after the acceleration is completed and again upsets the air/fuel ratio, influencing both driveability and emissions.

Removing deposits from valves has been shown to reduce both CO and NO_x emissions in one car model.[30]

Valve deposits can build up to such an extent that they interfere with the closing of the valve so that valve burning can result. The nature of the deposit varies with the fuel and the valve temperature so that sometimes they are soft and sticky and in other cases they are hard and brittle.[27] A number of investigations into the nature of valve deposits have been carried out[31,32,33,34] and there is a consistency of views that heavy aromatic compounds from reformate and heavy polar compounds formed by oxidation are largely responsible for deposit formation. The presence of alcohol in the fuel appears to increase deposit rate on intake valves and such blends may need additional additive treatment to overcome the higher deposit levels.[27] Where the deposits are brittle, pieces can break off and hold the valve open so that loss of power or even valve burning can occur. Pieces of hard valve deposit can also pass into the combustion chamber and become wedged in the ring belt or get lodged between the piston crown and valves or cylinder head. If sticky deposits extend up the valve stem, they can harden in cold weather while the vehicle is standing idle and cause failure to start from cold, and engine damage.[35]

As with all aspects relating to the interaction of fuel and engine factors, it is the latter which have the greatest effect so that some vehicles are entirely free from deposit problems while others can suffer very quickly. The lubricant can play a part, particularly with respect to valve deposits, because the valve stems are lubricated by oil flowing down them onto the valve undersides. The contribution of various additives in the lubricant has been assessed[36,37] and it has been shown that increasing amounts of VI improver increases valve deposits. Lead in gasoline was shown not to affect valve deposits.

The most potent fuel factor which influences deposit formation in the intake system is the presence of appropriate additives designed to keep the system clean and/or to clean up existing deposits. Different types of additive may be needed depending on the position in the fuel system where they form, and these are discussed in Chapter 9. There are additives such as some antioxidants which can worsen valve deposits if used at too high a concentration.[38]

8.3.4 Tests for Predicting Intake System Deposit Formation

With regard to other fuel factors that cause intake system deposits, there are no entirely satisfactory laboratory tests and, therefore, no fuel parameters that, when controlled to specified limits, will guarantee a satisfactory fuel from a deposit-forming viewpoint. It is primarily the olefin content and the types of olefin in the base fuel which are thought to have the greatest influence.[19,38,39] The most reactive of the olefins and, in particular, the heavier conjugated diolefins are generally regarded as the most likely to cause deposit problems. However, it is not always easy to identify precisely which olefins are present in a gasoline and so total olefin content, coupled with satisfactory levels for Existent Gum and Induction Period (see Chapter 4), is sometimes taken as a guide to likely stability in an engine. The Bromine number test (ASTM D 1159) gives a measure of the olefinic content but is not very reliable in predicting deposit-forming tendency because some olefins, and particularly the important diolefins, are not fully determined in the test.[40] Diene number is another test that has been used[41] in which the gasoline is refluxed with a known amount of maleic anhydride and the amount of unreacted maleic anhydride is determined. The maleic anhydride combines with conjugated dienes to form cyclic compounds by the Diels-Alder reaction and so the test gives a measure of the amount of this type of diolefin present. However, it does not determine the presence of other reactive di- and monoolefins and so, again, is not entirely satisfactory.

The Existent Gum test (D 381) is a measure only of the amount of gum actually present in a gasoline, not of how much can form under the conditions in an engine, and so is only of limited value in predicting the deposit-forming tendency of a gasoline. The Potential Residue test (also called the Potential Gum test - ASTM D 873) measures the total gum that can be formed in a fuel when it is oxidized under prescribed conditions in a bomb filled with oxygen, but this can be much more severe than occurs in an engine and, again, does not correlate well with field experience.[42] Other gum tests which accelerate the oxidation of olefins such as the Copper Beaker test can be valuable but do not separate out soluble (such as additives) from insoluble materials and cannot, therefore, be a reliable guide. Other tests have been developed such as the flash evaporation procedure[43] but none seems to have received wide usage.

The gum tests do give the possibility of actually measuring gummy materials formed from the fuel and so take into account other fuel factors that can be important. The presence of some nitrogen and sulfur compounds[44,45] can adversely influence engine deposit formation, but the extent will depend on a number of factors including temperature and so will vary in different parts of the fuel system and from model to model. Overall there does not seem to be an entirely satisfactory simple laboratory test for predicting deposit-forming tendency.

There are, however, engine tests which will give an indication of the relative deposit-forming tendency of different gasolines and the effectiveness of additives in overcoming them. The CRC (Coordinating Research Council) has developed carburetor deposit tests,[46,47,48] and in Europe, engine tests that have gained wide acceptance, mainly for the evaluation of additives, are the Renault 5,[49] the Opel Kadett test[50] and the Daimler Benz M102E test.[51] In the U.S., vehicle tests are frequently used to measure port fuel injector and inlet valve deposits and their effects on driveability, such as those developed by BMW[27] and CRC.[52]

There are many other engine tests, but most of them are "in-house" procedures developed by different oil or additive companies, using engines known to suffer from intake system deposit problems — or which have been modified to exaggerate the problems. These have been mainly developed to assess the effectiveness of additives.

Some blend components are clearly more likely to form deposits than others and it can be worthwhile to impose limits on the concentration of certain components in a gasoline based on engine test data. Some steam cracked naphthas and heavy cracked naphthas, for example, come into the category of "dirty" components,[53,54] but clearly the actual quality of individual streams will depend on the processing conditions and after-treatment that are used.

8.3.5 Combustion Chamber Deposits

These have been discussed in Chapter 6, Section 6.7, with regard to the influence of deposits on ORI, and also in Chapter 7, section 7.7, with respect to the effect of back-end volatility on deposits. In this section the overall importance of these deposits and the gasoline compositional aspects which influence them will be considered.

Probably the most important effect of combustion chamber deposits is that they increase ORI, as discussed earlier. However, they also influence exhaust emissions and carbon knock (mentioned in Chapter 5, Section 5.5), which will be discussed in more detail below. In addition they can cause surface ignition and, if they are on the spark plug, misfire. Each of these aspects is of significance—ORI because increases in octane requirement can give rise to catastrophic failure of the engine as also can surface ignition; emissions because of the stringent controls applied in the U.S. and elsewhere; and carbon knock because, although it does not appear to cause any serious damage, it does give real concern to the driver because of the noise it makes.

The amount and nature of combustion chamber deposits depend on the fuel, the lubricant, engine design and driving mode.[55]

Regarding the fuel, with leaded gasoline, regardless of the lead concentration, the deposits contain up to 70% of lead by weight, being present mainly as the halide, oxyhalide and sulfate, with the remainder being mostly carbonaceous. Of course, if there are metals present in the gasoline other than lead, such as manganese from the antiknock MMT, these will also appear in the deposits. If phosphorus is also in the fuel, it will modify the nature of the metallic deposits.

For unleaded gasolines the deposit forming tendency increases with the boiling point of the heaviest part of the fuel[56,57,58,59] but with aromatics giving the highest amount of deposits, olefins an intermediate level, and paraffins the least. The presence of certain types of multifunctional additive can increase combustion chamber deposits[60] as can the presence of carrier additives.[61]

Engine conditions and driving mode also play a large part in deposit formation. It has been found[62] that coolant temperature has the greatest effect and that air/fuel ratio is also important. Compression ratio and intake air temperature have only a very small influence on deposits.

The lube oil also contributes to combustion chamber deposits[61] and the amount depends on oil consumption,[63] particularly on the piston top,[64] and on the volatility[61,65] and sulfated ash content.[66]

The effect on emissions is to increase HC from the exhaust by an absorption/desorption mechanism of partially burned fuel on the deposits. NO_x is also increased because of the thermal insulation effect of the deposits.

8.3.6 Carbon Knock[67] (see also Chapter 5, Section 5.5)

This problem occurs when deposits build up on the top of the piston to such an extent that in engines with small squish clearances there is mechanical contact resulting in a knocking or rapping noise while the engine is warming up. The phenomenon is also called "Deposit Interference Noise" (DIN), "Carbon Rap" and "Combustion Chamber Deposit Interference" (CCDI). It started in the summer of 1991 in the U.S. and coincided with a general increase in unwashed gum level due to a higher use of Deposit Control Additives to overcome Intake Valve Deposits—the same engines prior to 1991 had no problems and there had been no problems in Japan with similar engines but where there was an unwashed gum specification maximum of 20 mg/100 mL.

Test work has shown that the factors which are important in combustion chamber deposit formation are also important for carbon knock, i.e., engine design, heavy aromatics in the fuel, certain types of additive, lubricant type and consumption, etc.

Although no damage is caused by carbon rap, it is of concern because the noise is upsetting to drivers and also because tight squish clearances of 1 mm or less are becoming more common since they help to reduce HC emissions. When the engine is cold the piston can tilt slightly towards the lag side, and this effectively reduces the squish clearance.

References

1. H.N. Giles, J.N. Bowden and L.L. Stavinoha, "Overview on Assessment of Crude Oil and Refined Product Quality During Long Term Storage," United States Department of Energy and U.S. Army Fuels and Lubricants Research Laboratory, Report DOE/FE0048, June 1985.

2. V.S. Azev, *et al.*, "Results of Tests of Automobile Gasolines by a Complex of Methods of Qualification Evaluation Following Their Long Term Storage in Salt," *Transport i Khranenie Nefteproduktov i Uglevodorod*, N. 2, pp. 1-4, 1977.

3. N. Por, "Stability Properties of Gasoline Alcohol Blends," 3rd International Conference on Stability and Handling of Liquid Fuels, London, September 1988.

4. A.C. Nixon, "Autoxidation and Antioxidants of Petroleum," Autoxidation and Antioxidants, W.O. Lundberg (Editor), Chap. 17, Interscience, N.Y., 1962.

5. P. Polsse, "What Additives Do for Gasoline," *Hydrocarbon Processing*, Vol. 52, N. 2, pp. 61-68, 1973.

6. V.E. Emel'yanov, V.P. Grebenshchikov, *et al.*, "The Effect of Hydroperoxides on the Knocking Stability of Gasolines," *Khimiya i Tekhnologiya Topliv i Masel*, No. 10, pp. 16-17, 1991.

7. 1989 Annual Book of ASTM Standards, Vol. 05.01, ASTM 1989.

8. D.L. Morris, J.N. Bowden and L.L. Stavinoha, "Evaluation of Motor Gasoline Stability," 3rd International Conference on Stability and Handling of Liquid Fuels, London, September 1988.

9. J.N. Bowden and D.W. Brinkman, "Stability Survey of Hydrocarbon Fuels," U.S. Department of Energy, Report BETC/17784, 1979.

10. G.H. Amberg and W.S. Craig, "Gasoline Detergents Control Intake System Deposits," SAE Paper No. 554D, August 1962.

11. A.C. Nixon, H.B. Minor and T.P. Rudy, "Induction System Reactions — Liquid or Vapor?" ASTM National Meeting, Sept. 1956.

12. A. Nersasian, "Effect of Sour Gasolines on Fuel Hose Rubber Materials," SAE Paper 790659, 1979.

13. G.A. Orloff, J.A.P. da Silva and A.A. Gentil, "Effects of Oxidated and Oxygenated Fuels on Rubber Hoses," SAE Paper No. 921501, 1992.

14. J.D. MacLachlan, "Automotive Fuel Permeation Resistance," SAE Paper No. 790657, 1979.

15. R.C. Tupa and C.J. Dora, "Gasoline and Diesel Fuel Additives for Performance/Distribution Quality II," SAE Paper 861179, 1986.

16. R.C. Tupa and D.E. Koehler, "Gasoline Port Fuel Injectors — Keep Clean/Clean Up with Additive," SAE Paper 861536, 1986.

17. A.F. Gerber and R.G. Smith, "Some Effects of Exhaust Gas Recirculation (EGR) Upon Automotive Engine Intake System Deposits and Crankcase Lubricant Performance," SAE Paper No. 710142, 1971.

18. E.L. Tandrup, "Evaluating Carburetor Detergent Performance," SAE Paper No. 660782, 1966.

19. J.D. Benson and P.A. Yaccarino, "The Effects of Fuel Composition and Additives on Multi-Port Fuel Injector Deposits," SAE Paper No. 861533, 1986.

20. B.Y. Taniguchi, *et al.*, "Injector Deposits — The Tip of Intake System Deposit Problems," SAE Paper No. 861534, 1986.

21. G.P. Abramo, A.M. Horowitz and J.C. Trewella, "Port Fuel Injector Cleanliness Studies," SAE Paper No. 861535, 1986.

22. D.L. Lenane and T.P. Stocky, "Gasoline Additives Solve Injector Deposit Problems," SAE Paper 861637, 1986.

23. R.C. Tupa, "Port Fuel Injector Deposits — Causes/Consequences/Cures," SAE Paper No. 872113, 1987.

24. A. Shiratori and K. Saitoh, "Fuel Property Requirements for Multiport Injector Deposit Cleanliness," SAE Paper No. 912380, 1991.

25. H.E. Alquist, G.E. Holman and D.B. Wimmer, "Some Observations of Factors Affecting ORI," SAE Paper No. 750932, 1975

26. L.B. Graiff, "Some New Aspects of Deposit Effects on Engine Octane Requirement Increase and Fuel Economy," SAE Paper No. 790938, 1979.

27. B. Bitting, *et al.*, "Intake Valve Deposits — Fuel Detergency Requirements Revisited," SAE Paper No. 872117, 1987.

28. J.P. Graham and B. Evans, "Effects of Intake Valve Deposits on Driveability," SAE Paper No. 922220, 1992.

29. G. Shibata, H. Nagaishi and K. Oda, "Effect of Intake Valve Deposits and Gasoline Composition on S.I. Engine Performance," SAE Paper No. 922263, 1992.

30. K.R. Houser and T.A. Crosby, "The Impact of Intake Valve Deposits on Exhaust Emissions," SAE Paper No. 922259, 1992.

31. B.G. Bunting, "An Analysis of Intake Valve Deposits from Gasolines Containing Polycyclic Aromatics," SAE Paper No. 912378, 1991.

32. P. Martin, F. McCarty and D. Bustamante, "Mechanism of Deposit Formation: Deposit Tendency of Cracked Components by Boiling Range," SAE Paper No. 922217, 1992.

33. P. Martin and D. Bustamante, "Deposit Forming Tendency of Gasoline Polar Compounds," SAE Paper No. 932742, 1993.

34. K. Ohsawa, Y. Nomura, *et al.*, "Mechanism of Intake Valve Deposit Formation Part III: Effects of Gasoline Quality," SAE Paper No. 922265, 1992.

35. S. Mikkonen, R. Karisson and J. Kivi, "Intake Valve Sticking in Some Carburetor Engines," SAE Paper No. 881643, 1988.

36. Y. Yonekawe, N. Okamoto and M. Kuroiwa, "The Influence of Lubricant Composition and Lead Content in Gasoline on Intake Valve Deposit," *J. Jap. Pet. Inst. 22*, No.2, pp. 98-104, March 1979.

37. J.B. Bidwell and R.K. Williams, "The New Look in Lubricating Oils," SAE Trans., 63, pp. 349-361, 1955.

38. T. Nishizaki, *et al.*, "The effects of fuel composition and fuel additives on intake system detergency of Japanese automotive engines," SAE Paper No. 790203, 1979.

39. K. Starke and E. Schwartz, "Effect of Fuel Components and Special Additives on Deposits in Intake Manifolds of Internal Combustion Engines," *Mineroeltechnik*, 28 (30), p. 12, 1983.

40. 1988 Annual Book of ASTM Standards, Method D1159, Annex A.2, 1988.

41. "Diene Value by Maleic Anhydride Addition Reaction," UOP Method 32682, Universal Oil Products Inc., 1965.

42. D.L. Morris, J.N. Bowden and L.L. Stavinoha, "Evaluation of Gasoline Stability," 3rd International Conference on Stability and Handling of Liquid Fuels, London, 1988.

43. S.R. Hills and M.J. Van der Zijden, "The Control of Gasoline Quality in Relation to Inlet System Deposits," *J. Inst. Pet.*, 50, No. 485, pp. 105-122, 1964.

44. E. Dimitroff and A.A. Johnston, "Mechanism of Induction System Deposit Formation," SAE Paper No. 660784, 1966.

45. L.M. Gibbs and C.E. Richardson, "Carburetor Deposits and Their Control," SAE Paper No. 790202, 1979.

46. "Carburetor Cleanliness Test Procedure — State-of-Art Summary Report: 1973-1981," Coord. Res. Counc. Inc. Report No. 529, 1983.

47. J.J. Malaker, J.B. Retzloff and L.M.M. Gibbs, "Throttle Body Deposits ... The CRC Carburetor Cleanliness Test Procedure," SAE Paper No. 831708, 1983.

48. "The Intake Manifold Deposit Engine Dynamometer Test Procedure. State-of-Art Summary Report 1973-78," Coord. Res. Counc. Report No. 505, 1979.

49. Coordinating European Council, "Evaluation of Gasolines with Respect to Maintenance of Carburetor Cleanliness," Tentative Test Method No. CEC F03T81, 1981.

50. Coordinating European Council, "The Evaluation of Gasoline Engine Intake System Deposition," Test Method No. CEC F04A87, 1987.

51. M. Gairing, "Zur Qualitat der Ottokraftstoffe aus der Sicht der Automobilindustrie: Vermeidung von Ablagerungen auf Einlassventilen," *Mineralolrundshau*, 34, No. 11, pp. 209-215, November 1986.

52. S.A. Bannon, G.O. Scherer and D.V. Swaynos, "Selection of an Engine Standard for Development of a CRC Intake Valve Deposit Test," SAE Paper No. 922260, 1992.

53. A.A. Gureev and S.M. Livshitz, "Decreasing the Deposit Forming Tendency of Automobile Gasolines," *Nevesti Neft. i Gas. Tekbn.*, Nefleporerab, i Neftekhem, No.5, 2426, 1961.

54. K. Owen and R.G.M. Landells, "Precombustion Gasoline Additives," Gasoline and Diesel Fuel Additives, Edited by K. Owen, John Wiley and Sons, London, 1989.

55. G.T. Kalghatgi, "Deposits in Gasoline Engines—A Literature Review," SAE Paper No. 902105, 1990.

56. L.B. Shore and K.F. Ockert, "Combustion Chamber Deposits—A Radio Tracer Study," <u>SAE Transactions</u>, Vol. 66, pp. 285-294, 1958.

57. C. Kim, S.-W.S. Cheng and S.A. Majorski, "Engine Combustion Chamber Deposits: Fuel Effects and Mechanisms of Formation," SAE Paper No. 912379, 1991.

58. M.K. Megnin and J.B. Furman, "Gasoline Effects on Octane Requirement Increase and Combustion Chamber Deposits," SAE Paper No. 922258, 1992.

59. P.J. Choate and J.C. Edwards, "Relationship Between Combustion Chamber Deposits, Fuel Composition and Combustion Chamber Deposit Structure," SAE Paper No. 932812, 1993.

60. P. Schreyer, K.W. Starke, *et al.*, "Effect of Multifunctional Additives on Octane Number Requirement of Internal Combustion Engines," SAE Paper No. 932813, 1993.

61. J.D. Benson, "Some Factors Which Affect Octane Requirement Increase," SAE Paper No. 750933, 1975.

62. S.S. Cheng and C. Kim, "Effect of Engine Operating Parameters on Engine Combustion Chamber Deposits," SAE Paper No. 902108, 1990.

63. H.E. Alquist, G.E. Holman and D.B. Wimmer, "Some Observations on Factors Affecting ORI," SAE Paper No. 750932, 1975.

64. B.D. Keller, G.H. Meguerin, *et al.*, "ORI of Today's Vehicles," SAE Paper No. 760195, 1976.

65. J.B. McNab, L.E. Moody and N.V. Hakala, "Effect of Lubricant Composition on Combustion Chamber Deposits," <u>SAE Transactions</u>, Vol. 62, pp. 228-242, 1954.

66. A. Maricante and P. Chiampo, "Influence of Lubricating Oil Ash on the ORI of Engines Running on Unleaded Fuel," SAE Paper No. 720945, 1972.

67. <u>Proceedings of the CRC Workshop on Combustion Chamber Deposits</u>, Nov. 15-17, 1993, at Orlando, Florida. Published by Coordinating Research Council, 1993.

Chapter 9

Gasoline Additives

Although lead is no longer an important additive in gasoline in many developed countries, it is still widely used elsewhere and sometimes at surprisingly high concentrations. Levels of 0.84 g/L (3.18 g/USG) or higher are relatively common in many developing countries.[1] Lead alkyls and other antiknock compounds have already been discussed in Chapter 6, and so we will cover here only additives utilized for purposes other than enhancing antiknock properties.

These other additives are becoming more important, partly for the same reasons as the reduction in the use of lead. Increasingly stringent restrictions on exhaust gas emissions make it essential that vehicles maintain their tune for long periods, so any deposit buildup must be minimized. In fact, many modern vehicles seem to be more critical to deposits in the engine than older vehicles, and require very little before driveability and other problems commence.

Fuels themselves may be less stable because the contribution of cracking processes in gasoline manufacture has expanded, bringing with it higher olefin contents. This started in many countries with the increase in crude price in the 1970s which changed the demand pattern and made it important to increase the yield of gasoline from crude. Although prices can still be rather volatile, it seems likely that the relatively high olefin contents associated with cracked components are here to stay although the lighter olefins may be restricted in some gasolines to avoid the presence of highly reactive hydrocarbons in evaporative emissions. The price increases of crude oil also emphasized the importance of fuel economy and, again, this brought with it many changes including the use of smaller engines operating at higher severities.

Another change in fuel composition that has occurred and has increased the need for additives is the use of oxygenated components to achieve required gasoline volumes, to meet octane specifications, and to reduce exhaust emissions of CO and HC, as discussed in Chapters 11 and 12.

Finally, there has been an increasing awareness of the value of additives in product differentiation. It can be difficult and expensive to modify octane, volatility or other parameters controlled during manufacture in order to show an advertisable quality advantage over competitive products. This is particularly true when product exchanges, to minimize transportation costs, mean that many different companies can be using the same base gasoline. However, additives can always be introduced into specific brands or grades to provide a very real advertisable benefit. There are many examples of valuable increases in market share of gasolines because of the use of additives supported by advertising campaigns.

A variety of units are used for additive treat rates. In the following text only ppm, i.e., parts per million by weight, has been used. This can also be written mg/kg. Another frequently used unit for additive treat rate is pounds per thousand barrels (ptb); 1 ptb is approximately equivalent to 4 ppm, depending on the density of the gasoline.

9.1 Additives to Improve Oxidation Stability

9.1.1 Antioxidants (also called oxidation inhibitors) function by terminating the free radical chain reactions involved in hydrocarbon oxidation (see Chapter 4, Section 4.3.5). The products of oxidation are gums that can cause a number of problems during storage and in use in an engine, as discussed in Chapter 8. Antioxidants are also valuable in protecting tetraethyl lead from decomposition,[2] and although this is unusual, it can occur under certain circumstances. Some gasolines containing no cracked stocks may not have antioxidant added, but if they also contain lead, then it is always desirable to use a low level of antioxidant, i.e., up to about 10 ppm, to avoid lead decomposition products forming and blocking filters, etc.

The type and amount of antioxidant to use will depend on such factors as the gasoline composition and the storage conditions. It is difficult to predict the optimum additive type and concentration, and to assess this, plant trials are best carried out using the components that will actually be in the gasoline blend. For cracked stocks it is important that the antioxidant is injected into the stream at the earliest possible time, and preferably in the rundown line from the process unit. Tests such as Induction Period and Existent Gum are used to establish that a gasoline has a satisfactory storage stability. These tests are described in Chapter 8, Section 8.2.

There are two main types of antioxidant in use — the aromatic di-amines and the alkylphenols. Aminophenols are also used to a smaller extent but have the disadvantage of being soluble in water and caustic soda, and so can be lost to water bottoms in tankage. They are also said to increase intake system deposits.[3]

Aromatic diamines such as the paraphenylenediamines are extremely active oxidation inhibitors and are usually used in the range 5 to 20 ppm. They are particularly useful in gasolines having a high olefin content. The compounds most frequently used have the general for-mula:

$$R\text{-}NH\text{-}C_6H_4\text{-}NH\text{-}R'$$

where R and R' can be the same or different and are often sec-butyl, isopropyl, 1,4 dimethylpentyl, or 1 methylpentyl. The compound NN'-di-sec-butyl-paraphenylenediamine is widely used.

These aromatic diamines are somewhat soluble in acidic tank water bottoms and so, if these conditions exist, there could be loss of addi-tive on storage and, hence, diminished protection. Tank water bot-toms are usually close to neutral or somewhat alkaline, but if upsets occur with such units as alkylation plants, then acidic water bottoms can result from unneutralized process water separating out of the product.

Alkylphenols are used mainly when the gasoline has a low olefin con-tent, although mixtures with phenylenediamines are also used for blends containing higher concentrations of olefins, as discussed be-low. The most commonly used compounds have sterically hindered hydroxyl groups as in 4-methyl-2,6-ditertiarybutylphenol. Quite often mixtures of various alkylphenols are used and these can have the ad-vantage of low freezing points and of being more cost-effective. They are used in the range of 5 to 100 ppm.

Mixtures of alkylphenols and phenylenediamines are frequently used and are claimed to be more effective than equivalent concentrations of either of the constituents alone. An approximate guide to the ratio of phenylenediamine type to alkylphenol type is given in Figure 9.1.[4]

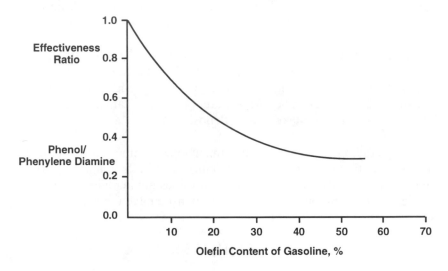

Fig. 9.1. *Effectiveness ratio of phenol/phenylenediamine antioxidants versus olefin content of gasoline.*[4]

9.1.2 Metal deactivators are used to prevent metals such as copper in a gasoline from catalyzing oxidation reactions. The dissolved metals usually arise from the action of acidic compounds such as mercaptans in the gasoline attacking metals in the distribution and vehicle fuel systems.

The most commonly used metal deactivator is probably N,N′-disalicylidene-1,2-propanediamine; it deactivates metals such as copper by forming a chelate having the structure shown in Figure 9.2.[4]

Metal deactivators are used in the range of 4 to 12 ppm and must be added downstream of any caustic wash unit because they are soluble in aqueous sodium hydroxide.

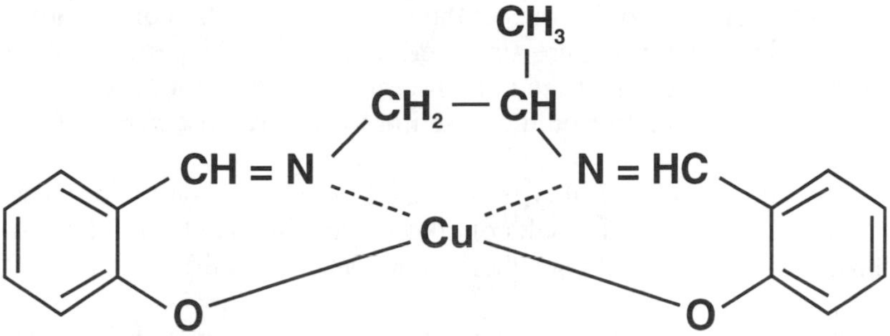

$$CH_3$$
$$CH_2-CH$$
$$CH=N \qquad N=HC$$
$$Cu$$
$$O \qquad O$$

Fig. 9.2. Copper chelate.[4]

9.2 Additives Used in Gasoline Distribution

9.2.1 Dyes and markers are often needed for legal reasons to distinguish one product or brand from another or to provide evidence in cases of theft, tax evasion, fuel adulteration, etc.

Azo dyes are normally used where red or orange colors are needed and anthroquinone dyes when blues are required. Concentrations are always very low (2 to 10 ppm) to avoid deposit formation on evaporation, and staining if they are spilled on the sides of light-colored vehicles. They are usually supplied as concentrated solutions.

Marker chemicals such as furfural or diphenylamine can be used for security reasons and do not color the gasoline. They are detected by chemical means.

9.2.2 Corrosion inhibitors are needed to protect pipelines and tanks from corrosion and, hence, to avoid rust being formed which will suspend in the gasoline and cause filters, etc., to be blocked.

These additives are surfactant materials which have a polar group at one end and an oleophilic/hydrophobic group at the other. The polar group attaches itself to a metal or other surface and the other group repels water and provides an oily layer to prevent rust from forming. They are used at low treat rates, often only 5 ppm, although treat levels as high as 20 ppm or more can be used in areas where severe corrosion is expected. In many cases the amount injected is controlled so that only 1 ppm or so emerges at the end of the pipeline or distribution system. These additives are also used to protect vehicle fuel systems (see Section 9.3.1) and more than the minimum amount for distribution system protection would be used in these circumstances.

A wide range of chemical types are used as anticorrosion additives including esters or amine salts of alkenyl succinic acids, alkyl orthophosphoric acids, alkyl phosphoric acids and aryl sulfonic acids.

Methods of test for gasolines containing anticorrosion additives are discussed in Section 9.3.1.

9.2.3 Biocides prevent microbial growth in the bottom of tanks (see Chapter 4, Section 4.2) which can cause suspended matter to accumulate in the water bottoms and in the gasoline. The microbial activity takes place mainly at the water/gasoline interface. It is less frequently encountered with gasoline than with middle distillate fuels because both lead and aromatics inhibit microbial growth.

The types of chemical that have been used to overcome this problem are boron compounds, quaternary ammonium salts of salicylic acid, glycol ethers, etc. Typical treatment levels range from 135 to 1000 ppm in the fuel.[4]

9.2.4 Antistatic additives are occasionally added to gasoline when it is being pumped at a high rate into a vessel or tank where there is a danger of an explosive mixture being formed. Gasolines, unlike middle distillates, generally give too rich a vapor/air mixture above the liquid in a tank for it to burn or explode.

The additives used increase the conductivity of the gasoline and prevent an electrostatic charge from building up and causing a spark. A totally organic type of additive is preferred for use in gasoline over the chromium-containing material often used in jet fuel. Some of the other additives or components used in gasoline (such as water soluble oxygenates) will themselves increase electrical conductivity and give some antistatic protection to the gasoline.

9.2.5 Drag Reducing Agents can increase pipeline capacity and postpone investment in new pipelines. They are high-molecular-weight polymers that shear readily and thus reduce drag.[5,6] Concentrations of up to 50 ppm have been used.

9.2.6 Demulsifiers and dehazers are occasionally added to a gasoline if a water haze or emulsion has formed which will not readily clear by allowing it to stand. Stable hazes and emulsions can form when a gasoline containing free water and a surfactant has been pumped or agitated in some way. Most gasoline surfactant additives are formulated to minimize these problems, but accidental overtreatment can occasionally cause problems. Hazes can form when there is a drop in temperature of the fuel, and may be stabilized if a surfactant additive is present.

Demulsifiers and dehazing additives are often complex mixtures[4,7] and function by promoting coalescence of the small droplets. They are themselves surfactants and so their use must be carefully controlled. Too much can make matters worse and, more importantly, they can interfere with other surfactants that may be present and reduce or eliminate their effectiveness.

It is important to be sure that any surfactant additive used in a gasoline will not cause haze and emulsion problems and will not suspend rust and other particulates from the distribution system. Water sensitivity tests should be carried out using a range of water samples of different pH. A number of test methods are used by the oil and additives industries, including the following:

1. The ASTM D 1094 water tolerance test[8] is a simple test that involves shaking 80 mL of fuel with 20 mL water and then, after 5 minutes standing, rating the fuel, the water and the interface for clarity and emulsions.

2. The 10 Cycle Multiple Contact test[9] gives a better simulation of field conditions than the ASTM D 1094 test because the water bottoms are contacted with several batches of gasoline, as normally happens in practice. In this test, 100 mL of fuel is shaken with 10 mL of water for 5 minutes and then allowed to stand for 24 hours. It is then rated for clarity and emulsion formation. The gasoline is carefully poured off leaving the water bottoms behind, and a fresh 100 mL of gasoline poured in, shaken as before, and rated after 24 hours. This is repeated for a total of ten cycles. The results give a good prediction of the performance of an additive in gasoline provided that a range of different water bottoms are used, varying in pH from about 4 to 9.

3. The Waring Blender test uses standard (but flame-proof) blend-ing equipment to mix 475 mL of gasoline and 25 mL water for 4 minutes. The fuel and water phases are rated as before at set time intervals.

4. The Particle Suspension test evaluates the tendency for a fuel to suspend particulate matter such as rust or dirt. A small amount (0.1 g) of dry precipitated iron oxide is added to 100 mL of fuel, shaken and then allowed to stand for a set period. Samples of gasoline are withdrawn periodically and filtered through a millipore filter to obtain a measure of the suspended matter. Sometimes a drop of water is added after shaking, followed by further shaking.

9.3 Additives Used to Protect Vehicle Fuel Systems

9.3.1 Corrosion inhibitors are important not only to protect the fuel distribution system but also to prevent corrosion of the fuel system in a vehicle (see also Section 9.2.3). If corrosion occurs, particles of corrosion products can block filters and jets, and leaks can develop in fuel tanks.

The same types of compound used to combat corrosion in the distri-bution system are used to protect the fuel system of vehicles.

Tests to evaluate gasoline corrosion inhibitors are of two types: dy-namic, to simulate use in vehicles or use in pipelines, and static, to simulate storage conditions.

The dynamic corrosion tests most frequently used are: ASTM D 665,[8] MIL-I-25017B and C,[10] the Colonial Pipeline test[11] and the National Association of Corrosion Engineers (NACE) test.[12] They all are somewhat similar and so the only one that will be described is the ASTM test.

This test (ASTM D 665) was originally developed to evaluate the antirust properties of steam turbine oils and has been modified for use with gasoline. A polished cylindrical steel spindle is immersed in a mixture of 300 mL of the gasoline and 30 mL water and stirred for a fixed period of time, often 24 hours, at a temperature in the range of ambient to 38°C. At the end of this period, the spindle is rated for the degree of rust that has formed. The water used in this test can be deionized to simulate rain water, it can be a synthetic sea water, or it can be at different pH levels to simulate the different types of process water that can find their way into gasoline.

The static corrosion test[4] often used consists of immersing a strip of carbon steel in a blend of 90 mL gasoline and 10 mL water in a glass bottle. After shaking, it is left for a fixed period, often 3 weeks or so, with estimates of the degree of rusting being made at set time intervals.

9.3.2 Anti-icing additives. The mechanism by which ice can form in a carburetor or throttle body has been described in Chapter 7, Section 7.6, together with the effects of the ice deposition and the methods of test.

Two types of additive are used to control icing — cryoscopic and surfactant.

Cryoscopic anti-icers function by depressing the freezing point of water and are usually alcohols or glycols. Materials that have been widely used are isopropyl alcohol, hexylene glycol and dipropylene glycol (DPG), as well as a number of complex mixtures of various glycols and other water-soluble oxygenated compounds. The treat rates used will depend on the additive and the likelihood and severity of problems in the vehicle population under consideration. Different compounds will require different treat levels to give the same protection and, for example, 2% of isopropyl alcohol will give about the same effect as 500 ppm of DPG.

Surfactant anti-icing additives function in much the same way as corrosion inhibitors, and good corrosion inhibitors are often also good anti-icers. They form a monomolecular layer on metal surfaces and thus prevent ice crystals from attaching themselves to the throttle plate and other surfaces so that they are swept into the combustion chamber. They also prevent ice crystals from sticking to each other and so stop the buildup of large crystals. Many types of surfactant are used including amine or imidazoline salts of carboxylic or alkenyl succinic acids.[4] As with all surfactant additives, care must be taken to be sure that they do not stabilize hazes and emulsions, as discussed in Section 9.2.6.

9.3.3 Carburetor detergents. Carburetor detergents are surfactant additives that function by the polar group at one end of the molecule attaching itself to a deposit or a particulate surface and the large non-polar, oleophilic group at the other dissolving in the fuel. The mono-molecular film that is formed around any particle effectively solubilizes it by forming a micelle which prevents aggregation of particles and allows the particulate matter to be carried into the com-bustion chamber with the fuel. Metal surfaces are protected against deposition in a similar way and deposits that have already been formed can gradually be removed. A higher treat rate or more power-ful additives are normally required for cleanup than for keep-clean performance.

Some early detergents were alkyl amine phosphates or fatty acid amides, but polymeric dispersants such as alkenyl succinimides, polybuteneamines and polyetheramines are commonly used nowa-days. Treat rates of the earlier additives were usually in the range of 20 to 100 ppm whereas the current materials, which are much more effective, are used at much higher concentrations (up to 1000 ppm), often together with a fluidizer to keep inlet valves clean (see Section 9.3.5).

Surfactant additives are normally multifunctional so that, although an additive might be primarily designed to be a carburetor detergent, it will often have other benefits as well such as anti-icing and/or anticor-rosion effectiveness.

The effectiveness of carburetor detergents is assessed by means of engine or road tests. A large number of engine tests have been devel-oped, many of them "in-house" procedures used by additive compa-nies to demonstrate the effectiveness of their products. However, a number of procedures have emerged over the years that have become accepted and are now widely used. These standard tests include the CRC Carburetor Detergency test[13] in the U.S. and the CEC Renault 5 Carburetor Cleanliness test[14] in Europe. The test conditions used are summarized in Table 9.1.

9.3.4 Port Fuel Injector (PFI) anti-fouling additives. There are two main types of fuel injector used — port fuel injectors and throttle body injectors (TBI). The use of fuel injectors instead of carburetors is growing rapidly because they give better control and are valuable in

Table 9.1. U.S. and European Carburetor Cleanliness Tests[15]

Test method	CRC	CEC
Engine type	Ford 3.9 L, 6 cyl.	Renault type 810-26
Test duration	20 hours	12 hours with 18-hour pause after 6 hours
Cycle	3 min at 700 rpm 7 min at 2000 rpm	2 min at 800 rpm 8 min at 1800 rpm
Cooling water temp.	88 - 90°C	80°C
Engine oil temp.	110°C max	78°C
Intake air temp.	66°C	40 - 65°C
EGR	None at idle Full at cruise	11%
Reference oil	REO-202-TI SAE 30 grade	CEC RL-51 Dispersant PMA 10W50 grade

helping to meet emissions and economy requirements; they also lend themselves more to computer feedback control. Deposits can, however, form in the pintle area (see Figure 9.3) of PFIs (TBFIs are relatively free from this type of problem) which reduce flow rate and modify fuel spray patterns,[16] as discussed in Chapter 5, Section 5.3. Even low deposit levels can have a significant influence on vehicle driveability.

These deposits became a serious problem in the U.S. in 1985-86 and a number of detergent additive solutions were developed which have largely overcome the problem. They are formed at much higher temperatures than carburetor deposits and so are more difficult to control. Polymeric dispersants and amine detergents[4] have been used successfully to control these deposits.[16,17,18,19,20,21,22]

Because the deposits are formed by heat from the engine evaporating any droplets of gasoline in the pintle area of the injector when the

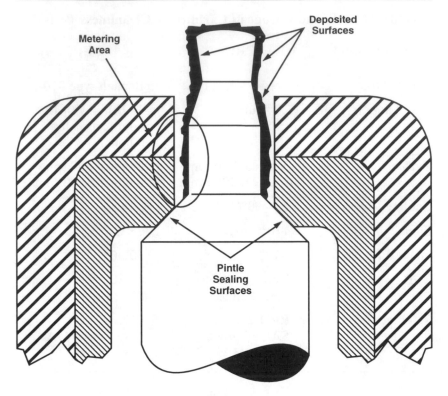

Fig. 9.3. Pintle tip deposits.[4]

engine is off, stop/start type driving is most likely to cause these deposits to form. Test procedures to evaluate the effectiveness of gasoline additive formulations in overcoming this problem (or the susceptibility of vehicles to it) all involve test cycles in which the engine or vehicle is driven at a moderate speed for 15 minutes or so and is then allowed to stand for 30 to 45 minutes with the engine off.[17,19] This cycle is repeated until driveability malfunctions begin to occur.

9.3.5 Additives to control inlet manifold, inlet valve and port deposits. Deposits in this region can cause many difficulties including driveability problems, increased exhaust emissions and poor fuel consumption[22] as discussed in Chapters 5 and 8.

In the manifold area deposits tend to accumulate on the "hot spot" region because the higher temperature there favors the gum-forming reactions. Surfactant additives such as carburetor detergents or anti-injector-fouling additives will usually provide adequate cleanliness. However, valve and port deposits require the addition of a high-boiling-point, thermally stable, oily material known as a *fluidizer, carrier oil* or *solvent oil*. These can be a petroleum-based lube oil base-stock, a synthetic oil or a polymeric material such as a polybutene or a polyetheramine.[23] They appear to function by dissolving, partially or completely, the deposit-forming compounds off the valve tulips (i.e., the undersides of the valves) or port area so that they are swept into the combustion chamber. They also appear to reduce the tendency for coking on the hot metal surfaces. These additives are used at relatively high concentrations, often as much as 1000 or 1500 ppm.[24]

Fluidizers alone without a detergent additive were quite effective in controlling valve and port deposits in older cars, but are inadequate in many modern designs where the combination with a detergent is necessary to overcome deposit problems. When ethanol is present in the gasoline, intake system deposits are much more difficult to control and may need up to 50% greater treat level to achieve adequate cleanliness.[25] Although reduced port deposits are claimed to reduce ORI (see Chapter 6, Section 6.7), increased combustion chamber deposits can result from too high a treat of these materials and this can itself raise ORI[26] and cause the lubricant to thicken. Problems of valve sticking have also been associated with some valve cleansing additives[27] in vehicles critical to this type of problem, so that it is necessary to check that an additive is free of this type of problem before using it.

Two tests in Europe have been developed to evaluate how fuel/additive combinations affect inlet valve deposits — the Opel Kadett test[28] and a Mercedes M102E test.[29] The test conditions are summarized in Table 9.2, although in the case of the Mercedes test, many variations are used. In the U.S., induction system deposit control has been investigated to some extent by the use of single- and multicylinder engine tests and fleet tests.[16] European vehicles have been used in some tests,[21] because these vehicles appear to be more critical to such problems under U.S. driving conditions.

9.3.6 Factory fill additives are used by some vehicle manufacturers if there is a likelihood that their vehicles will be stockpiled for a long period before being sold. They consist of a blend of a number of additives such as antioxidants, metal deactivators, corrosion inhibitors,

Table 9.2. Engine Test Conditions for Opel Kadett and Mercedes M102E Tests[15]

Test Conditions	Opel Kadett			Mercedes M102E			
Reference	CEC F-04-A-87			(22)			
Period of test	40 hours			150 hours			
Coolant outlet temperature, °C	92 max			90-95			
Engine oil temperature, °C	94 max			95-100			
Intake air temperature, °C	-			25-35			
Cycle	**Time (min)**	**Speed (rpm)**	**Power (kW)**	**Time (min)**	**Speed (rpm)**	**Load (Newtons)**	**Power (kW)**
	0.5	950	-	0.5	800	-	-
	1.0	3000	11.1	1.0	1300	31	4.0
	1.0	1300	4.0	2.0	1850	34	6.3
	2.0	1850	6.3	1.0	3000	37	11.0

biocides, etc., and are used in the factory fill gasoline at relatively high concentrations. While they are effective for factory fill purposes, they are not suitable for continuous use.

9.4 Additives that Influence Combustion

9.4.1 Antiknock additives are discussed in Chapter 6, Section 6.4. These additives can also help to overcome run-on and, in the case of lead, valve seat recession.

9.4.2 Anti-ORI additives reduce the increase in octane requirement (ORI) that occurs with all vehicles as they accumulate miles and lay down deposits in the combustion chamber and, to some extent, in the ports. The reduction of port deposits by the use of detergent additives/fluidizers has been discussed in Section 9.3.5 and these materials are claimed to reduce ORI by this mechanism,[30] although some fluidizers are known to increase combustion chamber deposits and possibly increase ORI.

In the early days when lead compounds were being developed, lead deposits in the combustion chamber caused a severe ORI problem. The use of halogen-based compounds such as ethylene dichloride provided a way of preventing the buildup of lead deposits and, hence, reduced ORI. Boron compounds such as glycol borates have also been used in the past to reduce lead deposits[31] and appear to work by making the deposits more friable so that they are physically removed by the mechanical action of the engine.

More modern additives used to reduce ORI are polyetheramines that are added to the fuel intermittently and at high concentrations.[32] They can rapidly reduce octane requirement by 1 or 2 numbers but it will immediately start to increase again so that the benefit is only obtained for a few thousand kilometers; a further additive treat is then needed, as shown in Figure 9.4.

9.4.3 Anti-preignition and anti-misfire additives have been used in leaded gasolines, particularly when lead levels were high. They consist of phosphorus compounds such as, for example, tricresyl phosphate which converts the lead halides, oxyhalides, oxides, etc., formed in the combustion chamber into lead phosphate. This has two effects: First, lead phosphate needs to be at a much higher temperature than the other lead compounds before it starts to catalyze the combustion of carbonaceous deposits and cause them to glow. Glowing deposits can be a source of preignition so that phosphorus compounds reduce the risk of preignition when using leaded gasolines. Second, lead phosphate has a much lower electrical conductivity at high temperatures than the halides, and so phosphorus additives are valuable in reducing misfire due to lead deposits on the plug conducting the charge away.

Phosphorus additives were used for both of the above purposes at 0.2 to 0.5 of the theoretical amount required to combine with all the lead in the fuel. As the use of lead has declined with lower concentrations in the gasoline, the use of these additives has diminished. Phosphorus also has an adverse effect on the efficiency of catalytic converters for improving exhaust gas quality, and so is banned in countries where they are used.

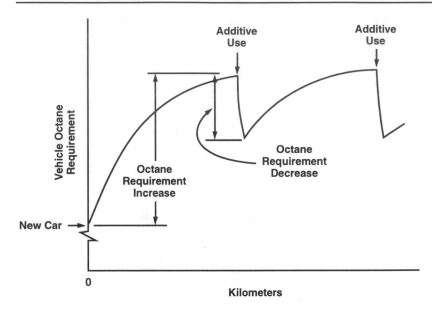

Fig. 9.4. Intermittent additive treats can reduce ORI.[32]

9.4.4 Spark-aider additives are designed to improve the spark so that when the air-fuel ratio is in the borderline condition such that it is almost too lean to burn, these additives will help prevent misfire and improve exhaust emissions.[22] This is most likely to happen in cold weather with vehicles that are designed to operate near the lean limit, because during the warm-up period it is sometimes not possible to get enough fuel vaporization to achieve a readily combustible mixture. The additive will therefore improve vehicle driveability under these conditions[33] with both leaded and unleaded gasolines.

The additive itself was introduced commercially for the first time in 1986 and is based on an organic gasoline-soluble potassium compound. The treat rate used gives a level of potassium in the fuel of only a few ppm. This is claimed to give a layer of deposit on the plug electrodes which has a lower electronic work function, so that more spark energy is put into igniting the air-fuel mixture. The initial flame is therefore stronger and there is a small increase in flame speed at the start of combustion. Other Group 1 and 2 metals also show this effect, to a greater or lesser degree.

Field problems have been experienced with this additive when used in leaded gasoline in that some antagonism was found between the halogen compounds used as scavengers and the potassium compound. This resulted in a viscous compound being formed which, in certain car models, caused sticking of valves with consequent engine damage.

9.4.5 Additives for improving fuel distribution between cylinders. When there is maldistribution of the air-fuel mixture between the cylinders of a multicylinder engine, different amounts of power are produced by each of the cylinders which can give rise to driveability problems, as well as cause a deterioration in exhaust gas quality and fuel economy. It is most likely to occur during cold weather with vehicles designed to operate on the lean side of stoichiometric.

Additives to reduce this problem appeared in the early 1970s, at a time when many U.S. vehicles suffered from this fuel maldistribution problem.[34] A commercial version was called HTA (hydrogenated tallow amine) and was composed of a mixture of tallow amines. It functions by forming a low surface energy coating on the internal surface of the inlet manifold, so that instead of a layer of liquid fuel being spread over the manifold surface, it is in the form of small discrete droplets which are much more readily entrained by the fuel-air stream.

9.4.6 Anti-valve-seat recession additives. Valve seat recession is a problem that can occur with engines having "soft" valve seats when operating on low-lead or unleaded gasoline.[35] It is most severe under high-speed/high-load conditions. The valve effectively grinds its way through the cylinder head so that the valve tappet clearance can be taken up in just a few thousand miles. All engines designed to operate on unleaded gasoline will have hardened valve seats, but many engines designed for leaded gasoline do not, because lead has a lubricating action on the valve seats and prevents them from wearing.

Lead is probably the best anti-valve-seat recession additive, but other materials are known to be at least partially effective. Phosphorus compounds such as tricresyl phosphate reduce this type of wear, but cannot be used in an unleaded gasoline that is also likely to be used in vehicles fitted with catalytic converters because the phosphorus has an adverse effect on catalysts. One commercial additive is called "Powershield"[36,37] and contains as the active ingredient a compound

239

containing sodium and sulfur. A dose rate of 1000 ppm, equivalent to 10 ppm of sodium and 9 ppm of sulfur, is claimed to be equivalent to about 0.13 g lead per liter. It cannot be used in countries where there is a restriction on the use of additives containing metallic elements. Other alkali metals are also known to be effective in reducing valve recession and even a totally organic additive has been claimed to have some effect.[38]

9.5 Additives that Improve Lubricant Performance

A small amount of gasoline additive finds its way into the lubricant by crankcase dilution during cold starts (see Chapter 7, Section 7.8), via unburnt fuel in the blowby gases, or because some gasoline is drawn into the cylinder after switching off the engine since it will continue to rotate for one or two cycles. The additive can help the lubricant especially in the ring/bore area as well as by replenishing lubricant additives such as detergents which might become depleted over the life of the lubricant. This class of additive was used in the 1950s and 1960s but is of limited use with today's highly developed lubricants, although there is some renewed interest because of a persistent "black sludge" problem in certain vehicles (see Section 9.5.4).

9.5.1 Upper cylinder lubricants were one of the first gasoline additives to be used and consisted of light mineral oils that were used at concentrations of up to 0.5 volume percent. They were valuable because they provided immediate lubrication to the ring and bore area after starting up. The early lubricants tended to drain away from the bores on switching off and it took some time before they reached them again on restarting. This caused considerable wear, particularly for vehicles carrying out a series of short runs every day. Nowadays, lube oils contain components that stick to metal surfaces and hold an oily film in place, so that this type of wear is no longer a serious problem.

9.5.2 Friction modifiers are an updated version of the upper cylinder lubricant and are designed to minimize engine friction in the piston ring area, and thus give a saving in fuel economy. Molybdenum-based additives have been used for this purpose as well as surfactants of the type used in lubricants and referred to in the previous section; these hold an oily film in contact with the cylinder walls.

9.5.3 Antiwear additives. Additives have been developed and used commercially that prevent corrosive wear in the cylinder bores and on the rings.[39] Halogen compounds such as lead scavengers, and sulfur compounds in the gasoline form corrosive gases on combustion which condense on the bores when the engine is switched off and begins to cool down. A coating of rust is formed on the metal surfaces and when the engine is restarted, this rust is swept away so that a clean metal surface is exposed for further corrosion to occur when the engine is again switched off. The additives used consisted of suitable surfactant anticorrosion agents, although the benefits with modern lubricants would be very marginal.

9.5.4 Antisludge additives were originally detergent/dispersants added to the gasoline so that the level of these additives in the lubricant was continually topped up to make up for materials lost by oxidation or used up in dispersing particulates. A black (or hot) sludge problem reappeared during the 1980s in both Europe and the U.S. when some vehicles were run under high-speed, high-temperature conditions.[40] Tests have shown that the cause can be due to both the fuel and the lubricant and also to fuel/lubricant interaction effects.[41] The fuel parameters that are important are fuel end point, the presence of heavy aromatic components, and the presence of a fuel detergent. Sump sludge is not affected by gasoline detergents although sludge in the rocker cover is influenced by these additives.[42]

References

1. "World Wide Survey of Motor Gasoline Quality 1993," The Associated Octel Company Limited, 1994.

2. F.G. Schwartz, C.S. Allbright and C.C. Ward, "Test Procedure for Predicting Gasoline Storage Stability," SAE Paper No. 690760, 1969.

3. J.H. Gilks, "Anti-Oxidants for Petroleum Products," *J. Inst. Pet.*, Vol. 50, No. 49, 1964.

4. R.C. Tupa and C.J. Dorer, "Gasoline and Diesel Fuel Additives for Performance/Distribution Quality II," SAE Paper No. 861179, 1986.

5. J.D. Culter and G.G. McClaffin, "Method of Friction Loss Reduction in Oleaginous Fluids Flowing Through Conduits," U.S. Patent 3,692,676, September 19, 1972.

6. Proceedings of the CRC Workshop on "Pipeline Drag Reducing Agents and Their Impact on Fuel Product Quality," Coordinating Research Council, November 1988.

7. A.A. Johnston and E. Dimitroff, "A Bench Technique for Evaluating the Induction System Deposit Tendencies of Motor Gasolines," SAE Paper No. 660783, 1966.

8. 1988 Annual Book of ASTM Standards, Vol. 05.01, 1988.

9. T.J. Sheahan, C.J. Dorer and C.O. Miller, "Detergent-Dispersant Fuel Performance and Handling," SAE Paper No. 690516, 1969.

10. U.S. Department of the Air Force, "Specification MIL-F-25017D for Fuel Soluble Corrosion Inhibitor," May 1981.

11. Colonial Pipeline Company, "Pipeline Rust Test (Modified ASTM D665)," November 1967.

12. National Association of Corrosion Engineers, "NACE Standard TM0172," 1972.

13. J.J. Malakar, J.B. Retzloff and L.M. Gibbs, "Throttle Body Deposits the CRC Carburetor Cleanliness Test Procedure," SAE Paper No. 831708, 1983.

14. Coordinating European Council. Tentative Test Method No. CEC F03T81 "Evaluation of Gasolines with Respect to Maintenance of Carburetor Cleanliness," 1981.

15. K. Owen and R.G.M. Landells, "Precombustion Gasoline Additives," *Gasoline and Diesel Fuel Additives*, Edited by K. Owen. Critical Reports on Applied Chemistry, Vol. 25, John Wiley & Sons, 1989.

16. B.Y. Taniguchi, *et al.*, "Injector Deposits — The Tip of Intake System Deposit Problems," SAE Paper No. 861534, 1986.

17. J.D. Benson and P.A. Yaccarino, "The Effects of Fuel Composition and Additives on Multi-Port Fuel Injector Deposits," SAE Paper No. 861533, 1986.

18. G.P. Abramo, A.M. Horowitz and J.C. Trewella, "Port Fuel Injector Cleanliness Studies," SAE Paper No. 861535, 1986.

19. R.C. Tupa and D.E. Koehler, "Gasoline Port Fuel Injectors — Keep Clean/Clean Up with Additives," SAE Paper No. 861536, 1986.

20. R.C. Tupa, "Port Fuel Injector Deposits — Causes/Consequences/Cures," SAE Paper No. 872113, 1987.

21. B. Bitting, *et al.*, "Intake Valve Deposits — Fuel Detergency Requirements Revisited," SAE Paper No. 872117, 1987.

22. C.D. Spink, P.G. Barraud and G.E.L. Morris, "A Critical Road Test Evaluation of Two High-Performance Gasoline Additive Packages in a Fleet of Modern European and Japanese Vehicles," SAE Paper No. 912393, 1991.

23. R.L. Sung, D.T. Daly and T.E. Hayden, "A Novel Gasoline Additive Package Removes Induction System Deposits and Reduced Engine Octane Requirement Increase," SAE Paper No. 891298, 1989.

24. "Proceedings of the CRC Workshop on Intake Deposits," Coordinating Research Council, August 22-24, 1989.

25. C.M. Shilbolm and G.A. Schoonveld, "Effect on Intake Valve Deposits of Ethanol and Additives Common to the Available Ethanol Supply," SAE Paper No. 902109, 1990.

26. P. Schreyer, K.W. Starke, *et al.*, "Effect of Multifunctional Fuel Additives on Octane Number Requirement of Internal Combustion Engines," SAE Paper No. 932813, 1993.

27. S. Mikkonen, R. Karlsson and J. Kivi, "Intake Valve Sticking in Some Carburetor Engines," SAE Paper No. 881643, 1988.

28. Coordinating Research Council. Tentative Test Method No. CEC F02T79, "The Evaluation of Gasoline Engine Intake System Deposition," 1979.

29. M. Gairing, "Zur Qualitaet der Ottokraftstoffe aus den Sicht der Automobilindustrie: Vermeidung von Ablagerungen auf Einlassventilen," *Mineroelrundshau*, Vol. 34, No. 11, pp. 209-215, 1986.

30. L.B. Graiff, "Some New Aspects of Deposit Effects on Engine Octane Requirement Increase and Fuel Economy," SAE Paper No. 790938, 1979.

31. E.C. Hughes, *et al.*, *Ind. Eng. Chem.*, Vol. 98, p. 1858, 1956.

32. J.A. Bert, *et al.*, "A Gasoline Additive Concentrate Removes Combustion Chamber Deposits and Reduces Vehicle Octane Requirement," SAE Paper No. 831709, 1983.

33. D.R. Blackmore, *et al.*, "Development of a Novel Gasoline Additive Package-Laboratory Test Work," Paper C311/86. Inst. Mech. Eng. Conference on Petroleum Based Fuels and Automotive Applications, London, 1986.

34. A.A. Zimmerman, L.E. Furlong and H.F. Shannon, "Improved Fuel Distribution — A New Role for Fuel Additives," SAE Paper No. 720082, 1972.

35. D. Godfrey and R.L. Courtney, "Investigation into the Mechanism of Exhaust Valve Wear in Engines Run on Unleaded Gasoline," SAE Paper No. 710356, 1971.

36. The Lubrizol Corporation, "Powershield. The Lead-Free Additive for Vehicles Designed to Operate on Leaded Gasoline," Technical Brochure, 1986.

37. R.C. Tupa, "Today's Gasoline Concerns — Injector Plugging and Valve Seat Wear," NPRA Paper AM8621, 1986.

38. O.L. Nelson, J.E. Larson, *et al.*, "A Broad-Spectrum, Non-Metallic Additive for Gasoline and Diesel Fuels: Performance in Gasoline Engines," SAE Paper No. 890214, 1989.

39. J.R. Hudnell, *et al.*, "New Gasoline Formulations Provide Protection Against Corrosive Engine Wear," SAE Paper No. 690514, 1969.

40. J.R.F. Lillywhite, P. Sant and S.B. Saville, "Investigation of Sludge Formation in Gasoline Engines," CEC Symposium, Paris, 1989.

41. I.R. Galliard and J.R.F. Lillywhite, "Field Trial To Investigate the Effect of Fuel Composition and Fuel-Lubricant Interaction on Sludge Formation in Gasoline Engines," SAE Paper No. 922218, 1992.

42. H. Moritani, *et al.*, "Effects of Fuel Composition on Sludge Formation," *J SAE Review*, Vol. 12, No. 3, July 1991.

Chapter 10

Other Gasoline Specification and Non-specification Properties

The major properties of gasoline relating to their combustion performance, their influence on volatility and their oxidation stability have been discussed in preceding chapters. The following covers a number of other aspects that can be important, not all of which appear in National and other specifications. The test methods mentioned all appear in the Annual Book of ASTM Standards.[1]

10.1 Appearance

Gasoline should always be clear and bright, without any suspended or other particulate matter and with no apparent free water. Particulate matter can block filters and orifices; water can block lines at ambient temperatures below freezing, increase intake system icing and worsen corrosion. The color of gasoline is not usually specified, except in the case of aviation gasolines where they are dyed different colors for easy identification of grade.

The standard procedures for evaluating the appearance of gasoline are contained in ASTM D 4176 in which Method A is for field use at ambient temperature and Method B is for use in the laboratory at a controlled temperature of 25°C. In both cases a 500 mL sample of the fuel is swirled in a clean glass jar and examined for visual sediment or water droplets just below the vortex formed by the swirling. Any sign of haziness or dullness is also noted. The results are recorded as a "Pass" or a "Fail."

If free water and/or sediment are evident, it may be worthwhile to determine them quantitatively by centrifuging, as in ASTM D 2705, or determining the total water by Karl Fischer titration (ASTM D 1744) and particulate contamination by a filtration test such as ASTM D 2276.

10.2 Composition

A number of compositional restrictions can be applied to gasoline for environmental or other reasons. The major ones are:

Benzene. The maximum allowable concentration of benzene is often specified because it is a highly toxic material and levels of 5% or even 3% have been imposed on motor gasolines in some countries. Even tighter restrictions are imposed on reformulated fuels where the maximum allowable limit can be 1% or even 0.8% in order to minimize both evaporative and exhaust emissions of this compound. It is usually determined by gas chromatography using a method such as ASTM D 3606 but other methods are available.[2]

Hydrocarbon composition. This is becoming increasingly important since it has been realized that the composition of a gasoline has a considerable influence on exhaust and evaporative emissions. The importance of benzene content has been discussed above but in addition the effect of aromatics content on benzene exhaust emissions, the photochemical activity of light olefins, and the impact of heavy hydrocarbons on exhaust HC concentration make the control of these compositional factors desirable.

In addition it is sometimes necessary to control composition in order to ensure, as much as possible, the severity of reference fuels for engine testing, particularly those for measuring deposit-forming tendency or the effectiveness of additives in overcoming deposits. Most of the specifications require only paraffins, aromatics and olefins to be measured and give only a loose control on composition because individual hydrocarbons within any of these categories can vary widely in deposit-forming tendency. Sometimes, as an extra control, specific active components such as dienes are also specified.

The most frequently specified procedure is the FIA (fluorescent indicator adsorption) method, ASTM D 1319. The sample is introduced into a special glass adsorption column packed with activated silica gel, a small layer of which contains a mixture of fluorescent dyes. Alcohol is used to desorb the sample and move it down the column.

The hydrocarbons are separated according to their tendency to be adsorbed into aromatics, olefins and saturates (or paraffins). Their position is identified on the column by the fluorescent dyes visible under ultraviolet light that separate selectively with the hydrocarbon types. This method is simple but does not give a reliable separation into different types. Thus, some diolefins, aromatics with olefinic side chains, and sulfur, nitrogen and oxygenated compounds all appear in the aromatics band. In the case of oxygenates, if they are present as components such as alcohols, they are best removed before carrying out this test.

Gas chromatography can also be used to determine gasoline composition in terms of the hydrocarbon types present, and the results can be presented in the same form as the FIA procedure, although it is difficult to get results that agree in view of the arbitrary separation involved in the FIA test. However, specific hydrocarbons can be identified by the GC method and this can be valuable, particularly in the control of streams going to gasoline blending.

Lead. Maximum lead contents are specified for all gasolines whether they are leaded or unleaded. In unleaded gasolines, the lead content is important because it may cause deposits in automotive pollution control equipment and also has a deleterious effect on exhaust gas catalysts. A variety of test methods are available for determining its concentration as summarized in Table 10.1.

In addition to the above, a method is also available (ASTM D 1949) for the separation and determination of tetraethyl lead and tetramethyl lead in a sample of gasoline.

Manganese. Where the manganese antiknock MMT (see Chapter 6, Section 6.4) is allowed to be used in gasoline, its maximum concentration is specified. The test procedure ASTM D 3831 uses atomic absorption and is suitable for estimating manganese concentrations in the range 0.001 to 0.12 gMn/USG (0.00025 to 0.03 gMn/L).

Oxygenates. Controls on the use of oxygenates in gasoline are summarized in Chapter 11, Section 11.1.

Phosphorus. Phosphorus is another element that will adversely affect automotive catalytic converters and so the maximum level is controlled in countries where this type of vehicle is present. The

Table 10.1 ASTM Test Methods for Determining
Lead in Gasoline

Expected Lead Concentration Range		ASTM Method[1]
gPb/USG	gPb/L	
0.12 - 6.0	0.026 - 1.3	D3341 - Iodine monochloride method
0.1 - 5.0	0.026 - 1.3	D2599 - X-ray spectrometry method
0.2 - 4.2	0.05 - 1.1	D2547 - Volumetric chromate method
0.01 - 0.50	0.0025 - 0.125	D3229 - X-ray spectrometry method
0.01 - 0.10	0.0025 - 0.025	D3237 - Atomic absorption spectroscopy
0.01 - 0.10	0.0025 - 0.025	D3348 - Rapid colorimetric field test
0.001 - 0.1	0.00025 - 0.025	D3116 - Spectrophotometric method

phosphorus can be present as an additive since phosphorus-containing additives have been used as anti-misfire and anti-preignition agents and as surfactants (see Chapter 9). It can be determined in gasoline by the ASTM method D 3231.

Sulfur. Even though the maximum allowed sulfur level in gasoline has been low compared with most other hydrocarbon fuels—generally it has been below 0.1% wt although some specifications allow up to 0.2% wt—for reformulated gasolines much lower levels can be required because of the adverse effect it has on the catalyst of controlled cars[3,4] (see also Chapter 12).

Apart from its effect on catalysts, its presence is controlled for a number of reasons including its contribution to the odor of the gasoline, its corrosivity, its antagonistic effect on lead and its general adverse environmental effect in that sulfur dioxide and, in some cases, the malodorous hydrogen sulfide[5] can be emitted from the exhaust system.

It has also been blamed as a factor in increasing the deposit-forming tendency of a gasoline[6] and for its adverse effect on fuel system elastomers.

Sulfur can be present in a number of forms. There can be very small amounts of mercaptans (thiols) present which are particularly undesirable because of the smell they give to the gasoline, although the various sweetening processes used during manufacture are designed to remove these compounds. It can also be present as elemental sulfur, sulfides, disulfides, polysulfides, thiophenes and in many partially oxidized forms such as sulfoxides.

Mercaptans can be determined qualitatively by the Doctor Test which, even though it is an old test, is sometimes still specified because it is simple to carry out and does not require expensive or complex equipment. A sample of gasoline is shaken up with sodium plumbite, and if the mixture slowly darkens it indicates the presence of both mercaptans and elemental sulfur. If it gives an immediate black precipitate, it indicates that hydrogen sulfide is present which would be highly undesirable and unusual and would necessitate washing a fresh sample with caustic soda solution and then starting again. If the mixture does not darken at all, a small quantity of flowers of sulfur is added and if, after reshaking, the mix darkens, the presence of mercaptans is indicated. This test will indicate the presence of about 5 to 10 ppm of mercaptan and the results are reported as a "pass" or "fail." Mercaptans can readily be measured quantitatively by a number of methods, including ASTM D 3227 which involves a potentiometric titration with silver nitrate.

Total sulfur in gasoline is usually determined by burning a known amount in a wick lamp and absorbing the combustion gases in hydrogen peroxide so that the sulfur dioxide is converted to sulfuric acid which is then determined by titration with standard caustic soda solution. This is the Lamp Method, ASTM D 1266. Other procedures that can be used are an X-ray Spectrographic method, ASTM D 2622, or an oxidative microcoulometry method, ASTM D 3120.

Water. Both dissolved and free water can be present in gasoline although free water is undesirable because it can freeze and block lines, promote corrosion, worsen intake system icing, and cause emulsion and haze formation under appropriate conditions. On the credit side, when injected under controlled conditions into the mixture stream, water will reduce the temperature of the end gas and, therefore, will reduce any tendency for detonation. It is possible to obtain more power from an engine by injecting water, provided the ignition timing is advanced into the region where knock would have occurred without water.[7]

Dissolved water in gasoline cannot usually be avoided because the components are almost always contacted with aqueous solutions during manufacture, and storage tanks invariably contain free water at the bottom. Haze formation can occur during a drop in temperature of the gasoline due to some of the water coming out of solution. Such hazes usually clear rapidly as the droplets coalesce and sink into the tank water bottoms, and this separation can be helped by the use of dehazing additives (see Chapter 9, Section 9.2.6). The amount of water that will dissolve will depend on the composition of the gasoline. Aromatics will increase water solubility and Figure 10.1 shows a fairly typical curve of how water solubility varies with temperature. The presence of certain oxygenated components in the gasoline can substantially increase the amount of water that will dissolve in a gasoline, as discussed in Chapter 11.

Free water can be determined by centrifuging, and total water by means of the Karl Fischer procedure as was discussed in Section 10.1.

10.3 Conductivity

The conductivity of a fuel is important in determining whether a charge of static electricity (see Chapter 9, Section 9.2.4) will build up during pumping operations. A high-conductivity fuel will dissipate an electrical charge more rapidly than a low-conductivity fuel and, hence, reduce the risk of an electrical discharge causing a fire. Fires/ explosions will only occur if such a discharge is in the presence of an air-fuel mixture that is within the combustible range. The conductivity affects the rate at which charged ions are removed and is measured in picosiemen/meter, one unit of which is equivalent to one conductivity unit (CU). The time taken for the charge to fall to half its value (half time value) and the time to drop to approximately 37% of its value (relaxation time) are sometimes used to indicate whether or not static charges are likely to be a hazard.

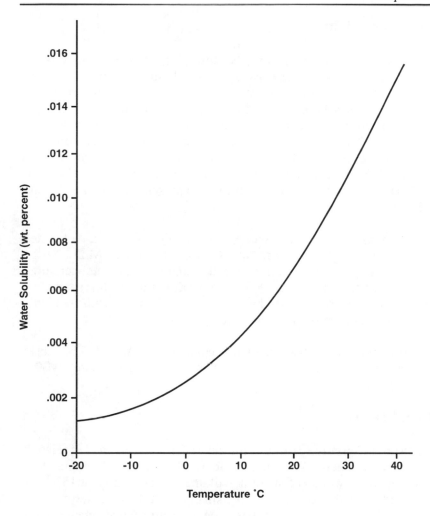

Fig. 10.1. Water solubility vs. temperature.

Values of conductivity for gasoline are difficult to quote because they depend on the hydrocarbon composition, and whether there are oxygenates and/or surfactants present. A pure dry hydrocarbon-only gasoline will have a conductivity of about 1 CU, but this can increase very rapidly with compositional changes and the use of certain additives. Standard test methods for determining conductivity are ASTM D 3114 and ASTM D 4308.

10.4 Corrosivity

Corrosion is a problem, not only because of the damage it does to equipment, but also because dissolved metals such as copper can catalyze oxidation reactions and give rise to excessive deposit formation (see Chapter 9, Section 9.1.2). In addition, the products of corrosion can block filters and orifices in a vehicle's fuel system and increase wear rates. The extent of corrosion is a function of the water content of a gasoline (Section 10.2), whether there are oxygenates present (see Chapter 11), the level and type of sulfur compounds present (Section 10.2) and whether corrosion inhibitors are present (Sections 9.2.3 and 9.3.1).

The corrosivity of gasoline to copper-containing parts of a fuel system is evaluated using the *Copper Strip Corrosion test* (ASTM D 130). A polished copper strip is immersed in the sample at fixed temperature and for a fixed period (usually 3 hours at 50°C for gasoline) and the degree of corrosion assessed by comparing the sample with standard ASTM colors representing various degrees of tarnish.

Corrosion of steel is assessed by using a modified ASTM D 665 test procedure as outlined in Section 9.3.1 or the NACE Standard TM-01-72.

10.5 Density

The SI term "relative density" replaces specific gravity and is related to the measurement of the ratio of the weight of a given volume of gasoline to the weight of the same volume of water, both at 15°C and at a pressure of 101.325 kPa . To a small extent the older term "degrees API" is also used and is based on an arbitrary hydrometer scale that is related to specific gravity as follows:

$$\text{Degrees API} = \frac{141.5}{\text{Sp. Gr. (60°F/60°F)}} - 131.5$$

Relative density is useful for converting volumes to weight and is sometimes used as a way of identifying gasolines. It is important in long-trip fuel economy (see Chapter 7) because it is a guide to the heat content of a gasoline (see Section 10.8). Measurement is usually by the Hydrometer method, ASTM D 1298.

When considering the flow of fuel through a carburetor jet, the coefficient of discharge for normal hydrocarbon fuels is virtually constant above a critical value of the Reynolds number. Under these conditions, the mass flow is a function of the density of the fuel so that increasing density increases mass flow and lowers the air-fuel ratio of the mixture. This is partly offset by the lower level in the float chamber that occurs with heavier fuels. There is no such compensating factor with injection vehicles where the mass of fuel injected will depend directly on the density of the fuel. Variations in density between commercial fuels will therefore influence the air/fuel ratio unless there is an engine management system to hold it at the optimum level.

Most gasolines have a density between about 0.72 and 0.78, equivalent to an API range of 65 to 50. The density will depend on the types of components used in blending the gasoline — aromatic compounds have the highest density, with the olefins being intermediate and the paraffins having the lowest density, when compounds having the same number of carbon atoms are compared.

10.6 Flash Point

The flash point is the lowest temperature of a sample at which application of an ignition source causes the vapor of the sample to ignite under the specified conditions of the test. For gasoline it is so low (normally below -40°C) that it cannot be determined by any standard methods.

10.7 Freezing Point

This is also not very relevant for motor gasolines, although a maximum temperature of -60°C is usually specified for aviation gasolines using ASTM Method D 2386. It is the temperature at which crystals of hydrocarbons formed on cooling the fuel just disappear when the temperature is allowed to rise. Motor gasoline has a freezing point of about -70°C[8] depending on its composition.

10.8 Heat of Combustion (also called Heating Value or Calorific Value)

The Gross Heat of Combustion is the heat released by the combustion of a unit mass of fuel in a constant volume bomb with substantially all the water condensed to the liquid state. The Net Heat of Combustion is the heat released by the combustion of a unit mass of fuel at a constant pressure of 1 atmosphere (0.1 MPa) with the water remaining in the vapor state. It is the Net or Lower Heat of Combustion that is normally used when considering automotive fuels because exhaust gases leave the combustion chamber at a high temperature carrying the uncondensed water vapor with them. It is usually measured in terms of kJ per kg, or British Thermal Units per pound, where 1 Btu equals 1055.06 J.

The heat of combustion is clearly important in that it is a measure of the energy content of the gasoline and, hence, will influence the fuel economy that can be achieved. There is an empirical linear relationship between density and heating value for gasoline as shown in Figure 10.2.[13]

Heat of combustion of gasoline can be measured using ASTM D 240 or ASTM D 2382, both of which involve combusting some sample in a bomb calorimeter and measuring the temperature increase. It is often more convenient to measure the density of the fuel and relate it to the heat content because there is a direct relationship between both parameters. Gasolines which contain no oxygenates typically have heating values of about 43 kJ per kg.

10.9 Surface Tension

Surface tension has an effect on flow through jets because increased venturi suction is required to overcome the surface tension at the discharge nozzle.[9] Increasing surface tension will reduce mass flow and air-fuel ratio. It increases as the temperature is reduced (as does density and viscosity). Atomization of the fuel is also influenced by surface tension so that the droplet size distribution and the mean droplet size can be changed significantly by changes in surface tension.

Measurement can be made by a variety of methods including ASTM D 3825. Units of measurement are Newtons/meter or dynes/cm and a typical value for gasoline at 20°C is 20 dynes/cm or 20×10^{-3} N/m.

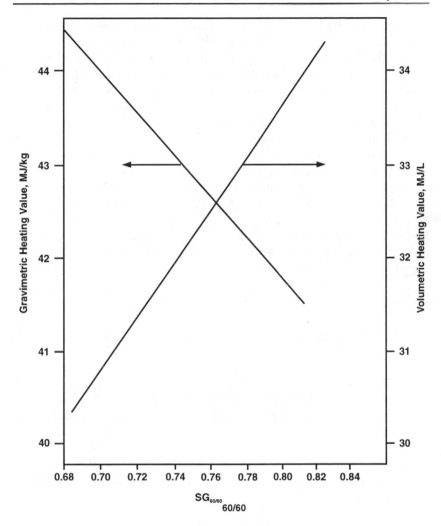

Fig. 10.2. Correlation between lower heating value and specific gravity for gasolines.[13]

10.10 Viscosity

Viscosity can also influence flow through metering orifices since the Reynolds number is an inverse function of fuel viscosity. However, the coefficient of discharge is fairly constant above a critical value of the Reynolds number[10] and, because for most vehicles the Reynolds number is usually above this critical value, normal variations in viscosity have a relatively small but significant influence on flow. However, if the fuel viscosity is above the normal range, the Reynolds number may fall below the critical level so that non-turbulent flow occurs. Under these circumstances, increasing fuel viscosity will then decrease fuel flow and so the air-fuel ratio will increase. Thickening agents have been proposed that will effectively lean off the mixture strength in this way for an improvement in fuel economy[11,12] and, although this might have been feasible at one time when many vehicles were operating on the rich side of stoichiometric in order to maximize power, most of today's vehicles would suffer severe driveability problems if viscosity was significantly increased.

Viscosity is measured by ASTM D 445 in which the flow under gravity through a calibrated capillary tube is measured. A typical range for gasoline is 0.5 to 0.6 centistokes.[9]

References

1. 1993 Annual Book of ASTM Standards, Volumes 05.01 to 05.03, 1993.

2. R.E. Pauls, G.J. Weight and P.S. Munowitz, "A Comparison of Methods to Determine Benzene in Gasoline Boiling Range Material," *Journal of Chromatographic Science*, Vol. 30, No. 1, January 1992.

3. J.C. Summers, J.F. Skowron, *et al.*, "Fuel Sulfur Effects on Automotive Catalyst Performance," SAE Paper No. 920558, 1992.

4. J.D. Benson, V. Burns, *et al.*, "Effects of Gasoline Sulfur Level on Mass Exhaust Emissions—Auto/Oil Air Quality Improvement Research Program," SAE Paper No. 912323, 1991.

5. D.C. Trimm, "Hydrogen Sulphide in Exhaust Gases," API Abstracts, Health and Environment, November 17, 1986.

6. V.S. Azev, *et al.*, "Results of Tests of Automotive Gasolines by a Complex of Methods of Qualification Evaluation Following Their Long Term Storage in Salt Caverns," *Transport i Khranenie Nefteproduktov i Uglevodorod*, N. 2, pp. 1-4, 1977.

7. E.M. Goodger, <u>Hydrocarbon Fuels</u>, The Macmillan Press, p. 178, 1975.

8. Ibid. p. 133.

9. J.A. Bolt, S.J. Derezinski and D.L. Harrington, "The Influence of Fuel Properties on Metering in Carburetors," SAE Paper No. 710207, 1971.

10. A. Lichtarowicz, R.K. Duggins and E. Markland, "Discharge Coefficients for Incompressible Non-Cavitating Flow Through Long Orifices," *J. Mech. Eng. Sci.*, Vol. 7, No. 2, pp. 210-219, 1965.

11. R.M. Reuter and G.W. Eckert, US Patent No. 3164138, 5 January 1965.

12. D.R. Blackmore and A. Thomas, <u>Fuel Economy of the Gasoline Engine</u>. The Macmillan Press, p. 74, 1977.

13. Ibid. p. 16.

Chapter 11

Oxygenated Blend Components for Gasoline

11.1 Introduction

The use of oxygenated compounds such as alcohols in gasoline started in the 1920s when the high-octane quality of methanol and ethanol made them extremely valuable blend components at a time when only relatively low-octane components were available from refinery processing. Although the blending of alcohols into gasoline continued to a small extent in a limited number of brands after World War II, their use expanded considerably in the 1970s when, in relative terms, they became more economical as a result of the rapid increase in crude oil prices. Ethers also began to be used at that time. The importance of oxygenates increased as refiners began to appreciate their potential benefits in providing a means of extending the gasoline pool, of overcoming an octane shortage due to the phase-out of lead antiknock compounds, and of giving additional flexibility for meeting ever-increasing demands on quality. They also reduce the dependency on crude oil and, therefore, provide a stabilizing influence in the market. They are of most interest to gasoline blenders when crude oil prices are high. More recently the benefits of oxygenates in reducing exhaust emissions of CO and HC have been appreciated and in some areas there is a mandatory minimum level.

Only two types of oxygenated compounds — alcohols and ethers — are used to any significant extent in gasoline as components (i.e., at concentrations greater than one or two percent), and of these the most important are:

Alcohols: Methanol (MeOH)
Ethanol (EtOH)
Isopropanol (IPA)

261

t-Butanol (TBA)
Mixed C1 to C5 alcohols

Ethers: Methyl tertiary butyl ether (MTBE)
Tertiary amyl methyl ether (TAME)
Ethyl tertiary butyl ether (ETBE)
Mixed ethers

Because of its poor solubility in gasoline when water is present, methanol is used with a cosolvent such as TBA and, in fact, all gasolines containing alcohols require careful handling to avoid or minimize water contact. Ethers tend to be relatively trouble-free as gasoline blend components.

The manufacture of the oxygenates commonly used in motor gasoline has been described in Chapter 3, Section 3.9.

There are two major problems which effectively limit the amount of oxygenate that can be used in gasolines marketed in areas where the vehicle population is designed for conventional hydrocarbon gasolines. These are the chemical leaning effect arising because of the oxygen content of the compound and the adverse effect on vehicle fuel system materials. They will be discussed in more detail later, but the overall restrictions that have been imposed because of these factors are as follows:[1]

U.S.A.

The EPA has ruled that aliphatic alcohols and glycols, ethers and polyethers may be added to the fuel provided that the amount of oxygen in the finished fuel does not exceed 2.7% wt (revised from 2.0% in 1991). Methanol is excluded from this approval. This is known as the "substantially similar" ruling because these compounds are considered to be substantially similar to fuels in widespread use before the requirement for EPA approval. There are also a number of specific proposals that have been approved (see Appendix 7 for more details) including:

- "gasohol" consisting of gasoline with 10% ethanol. This contains 3.5% oxygen.

- a mixture of TBA and methanol up to a maximum concentration of 3.7% oxygen provided that methanol does not form more than 50% of the mixture.

- methanol up to 5.0% volume plus at least 2.5% volume cosolvent (ethanol, propanols or butanols) plus corrosion inhibitor, with a maximum oxygen content of 3.7% wt. This is known as the "Dupont waiver."

- MTBE up to 15%.

European Community

The EC Directive allows the use of monoalcohols and ethers with atmospheric boiling points lower than the final atmospheric boiling point laid down in the national gasoline standards. The ethers must also have molecules containing 5 or more carbon atoms.

Member States must permit fuel blends containing levels of oxygenates not exceeding the level set out in column A of Table 11.1. If they so desire, they may authorize proportions of oxygenates above these levels. However, if the levels so permitted exceed the limits set out in column B, the dispensing pumps must be clearly marked accordingly, in particular to take into account the calorific value of such fuels.

The use of higher levels of oxygenate in gasoline than those currently allowed and summarized above necessitates the use of modified vehicles. These would need changes to the carburetor/fuel injection system to compensate for the oxygen content of the fuel, and to the fuel system construction materials to ensure that they are not attacked by the fuel. A large scale program was carried out in Germany from 1979 to 1982 mainly directed at blends containing 15% methanol.[2,3,4] Some tests were also carried out with concentrations of methanol as high as 60%. The work demonstrated that suitably modified vehicles can perform satisfactorily in all respects on such medium to high levels of methanol. The use of neat and near neat alcohols is discussed in Chapter 21.

11.2 Chemical and Physical Properties of Oxygenates

Table 11.2[5,6,7] summarizes some of the important characteristics of the most commonly used oxygenated gasoline components as compared with a conventional gasoline.

Table 11.1. EC Oxygenates Limits as Set Out
in Directive 85/536/EEC[1]

	A (% vol)	B (% vol)
Methanol, suitable stabilizing agents must be added [a]	3%	3%
Ethanol, stabilizing agents may be necessary [a]	5%	5%
Isopropyl alcohol	5%	10%
TBA	7%	7%
Isobutyl alcohol	7%	10%
Ethers containing 5 or more carbon atoms per molecule [a]	10%	15%
Other organic oxygenates defined in Annex section I	7%	10%
Mixture of any organic oxygenates defined [b] Annex section I	2.5% oxygen weight, not exceeding the individual limits fixed above for each component	3.7% oxygen weight, not exceeding the individual limits fixed above for each component

Notes:

   (a) In accordance with national specifications or, where these do not exist, industry specifications
   (b) Acetone is authorized up to 0.8% by volume when it is present as a by-product of the manufacture of certain organic oxygenate compounds

Not all countries permit levels exceeding those in column (A) even if the pump is labeled. Fuller details are included in Appendix 7.

Table 11.2. Some Properties of Oxygenates and Gasoline[5]

Property	Methanol	Ethanol	IPA	TBA	MTBE	ETBE[8]	TAME	Gasoline
Formula	CH_3OH	C_2H_5OH	$(CH_3)_2CHOH$	$(CH_3)_3COH$	$(CH_3)_3COCH_3$	$(CH_3)_3COC_2H_5$	$(CH_3)_2(C_2H_5)COCH_3$	C_4 to C_{12}
Molecular weight	32.04	46.07	60.09	74.12	88.15	102.18	102.18	100-105
Composition, weight %								
Carbon	37.5	52.2	60.0	64.8	68.1	70.5	70.5	85-88
Hydrogen	12.6	13.1	13.4	13.6	13.7	13.8	13.8	12-15
Oxygen	49.9	34.7	26.6	21.6	18.2	15.7	15.7	0
Specific gravity, 60°F/60°F	0.796	0.794	0.789	0.791	0.744	0.742	0.77	0.72-0.78
Density, lb/gal @ 60°F	6.63	6.61	6.57	6.59	6.19	6.25	6.41	6.0-6.5
Boiling temperature, °F	149	172	180	181	131	164	187	80-437
Reid vapor pressure, psi	4.6	2.3	1.8	1.8	7.8	4.4	1.5	8-15

Table 11.2. Some Properties of Oxygenates and Gasoline[5] (Cont)

Property	Methanol	Ethanol	IPA	TBA	MTBE	TAME	Gasoline
Water solubility, @ 70°F							
Fuel in water, volume %	100	100	100	100	4.3	—	Negligible
Water in fuel, volume %	100	100	100	100	1.4	0.6	Negligible
Viscosity, Centipoise @ 68°F	0.59	1.19	2.38	4.2 @ 78°F	0.35	—	0.37-0.44
Centipoise @ –4°F	1.15	2.84	9.41	Solid	0.60	—	0.60-0.77
Flash point, closed cup, °F	52	55	53	52	–14	—	–45
Autoignition temperature, °F	867	793	750	892	815	—	495
Flammability limits, volume %							
Lower	7.3	4.3	2.0	2.4	1.6	—	1.4
Higher	36.0	19.0	12.0	8.0	8.4	—	7.6

Table 11.2. Some Properties of Oxygenates and Gasoline[5] (Cont)

Property	Methanol	Ethanol	IPA	TBA	MTBE	TAME	Gasoline
Latent heat of vaporization,							
Btu/gal @ 60°F	3,340	2,378	2,100	1,700	863	—	900 (approx.)
Btu/lb @ 60°F	506	396	320	258	138	—	150 (approx.)
Heating value, Lower (liquid fuel-water							
vapor) Btu/lb	8,570	11,500	13,300	14,280	15,100	15,690	18,000-19,000
Btu/gal @ 60°F	56,800	76,000	87,400	94,100	93,500	100,600	109,000-119,000
Stoichiometric air-fuel, weight	6.45	9.00	10.3	11.1	11.7	12.1	14.7
Ratio moles product/moles $O_2 + N_2$	1.21	1.12	1.10	1.10	1.10	1.09	1.08

The higher the oxygen content, the lower the stoichiometric air-fuel ratio and the leaner the mixture strength when used in blends with gasoline. The lower heat content of oxygenates may adversely influence fuel economy. The ability of alcohols to form azeotropes with gasoline components means that the volatility characteristics of the gasoline into which they are blended will be severely modified. It is these factors that give rise to differences in performance between hydrocarbon-only gasolines and gasolines containing oxygenates.

Fungibility (i.e., interchangeability) of gasolines containing alcohols with hydrocarbon-only gasolines is a problem, and in many countries alcohol blends are regarded as non-fungible. This is because of their effect on volatility and their water sensitivity.

11.3 Influence of Oxygenates on Gasoline Quality

In this section the effect of oxygenates in meeting specification levels will be discussed. The performance of such blends in vehicles on the road may differ substantially from that of purely hydrocarbon blends, and this aspect is discussed in Section 11.5.

11.3.1 Octane

It is not possible to give absolute values for the octane quality of the various oxygenates because the way they blend into gasoline depends on the other components that are present. However, Table 11.3 summarizes a range of octane blending data that have been obtained with different gasoline blends and demonstrates that these components can be valuable in helping to meet required octane levels, particularly in an unleaded situation.

It will be seen that the Sensitivities (RON minus MON) of the oxygenates are high and this means that in a situation where MON is a critical specification point, it may be difficult to meet MON specifications using these materials without going over the top of the RON specification, i.e., without giving quality away. In this situation, blending components with low sensitivities such as alkylate or isomerate may be needed.

Table 11.3. Approximate Octane Blending Values

	Blending RON	Blending MON
Methanol	127 - 136	99 - 104
Ethanol	120 - 135	100 - 106
Tert. butanol	104 - 110	90 - 98
Methanol/TBA 50/50	115 - 123	96 - 104
MTBE	115 - 123	98 - 105
TAME	111 - 116	98 - 103
ETBE	110 - 119	95 - 104

The standard tests for RON and MON (i.e., ASTM D 2699 for RON and ASTM D 2700 for MON) are equally applicable to fuels containing oxygenates provided that the concentration does not exceed the restrictions given in Section 11.2. For gasolines containing high levels of oxygenates or for pure oxygenates, these procedures are not suitable (see Chapter 20).

11.3.2 Volatility

The presence of alcohols, and particularly of methanol, can have a dramatic effect on the RVP of a gasoline to which it is added, as illustrated in Figure 11.1.[9]

The increase in RVP when methanol is added to a hydrocarbon blend can adversely affect the economics of blending because where there is a restrictive RVP specification, butane, which has a high vapor pressure, may have to be backed out of the blend.

The procedure used for measuring the RVP of a gasoline blend containing alcohols can affect the result. The standard RVP procedure (ASTM D 323) measures the vapor pressure in the presence of water-saturated air and this gives rise to a lower value than if a dry procedure is used. Some specifications will allow the "wet" procedure to be used and some require the use of a "dry" procedure. Fig. 11.2[4] shows the differences that can occur between the two procedures.

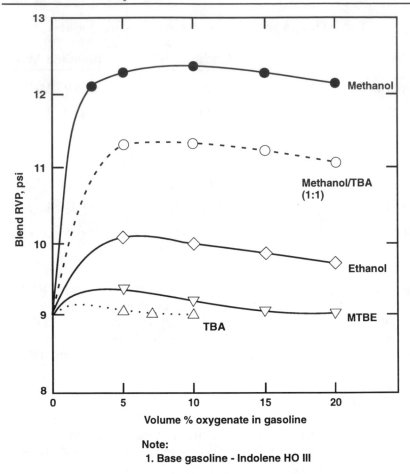

Fig. 11.1. Effect of oxygenate concentration on blend vapor pressure.[5,9,10]

The blending RVP of an oxygenate will depend on the oxygenate type, the amount present, and the RVP and nature of the base blend. Figures 11.1 and 11.3[4] illustrate the effect of concentration and type in different hydrocarbon blendstocks and show that low concentrations of methanol, in particular, can have very high RVP blending numbers.

The presence of TBA as a cosolvent with methanol reduces the effect on RVP as shown in Figures 11.1 and 11.4.[4] Figure 11.1 is based on

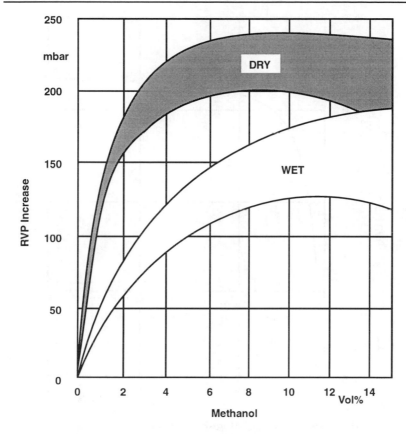

Fig. 11.2. RVP of methanol blends depending on different methods and different base gasolines.[4]

the use of a base gasoline having a relatively low RVP (9 psi), whereas the data in Figure 11.4 are based on a gasoline with a higher RVP (0.7 bar, 10.2 psi).

The blending vapor pressure of MTBE is below that of finished gasoline and varies from about 50 to 65 kPa depending on the other blend components and the concentration used. This may allow additional butane to be used in a blend without exceeding the RVP specification limit. For ETBE the blending RVP is even lower (between about 20 and 32 kPa[8,11,12]) and so allows even more butane to be used.

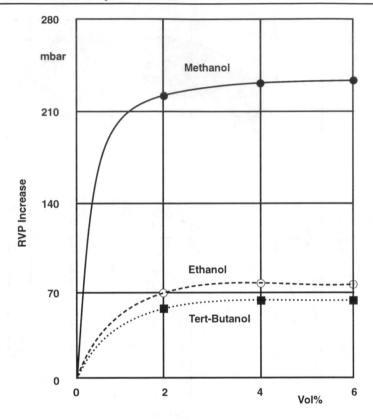

Fig. 11.3. RVP increase of alcohol-gasoline blends.[4]

Figure 11.5[5,10] shows the influence on the ASTM D 86 distillation curve of adding various oxygenates to a gasoline; it can be seen that the curve can be highly distorted by the presence of alcohols.

As the molecular weight of the alcohol increases, the effect becomes less and the temperature range of the section of the distillation curve that is influenced moves up the curve.

Ethers do not have the same effect of changing just a part of the distillation curve but they can make the whole gasoline lighter or heavier depending on the volatility of the base gasoline and the specific ether being used.[8,11,12] An example of the effect of MTBE on the distillation curve is also shown in Figure 11.5.

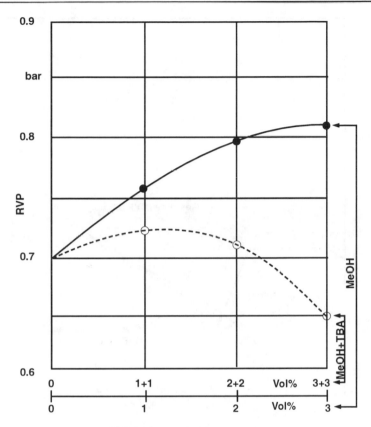

Fig. 11.4. TBA lowers RVP increase.[4]

Vapor-liquid ratio is increased by the use of alcohols as would be expected from a consideration of the effect on RVP and the distillation curve. This is shown in Figure 11.6.[9] ETBE behaves in much the same way as a pure hydrocarbon of the same volatility in the way it affects RVP and distillation characteristics.[12]

11.3.3 Density

When certain alcohols are blended into gasoline, non-ideal volume blending occurs giving a gain in volume. For example, at the 5% addition level, TBA has a blending relative density value of approximately 0.75 compared with its actual relative density of 0.785.[13] This volume increase, which is slightly less for blends of methanol

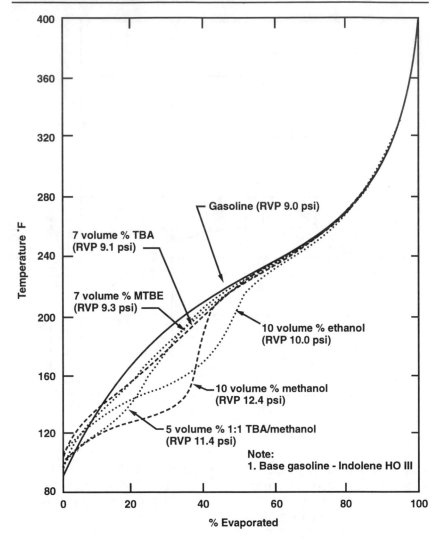

Fig. 11.5. Effect of oxygenates on distillation.[5,10]

and TBA, will give an economic advantage in areas where gasoline is blended by weight and sold by volume. It needs to be taken into account when blending in tankage. The effect does not occur with ethers.

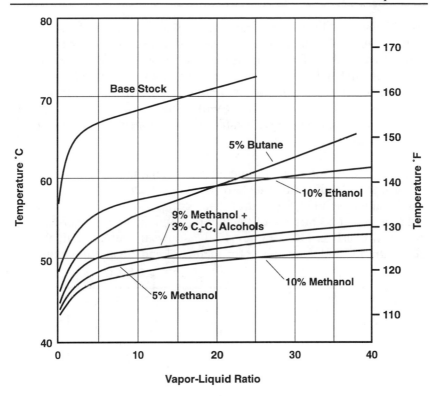

Fig. 11.6. Vapor-liquid ratio versus temperature for blends in debutanized gasoline.[9]

11.4 Storage and Handling of Oxygenate Blends

In general, the storage and handling of blends containing ethers present no difficulties over those of purely hydrocarbon blends although, because of its slight water solubility, there have been concerns over losses and contamination of ground water when it has been stored in caverns over natural water bottoms. However, fuels containing alcohols, and methanol in particular, can cause problems unless care is taken during blending, storage and distribution.

11.4.1 Water Sensitivity of Alcohol Blends

Both methanol and ethanol are completely miscible in water and so are partially extracted from gasoline if it is contacted with water. The

275

gasoline/alcohol blend can separate into two layers under these conditions — an alcohol-rich aqueous layer and a gasoline-rich upper layer. This has a number of undesirable effects: First, it depletes the gasoline of alcohol and so it may not meet the octane specifications. Second, the lower phase will have increased in volume due to the alcohol and have a lower density so that it will be more readily entrained and carried over into the motorist's tank. If this happens, it can stop a vehicle completely.

The water tolerance of a gasoline/alcohol blend, i.e., the amount of water that a blend can tolerate before it breaks into two phases, depends on temperature, the hydrocarbon composition of the gasoline, the type and concentration of the alcohol, and the presence of cosolvents such as TBA. Figure 11.7[9] shows the effect of alcohol concentration and temperature for methanol and ethanol in an un-

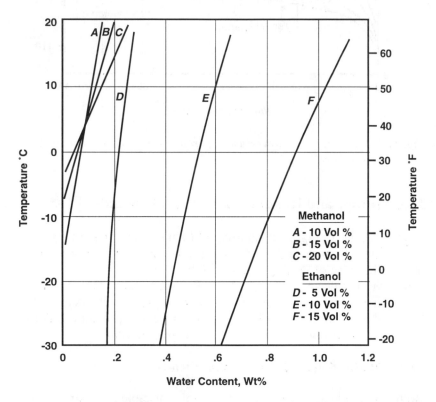

Fig. 11.7. Water tolerance of methanol and ethanol blends in gasoline of 26 vol. % aromatic content.[9]

leaded gasoline having an aromatics content of 26% by volume. No cosolvent is present.

It can be seen that ethanol/gasoline blends have a much greater water tolerance than methanol/gasoline blends. It is also seen that water tolerance tends to increase with methanol content at temperatures above about 2°C, and to decrease with increasing methanol concentration below this temperature.[9]

Figure 11.8 shows how water tolerance varies with aromatic content and methanol-ethanol ratio in blends containing 10% alcohol.[9] Figure 11.9 illustrates how the use of a higher alcohol acting as cosolvent increases the water tolerance.[9] Water tolerence curves for TBA and methanol mixtures, which have been widely used in practice, are shown in Figure 11.10.[14]

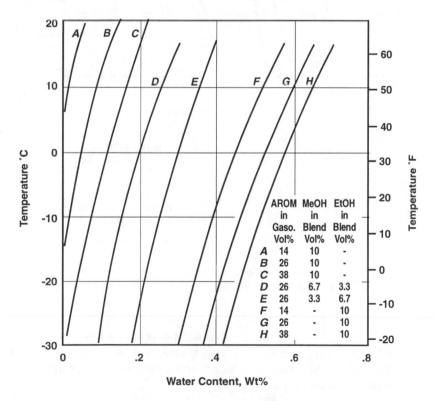

Fig. 11.8. Water tolerance of blends containing 10 vol. % alcohol in gasolines of different aromatic content.[9]

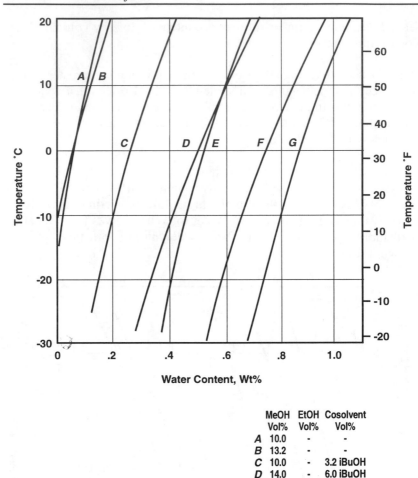

	MeOH Vol%	EtOH Vol%	Cosolvent Vol%
A	10.0	-	-
B	13.2	-	-
C	10.0	-	3.2 iBuOH
D	14.0	-	6.0 iBuOH
E	-	10.0	-
F	-	13.2	-
G	-	10.0	3.2 iBuOH

Fig. 11.9. Effect of cosolvents on water tolerance of methanol and ethanol blends in 26% aromatic gasoline.[9]

The use of certain surfactant additives to prevent or reduce the tendency for phase separation in alcohol mixtures with gasoline has been investigated[15] and found to reduce significantly the temperature at which phase separation occurs.

Aromatics concentration in base fuel = 26% volume

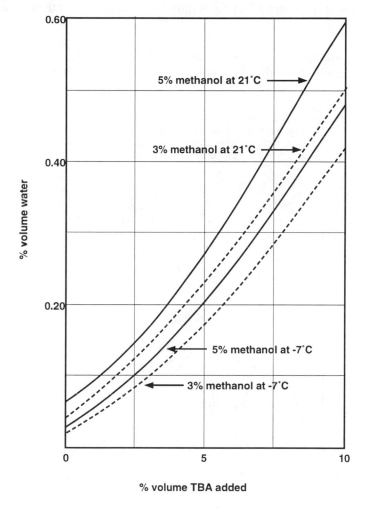

Fig. 11.10. Effect of TBA on water tolerance of fuels containing up to 5% volume methanol.[14]

Water bottoms growth is a feature of alcohol blends in storage tanks. The water in the bottom of a gasoline tank is normally left behind every time gasoline is discharged, so that if alcohol is present in the gasoline, the level of water bottoms will increase as alcohol enters the water phase from successive batches of gasoline. Eventually, the aqueous phase will contain so much alcohol that it becomes soluble in the gasoline so that the water bottoms will gradually disappear. This is illustrated in Figure 11.11 for a number of blends of methanol and TBA.[13]

Once the storage tank has been "dried out" in this way, no further losses of methanol will occur, but care must be taken when this drying

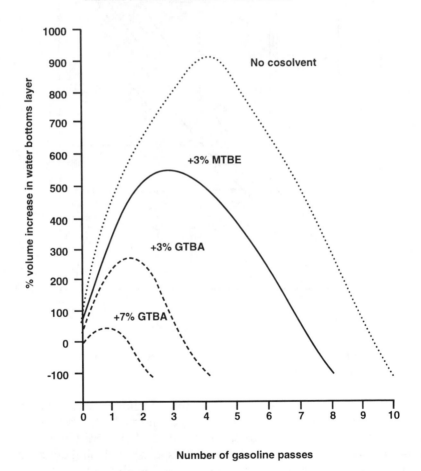

Notes:
GTBA is gasoline-grade tertiary butyl alcohol
Multi-passes of gasoline through tank
Original tank water = 0.5% of gasoline volume
Gasoline aromatics content = 26% volume

Fig. 11.11. Effect of cosolvents on water bottoms growth for gasoline containing 3% methanol.[13]

out process is taking place to avoid discharging some of the aqueous phase at the same time as the gasoline.

Haziness is another form of phase separation in which separated water droplets are in suspension. If water is forced out of solution for any reason such as a reduction in ambient temperature or the addition during blending of a component having a low aromatics content, it will usually appear as a haze which will gradually disperse as the droplets agglomerate and sink to the bottom of the tank. If surfactant additives are present, such hazes can be very stable and require the use of demulsifiers to clear them (see Section 9.2.6).

Only a small amount of water usually accumulates in a vehicle's fuel tank, and this is much less of a problem when alcohol is present because the alcohol will eventually dry out the tank completely. Sometimes alcohol is added to a vehicle's fuel tank solely for this purpose. Gross amounts of water in the fuel tank, however, can cause problems if alcohol fuels are used because the increase in water bottoms may cause the engine to take in a blend of methanol and water.

11.4.2 Safety and Fire Protection

Underground leaks of gasoline/alcohol blends are particularly undesirable because the alcohol will separate from the gasoline and dissolve in ground water. It cannot be separated from the water using conventional gasoline techniques such as gravity separation. There is evidence, however, that the alcohol is biodegraded.[16]

In general, the handling of blends containing alcohols or ethers from a safety standpoint is no different from that of a conventional gasoline (see Chapter 4, Section 4.5). Care should be taken to minimize skin contact and the breathing of vapors, and to avoid ingestion. Any spills should be cleaned up promptly and work areas should be well ventilated.[17]

All types of gasoline/alcohol or gasoline/ether fires can be extinguished with dry chemical, carbon dioxide or Halon 1214 and 1301 fire extinguishers. Suitable fire fighting foams are also available.[18]

11.5 Effect of Gasoline/Oxygenate Blends on Vehicle Performance

11.5.1 Road Antiknock Performance

The high-octane blending values of oxygenates make them attractive to refiners and it is important, therefore, to establish if these octane benefits are seen by vehicles on the road. There has been some question as to whether the conventional octane parameters of RON and MON give a reliable guide to road octane performance when some oxygenates, and particularly ethanol, are used in gasoline.[19,20] However, other work has shown a satisfactory correlation.[21]

A CRC program[21] in the U.S. showed that in regular-grade fuels at full throttle, all the oxygenates tested, which included both alcohols and ethers, improved Road Octane performance relative to hydrocarbon blending components, although no trend was readily apparent in premium fuels. High-speed road octane performance with some oxygenates, and particularly methanol, has been shown to be relatively poor.[22] MTBE does not appear to show this deterioration in performance at high speed.[23]

A comprehensive study in Europe using a wide range of different oxygenates in gasolines blended to constant RON and MON levels showed the following:[24]

- At low olefin levels (10%), accelerating knock performance of all oxygenate-containing fuels was better than that of the hydrocarbon-only fuel of the same RON and MON. For most oxygenates, when used at levels near the maximum oxygen content allowable, there was a reduction in octane requirement of around 0.5 number. However, at high olefin levels (20%), the reverse was true.

- At constant speed the oxygenated fuels were generally inferior to the corresponding hydrocarbon fuel, increasing octane requirements by an average of about one number. Ethanol and methanol give the biggest increase, and MTBE and mixtures of ethers appear to have only a marginal effect on octane requirement.

It is only possible to generalize on the effect of oxygenates because, as with all octane work, the actual effect will depend on the vehicle, the composition of the fuel including whether it is leaded or unleaded, and the method of test.

11.5.2 Driveability

Because oxygenate-containing fuels effectively lean the mixture strength and, thus, directionally bring it closer to the lean limit of combustion, for most vehicles oxygenates somewhat worsen cold starting and cold weather driveability.[25]

Cold starting depends on vaporization of the gasoline front end, and when an alcohol such as methanol is present, the vapor contains a greater concentration of alcohol than would be expected from the vapor pressure of the alcohol or its concentration in the gasoline. In addition, the heat of vaporization is higher for alcohols than for hydrocarbons, so more heat is needed to vaporize blends containing them. All these factors indicate that cold starting of vehicles on alcohol blends should be more difficult than with conventional gasolines, and test work confirms this.[26] The addition of ethers, on the other hand, appears to slightly improve cold starting characteristics.[8,11]

A considerable number of test programs have been carried out to assess the influence of oxygenates on cold and moderate temperature driveability because drivers readily notice a worsening of performance under these conditions.

Vehicles not fitted with exhaust emission controls such as catalytic converters will show an inferior driveability performance on gasolines containing oxygenates, particularly if they are set to operate as lean as possible for fuel economy and emissions reasons. The results of such tests are shown in Figure 11.12, where there is an increase in warm-up time (see Chapter 7, Section 7.3) at the same volatility when oxygenates are present.[24] The extent of the increase will depend on vehicle design, oxygenate content, ambient temperature, fuel volatility and test method.

In tests carried out in the U.S. with both open-loop (where the fuel delivered to the engine is proportional to the air consumption and there is no system to compensate for fuel differences) and closed-loop

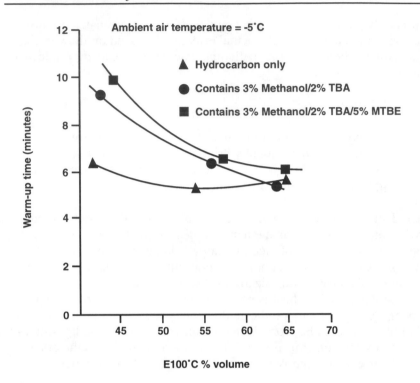

Fig. 11.12. Cold weather performance in terms of warm-up time (carburetted vehicles).[24]

(where there is a feedback system to maintain engine operation at the required air-fuel ratio) fuel control systems, poorer driveability was obtained with blends containing alcohol,[22,27,28] even when the volatilities were matched. As the molecular weight of alcohols and ethers increases, their adverse effect on driveability diminishes because their properties get closer to those of the corresponding hydrocarbons of the same carbon number. MTBE has a blending RVP of about 9 and its effect on the distillation curve is quite small. ETBE even seemed to improve cold weather driveability in one test[11] but in another not to have a noticeable effect.[8]

Excessive front-end volatility can cause poor hot weather driveability problems such as vapor lock, and it was shown in Section 11.3.2 that both RVP and E70 are increased if alcohols such as methanol are added to gasoline, but that the RVP test procedure used influences the result obtained. In most cases, the front-end volatility is controlled to match the specification requirements, and in cases where methanol is used, this can mean backing out butane. The other parameter used to predict gasoline performance at high temperatures is V/L ratio. However the calculated V/L ratio using RVP and distillation characteristics (see Chapter 7, Section 7.1.3) does not agree with measured values when alcohols are present mainly because of the distortion of the distillation curve, although the correlation is quite good with MTBE.[29]

Test work has shown that where the volatility of the gasoline/oxygenate blend is matched with that of a hydrocarbon-only gasoline, there is no significant difference in vapor lock performance between them.[27,30,31,32] Tests with European vehicles also confirm that, for fuels of constant front-end volatility, there is no significant difference in hot weather driveability performance regardless of the oxygenate used within the allowable limits, with the exception of methanol when used without cosolvent at a concentration of 5 percent by volume.[24]

11.5.3 Materials Compatibility

Corrosive attack on some construction materials used in the gasoline distribution system is a real possibility unless corrosion inhibitors are used (see Section 9.2.3). Fiberglass-reinforced plastic tanks may cause problems depending on the type of resin used during fabrication.[16] The materials used in vehicle fuel systems are very diverse and the potential for attack is quite high unless they are selected carefully.

Such materials as terne plate (lead/tin-coated steel used in fuel tank construction), zinc die castings, and aluminium fuel system components are all attacked by alcohols and need corrosion inhibitors to minimize this effect. Corrosion of steel is accelerated by the presence of alcohols in the fuel, partly because of the increased water content of the fuel and partly because of the organic acids that can be present in commercial oxygenates.

Both methanol and ethanol in gasoline can cause elastomers to swell and lose tensile strength. This can give rise to failures of critical components such as fuel pumps, accelerator pumps, and hoses.[33,34,35,36] The aromatic content of a gasoline alcohol blend is important because the higher the aromatics, the more the swelling for most elastomers.[37,38,39] An example of this is shown in Figure 11.13[38] for

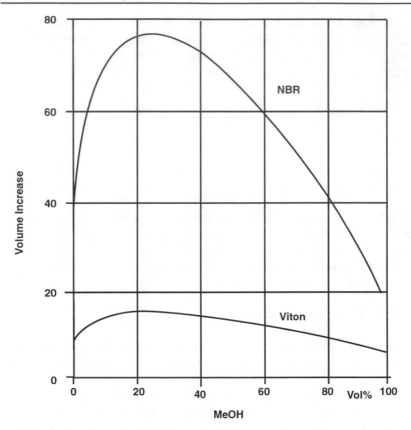

Fig. 11.13. Swelling effect of methanol blended in gasoline.[38]

methanol which also shows that as the alcohol content of the blend goes beyond its allowable concentration as a gasoline component, the amount of swelling starts to decrease.

Ethers cause many fewer materials compatibility problems—no significant corrosion, and the swelling of polymeric and elastomeric parts appears to be no greater than with typical fuels in current use.[8]

11.5.4 Intake System Deposits

Ice formation in the intake system (see Chapter 7, Section 7.7 and Chapter 9, Section 9.3) can be strongly influenced by oxygenates. Alcohols act as cryoscopic anti-icing additives and will normally prevent icing completely, although the increase in volatility that occurs with methanol could act in the opposite direction. Some work has been carried out[40] which shows that ice deposition is still possible even when methanol is present. Ethers, on the other hand, have so low a solubility in water that they have virtually no effect as anti-icing additives except, perhaps, due to any change in volatility that may occur as a result of their addition. MTBE may increase the volatility of a low-volatility gasoline and if this is not compensated for in the blend, then such a gasoline may be more susceptible to icing problems.

There is evidence that intake system deposits, particularly in the valve and manifold area, are increased by the use of oxygenates[4,24,41,42] as shown by Figure 11.14, which also shows that additives are effective in overcoming these deposits. Alcohols are known to require a higher additive treat to overcome the deposits than would be required for a hydrocarbon-only gasoline.[43,44]

Some ethers might be expected to worsen fuel stability in view of their well-known tendency to form potentially dangerous peroxides on exposure to air. However, neither MTBE nor TAME have shown any tendency to peroxide formation[43,45] although DIPE (di-iso-propylether) on its own is known to form peroxides and it would be expected that ETBE would also be likely to do so because of the labile methylene hydrogen atoms.[8] Gum and peroxide formation, however, both in storage and in use, are of concern in that decreased induction periods and higher unwashed gums have been reported for gasoline/ ether blends,[43] although the use of additives such as antioxidants overcomes this problem.

Increases in gum on storage have been reported for alcohol blends as well as for ethers.[46] These blends also increase the solubility of the gum in the gasoline and this may partly account for the increased intake system deposit-forming tendency. Gum-bound deposits of rust and other types of sediment can be loosened by dissolving the gum, thus giving rise to suspended material which can block filters. This is a particular problem when existing fuel tankage is converted from hydrocarbon-only duty to the storage of alcohol/gasoline blends.

11.5.5 Additives for Use with Oxygenated Fuels

There is evidence to indicate that not only are oxygenated fuels more likely to form deposits, but also that surfactant additives to overcome

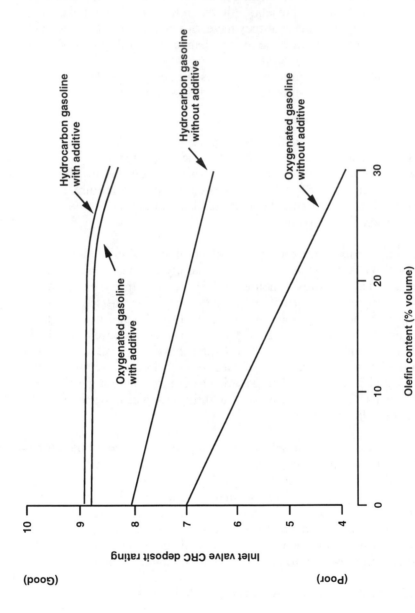

Fig. 11.14. Influence of oxygenates and additives on inlet system cleanliness (unleaded gasoline).[24]

these deposits are less effective than in hydrocarbon-only fuels. This is probably because surfactant additives function by fastening themselves to polar surfaces and, because the presence of oxygenates increases the polarity of the fuel, there is a greater tendency for them to remain in solution rather than to migrate to the surfaces.

The corrosive properties of gasolines containing alcohol have already been mentioned and, although higher treats than normal of surfactant anticorrosion additives are required for protection, such additives are generally effective.

The types of polar groups used in surfactant additives that are suitable as intake system cleansers are the carbamides, carbimides and alkylamines, although they need to be formulated with auxiliary additives to ensure they are satisfactory under all conditions.[40] Polybuteneamines and polybutenesuccinimides, in particular, have been shown to be effective together with polyethers to keep intake valves clean, although some additives designed to keep fuel injectors clean have been shown to worsen valve deposits when ethanol is present in the gasoline.[44]

Many additive manufacturers will specify specific additive packages designed especially for oxygenated fuels.[47]

11.5.6 Environmental Aspects of Fuels Containing Oxygenates

Both exhaust and evaporative emissions can be influenced by the use of oxygenates in gasoline, and a number of studies have been carried out[48,49,50,51,52] to establish the extent of the effect which depends on the type of vehicle and the emission-control equipment that is fitted.

The leaning effect on the air-fuel ratio when low levels of oxygenates are used can mean that, for vehicles with no exhaust gas treatment, both CO and to a small extent hydrocarbons emitted from the exhaust are reduced with no significant change in NO_x. A typical reduction in CO of 30% has been found with a fuel oxygen content of 2%.[49] However, if the vehicle is designed to operate close to the lean limit of combustion on hydrocarbon fuels, misfire may occur when oxygenates are present under certain driving regimes with a consequent increase in unburnt hydrocarbons.

With vehicles equipped with oxidation catalysts, the effect of oxygenates found in one study indicated reductions in CO and HC and a slight increase in NO_x, as shown in Figure 11.15.[49] These data were obtained using 1978 to 1980 U.S. vehicles and a wide range of different oxygenates.

For vehicles fitted with 3-way catalyst, closed-loop systems, substantial reductions in CO have been found, as shown in Table 11.4, for tests on a range of U.S., European and Japanese vehicles operating on fuels with and without MTBE:[49]

Table 11.4 Reduction in Emissions of Catalyst Cars with MTBE

Vol % MTBE	% Reduction in:		
	CO	HC	NO_x
11	20	6	8
16	26	14	7

The above results are a little surprising since a closed-loop system would be expected to compensate for changes in air-fuel ratio. They are probably explained by the fact that in test cycles these systems are open-loop until the catalyst has warmed up and during some accelerations, so that the effect on emissions may depend on the type of drive cycle and the testing time. It is also possible that combustion is improved by the presence of oxygenates. CO emissions tend to decrease in proportion to the oxygen content of the fuel up to 3.5% oxygen.[53]

The reduction in emissions obtained by the use of oxygenates in gasoline has been used in the U.S. to improve air quality by requiring some gasolines to have minimum oxygen contents depending on the time of year.[54]

Other unregulated exhaust gas pollutants that have been studied and that are influenced by oxygenates in the fuel include 1,3 butadiene, benzene, and other aromatics including polynuclear aromatics (PNAs), formaldehyde and acetaldehyde.[55] Benzene from the exhaust is a direct function of the benzene and other aromatics content of the gasoline[56,57] and so the use of oxygenates improves this situa-

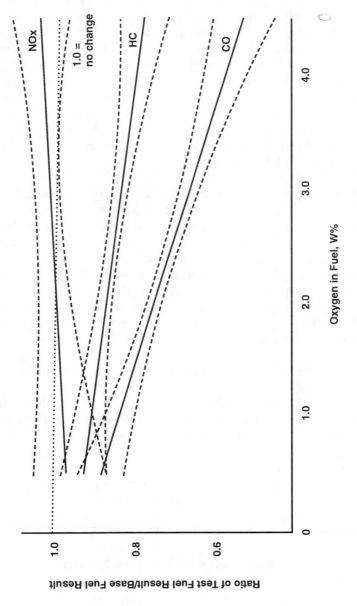

Fig. 11.15. Effect of oxygen content of fuel on exhaust emissions of oxidation catalyst vehicles.[49]

tion by dilution as well as by the general reduction in exhaust hydro-carbons. PNAs and particulates are very low from spark ignition engines, particularly when fitted with a catalyst, but reductions varying from 12 to nearly 50% have been demonstrated[58] by the use of oxygenates in gasolines. Aldehydes are increased in the exhaust gases when oxygenates are present in the gasoline[50,55,59] mainly because of the leaning effect, so that the increase is proportional to the oxygen content of the fuel. Oxygenates can be regarded chemically as immediate precursors of aldehydes and this also plays a part in giving higher aldehyde yields than with hydrocarbon-only gasolines. For fuels containing up to about 3.7% of oxygen, the increase in aldehydes is typically between 5 and 20%.

Regarding evaporative emissions, these are strongly dependent on the RVP of the fuel[60,61,62] and any increase due to the presence of oxygenates such as methanol or ethanol will give a corresponding increase in evaporative emissions.[48,63,64] However, fuels of the same volatility, whether they contain oxygenates or not, appear to give about the same or lower evaporative emissions.[65,66] Alcohol and ether vapors are preferentially adsorbed by charcoal as compared to hydrocarbon vapors.[67] Work has been carried out which suggests that the working capacity of a canister is reduced,[68] although there are divergent views on the long-term effects.[64,69,70]

11.5.7 Effect of Oxygenated Fuel Blends on Fuel Economy

Because fuels containing oxygenated components have lower calorific values than conventional gasolines, it would be expected that they would show an inferior fuel economy. In practice, losses of fuel economy for this reason are at least partly offset by a gain in energy efficiency mainly due to the leaning of the air-fuel ratio. In addition, the higher blending density of most types of oxygenate tends to counter possible fuel economy penalties so that for oxygen contents within the allowable limits for unmodified vehicles, losses in economy are very small.[22,24,71,72,73,74]

11.5.8 Effect of Oxygenated Fuel Blends on Lubricant Performance

Some concerns have been expressed as to whether the use of alcohols in fuels would increase engine wear, particularly as methanol on its own is known to cause a substantial increase in wear.[75,76] Test work, however, has shown[4,77] that gasoline containing methanol and TBA up to a 3.5% oxygen level does not give a perceivable increase in engine wear with commercially available lubricants. It is concluded that specially formulated lubricants are not necessary up to this level of oxygenate addition.

References

1. CONCAWE, "Trends in Motor Vehicle Emission and Fuel Regulations — 1994 Update," Report 4/94, CONCAWE, The Hague, 1994.

2. H. Menrad, "The Part of Volkswagen in the German Program for Research on Alcohol Fuels," Fifth International Automotive Progression Systems Symposium, Dearborn, Michigan, April 1980.

3. K. Reders, "M 15 Kraftstoffqualitatsuberwachung im Flottentest," <u>Entwicklungslinien in Kraftfahrzeugtechnik und Strassenverkehr</u>, Verlag TUV Rheinland, pp. 86-91, 1980.

4. H. Menrad and B. Nierhauve, "Engine and Vehicle Concepts for Methanol-Gasoline Blends," SAE Paper No. 831686, 1983.

5. American Petroleum Institute, "Alcohols and Ethers ... A Technical Assessment of Their Application as Fuels and Fuel Components," API publication N. 4261, July 1988.

6. SAE Information Report, "Alternative Automotive Fuels," SAE J1297, Table 2, June 1985.

7. K. Owen, "Fuels for Spark Ignition Engines," <u>Modern Petroleum Technology</u>, edited by G.D. Hobson, p. 816, John Wiley & Sons, 1984.

8. C.M. Shiblom, G.A. Schoonveld, *et al.*, "Use of Ethyl-t-Butyl Ether (ETBE) as a Gasoline Blending Component," SAE Paper No. 902132, 1990.

9. J.L. Keller, "Methanol and Ethanol Fuels for Modern Cars," API Symposium on Fossil Fuels for the 1980s, 1979.

10. R.L. Furey, "Volatility Characteristics of Gasoline-Alcohol and Gasoline-Ether Blends," SAE Paper No. 852116, 1985.

11. J. Kivi, A. Niemi, *et al.*, "Use of MTBE and ETBE as Gasoline Reformulation Components," SAE Paper No. 922379, 1992.

12. R.L. Furey and K.L. Perry, "Volatility Characteristics of Blends of Gasoline with Ethyl Tertiary-Butyl Ether (ETBE)," SAE Paper No. 901114, 1990.

13. ARCO Chemical Europe Inc, "Technical Bulletin, Oxinol 50, Gasoline Blending Component," Atlantic Richfield Company, 1985.

14. ARCO Chemical Europe Inc, "Technical Bulletin, GTBA, Gasoline Grade Tertiary Butyl Alcohol," Atlantic Richfield Company, 1986.

15. E.J. Smith and D.R. Jordan, "The Use of Surfactants in Preventing Phase Separation of Alcohol Petroleum Fuel Mixtures," SAE Paper No. 830385, 1983.

16. J.T. Novak, C.D. Goldsmith, R.E. Benoit and J.H. O'Brien, "Biodegradation of Methanol and Tertiary Butyl Alcohol in Subsurface Systems," International Seminar on Degradation, Retention and Dispersion of Pollutants in Groundwater, Copenhagen, 1984.

17. American Petroleum Institute, "Storage and Handling of Gasoline-Methanol/Co-Solvent Blends at Distribution Terminals and Service Stations," API Recommended Practice 1627, August 1986.

18. National Fire Protection Association, "Foam Extinguishing Systems and Combined Agent Systems," NFPA 11.

19. P.D. Histon and R.T. Roles, "The Road Antiknock and Preignition Characteristics of Gasoline Containing Oxygenates," 5th Alcohol Fuel Symposium, New Zealand, 1982.

20. K. Campbell and T.J. Russell, "The Effect on Gasoline Quality of Adding Oxygenates," Associated Octel Publication OP82/1, April 1982.

21. Coordinating Research Council, "Fuel Rating Program — Road Octane Performance of Oxygenates in 1982 Model Cars," CRC Report No. 541, July 1985.

22. N.D. Brinkman, N.E. Gallapoulos and M.W. Jackson, "Exhaust Emissions, Fuel Economy and Driveability of Vehicles Fueled with Alcohol-Gasoline Blends," SAE Paper No. 750120, 1977.

23. B. Tanaguch and R.T. Johnson, "MTBE for Octane Improvement," *Chemtech*, pp. 502-510, August 1979.

24. G.J. Lang and F.H. Palmer, "The Use of Oxygenates in Motor Gasolines," *Gasoline and Diesel Fuel Additives*, Edited by K. Owen, Critical Reports on Applied Chemistry, No. 25, pp. 147 - 152, John Wiley & Sons, 1989.

25. K. Kanehara, N. Sasajima, *et al.*, "Analyzing the Influence of Gasoline Characteristics on Transient Engine Performance," SAE Paper No. 912392, 1991.

26. M.F. Bardon, D.P. Gardiner and V.K. Rao, "Cold Starting Performance of Gasoline/Methanol M10 Blends in a Spark Ignition Engine," SAE Paper No. 850214, 1985.

27. Coordinating Research Council, Inc., "Performance Evaluation of Alcohol-Gasoline Blends in 1980 Model Automobiles — Phase II — Methanol-Gasoline Blends," CRC Report No. 536, January 1984.

28. Coordinating Research Council Inc., "1984 CRC Intermediate Temperature Driveability Program Using Gasoline-Alcohol Blends," CRC Report No. 554, August 1987.

29. P. Dorn, A.M. Mourao and S. Herbstman, "The Properties and Performance of Modern Automotive Fuels," SAE Paper No. 861178, 1986.

30. Coordinating Research Council Inc., "Performance Evaluation of Alcohol-Gasoline Blends in 1980 Model Automobiles — Phase I — Ethanol-Gasoline Blends," CRC Report No. 527, July 1982.

31. P.A. Yaccarino, Hot Weather Driveability and Vapor-Lock Performance with Alcohol-Gasoline Blends, SAE SP-638, October 1985.

32. Coordinating Research Council Inc., "1983 CRC Two-Temperature Vapor Lock Program Using Alcohol-Gasoline Blends," CRC Report No. 550, October 1986.

33. J.R. Dunn and H.A. Pfisterer, "Resistance of NBR-Based Fuel Hose Tube to Fuel-Alcohol Blends," SAE Paper No. 800856, 1980.

34. A. Nersasian, "The Volume Increase of Fuel Handling Rubbers in Gasoline/Alcohol Blends," SAE Paper No. 800789, 1980.

35. A. Nersasian, "The Use of Toluene/Isooctane/Alcohol Fuels to Simulate the Swelling Behavior of Rubber in Gasoline/Alcohol Fuels," SAE Paper No. 800790, 1980.

36. B. Spoo, "High Performance Fuel Line for Emerging Automotive Needs," SAE Paper No. 800787, 1980.

37. I.A. Abuisa, "Elastomer-Gasoline Blends Interactions I. Effects of Methanol-Gasoline Mixtures on Elastomers," *Rubber Chemistry and Technology*, Vol. 56, No. 1, p. 155, 1983.

38. I.A. Abuisa, "Elastomer-Gasoline Blends Interactions II. Effects of Ethanol-Gasoline and Methyltbutyl Ether-Gasoline-Mixtures on Elastomers," ibid., p. 169.

39. I.A. Abuisa and M.E. Myers, Jr., "Elastomer-Solvent Interactions III. Effects of Methanol on Fluorocarbon Elastomers," American Chemical Society Symposium on Solubility Parameters, Philadelphia, Pennsylvania, August 1984.

40. F. Hovemann, "Additives for Oxygenated Fuels," EFOA Second Conference, Rome, 1987.

41. A. Shiratori and K. Saitoh, "Fuel Property Requirements for Multiport Fuel Injector Deposit Cleanliness," SAE Paper No. 912380, 1991.

42. K. Ohsawa, Y. Nomura, *et al.*, "Mechanism of Intake Valve Deposit Formation Part III: Effects of Gasoline Quality," SAE Paper No. 922265, 1992.

43. J. Chase and H. Woods, "Processes for High Octane Oxygenated Gasoline Components," Symposium on Octane in the 1980s, ACS, Division of Petroleum Chemistry, Florida, 1978.

44. C.M. Shilbolm and G.A. Schoonveld, "Effects on Intake Valve Deposits of Ethanol and Additives Common to the Available Ethanol Supply," SAE Paper No. 902109, 1990.

45. R. Reynolds, J. Smith and I. Steinmetz, "Methyl Ethers as Motor Fuel Components," ACS National Meeting, Division of Petroleum Chemistry, Atlantic City, New Jersey, 1974.

46. J.L. Keller, G.M. Nakaguchi and J.C. Ware, "Methanol Fuel Modification for Highway Vehicle Use — Final Report," HCP/W368318, U.S. Dept. of Energy, July 1978.

47. P. Polss, "Du Pont Additives for Gasoline-Alcohol Fuels," Du Pont Petroleum Chemicals Division, Report No. PLMR2584, October 1984.

48. R.L. Furey and J.B. King, "Evaporation and Exhaust Emissions from Cars Fueled with Gasoline Containing Ethanol or MTBE," SAE Paper No. 800261, 1980.

49. M.A. Mays, "Exhaust and Evaporative Emission Studies with Fuels Containing Oxinol Blending Component and MTBE," EFOA Second Conference on Fuel Oxygenates, Rome, October 1987.

50. P. Garibaldi, "Fuel Oxygenates: The Right Solution for Cleaner Cars and a Better Environment," EFOA Second Conference on Fuel Oxygenates, Rome, October 1987.

51. F.D. Stump, K.T. Knapp, *et al.*, "The Seasonal Impact of Blending Oxygenated Organics with Gasoline on Motor Vehicle Tailpipe and Evaporative Emissions—Part II," SAE Paper No. 902129, 1990.

52. M.T. Noorman, "The Effect of MTBE, DIPE and TAME on Vehicle Emissions," SAE Paper No. 932668, 1993.

53. K. Nelson, R. Ragazzi and G. Gallagher, "The Effects of Fuel Oxygen Concentration on Automotive Carbon Monoxide Emissions at High Altitudes," SAE Paper No. 902128, 1990.

54. State of Colorado Regulation No. 13, "The Reduction of Carbon Monoxide Emissions from Gasoline Powered Motor Vehicles Through the Use of Oxygenated Fuels."

55. M.A. Warner-Selph and C.A. Harvey, "Assessment of Unregulated Emissions from Gasoline Oxygenated Blends," SAE Paper No. 902131, 1990.

56. CRCAPRAC Project No. CAPE3583, "Automotive Benzene Emissions," September 1985.

57. N. Zaghini, *et al.*, "Polynuclear Aromatic Hydrocarbons in Vehicle Exhaust Gas," SAE Paper No. 730836, 1973.

58. Colorado Dept. of Health, Air Pollution Control Division, Aurora Vehicle Emissions Technical Center, "The Effect of Two Different Oxygenated Fuels on Exhaust Emissions at High Altitude," January 1986.

59. J.M. Biren, A. Milhau and O. Boulhol, "The Evaluation of the Environmental Impact of Fuel Oxygenates in Gasoline," EFOA Second Conference on Fuel Oxygenates, Rome, October 1987.

60. B.H. Eccleston, B.F. Noble and R.W. Hurn, "Influence of Volatile Fuel Components on Vehicle Emissions," U.S. Bureau of Mines Report No. 7291, 1970.

61. B.H. Eccleston and R.W. Hurn, "Effect of Fuel Front-End and Mid-Range Volatility on Automobile Emissions," U.S. Bureau of Mines Report No. 7707, 1972.

62. J.S. McArragher, *et al.*, "Evaporative Emissions from Modern European Vehicles and Their Control," SAE Paper No. 880315, 1988.

63. K.R. Stamper, "Evaporative Emissions from Vehicles Operating on Methanol/Gasoline Blends," SAE Paper No. 801360, 1980.

64. T.M. Naman and J.R. Allsop, "Exhaust and Evaporative Emissions from Alcohol and Ether Fuel Blends," SAE Paper No. 800858, 1980.

65. E.N. Cantwell and W.E. Smith, "Volatility Characteristics of Alcohol/Gasoline Blends and Automotive Evaporative Emissions," Du Pont de Nemours Report No. PLMR1284, 1984.

66. S.R. Reddy, "Evaporative Emissions from Gasolines and Alcohol-Containing Gasolines with Closely Matched Volatilities," SAE Paper No. 861556, 1986.

67. R.L. Furey and J.B. King, "Evaporative and Exhaust Emissions from Cars Fueled with Gasoline Containing Ethanol or Methyl Tertbutyl Ether," SAE Paper No. 800261, 1980.

68. P.A. Gabele, J.O. Baugh and F. Black, "Characterization of Emissions from Vehicles Using Methanol and Methanol-Gasoline Blended Fuels," *J. of the Air Pollution Control Ass.*, Vol. 35, pp. 1168-1175, 1985.

69. K.R. Stamper, "Evaporative Emissions from Vehicles Operating on Methanol/Gasoline Blends," SAE Paper No. 801360, 1980.

70. J.M. DeJovine, *et al.*, "Material Compatibility and Durability of Vehicles with Methanol/Gasoline Grade Tert. Butyl Alcohol Gasoline Blends," ARCO Chemical Company, October 1984.

71. E.E. Wigg and R.S. Lunt, "Methanol as a Gasoline Extender — Fuel Economy, Emissions and High Temperature Driveability," SAE Paper No. 741008, 1974.

72. J.C. Ingamells and R.H. Lindquist, "Methanol as a Motor Fuel or a Gasoline Blending Component," SAE Paper No. 750123, 1975.

73. A.W. Preuss, "Energy Efficiency of Oxygenates from Their Production to Their Engine Use," SAE Paper No. 830384, 1983.

74. T.O. Wagner and H.L. Muller, "Experience with Oxygenated Fuel Components," API Paper 82000041, 49th Mid-year Refining Meeting, American Petroleum Institute, May 1984.

75. E.C. Owens, *et al.*, "Effects of Alcohol Blends on Engine Wear," USAFLRL, SWRI, SAE, June 1980.

76. W.H. Baisley and C.F. Edwards, "Wear Characteristics of Fleet Vehicles Operating on Methyl Alcohol," University of Aukland and University of Santa Clara, SAE, October 1981.

77. J.M. DeJovine, D.A. Drake and M.A. Mays, "Test Stand Evaluations of Commercial Lubricants with Methanol and Tertiary Butyl Alcohol Blends," SAE Paper No. 830242, 1983.

Further Reading

1. American Petroleum Institute, "Alcohols and Ethers ... A Technical Assessment of Their Application as Fuels and Fuel Components," API publication N. 4261, July 1988.

2. G.J. Lang and F.H. Palmer, "The Use of Oxygenates in Motor Gasolines," *Gasoline and Diesel Fuel Additives*, Edited by K. Owen, Critical Reports on Applied Chemistry, No. 25, pp. 147 - 152, John Wiley & Sons, 1989.

3. A.F. Talbot, "Alkyl Ethers as Motor Fuels," API Symposium, "Update on Octane Improvement," 1979.

4. N.E. Gallopoulous, "Alcohols for Use as Motor Fuels and Motor Fuel Components," CEC Second International Symposium on The Performance Evaluation of Automotive Fuels and Lubricants, Wolfsburg, West Germany, June 1985.

Chapter 12

Influence of Gasoline Composition on Emissions

12.1 Legislated Changes

In the U.S. the first legislated national exhaust emission limits for cars were set by the Clean Air Act of 1968, and these were first amended in 1970 and again in subsequent years to become progressively more severe. In California the limits are even more stringent and require the gradual introduction of Transitional Low Emission (TLEV), Low Emission (LEV), Ultra Low Emission (ULEV) and, by the end of the century, Zero Emission (ZEV) vehicles. The requirements for these vehicles and the legislation governing vehicle emissions for the U.S. and other countries are summarized in Appendix 3, with details of the test cycles and procedures used in Appendix 4.

Until comparatively recently in the U.S. it has fallen mainly on the motor industry to meet the emissions standards for new cars, but now there are mandatory requirements for gasoline composition, in addition to the early controls on lead content, to help meet the latest clean air requirements. This regulation on gasoline composition for environmental reasons is spreading to other countries.

The ever increasing importance of this subject means that work is still being actively carried out, so that it is impossible to give a final and definitive view. However, much of the really fundamental work, from a fuel effect on emissions viewpoint, has already been carried out and investigations now in progress are directed more at finding the relative importance of each of the fuel variables on emissions in different engine designs and different vehicle populations.

It is interesting that one of the unexpected benefits of automotive emission control has been the reduction in fatalities by motor vehicle exhaust gas in the U.S. The decrease in accidents and avoided suicides together with the projected cancer risk avoidance show that, in terms of mortality, the unanticipated benefits could outweigh the intended benefits.[1]

12.2 Overall Fuel Effects on Emissions

Previous chapters have already discussed many of the gasoline parameters that influence both exhaust and evaporative emissions from vehicles. Thus Chapter 7, Section 7.5 discusses the influence of fuel volatility on evaporative emissions, and Section 7.6 covers its effect on exhaust emissions. Chapter 9, Sections 9.3 and 9.4 discuss additives that can influence emissions, and Chapter 11, Section 11.5.5 briefly covers the effect of oxygenates on emissions. In this chapter the effect of compositional factors will be discussed. These are summarized in Table 12.1 on a "well to wheels" basis, i.e., taking into account all the emissions and costs associated with making changes to gasoline composition including the effect of increased energy consumption and CO_2 emissions at refineries due to extra conversion processes, etc.:[2]

Table 12.1. Compositional Effects on Exhaust/ Evaporative Emissions[2]

	CO	NO_X	Hydrocarbons Exhaust	Hydrocarbons Evaporative	Ozone	Air Toxins Benzene	Air Toxins Butadiene	Air Toxins Aldehydes	Total CO_2	Cost
Add oxygenates	↓↓	↑0	↓	↑0	0	↓	0	↑	↑↑↑	↑↑↑
Reduce Benzene	0	0	0	0	0	↓↓	0	0	↑	↑
Reduce Aromatics	↓	0?	0	0	0↑	↓	↑	↑	↑↑↑	↑↑↑
Reduce Olefins	0	↓	↑	↓	↓	0	↓	0	↑↑	↑↑
Reduce Sulfur**	↓	↓	↓↓	0	↓	↓	↓	↑	↑↑	↑↑
Reduce T90E	0	0	↓	0	↓	↓	↓	↓	↑	↑↑
Reduce RVP	0	0	↑	↓↓	↓	0	0	0	↑	↑
Increase Octane	0	0	0	0	0	0	0	0	↑*	↑

↓ = Decrease   ↑0 = Small or no effect   0 = No effect   ↑ = Increase

* = Above optimum value   ** = Catalyst cars only

It can be seen that whatever is done to the fuel, emissions of CO_2 increase and so do the manufacturing costs. It is also apparent that improvements in one type of emission will often lead to a worsening of other types of emission.

There are indications that advances in engine technology may reduce the effect that the fuel has on emissions.[2] For example, engine management equipment with adaptive learning systems can reduce the sensitivity to fuel effects. Also, since most emissions occur during the first few minutes of operation before the catalyst has warmed up enough for it to be effective, developments aimed at rapid light-off of the catalyst,[3,4,5] have been shown to have a very substantial effect in reducing vehicle emissions.

12.3 U.S. Studies—The Auto/Oil AQUIRP Program

In the U.S. a cooperative test program between 14 oil companies and 3 domestic automakers was set up in 1989 in order to develop data that would help meet the nation's clean air goals.[6,7,8] This large and comprehensive program, known as AQUIRP (Air Quality Improvement Research Program), involved extensive measurements of vehicle emissions using a carefully designed fuel matrix; air quality modeling studies to predict the effects of emissions on ozone formation; and the economic analysis of the fuel/vehicle systems.

Phase I of this program was designed to study gasoline compositional effects in order to establish some of the key factors that influence emissions. Phase II will investigate areas requiring further study that were identified in Phase I, but this work had not been completed at the time of writing. The major effects found in Phase I are summarized below.

12.3.1 Regulated Emissions (HC, CO, NO_x)

Two test fleets of cars were used in this investigation: a "current" fleet consisting of twenty 1989 model vehicles fitted with new emission technology and an "old" fleet of fourteen 1983-1985 model year vehicles mainly fitted with carburetted engines. The average effects on regulated exhaust emissions are given in Figure 12.1.

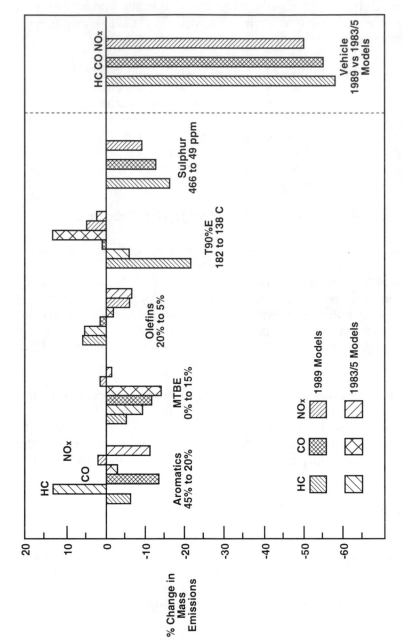

Fig. 12.1. U.S. auto/oil test program showing effects of fuel variables on regulated emissions.[2] *(Courtesy: CEC)*

Clearly the engine design and exhaust gas treatment system has the greatest effect, as shown by the 50-60% lower emissions of the 1989 vehicles versus the older 1983-85 models.

It was also found that decreasing the RVP of gasoline by 1 psi (from 9 to 8 psi) reduced evaporative emissions by 34%, exhaust CO by 9% and exhaust HC by 4%.

12.3.2 Air Toxins

In addition to the regulated emissions, air toxins such as benzene, 1,3 butadiene, formaldehyde and acetaldehyde were also determined. It was found that decreasing aromatic level had the greatest effect, reducing air toxins by 28% in the 1989 vehicles (of which benzene was 74%) and by 23% in the older models (where benzene comprised 56% of these emissions).

Adding oxygenates to the fuel decreased benzene emissions and increased aldehydes. However, the effect of MTBE on total exhaust toxins was not significant.

Olefin reduction from 20% to 5% reduced 1,3 butadiene emissions by 30% although the overall effect on air toxins was again not significant.

Lowering T90 from 360°F to 280°F dropped toxic emissions by 16% in the current fleet but was not significant in the older fleet.

Finally, sulfur reduction from 450 ppm to 50 ppm reduced total air toxins by 10%.

12.3.3 Air Quality (Ozone)

Air-quality modeling studies were conducted for New York City, Los Angeles, and Dallas-Fort Worth, taking into account changes in fuel/vehicle systems on exhaust, evaporative and running loss emissions as well as stationary source emissions associated with refueling and storage. Future projections for 1995 and 2005/2010 were also made.

It was predicted that there would be a drop in the peak ozone contribution of light-duty vehicles from between 28 and 37% in the 1980s to only 5 to 9% by 2005/2010. The fuel compositional changes which were foreseen to have the greatest effect were the reduction of olefins and T90 levels. Reducing RVP and sulfur also reduced predicted ozone.

12.4 European Program on Emissions, Fuels, and Engine Technologies (EPEFE)

In Europe a dialogue has now started between the European Auto and Oil Industries and the European Commission to find cost-effective solutions to overcome the air pollution resulting from motor vehicles and their fuels.[9] A European auto/oil research program has been designed that will extend the benefits of the American AQUIRP program to the European situation, since in Europe not only are fuels and engines quite different from those in the U.S., but also the test procedures. For example, the average aromatics content of U.S. gasolines in 1992 was about 32% versus about 40% in Europe and for olefins the figures were approximately 7% in Europe and 12% in the U.S. Test work is in progress on this program but no conclusions were available at the time of writing.

12.5 Other Studies on Fuel Effects on Emissions

There have been a great number of studies carried out on the influence of fuel quality on emissions and many of them have already been mentioned in the earlier chapters on volatility, oxygenates, additives, and so on. Not all of the studies agree with each other and this may be for a variety of reasons—as already discussed, different cars can respond to fuel changes in different ways, ambient temperature has an effect, and test procedures are important. However, in general terms, there are large areas of agreement with the AQUIRP studies.

Thus the use of oxygenates is consistently shown to improve CO and HC emissions[10,11,12,13,14,15] as does the reduction of aromatics[10,16,17] and of sulfur.[10] Increasing ambient temperature also decreases exhaust CO and HC.[11,18] High aromatics increases engine-out (i.e., before the catalytic converter) HC and NO_x emissions but reduced the tailpipe emissions of these gases.[10] In another test only the oldest vehicles showed a reduction in NO_x when aromatics were reduced from 30% to 10%—newer vehicles showed no significant response.[19]

Regarding effects on unregulated emissions such as benzene, 1,3 butadiene, etc., again there have been a number of different studies. MTBE in the fuel increased formaldehyde emissions, and ETBE increased acetaldehyde emissions but ethylene, 1,3 butadiene and N_2O emissions were not statistically different from those from a fuel containing aromatics rather than oxygenates.[20,21] Aromatics are the main precursors of exhaust benzene, due to dealkylation of substituted aromatics, but they reduce exhaust formaldehyde and acetaldehyde and increase aromatic aldehydes.[16] Increasing the olefin content of the fuel increases engine-out nitrogen oxide and 1,3 butadiene emissions.[16]

With respect to ozone forming potential, olefins have been shown to be more photochemically reactive than paraffins[22] so that from an evaporative emissions viewpoint, it is important to keep the light olefin content low. However, the situation regarding exhaust emissions is more complex since, although olefins are a major combustion product of paraffins,[23] they are more easily converted by the catalyst.[24,25] Tailpipe emissions from highly olefinic fuels do not necessarily therefore increase ozone forming potential over those found from paraffin fuels[26] although olefins are second to aromatics[27,28] in being the largest contributors to the specific ozone reactivity of tailpipe emissions.

12.6 Reformulated/Clean Gasolines

Reformulated gasolines, as they are generally called in the U.S., or Clean gasolines as they are sometimes termed in Europe and elsewhere, are designed to reduce both exhaust and evaporative emissions from vehicles. They were first introduced by ARCO who, in September 1989, started marketing an "Emission Control" gasoline, EC-1 in Southern California. It was intended as a fuel for vehicles not equipped with catalytic converters, of which there were an estimated 1.2 million in the South Coast Air Basin of Southern California, representing at that time about 15% of the cars and trucks in the area, but causing more than 30% of the vehicular air pollution.

This EC-1 gasoline was unleaded although it was designed to replace leaded regular gasoline and had the same octane specification as the gasoline it replaced. Its effect in reducing emissions was achieved by several different routes, including reduced vapor pressure to minimize evaporative emissions; lower aromatics and olefins to reduce smog-forming hydrocarbons and benzene exhaust emissions; a low benzene content; the use of oxygenates to improve CO and HC exhaust emissions; and a reduced sulfur content.

Tests carried out on ARCO's EC-1,[29,30] based on a comparison with leaded regular gasoline in a 20-car fleet test, indicated statistically significant reductions in regulated and other emissions, as shown in Table 12.2.

Table 12.2. Effect of EC-1 in Reducing Vehicle Emissions

	Approx. Reduction
Exhaust hydrocarbons	5%
Evaporative emissions	22%
Carbon monoxide	10%
NO_x	6%
Exhaust benzene emissions	43%

Soon after the introduction of EC-1, many other U.S. companies followed suit with their own versions of reformulated gasoline, most of which were marketed only in restricted areas of the U.S. There were considerable variations between the specifications which each company imposed on itself.

The U.S. Congress passed S.1630, the Clean Air Act Amendments of 1990, at the end of October 1990 and this included a requirement for "Reformulated Gasoline" to be sold in over 40 ozone and carbon monoxide non-attainment areas starting in November 1992. The primary targets for these fuels were the nine cities which had the worst record of noncompliance with the U.S. National Ambient Air Quality Standards set by the 1970 Clean Air Act. This Act set the major requirements for reformulated gasoline, but it was left to the Environmental Protection Agency (EPA) to define the detailed specifications.

A "regulation-negotiation" (reg-neg) has now been completed with industry by the U.S. EPA in order to define "Reformulated Gasoline" and the outcome is summarized in Table 12.3.

The United States has some of the most severe regulations controlling automotive emissions in the world and although there are many cities and areas in other countries where air quality is unacceptable, the direction that legislation is going worldwide is largely led by the ac-

Table 12.3. U.S. Reformulated Gasoline—"Reg-Neg" Agreement[31]

Specification Item	Target	Refinery Average Limit	Absolute Limit
RVP Class A*	7.2 psi/49.7 kPa	7.1 psi/49.0 kPa	7.4 psi/51.1 kPa max
RVP Class B	8.1 psi/55.9 kPa	8.0 psi/55.2 kPa	8.3 psi/57.3 kPa max
Oxygen	2.0% mass	2.1% mass	1.5% mass min
Benzene	1.0% vol	0.95% vol	1.3% vol max
Heavy Metals	None without waiver from the EPA		
T90E	Average no greater than refiner's 1990 average		
Sulfur	As above		
Olefins	As above		
Detergent additives	Compulsory, but not yet defined by the EPA		

* Applies for 1995–96 only, for 1997-99 VOC emissions must be reduced by at least 16.5% relative to 1990 average gasoline.

tions taken in the U.S. In particular, the state of California has always been in the forefront of emission control legislation mainly because of the poor air quality and atmospheric smog experienced in the Los Angeles area—although there has been a considerable improvement over the past several years.

The major advantage of reformulated fuels is that they have an immediate beneficial effect, especially on the emissions of non-controlled, and hence, the most polluting vehicles in the car and truck population. In addition, emitted hydrocarbons from these fuels will have lower photochemical reactivities. It can take many years before the older vehicles disappear completely and are replaced by newer designs and so it can take a long time for improvements in air quality to take place if reliance is placed on vehicle hardware changes only.

Even with vehicles that have emission controls there are still advantages to using reformulated gasolines, although these benefits will diminish as control systems get more and more sophisticated. At present many of the feedback controls that keep a three-way catalyst system operating at the optimum air/fuel ratio do not operate under some driving regimes—during warm-up for example and under full throttle accelerations in some vehicles. During these periods there is clearly a benefit in using reformulated gasoline. In addition the improved exhaust quality before it goes to the catalyst, and the reduced amount of evaporative losses will place a lower burden on the emission control equipment and result in improved emissions.

Some reformulated gasolines have appeared in Europe, although the studies mentioned above in Section 12.4 are being carried out to ensure that to meet a given emissions level, any legal constraints applied to the fuel or to engines will minimize the costs to the consumer and to the economy in general.

Tests with reformulated gasolines all show very significant benefits, although these will depend on the vehicles used and the test conditions.[10,32,33] The principle of diminishing returns seems to apply in that little additional benefits seem to be gained by moving to extreme reformulation such as, for example, by having a very low aromatics level.[33] Modest reformulation with only minor changes from the standard gasoline, such as using 2% oxygenates and reducing aromatics to meet the required octane, have shown very important emissions benefits—in one test a 23% reduction in CO and an 11% reduction in both HC and NO_x.[34]

References

1. M. Shelef, "Unanticipated Benefits of Automotive Emission Control: Reduction in Fatalities by Motor Vehicle Exhaust Gas," SAE Paper No. 922335, 1992.

2. M. Booth, J.S. McArragher and J.M. Marriott, "The Influence of Motor Vehicle Emission Control Legislation on Future Fuel Quality," Paper CEC/93/EF05, Fourth International Symposium on the Performance Evaluation of Automotive Fuels and Lubricants, Birmingham, U.K., May 1993.

3. R.J. Farrauto, R.M. Heck and B.K. Speronello, "Environmental Catalysts," *Chemical and Engineering News*, 70, No. 36, July 1992.

4. K.H. Hellman, G.K. Piotrowski and R.M. Schaefer, "Evaluation of Different Resistively Heated Catalyst Technologies," SAE Paper No. 912382, 1991.

5. J.E. Kubsh, P.W. Lissiuk and I. Gottberg, "Electrically Heated Catalysts and Reformulated Gasolines," SAE Paper No. 930385, 1993.

6. Auto/Oil Air Quality Improvement Research Program, SAE SP-920, 1992.

7. Auto/Oil Air Quality Improvement Research Program, Volume II, SAE SP-1000, 1993.

8. Auto/Oil Air Quality Improvement Research Program, Phase I Final Report, May 1993.

9. W.G. Luding, "The Auto/Oil Industries and Clean Fuels, Vehicles," The European Conference on New Fuels and Vehicles for Clean Air, Amsterdam, June 1993.

10. T.D.B. Morgan, G.J. den Otter, *et al.*, "An Integrated Study of the Effects of Gasoline Composition on Exhaust Emissions Part I: Program Outline and Results on Regulated Emissions," SAE Paper No. 932678, 1993.

11. H.M. Doherty and W.H. Douthit, "Effect of Oxygenates and Fuel Volatility on Vehicle Emissions at Seasonal Temperatures," SAE Paper No. 902130, 1992.

12. F.D. Stump, K.T. Knapp, *et al.*, "The Seasonal Impact of Blending Oxygenated Organics with Gasoline on Motor Vehicle Tailpipe and Evaporative Emissions—Part II," SAE Paper No. 902129, 1990.

13. M.T. Noorman, "The Effect of MTBE, DIPE and TAME on Vehicle Emissions," SAE Paper No. 932668, 1993.

14. K. Nelson, R. Ragazzi and G. Gallagher, "The Effects of Fuel Oxygen Concentration on Automotive Carbon Monoxide Emissions at High Altitudes," SAE Paper No. 902128, 1990.

15. J. Kivi, A. Niemi, *et al.*, "Use of MTBE and ETBE as Gasoline Reformulation Components," SAE Paper No. 922379, 1992.

16. A. Petit and X. Montagne, "Effects of the Gasoline Composition on Exhaust Emissions of Regulated and Speciated Components," SAE Paper No. 932681, 1993.

17. H. Hoshi, M. Nakada, *et al.*, " Effects of Gasoline Composition on Exhaust Emissions and Driveability," SAE Paper No. 902094, 1990.

18. J. Doyon, K. Mitchell and T. Mayhew, "The Effect of Gasoline Composition on Vehicle Tailpipe Emissions at Low Ambient Temperature," SAE Paper No. 932669, 1993.

19. J.A. Gething, S.K. Hoekman, *et al.*, "The Effect of Gasoline Aromatics Content on Exhaust Emissions: A Cooperative Test Program," SAE Paper No. 902073, 1990.

20. M.A. Warner-Selph and C.A. Harvey, " Assessment of Unregulated Emissions from Gasoline Oxygenated Blends," SAE Paper No. 902131, 1990.

21. R.A. Gorse, J.D. Benson, *et al.*, "Toxic Air Pollutant Vehicle Exhaust Emissions with Reformulated Gasolines," SAE Paper No. 912324, 1991.

22. A. Lowi, Jr. and W.P.L. Carter, "A Method for Evaluating the Atmospheric Ozone Impact of Actual Vehicle Emissions," SAE Paper No. 900710, 1990.

23. N.M. Dempster and P.R. Shore, "An Investigation into the Production of Hydrocarbon Emissions from a Gasoline Engine Tested on Chemically Defined Fuels," SAE Paper No. 900354, 1990.

24. W.R. Leppard, L.A. Rapp, *et al.*, "Effects of Gasoline Composition on Vehicle Engine-out and Tailpipe Hydrocarbon Emissions—The Auto/Oil Air Quality Improvement Research Program," SAE Paper No. 912320, 1991.

25. S. Kubo, M. Yamamoto, *et al.*, "Speciated Hydrocarbon Emissions of SI Engine During Cold Start and Warm-up," SAE Paper No. 932706, 1993.

26. W.M. Studzinski, D.R. Lachowicz, *et al.*, "Paraffinic versus Olefinic Refinery Streams: An Exhaust Emissions Investigation," SAE Paper No. 922377, 1992.

27. Y. Takei, H. Hoshi, *et al.*, "Effect of Gasoline Components on Exhaust Hydrocarbon Components," SAE Paper No. 932670, 1993.

28. S. Yamazaki, S. Kubo, *et al.*, "Effects of the Gasoline Composition and Emission Control Systems on Exhaust HC Emission," SAE Paper No. 922182, 1992.

29. L.K. Cohu, L.A. Rapp and J.S. Segal, "EC-1 Emission Control Gasoline," ARCO Products Co., Anaheim, California, U.S.A., September 1989.

30. R.A. Corbett, "Substitute Transport Fuels, Emissions and Air Quality 1990," *Oil and Gas Journal Special*, June 18, 1990.

31. "Motor Vehicle Emission Regulations and Fuel Specifications— 1992 Update," CONCAWE, Report No. 2/92, November 1992.

32. G.A. Schoonveld and W.F. Marshall, "The Total Effect of a Reformulated Gasoline on Vehicle Emissions by Technology (1973 to 1989)," SAE Paper No. 910380, 1991.

33. G.J. den Otter, R.E. Malpas and T.D.B. Morgan, "Effects of Gasoline Reformulation on Exhaust Emissions in Current European Vehicles," SAE Paper No. 930372, 1993.

34. B. Hery, "Introducing Clean Gasoline in Europe," The European Conference on New Fuels and Vehicles for Clean Air, Amsterdam, June 1993.

Chapter 13

Racing Fuels

13.1 General Considerations

A primary objective in racing is to maximize engine power output consistent with any imposed restrictions on fuel quality and the ability of the engine to survive for the duration of the race. Power output is mainly increased by engine design features, but the use of specialized fuels to maximize power output can play an important part in achieving higher outputs than can be obtained by mechanical means alone.

The factors that are important in a racing fuel are as follows:

- **Heating value.** The greater the heat content of the fuel, the higher the possible energy output.

- **Stoichiometry.** The lower the stoichiometric air-fuel ratio (or oxygen-fuel ratio if another oxidant than air is used), the more fuel can be introduced into the combustion chamber.

- **Ratio of products to reactants.** The higher this ratio, the higher the combustion pressure in the cylinder and, hence, potentially the higher the power output.

- **Resistance to detonation and preignition** because these can swiftly destroy an engine under racing conditions. The relatively severe driving conditions used in racing may mean that Motor Octane Number (MON) is generally more important than Research Octane Number (RON) as a guide to the antiknock performance of the fuel, although for turbo or supercharged vehicles the reverse may be true. However, under the rather extreme conditions in a racing engine, the RON and MON of a fuel may not have very much meaning in terms of predicting whether knock, or preignition resulting from knock, will occur.

- **Flammability limits.** The flammability limits should be such that fuel-rich air-fuel ratios can be used that maximize power.

- **Flame speed.** A fast burn rate is important in a racing engine since it helps to maintain good combustion under varying conditions of speed, load, air/fuel ratio, etc., and also because combustion needs to be virtually complete before the combustion stroke is over, in order to maximize the thermal efficiency.[1] In an engine operating at over 10,000 rpm, the time available for complete combustion is rather limited.

- **Heat of vaporization.** A high heat of vaporization will increase the volumetric efficiency of an engine by cooling the intake charge as the fuel evaporates, and increase the energy density.

- **Volatility.** The fuel must have a boiling range that allows it to be transported readily in the liquid form and yet volatilize satisfactorily in the intake system.

- **Safety.** The fuel must be sufficiently stable to ensure that it does not cause an explosion hazard during use or when being transported and handled under the normal conditions for automotive fuels. Its toxicity should be such that exposure of personnel handling the fuel is within the limits accepted for the components present in the fuel.

It is not possible to find a fuel that has the optimum level of all the characteristics listed above, and so racing fuels must always be a compromise. Because the characteristics of individual fuels vary widely from one to another, engines must be designed for a specific fuel and will not normally run efficiently on other blends.

One important parameter that provides a method of comparing the heat release of different fuels in an engine is the Specific Energy (SE). The *theoretical* SE is calculated by dividing the heating value (usually the lower heating value) by the air-fuel ratio so that it represents the fuel energy delivered to the combustion chamber per unit mass of air inducted. It is often calculated for stoichiometric air-fuel ratios. The *actual* SE is calculated from the actual heat of combustion and the actual air-fuel ratio. SE changes with changing air-fuel ratio and is only valid within the flammability limits of the fuel.

Specific Energy Ratio is also used to compare fuels and is the ratio of the SE of the fuel to that of a standard reference fuel, usually isooctane since this is similar to a commercial gasoline.

Table 13.1[2] compares some characteristics of nitromethane and methanol, which are both commonly used as racing fuels, with isooctane. It can be seen that nitromethane has an SE of 2.3 times that of isooctane and that methanol is marginally better than isooctane because, in both cases, the air-fuel ratio at stoichiometric is much lower than that of isooctane, despite the fact that isooctane has the highest heating value.

Table 13.1. Properties of Some Common Racing Fuels

Property	Nitromethane	Methanol	Isooctane
Formula	CH_3NO_2	CH_3OH	C_8H_{18}
Mol. Wt.	61	32	114
Oxygen content, wt%	52.5	49.9	0
Stoichiometric Air-Fuel Ratio	1.7:1	6.45:1	15.1:1
Lower Heating Value, MJ/kg	11.3	19.9	44.3
Specific Energy at stoichiometric	6.6	3.1	2.9
SE Ratio	2.3	1.06	1.00
Heat of vaporization, MJ/kg	0.56	1.17	0.27

13.2 Hydrocarbon Racing Fuels

Gasoline is the most commonly used racing fuel in view of its ready availability and low cost. Its greatest limitations are its relatively low specific energy and its tendency to knock at the high compression ratios necessary to maximize power output.

In the period between the wars, blends of the then-poor octane quality gasoline were used together with benzol and higher aromatics to boost octane level. General Motors patented an aviation fuel called "Hector Fuel," consisting of an 80:20 blend of cyclohexane and benzole[3] soon after World War I and this would have had an excellent octane quality and a high relative density. Blend components containing benzene would not be acceptable today because of the toxicity of this compound and, in any case, such blends which boil over a restricted temperature range do not always give good driveability.

Nowadays, gasolines having high octane levels are not normally available commercially because of restrictions on the use of lead, although fuels with Research Octane levels of 100 or more and with Motor Octane levels of at least 88 have been marketed in the past. They were often highly leaded blends of butane, light catalytically cracked naphtha and heavy catalytic reformate.

A fuel having a RON in the range of 110-120 and a MON from 105-110 has been suggested as being capable of satisfying most naturally aspirated piston engines operating under the most severe conditions likely to be encountered in racing.[2] Leaded blends of selected compounds would be needed to meet such a target. Specific hydrocarbons and, in particular, the highly branched paraffins such as isooctane (2,2,4 trimethylpentane) and triptane (2,2,3 trimethylbutane), which have excellent antiknock properties, have been used in racing fuel blends to give acceptable antiknock performance. Table 13.2 summarizes the octane quality of some of the compounds that have been used in such blends.[2]

Although cyclohexane appears to have a relatively low octane quality, it has, in fact, quite high blending values depending on the other components present. Triptane was investigated as a possible aviation gasoline component during World War II[4] in view of its exceptionally high knock resistance, but it has the disadvantage of a high production cost.

Table 13.2. Octane Qualities of Some Pure Hydrocarbons

Compound	Formula	RON	MON
Isooctane	C_8H_{18}	100	100
Triptane	C_7H_{16}	112	101
Isodecane	$C_{10}H_{22}$	113	92
Cyclopentane	C_5H_{10}	101	85
Cyclohexane	C_6H_{12}	83	77
Toluene	C_7H_8	120	109
Xylene	C_8H_{10}	118	115

Racing gasolines blended from refinery streams usually consist of a large proportion of alkylate together with some highly aromatic stream such as catalytic reformate, and treated with the maximum amount of antiknock additive allowed.

Because of the toxicity of TEL, this antiknock additive is not generally available as a blending compound, so many racing fuel blenders have utilized aviation gasoline as a blending source of TEL.[2] The Avgas grades with relatively high lead levels are the most valuable for blending such as 115/145 and 100/130. The first number refers to the antiknock rating performed by the lean mixture method (ASTM D 2700) and the second refers to the rating performed by the super-charge method (ASTM D 909). The 100/130 grade contains a maximum of 0.85 g lead/L and the 115/145 a maximum of 1.28 g lead/L. A further grade designated as 100LL is less valuable as a blend component in that it is a low-lead grade with a maximum content of 0.56 g lead/L.

Aviation fuel consists mainly of alkylate and can be blended with aromatics such as toluene, xylene or heavy catalytic reformate to give a satisfactory fuel for many racing applications.[2] Another reported practice is to blend it with a premium unleaded automotive gasoline containing a high proportion of alkylate and/or isomerate since this will give a blend having a higher antiknock quality than either of the two components. This is because TEL gives a better octane response in paraffinic hydrocarbons such as alkylate than it does in aromatic-type blend components.

For Formula 1 racing, the regulations restrict octane quality and do not allow the use of methanol and nitro-methane based fuels.[1] Because of this a considerable amount of work has been carried out to develop fuels that will give enhanced performance under the exacting conditions of a racing engine. These fuels can be unleaded and contain only a relatively few components as compared with a pump fuel. The components can include aromatics—one racing fuel is reported to contain 84% toluene[5]—to give good octane quality and high energy content, as well as other components such as olefins to increase flame speed,[6] and isoparaffins and oxygenates. Because conditions in a racing engine are so different from those of a CFR engine, conventional octane parameters such as RON and MON are not necessarily applicable and these high-energy fuels tend to have a lower conventional octane quality, as well as higher flame speeds, lower stoichiometric air/fuel ratio, and increased density relative to commercial fuels.[1]

13.3 Alcohols as Racing Fuels

Straight methanol is well known as a racing fuel, having many advantages over hydrocarbon fuels. It is somewhat better than gasoline in terms of specific energy (see Table 13.1) since its low heating value is more than offset by the low stoichiometric air-fuel ratio. In addition, maximum power is obtained at the very rich air-fuel ratio of 4:1 so that under these conditions the SE is considerably higher than that of gasoline where maximum power is obtained only at about 20% richer than stoichiometric. Methanol has excellent antiknock properties allowing high compression ratios to be used — up to about 16:1 have been reported.[7] It also has a high heat of vaporization giving good volumetric efficiency. Finally, it is readily available and relatively inexpensive.

The main problem is the high fuel consumption, particularly when operating at the air-fuel ratio for maximum power, and this makes it necessary to carry a large fuel supply. At an air-fuel ratio of 4:1, the engine is quite cool running and, as one leans off, the engine will give better fuel economy but will run hotter and show some reduction in performance. Cold starting can be a problem with straight methanol because it has a low vapor pressure and small amounts of hydrocarbons or ether are sometimes added for this reason. Because methanol boils at a single temperature, carburetion can be difficult and result in poor driveability which can be overcome by blending in some gasoline or other hydrocarbons. Fuel injection can avoid or minimize these driveability problems.

Ethanol has also been used as a racing fuel and is intermediate in properties between methanol and gasoline, as can be seen from Table 13.3, where some of the important characteristic of methanol and ethanol are summarized and compared with those of gasoline. The air-fuel ratio for maximum power is approximately 7:1 and the relatively high heat of vaporization makes it a useful racing fuel. As one goes up the aliphatic alcohol series, the oxygen content reduces and the closer one approaches the combustion characteristics of gasoline.

Alcohols have a poor lead response and so treatment with lead alkyls is not very useful.

13.4 Antiknock Additives

The use and mode of action of the most common antiknock additives have already been discussed in Section 6.4 and the benefits of lead antiknocks in racing fuels are referred to above. In particular, the

Table 13.3. Properties of Methanol and Ethanol

	Methanol	Ethanol	Gasoline
Oxygen content, wt%	50.0	34.8	0
Boiling point, °C	65	78	35-210
Lower heating value, MJ/kg	19.9	26.8	42.7 approx
Heat of vaporization, MJ/kg	1.17	0.93	0.18 approx
Stoichiometric air-fuel ratio	6.45:1	9.0:1	14.6:1 approx
Specific energy	3.08	3.00	2.92 approx
Blending RON	115-130	112-120	90-100
Blending MON	95-103	95-106	80-90

toxicity of lead antiknock compounds means that they are not generally available except to refiners and blenders with the appropriate handling equipment. The response of different types of hydrocarbon to the addition of lead alkyls is variable, with the lowest response being found with aromatic components and the best response with paraffinic components such as alkylate and isomerate. It is partly for this reason that these two components appear so frequently in racing gasolines, and partly because branch chain paraffins tend to have good Motor Octane levels and low Sensitivities.

Methyl cyclopentadienyl manganese tricarbonyl (MMT) (see Section 6.4) has also been used to a limited extent in racing gasolines.[2] Although it is classed as a Class B poison according to U.S. shipping regulations and is toxic by all exposure routes, the handling of MMT is much less restrictive than required for lead compounds because of its low vapor pressure and high thermal stability. In unleaded gasolines at a treat level of 0.07 g Mn/L, it is capable of giving octane gains of 2 to 3 units for MON and over 4 for RON, but such high treat levels are not acceptable for normal use on the road. At higher treat levels, spark plug fouling and valve wear problems have been reported.[8] It also has a synergistic effect when used with lead so that octane gains of up to one number are possible when it is added to leaded gasoline. Ferrocene has also been used to a small extent to enhance octane quality in racing fuels.

Methanol and ethanol can be regarded as antiknock additives, although strictly they are high-octane blend components. They have both been widely used to upgrade gasoline for use as a racing fuel,[7] and in the 1920s, European factory teams used blends of ethanol (E), benzol (B) and aviation gasoline (G) in the ratios of E:B:G of between 20:20:60 to 80:10:10. These blends had a good water tolerance (see Section 11.4.1) but were somewhat prone to preignition. Methanol (M) took over to some extent in the 1930s due to its high heat of vaporization and ability to combust at lower air-fuel ratios, with M:E:B:G blends varying from 30:30:20:20 to 60:10:25:5. M:E:G blends were also used and ranged from 20:60:20 to 80:10:10.[7]

The present-day use of ethanol and methanol in normal commercial gasolines is discussed in Chapter 11, but for this purpose the amount that can be used is severely restricted because vehicles need to be able to run both on these gasolines as well as on conventional gasolines.

This restriction does not apply to racing vehicles where the whole system can be tailored to a specific fuel.

It can be seen from Table 13.3 that alcohols are excellent blend components in terms of octane quality, particularly RON, and that they are all somewhat better than gasoline from a specific energy viewpoint. The actual blending numbers will depend on the other components present in the blend and the amount of alcohol used.

The high heat of vaporization of the alcohols, especially methanol, will cool the intake charge and improve volumetric efficiency, as it does when neat alcohol is used.

Another advantage of the use of methanol as a blend component is that, as discussed above, maximum power is achieved at an air-fuel ratio of 4:1, some 40% richer than the stoichiometric air-fuel ratio. This tends to keep the engine cool and allows the possibility of leaning off under lighter load conditions to improve economy.

Disadvantages of the alcohols are that they increase RVP when blended with gasoline (see Section 11.3.2), causing hot driveability problems, and that they attack metals and elastomers in the fuel system. When the engine is not in use, it is advisable to drain off the fuel if it contains a high amount of alcohol, and especially methanol. Alcohols also absorb moisture from the atmosphere and such water can cause phase separation (see Section 11.4.1).

Ethers such as MTBE or TAME (see Chapter 11) can also be used as blend components in racing fuels in order to improve octane quality. They are both relatively trouble-free but do not have a significant cooling effect on the intake charge because of their relatively low heats of vaporization as compared with gasoline. Their characteristics are summarized in Table 13.4.

Although water is strictly not an antiknock additive, it has been known for many years that its use will suppress knock.[9] It can be introduced into the engine in many different ways, such as by injection into the inlet manifold, ports or directly into the cylinder. It can even be incorporated into the fuel in the form of an emulsion.[10] In addition to suppressing knock, water can also improve performance and substantially reduce exhaust emissions of nitrogen oxides.[11]

Table 13.4. Characteristics of MTBE and TAME as Racing Fuel Blend Components

	MTBE	TAME
Oxygen content, wt%	18.2	15.7
Boiling point, °C	55	86
Lower heat of combustion, MJ/kg	35.1	37.7
Heat of vaporization, MJ/kg	0.32	0.32
Stoichiometric air-fuel ratio	11.7:1	11.9:1
Blending RON (approx.)	115	111
Blending MON (approx.)	104	100

Emissions of hydrocarbons are increased, however, but CO is only slightly affected. Adverse effects are that the lean limit of combustion is narrowed and that corrosion and wear are increased. The introduction of 40% wt of water into a gasoline in the form of an emulsion increased the RON of a gasoline blend from 91.2 to 100,[11] although such an emulsion would be difficult to use because of its high viscosity and the difficulty of preventing the emulsion from breaking. Lower amounts injected separately will have a very real effect.

Finally, it is perhaps worth mentioning that nitrogen compounds such as N-methyl aniline and aniline are antiknock agents and have been used to boost the octane quality of gasolines. Rather large amounts, on the order of 1 or 2 vol. percent, are needed to gain 1 or 2 octane numbers, and at this concentration they tend to increase the deposit-forming tendency of the gasoline. They are unpleasant and toxic materials which are relatively expensive, so they have not enjoyed widespread use.

13.5 Nitroparaffins as Racing Fuels

One of the limiting factors in increasing power output of the I.C. engine is volumetric efficiency. Hydrocarbon fuels require a rather narrow range of air-fuel ratio in order to burn efficiently and give maximum performance, and so there is little scope for increasing this.

However, by using fuels which require less oxygen for complete combustion, it is possible to increase the available energy, provided, of course, that the heating value of the fuel is not so low that it more than offsets the gain due to the lower air-fuel ratio.

Table 13.1 showed that nitromethane has an SE ratio almost 2.3 times higher than isooctane so it is potentially an excellent fuel from this viewpoint. Nitromethane combusts in oxygen or air according to the following equation:

$$4CH_3NO_2 + 3O_2 \rightarrow 4CO_2 + 6H_2O + 2N_2$$

Thus, seven moles of reactant give twelve moles of product, or, taking atmospheric nitrogen into account, 18.29 moles of reactant give 23.29 moles of product, a ratio of 1.27:1 as compared with 1.00:1 for a hydrocarbon fuel such as isooctane. In fact, nitromethane will "combust" in the complete absence of air according to the following equation:[12]

$$4CH_3NO_2 \rightarrow CO_2 + 3CO + 3H_2O + 3H_2 + 2N_2$$

Nitromethane will therefore act as a monopropellant and will combust over a wide range of air-fuel ratios from 100% fuel to extremely lean fuel-air mixtures. Other nitroparaffins can be used as engine fuels, including nitroethane, nitropropane and 2,2 dinitropropane.[12]

Table 13.1 also shows that the heat of vaporization for nitromethane is quite high so that when the increased fuel charge resulting from the low air-fuel ratio is taken into account, a cooling effect about twice that of pure methanol is obtained. Of course, not all of this cooling effect is achieved in practice, mainly because of incomplete vaporization and heat transfer.

Nitro fuels are fast burning and particularly suited to engines running at high speeds where the slower-burning methanol cannot release all its energy at the peak of piston travel. However, because of severe engine stress, nitromethane is only used as a primary fuel in short duration runs such as in drag racing, where the engine is usually run at full power for only a few seconds at a time.[13] The tendency to produce severe knock and preignition is usually offset by operating at very rich air-fuel ratios and by using low compression ratios and a retarded spark timing.

In practice, in order to reduce peak flame temperatures, mixtures with methanol are used. The amount of nitromethane used in such a blend with methanol usually varies from 10 to 90%. Fifty percent of nitromethane in methanol was shown to increase the power output over that of pure methanol by almost 45%, although knock increased by about 20%.[13] Work has also been carried out using nitroethane or nitropropane as cosolvent instead of methanol and it has been found that with these blends increasing nitromethane concentration increases power and reduces knock.[13]

The addition of TEL or MMT to nitromethane does not reduce significantly the tendency for the engine to knock.

Nitromethane is highly corrosive to aluminum and magnesium so that it is important to drain and purge the fuel system to avoid damage.

13.6 Hydrazine as a Racing Fuel Component

Hydrazine (N_2H_4) is a liquid boiling at 113°C and has been used as a fuel additive in nitromethane and methanol blends where the fuel properties and composition are not restricted by regulations. It forms an explosive salt with nitromethane that requires only the oxygen in the nitromethane for it to combust. High rates of pressure rise are achieved in the cylinder with dramatic increases in power but with an increased risk of damage to the engine. Very little is needed and less than 1% of hydrazine by volume gives a significant power increase.[2] The mixture is so unstable that it causes a severe safety hazard.

13.7 Nitrous Oxide as an Oxidant for Racing Fuels

Nitrous oxide (N_2O) is a gas containing 36% oxygen by weight as compared with 23% for air, so it has some attraction as an oxidant because it enables an increase in specific energy for any given fuel.[2] It is stable and readily available but it is an anaesthetic and must be handled with care. It is normally supplied in gas cylinders so that storage on racing vehicles limits its use.

It is usually used for short durations in view of the storage problems, the high combustion temperatures achieved and the high fuel flow rates required. It has been calculated[2] that a gaseous stoichiometric mixture of isooctane, air and 5% nitrous oxide would produce a 9% increase in available specific energy and a 5% increase in the product/

reactants mole ratio. In racing engines, the use of nitrous oxide in short bursts has resulted in estimated increases in power of about 20%.[14]

References

1. A.R. Glover, A. Yasuoka, *et al.*, "Optimizing Formula 1 Engine Performance through Fuel and Combustion System Development," SAE Paper No. 938094, 1993.

2. G.J. Germane, "ATechnical Review of Racing Fuels," SAE Paper No. 852129, 1985.

3. United States Advisory Committee for Aeronautics (NACA-NASA), T.R. Nos. 47, 89 and 90.

4. J.R. Branstetter, "Comparison of Knock-Limited Performance of Triptane with 23 Other Purified Hydrocarbons," NACA MR No. E5E15, 1945.

5. "Honda Experiments with Engine Technologies in F1," *Motor Trend*, pp. 103-104, April 1992.

6. E.M. Goodger, Hydrocarbon Fuels, p. 112, The MacMillan Press Ltd., 1975.

7. T. Powell, "Racing Experiences with Methanol and Ethanol Based Motor Fuels," SAE Paper No. 750124, 1975.

8. Ethyl Corporation, "Information for the National Research Council Concerning Methylcyclopentadienyl Manganese Tricarbonyl," October 1971.

9. B. Hopkinson, "A New Method of Cooling Gas Engines," *Proc. Inst. Mech. Engineers*, pp. 679-715, 1913.

10. B.D. Peters and R.F. Stebar, "Water-Gasoline Fuels — Their Effect on Spark Ignition Engine Emissions and Performance," SAE Paper No. 760547, 1976.

11. J.A. Harrington, "Water Addition to Gasoline — Effect on Combustion, Emissions, Performance and Knock," SAE Paper No. 820314, 1982.

12. E.S. Starkman, "Nitroparaffins as Potential Engine Fuel," *Ind. and Eng. Chemistry*, Vol. 51, No. 12, pp. 1477-1480, 1959.

13. K.C. Bush, G.J. Germane and G.L. Hess, "Improved Utilization of Nitromethane as an Internal Combustion Engine Fuel," SAE Paper No. 852130, 1985.

14. J. Fuchs, "N_2O injection." *Hot Rod*, p. 136, April 1973.

Chapter 14

The Diesel Engine and its Combustion Process

14.1 Introduction

Chapter 2, Section 2.2 describes how the compression ignition engine came into existence as the successful outcome of the independent efforts by Rudolf Diesel in Germany and Herbert Akroyd-Stuart in England to overcome particular limitations of the spark-ignited engines.

Both inventors designed their engines to breathe pure air, and fuel was only injected towards the end of the compression stroke, thus avoiding the problem of preignition during compression of an explosive mixture. However, whereas Ackroyd-Stuart's engine had a relatively low compression ratio and relied on a heated vaporizer for ignition, Diesel's scientifically based approach was to achieve greater thermal efficiency by the utilization of a higher expansion ratio, by increasing the compression ratio which, at the same time, produced a temperature high enough to cause the fuel to ignite spontaneously when injected into the hot compressed air.

Another difference between the initial designs of these two pioneers was in the fuel injection systems. Akroyd-Stuart used a mechanical pump to spray a liquid fuel into the combustion space, while Diesel relied on a blast of compressed air to inject the fuel which, in the first design, was finely powdered coal dust. Air blasting was subsequently abandoned in favor of the mechanical pump "solid injection" system used by Akroyd-Stuart which is common to all modern automotive diesel engines.

331

Akroyd-Stuart's vaporizer model was the first oil engine to run without spark-induced ignition and it was put on exhibition in 1891, two years before Diesel's first prototype was completed. However, final credit has rightly been given to Diesel as the originator of the modern engines, since they all embody his fundamental concept of compression ignition.

14.2 The Diesel Compression Ignition Engine

There are two broad categories of diesel engine based on their combustion chamber arrangement — Direct Injection (DI) and Indirect Injection (IDI).

With DI engines, the fuel is injected directly into a combustion chamber formed by the confined space between the fixed cylinder head and the crown of the piston as it reaches its highest position. A typical compression ratio for a DI engine is 18:1. The fuel injector, with multiple spray orifices in its nozzle, is mounted in the cylinder head. In many of the smaller engines a shaped cavity is forced in the piston crown to give improved fuel-air mixing. Figure 14.1[1] illustrates the arrangement of the two types of combustion chamber.

IDI engines, which were developed to enable steady and progressive burning of the fuel, have a combustion prechamber within the cylinder head, connected to the cylinder by a narrow passage through which air is forced during the compression stroke. The fuel injector, usually a single hole pintle nozzle, sprays fuel into the prechamber, where it mixes with the incoming compressed air and ignites. The hot burning gases expand through the narrow connecting passage into the cylinder, forcing down the piston on the power stroke. Good mixing of fuel and air is achieved with the high turbulence created by the rapid transfer of air from the cylinder into the prechamber during compression. In most small passenger car engines, the recess in the piston crown is shaped to give high air swirl. Compared with DI engines, the IDI system allows operation at higher speeds and the use of simpler and cheaper injection equipment.

Combustion chambers of IDI engines have larger surface-to-volume ratios than those of DI engines and, therefore, higher compression ratios, typically around 22:1, are used to compensate for the greater

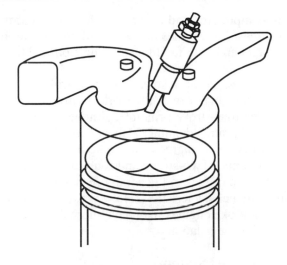

Direct Injection Diesel Engine

Indirect Injection Diesel Engine

Fig. 14.1. The two types of combustion chamber.[1]

333

heat losses during compression and ensure an adequate ignition temperature. Even so, most IDI engines are equipped with heater plugs or glow plugs in the prechamber to facilitate starting from cold.

As with gasoline engines, diesels also exist in two-stroke and four-stroke versions. The vast majority of the world's current automotive engines operate on the four-stroke principle, although two-strokes, with their power-weight ratio advantage, have become well established in the U.S., where high engine power output is of greater consequence than fuel economy. Two-stroke and four-stroke diesel engines operate on the same type of fuel and generally respond in a similar way to variations in fuel characteristics. However, as environmental controls tighten, two-stroke engines may have more difficulty in meeting some emissions legislation.

14.3 Diesel Vehicle Fuel Systems

The diesel vehicle fuel system consists of a low-pressure section and a high-pressure section. In the low-pressure section, a transfer or lift pump draws fuel from the tank and delivers it, at a relatively low pressure of 1 or 2 bar, to the injection pump which, together with the fuel injectors, comprise the high-pressure section. Typical fuel system arrangements are shown in Figure 14.2.

14.3.1 Strainers and Filters

Some vehicle tanks are fitted with a coarse mesh strainer to prevent any large pieces of rust, dirt or other foreign matter which gets into the tank from being drawn into the fuel lines. Drain plugs allow regular draining of water and sediment from fuel tanks but these are becoming less common. A water sensor has been fitted in the fuel tanks of some diesel vehicles to give a visual signal on the instrument panel when a critical level has been reached.

Where a fuel transfer pump is fitted, it is sometimes protected by a prefilter fitted in the line from the tank. Woven wire or nylon mesh screen filters of about 250 micrometers (microns) aperture size are considered adequate for diaphragm and plunger pumps, but finer porosity material of cloth or paper is required for gear pumps.

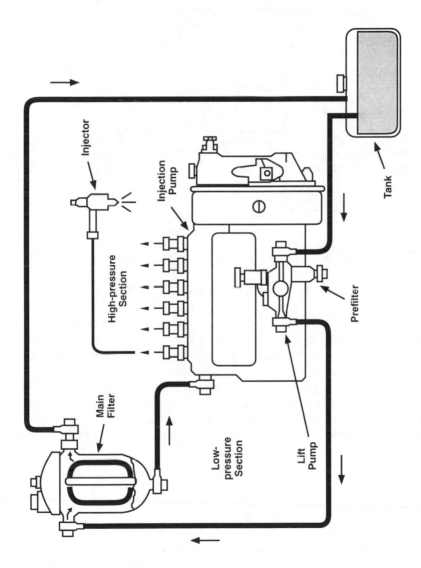

Fig. 14.2. (a) European diesel fuel system with in-line fuel injection pump.

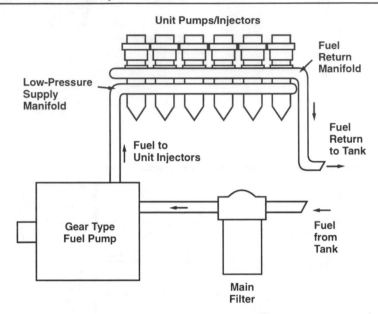

Fig. 14.2. (b) U.S. diesel fuel system with unit injection pumps/injectors.

The small clearances between the pumping elements of injection pumps necessitate the use of fine porosity filters to protect against damage or excessive wear. These are paper, cloth or felt main filters, capable of retaining particles larger than 5 or 10 micrometers. Unlike the simple woven mesh screens, these are depth filters which become more restrictive with use due to entrapment of dirt particles, so they need to be replaced periodically before becoming completely choked. To avoid high pressure drops because of their fine pore size, main filters generally have a large surface area, 3500 cm^2 being typical for the spirally wound paper element fitted in many European passenger car diesel systems. Some vehicles are fitted with two main filters, either connected in parallel to increase the total surface area or in series to provide better protection for the injection pump.

14.3.2 Transfer or Lift Pumps

Various types of transfer pumps such as diaphragm, plunger-in-barrel, or gear type are used in diesel systems to lift fuel from the tank and supply it to the injection pump. Most transfer pumps are capable of delivering more fuel than the full load requirement of the engine, so the excess fuel is recycled back to the fuel tank. The exceptions are some diaphragm pumps fitted with an internal pressure relief valve which opens to prevent further delivery when the injection pump has taken the fuel it needs. In some designs the recycled fuel is returned to the bottom of the tank near the suction line, while in others it is returned through a short pipe in the top of the tank, sometimes remote from the suction line. Fuel recirculation ratios vary widely, depending on the requirements of the engine designer or the recommendations of the fuel injection equipment (f.i.e.) manufacturer.

14.3.3 Injection Pumps and Injectors

A high proportion of diesel engines originating in the U.S. are equipped with the engine makers' own design of individual jerk fuel injection pump/injector unit fitted into each cylinder head and driven by a push rod and rocker from an engine-mounted camshaft in the same way as the inlet and exhaust valves.

Diesel engine manufacturers in Europe and most other countries fit separate fuel injection pumps and injectors that are usually bought-in components from companies such as Robert Bosch and Lucas Diesel Systems (formerly Lucas CAV), whose equipment was not as readily available in the U.S. during the 1920s, when Cummins and other American pioneer diesel engine builders were getting started.[2] Figure 14.3 is a photograph of a Cummins engine, showing the main fuel filter and the pressure-time (PT) fuel pump and control unit, with the low-pressure fuel pipe to the manifold in the cylinder head. Figure 14.4 illustrates the PT system developed by the Cummins Engine Company in the U.S.

A transfer pump delivers fuel to a manifold in the cylinder head to which all the pump/injector units are connected. The fuel pressure varies according to engine speed and load. Fuel is admitted to the injector as the plunger lifts, the time element (T) relating to the duration of opening, which is a function of engine speed, while the fuel supply pressure (P) determines the quantity admitted in unit time. The fuel is injected into the combustion chamber through a multihole nozzle by the action of the push rod and rocker on the plunger.

The fuel pump/injector of the Detroit Diesel two-stroke engine uses a variable spill system to control the amount of fuel admitted, but otherwise has a similar push rod and rocker pumping arrangement to that

Fig. 14.3. Photo of Cummins engine.

of the Cummins PT system. With combined pump/injectors, no external high-pressure pipes and connections are required but high rates of recirculation back to the tank are necessary to avoid fuel overheating in the manifold to the pump/injector units.

Using the terminology given in the international standard, ISO 7876/ 1-1984 (Fuel injection equipment - Vocabulary - Part I: Fuel injection pumps), the fuel injection pump supplying more than one cylinder of an engine may be an in-line block pump, a rotary pump or a distributor pump. They are all jerk pumps, in which the force required to move the plunger is received directly from the prime motivator, but there are differences in the way they operate.

The separate injection pump is externally mounted on the side of the engine and coupled to an auxiliary drive shaft that runs at half engine speed for four-stroke engines. Operation of the pumping elements is

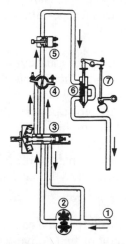

KEY:
1. Fuel from tank
2. Gear pump
3. Governor and pressure
   regulating component
4. Throttle
5. Shutdown valve
6. Injector
7. Injector actuating rocker,
   pushrod and cam follower

Fuel is drawn by a gear pump from the tank, filtered and delivered through the governor, pressure regulator and flow control valve, into the manifold supplying the pump/injector unit in each cylinder head. The supply pressure is a function of engine speed and load. High pressure to inject fuel into the engine is developed within the injector, as the plunger is driven downward by the cam-operated push rod.

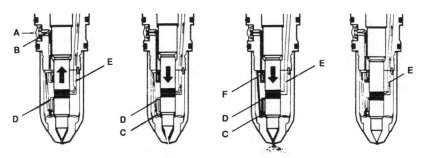

At the start of the up stroke, fuel entering through the inlet orifice B, flows through passages D and E and returns to the tank. As the plunger moves up, D is covered and the metering orifice C is uncovered, allowing fuel to enter the injector cup. On the down stroke, the metering orifice is closed and pressure builds up, forcing fuel out of the small holes in the cup as a fine spray.

Fig. 14.4. Cummins PT fuel system (source: Cummins Engine Company Ltd.).

effected by cams within the pump. The high-pressure fuel from the injection pump is delivered through external pipes of rigid steel to individual injectors in each cylinder head.

An early design of separate injection pump for multicylinder diesel engines was the Bosch in-line block pump, which has its own camshaft to operate the plunger and barrel pumping elements supplying fuel to each injector in turn. Most in-line block pumps have oil lubrication of the cams and followers but the pumping elements have to rely predominantly on the fuel being pumped for lubrication of the sliding surfaces. A cutaway view of the internals of the Minimec in-line block fuel injection pump, manufactured by Lucas Diesel Systems, is shown in Figure 14.5.

In-line block pumps have a straight camshaft with a cam lobe for each pumping element. Control of the amount of fuel delivery to govern the engine is by variation of the effective length of the plunger stroke. This is achieved by changing the relative angular position of the plunger and the barrel which contains the fuel entry helix and the exit port. An internal linkage ensures that changes in fuel delivery apply to all engine cylinders.

The in-line injection pump is still widely used for the larger engines in commercial vehicles, but many of the small, faster-revving engines for passenger cars and light delivery vehicles are now equipped with rotary or distributor injection pumps which are better able to operate at high speeds.

A high-pressure distributor system for the output from a single pump element was under development by the American Bosch Company during the 1930s but did not reach maturity. However, in 1939, V. Roosa of Hartford, Connecticut, developed a rotary injection pump which was marketed as the Roosa Master. The design had opposing plungers in a body rotating within a cam ring and the basic principles of this design have set the pattern for many of the small automotive pumps in current use. Figure 14.6 gives a cutaway view of a Lucas Diesel Systems DPA pump with a mechanical governor.

Rotary pumps made by Lucas Diesel Systems have an internally lobed cam ring to actuate radially mounted pumping plungers carried in a rotor, which also distributes the fuel to each injector in turn. The

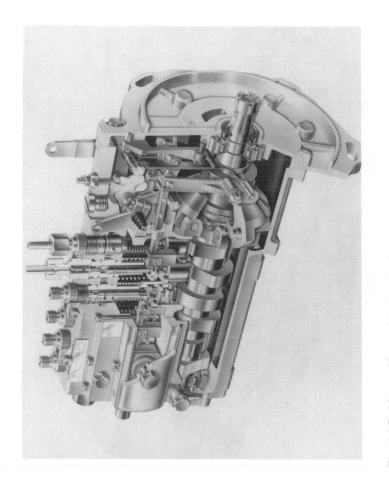

Fig. 14.5. Photo of Lucas Minimec fuel injection pump (source: Lucas Diesel Systems).

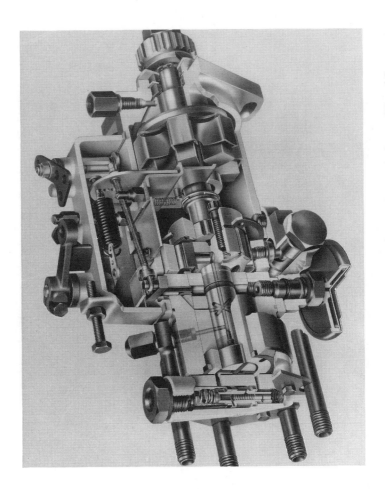

Fig. 14.6. Lucas DPA distributor-type fuel pump (source: Lucas Diesel Systems).

roller and cam action is shown in Figure 14.7. The pump relies entirely on the fuel for lubrication and an incorporated vane pump draws
fuel from the tank and floods the interior of the pump housing with
fuel at a transfer pressure that varies with engine speed. The vane
pump provides servo-pressure for variation of injection timing and for
the hydraulic governor version of the pump. Alteration of the plunger
stroke to regulate the engine speed is achieved by limiting the quantity of fuel entering the pumping chamber by means of a metering
valve controlled by the movement of the accelerator pedal and the
governor. Depending on whether the governor is hydraulic or mechanical, the metering valve is either lifted or rotated to vary the opening of the metering orifice. The amount of fuel admitted through the
metering orifice will depend on the transfer pressure, the time available when the stator and rotor feed ports coincide, as well as the size
of the metering orifice.

The Bosch distributor pump employs a rotating face cam to give a
reciprocating movement to a single, axially mounted pump plunger
which also rotates to distribute the fuel. Like the rotary pump, it is
entirely fuel-lubricated and has an incorporated vane pump. In the
Bosch VA distributor pump, the amount of fuel injected is regulated
by a hydraulic governor consisting of a constant-volume governor
pump, a spool control valve with a governor spring and a control
throttle linked to the accelerator pedal. The axial pump plunger in the
VA pump version is stepped, the smaller part acting as the high-pressure injection pump, while the larger part functions as the hydraulic
governor pump. Fuel pressure from the governor pump acts on the
spool control valve, causing it to move against its spring in synchro-
nism with the injection pump plunger and allowing fuel to pass
through the port helix of the spool into the pressurizing chamber of
the injection pump. Variations in the restriction of internal fuel passages by the pedal-linked control throttle influence the return speed of
the control spool, thereby determining the time available for fuel to be
admitted to the injection pump. As engine speed increases, the time
available for the spool to complete its return stroke reduces, so less
fuel is injected and the engine speed is regulated.

The Bosch VE distributor injection pump has a mechanical governor.
As engine speed increases, flyweights move a sliding sleeve against a
pivoted lever to adjust the position of a regulating collar on the dis-
tributor/plunger and control the effective stroke of the pump plunger

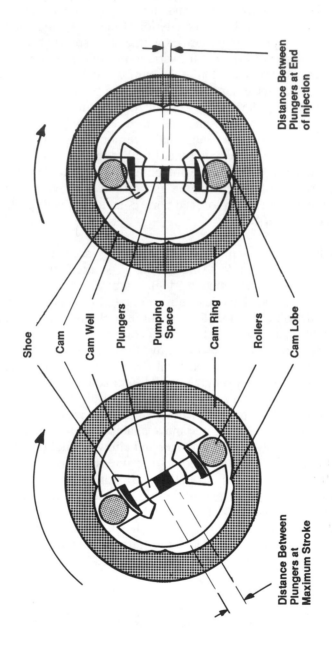

Shoe

Cam

Cam Well

Plungers

Pumping Space

Cam Ring

Rollers

Cam Lobe

Distance Between Plungers at End of Injection

Distance Between Plungers at Maximum Stroke

to reduce fuel delivery. Figures 14.8a and 14.8b show drawings of two European diesel engines with separate injection pumps, one an in-line block pump and the other a distributor-type pump.

As mentioned above, a feature of rotary and distributor pumps is the incorporated vane pump that delivers fuel to the interior of the pump housing. The good suction capabilities of vane pumps enable them to draw fuel from the vehicle tank and, because of this, many engines do not have a separate transfer pump to deliver fuel at a low positive pressure to the main filter protecting the injection pump.

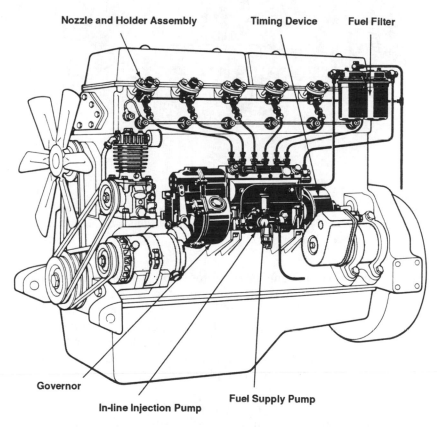

Nozzle and Holder Assembly   **Timing Device**   **Fuel Filter**

Governor

In-line Injection Pump   **Fuel Supply Pump**

Fig. 14.8. (a) Six-cylinder diesel engine with in-line fuel injection pump (source: Robert Bosch Corporation).

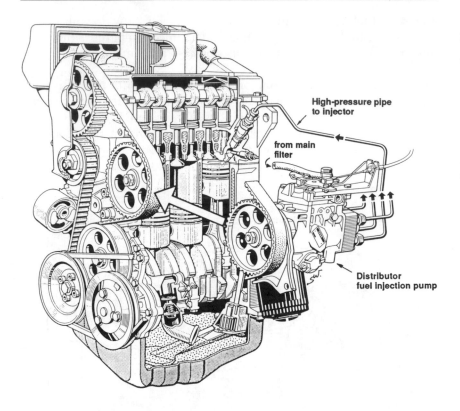

High-pressure pipe
to injector

from main
filter

Distributor
fuel injection pump

*Fig. 14.8. (b) Four-cylinder diesel engine with distributor fuel injection pump
(source: Volkswagen A.G.).*

Direct injection (DI) engines, except for small stationary engines, are
usually equipped with multi-hole fuel injectors, while pintle injectors
with a single annular hole are used in engines with indirect injection
(IDI) systems, where there is a higher degree of swirl. Figure 14.9
shows examples of the two types of injector nozzle. The upper drawings show a single-hole and multi-hole nozzle, as used in DI engines,
while the lower drawings are of pintle nozzles for IDI engines.

The function of the injector nozzle is to produce a finely atomized
spray of fuel droplets that will mix readily with the air and ensure
complete combustion in the time available. The nozzle outlet is controlled by a needle valve held against its seat by a stiff spring, which
determines the fuel pressure needed to lift the valve off its seat.

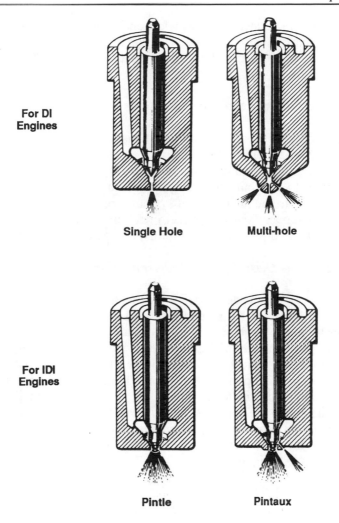

For DI Engines

Single Hole **Multi-hole**

For IDI Engines

Pintle **Pintaux**

Fig. 14.9. Injector nozzles (source: Lucas Diesel Systems).

Depending on the engine size, multi-hole nozzles for DI systems usually have four, six, or more orifices to ensure good fuel-air mixing. In IDI engines, the pintle on the lower end of the needle valve protrudes through the single orifice, so that the nozzle aperture increases progressively as fuel pressure lifts the needle valve off its seat. The initial restriction of fuel flow by the pintle controls the rate of combustion pressure rise, resulting in smooth engine running. A variant design to further control the rate of pressure rise is the Pintaux nozzle, in which an auxiliary spray aperture gives additional control over combustion during the early stages of injection. Typical peak injection pressures range between 300 bar for IDI engines to as much as 1000 bar for DI engines.[3]

14.3.4 Electronic Management Systems

The need for ever closer control on exhaust emissions has led to the development of sophisticated electronic management systems. Computers continuously monitor engine operating conditions and use this information to determine both the timing and quantity of fuel to be injected. Electronic management systems are made by the two principal European manufacturers of fuel injection equipment, Lucas Diesel Systems Ltd. and Robert Bosch GmbH, and by Stanadyne in the U.S.

14.3.4.1 Lucas Diesel Systems Electronic Controls

Lucas Diesel Systems developed their Electronically Programmed Injection Control (EPIC) system for direct and indirect injection diesel engines to be marketed in low emissions areas.[4] Precise control of the timing and quantity of fuel injected is achieved by means of an electronic control unit (ECU) and sensors which continuously monitor vehicle, engine, and pump operating conditions. Signals from the various sensors are entered in digital form into memory maps in the ECU, where they are converted into signals to the distributor-type injection pump actuators to give the desired control. A schematic diagram of the EPIC fuel control system for the pump of an IDI engine is shown in Figure 14.10.

The system, which has a self-diagnostic performance capability, gives improved idling and driving vehicle performance as well as reducing exhaust emissions. Features available as additional options to the EPIC system include exhaust gas recirculation, turbocharger control, cruise control, and an anti-theft device.

Lucas Electronic Unit Injector (EUI) systems were developed for use in markets such as California and Sweden, where stringent emissions controls are imposed. Initial applications of the combined fuel injection pump and injector units were for premium truck engines and they are currently used on Caterpillar and Volvo 12-liter engines. Lucas Electronic Unit Injectors are now available in a number of sizes to suit

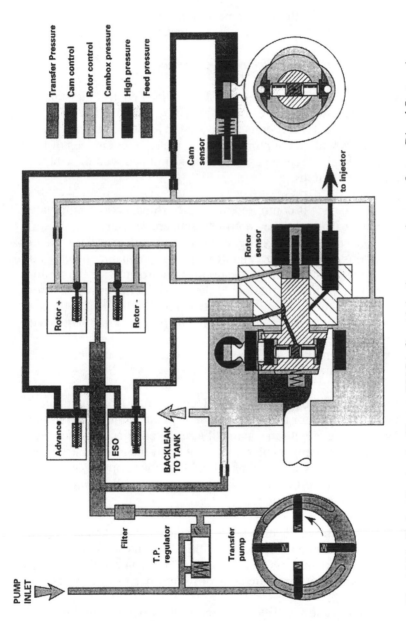

Fig. 14.10. Hydraulic circuit of Lucas EPIC indirect injection pump (source: Lucas Diesel Systems).

engines ranging in size from 2.5 liters per cylinder down to small high-speed direct injection (HSDI) engines of 0.4 liter per cylinder. Figure 14.11 shows the arrangement of electronic control unit, sensors, and the pump/injector units for a four-cylinder engine.

14.3.4.2 Robert Bosch Electronic Controls

Bosch Electronic Diesel Control (EDC) units[5] control the start of injection and the quantity of fuel injected by the injection pump. Versions of the Bosch EDC units are available for large commercial vehicle engines with inline pumps and for the smaller engines of light commercial vehicles and passenger cars equipped with distributor-type injection pumps.

The timing device is controlled by an electromagnetic timing valve and input variables for the control system are measured by sensors and engine speed; road speed; air, water and fuel temperatures; ambient and boost pressures; accelerator pedal position and injector nozzle opening. Signals are sent by the control unit to activate the timing device and adjust the input data to the map of set point values stored in the memory of the electronic control unit. Figure 14.12 gives a sectional view of a Bosch distributor-type fuel injection pump with electronic governor and Figure 14.13 shows the fuel system for a four-cylinder engine with a Bosch electronically controlled distributor-type fuel injection pump.

The schematic arrangement of the electronic fuel injection control system developed for the Ford Mondeo engine is illustrated in Figure 14.14. Compared with mechanically governed injection, the electronic control system enables optimization of the torque curve, reduces exhaust emissions and part-load fuel consumption, gives better ride comfort, and offers additional functions as well as improvement of existing control functions.

Other Robert Bosch developments to reduce diesel exhaust emissions and fuel consumption include electronically controlled unit injectors, an electronic exhaust gas recirculation (EGR) control, an electrostatic soot filter and boost pressure control for turbocharged engines.

In the electrostatic soot filter, ionized soot particles coagulate on the filter walls and are separated in a cyclone. The clean gas stream

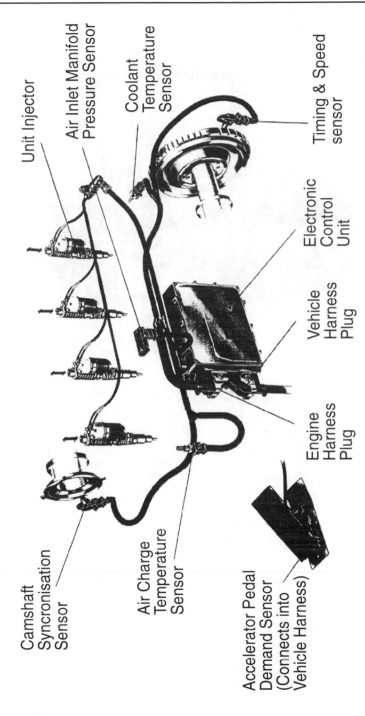

Unit Injector

Air Inlet Manifold
Pressure Sensor

Coolant
Temperature
Sensor

Timing & Speed
sensor

Camshaft
Syncronisation
Sensor

Air Charge
Temperature
Sensor

Accelerator Pedal
Demand Sensor
(Connects into
Vehicle Harness)

Engine
Harness
Plug

Electronic
Control
Unit

Vehicle
Harness
Plug

Fig. 14.11. The Lucas EUI system (source: Lucas Diesel Systems).

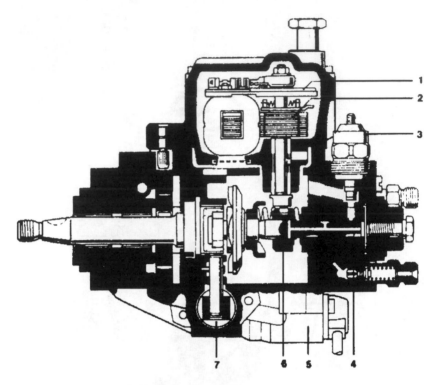

1. Control-collar position sensor
2. Injected fuel quantity actuator
3. Electromagnetic shut-off valve (ELAB)
4. Supply plunger
5. Solenoid valve for start of injection
6. Control collar
7. Timing device

Fig. 14.12. Section through Bosch distributor-type fuel injection pump with electronic governor (source: Robert Bosch Corp.).

leaves by the exhaust pipe and the soot-laden gas stream can be recirculated into the engine intake manifold for the particles to be burned.

14.3.5 Starting Aids

At low temperatures, diesel engines require starting aids to compensate for the reduced compression temperature when breathing cold air.

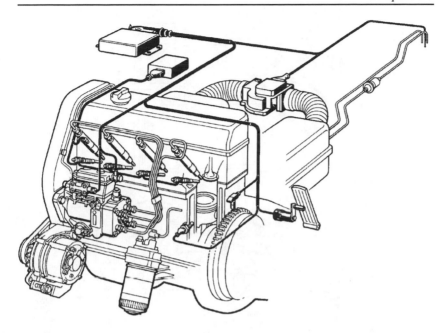

Fig. 14.13. Arrangement of fuel injection system with electronically controlled distributor-type pump (source: Robert Bosch Corp.).

For IDI engines the usual starting aid is an electrically heated glow plug protruding into the prechamber, onto which fuel droplets from the injector will impinge and be ignited. It contains an enclosed coil of resistance wire which is heated to over 1000°C when current from the battery is applied. Depending on the type, glow plugs are heated for periods of up to 30 seconds before operating the starter motor. The current is switched off when the engine starts, but combustion will keep the plug glowing.

Cold starting of DI engines is sometimes assisted by devices to preheat the intake air. Flame heaters use fuel from the vehicle tank which is vaporized and ignited in the intake manifold by a sheathed electric element. Alternatively, electrically heated glow tubes are fitted between the air filter and cylinder head or, as with some small DI engines, electric glow plugs are fitted in each cylinder head.

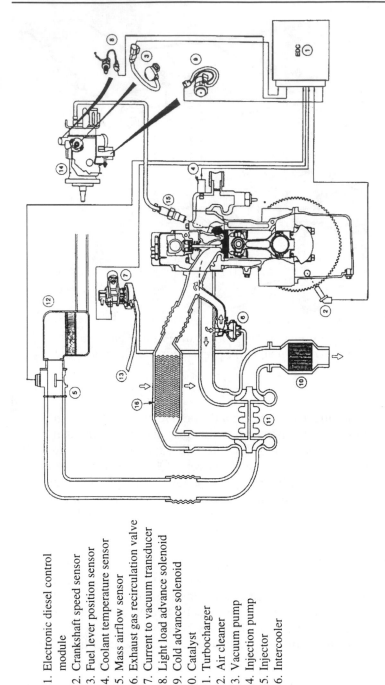

1. Electronic diesel control module
2. Crankshaft speed sensor
3. Fuel lever position sensor
4. Coolant temperature sensor
5. Mass airflow sensor
6. Exhaust gas recirculation valve
7. Current to vacuum transducer
8. Light load advance solenoid
9. Cold advance solenoid
10. Catalyst
11. Turbocharger
12. Air cleaner
13. Vacuum pump
14. Injection pump
15. Injector
16. Intercooler

Fig. 14.14 Schematic of electronic diesel control system for Ford Mondeo passenger car (source: Ford Motor Company Ltd.).

14.4 The Diesel Combustion Process

Combustion in an engine requires a flammable fuel-air mixture and a source of energy to ignite the mixture and initiate the combustion process. In the diesel engine, the flammable mixture is obtained by injecting the fuel in the form of a finely dispersed spray that mixes readily with the rapidly swirling air in the combustion chamber. The necessary energy comes from the high temperature of air compressed in the cylinder, which causes self-ignition of the fuel when it is injected near the end of the compression stroke. Temperatures sufficient to ensure autoignition are achieved by the use of high compression ratios, typically in the range 15:1 to 23:1 for automotive diesel engines.

Combustion in the diesel engine proceeds in phases which involve physical and chemical processes. The physical processes produce the turbulence to bring the fuel and air into an intimate mixture in the combustion chamber, and create the pressure and temperature environment needed for the chemical reactions to occur. The chemical processes lead to autoignition of the fuel and the ensuing combustion releases the energy in the fuel. A pressure diagram from a naturally aspirated engine, Figure 14.15 illustrates the progression of the combustion phases.[2]

The first phase commences with injection of the first droplets of fuel, about fifteen degrees before the piston reaches top dead center (TDC). On the compression stroke, the air in the combustion chamber is heated to over 500°C, which is sufficient to cause spontaneous ignition of fuel injected into this environment. However, ignition will not occur immediately; a small but finite time, known as the ignition delay period, is required for the fuel spray to be broken into finer droplets to form a combustible mixture with the air, and for preflame reactions which lead to ignition to occur.

Some ignition delay, typically about 1,000 microseconds (0.001 s), is an inherent part of the diesel combustion process, but the actual duration will depend on a number of factors: how readily the fuel oxidizes, the air temperature, the size of the injected fuel droplets and the extent of fuel-air mixing.[2,6,7] During this first phase, the cylinder pressure tends to follow the compression line until ignition occurs.

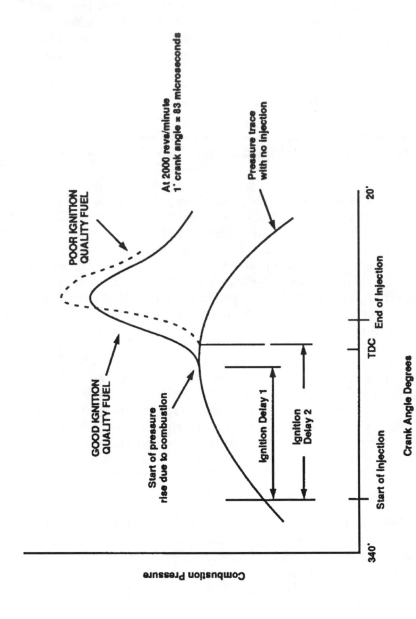

Fig. 14.15. Diesel engine pressure diagram (source: Exxon Chemical International).

A phase of uncontrolled burning and rapid pressure rise follows, as the flame spreads through the fuel-air mixture formed before ignition occurred. The rate and extent of pressure increase will depend on the amount of fuel present in the combustion chamber, which in turn will be influenced by the length of the ignition delay and the amount of fuel injected during the delay period.

In the third and final phase, combustion will be controlled largely by the rate of fuel injection and mixing, but it will continue after injection stops and until all the available fuel has been consumed. Cylinder pressure will rise more gradually and then fall as the piston descends on its power stroke.

The solid line in Figure 14.15 shows how the pressure varies with a fuel of good ignition quality. The dotted line indicates the different situation when a fuel of lower ignition quality is used. The ignition delay period is increased as more time is needed to cause the poorer fuel to self ignite. As a result of the longer delay period, more fuel has been injected and the air temperature is higher due to continuation of the compression stroke towards top dead center (TDC). When ignition does occur, the pressure rises faster due to uncontrolled burning of the fuel-air mixture, possibly resulting in a higher peak pressure. However, the work done, represented by the area under the pressure-time curve, is not appreciably different from that obtained with the better ignition quality fuel.

The most readily discernible effects associated with poorer ignition quality fuels are longer cranking times and increased output of white smoke when starting at low temperatures, higher combustion noise due to "diesel knock" and rougher running, especially at light loads. There may also be an increase in the levels of gaseous emissions. The influence of fuel characteristics on exhaust emissions will be discussed in the following sections.

14.5 Diesel Environmental Considerations

Increasing government and industry efforts are being directed towards environmental protection and more efficient use of the non-renewable petroleum resources. In the area of road transport fuels, this has contributed to the phased removal of lead antiknock additives from gasolines in North America, Europe and Japan, and the steady growth in popularity of the diesel engine for the private car market, particularly in Europe since the 1970s. In Europe and Japan, diesels are used almost exclusively for road transportation of freight as well as for city and long-distance buses, and Europe has a higher proportion of diesel-engined road vehicles than any other world market. The U.S. has a high proportion of gasoline trucks but the diesel engine is taking over in the medium- to heavy-duty sector.[8]

With diesel vehicles, attention focused initially on the visual evidence of sooty exhaust emissions from overloaded or poorly maintained engines and this lead to the introduction of controls on smoke levels. Regulation of gaseous emissions from diesel engines was first introduced in the U.S. in 1970. Recent legislation in both the U.S. and Europe has defined further limits on gaseous and particulate pollutants emitted by diesel engines.

The situation is continuing to evolve but serious attempts have been made to establish emissions standards that are technically realistic and also achievable within a well-defined time frame. In the U.S., this resulted in the Environmental Protection Agency (EPA) issuing an emissions timetable designed to produce significant reductions in emissions from heavy-duty diesel (HDD) vehicles by the year 1994,[9] bringing the 1991 HDD particulates limit of 0.25 g/BHP hour down to 0.10 g/BHP hour for urban buses. California HDD particulates limits are to fall to 0.18 or 0.15 (depending on vehicle weight) by 1995; to 0.12 by 1998 and, with the phasing-in of low emission (LEV) and ultra low emission (ULEV) vehicles between 1998 and 2003, particulates levels will be further brought down to 0.06 g/BHP hour.

A European Program on Emissions, Fuels and Engine Technologies (EPEFE) was set up as a result of the dialogue between the EC and the relevant motor and oil industry associations, ACEA and EUROPIA assisted by CONCAWE. The objectives of the Auto/Oil program are to provide test data on the influence of engine/exhaust clean-up technologies and fuel properties on gasoline and diesel vehicle emissions. As reported in Section 12.4, test work on EPEFE is in progress but no conclusions were available at the time of writing.

Other aspects which are again under close scrutiny by environmental groups relate to health risks associated with diesel engines and the short- and long-term effect of their emissions on environmental concerns such as acid rain, damage to buildings and the greenhouse effect.

14.5.1 Sources of Exhaust Emissions

Diesel exhaust emissions that have to be controlled are: hydrocarbons (HC) and carbon monoxide (CO) which increase as a result of incomplete combustion; nitrogen oxides (NO_x) which increase with the combustion temperature; particulates and smoke which are affected by several fuel characteristics, while ignition quality is the fuel property having most impact on combustion noise. Other factors contributing to the type and level of exhaust emissions are the lubricant and the design and operation of the engine.

Reductions in emissions can be achieved in a number of ways, some of which may be used in combination. These include turbocharging, after-cooling, exhaust gas recirculation, particulate traps and exhaust catalyzers. Control of the amount of lubricating oil that gets into the combustion chamber also has to be considered. However, close attention to the fuel injection system, employing higher pressures and carefully scheduled timing, may be necessary additional improvements to ensure meeting future emissions standards.[10,11]

The modern high-speed diesel engine has attained a high standard of performance and reliability with excellent fuel economy. The principal challenge to its continuing success is control of its emissions to conform to the tightening legislation to protect the environment. If limits on diesel emissions become too stringent, the automotive diesel, for both commercial and private use, may have a very uncertain future and the inherent benefits of its good fuel efficiency may be wasted.

14.5.2 Emissions Legislation

Legislation on emissions from automotive diesel engines is at different stages and moving at different rates in various parts of the world. This tends to make direct comparisons difficult, with the almost continuous introduction of new legislation. However, interchange of ideas and also of standards is becoming more common, which facilitates assessment of the approaches being taken to identify and control noxious exhaust emissions.

The influence on exhaust emissions of changes in various fuel properties is shown by the engine and fuel interactions in Table 14.1.[12] In most cases, a change in any one fuel property tends to have the same

Table 14.1. Interactions Between Engines and Diesel Fuel[12]

FUEL PROPERTY	PERFORMANCE		OPERABILITY			EXHAUST EMISSIONS						
			Starting		Cold	Smoke[2]		Gases			Particu-	Noise
	Max Power	Specific[1] Fuel Cons	Cold	Hot	Op	White	Black	HC	CO	NOx	lates	
DENSITY												
up	+	-				-						
down	-	+				+						
DISTILLATION												
IBP up						-	+			+		
10% up						-	-	+				
50% up							-	+		+	-	
90% up							-			-	-	
FBP up							-			-	-	
CETANE NUMBER												
up	+	+	+		+	+		+	+	+	+	+
down	-	-	-		-	-		-	-	-	-	-
VISCOSITY												
up	+	-	-	+							-	-
down	-	+	+	-			+				+	+

NOTES

  +   Positive/beneficial effect on engine or environment
  -    Negative/undesirable effect on engine or environment
  (1) At maximum load
  (2) Cold engine

Reprinted by permission of the Council of the Institution of Mechanical Engineers from Proceedings of an International Conference on Petroleum Based Fuels and Automotive Applications, London, England, 25-26 November 1986. On behalf of the Institution of Mechanical Engineers.

directional effect on all the emissions, but there can sometimes be conflicts as a result of the complex relationships involved.[12]

The evolution of emissions legislation in the U.S., from 1969 into the 1990s, is given in Appendix 3,[13] together with standards and directives issued by the European Economic Community, Japan and other countries. Direct comparisons are not always possible because emissions measurement units vary, depending on the class of vehicle and the legislating authority.

It must be pointed out that, while the information contained in Appendix 3 was correct at the time of its publication in 1994, the legislation is under constant review and it is possible that revisions may have been incorporated subsequently. A regular (and generally annual) update of emissions legislation around the world is published by CONCAWE, the oil companies' European organization for environmental and health protection, who compiled the information given in Appendix 3.

At present, over 30 countries have specific measures in place to control diesel smoke, noise, gaseous and particulate emissions. Pertinent aspects of exhaust emissions legislation relating to diesel fuel properties will be covered in the following sections.

14.5.3 Sulfur

Sulfur content was the first fuel property to be controlled as a means of limiting harmful diesel exhaust emissions. Sulfur is present to a greater or lesser extent in all crude oils and, as a result of the distillation process, some sulfur compounds which boil in the same range are present in streams used for diesel fuel blending. During combustion they burn to form acidic by-products (sulfur dioxide, SO_2, and trioxide, SO_3), together with other gases and also solid compounds, such as sulfates. The gases will have an influence on the exhaust odor, while the sulfates will contribute to the particulates burden in the exhaust, as shown by Figure 14.16.[14]

While control of sulfur-containing emissions is by means of the fuel sulfur specification, amendments to the clean air act in the U.S. require measurement of sulfur compounds as unregulated exhaust pollutants. The U.S. EPA adopted a maximum fuel sulfur limit of 0.05%

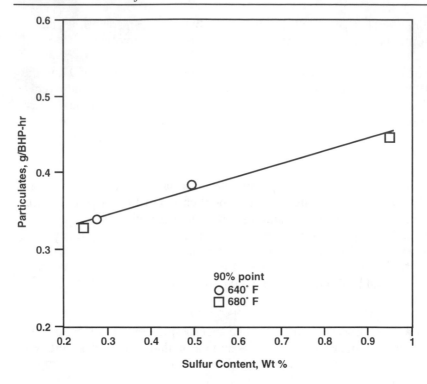

Fig. 14.16. Sulfur content effect on particulates.[14]

mass, effective October 1, 1993, although small refiners will have until 1995 to comply.

In Europe, the council of the EC in 1987 adopted a maximum limit of 0.3%, for implementation by January 1, 1989, although member states were allowed to set a lower limit of 0.2% in heavily polluted areas. From October 1, 1994, a limit 0.2% will apply throughout the EC/EU and a further lowering of the limit to 0.05% mass will be implemented by October 1, 1996. As it is believed that sulfur in the fuel can provide benefits in the form of extreme pressure (E.P.) lubrication properties, there is some concern within the motor industry that too much reduction in sulfur levels could give rise to increased engine bore and ring wear and also cause problems with fuel injection equipment.

14.5.4 Smoke

The appearance of diesel exhaust smoke depends on whether the engine is being started from cold or being driven under load after reaching normal operating temperature. White smoke is the term generally used to describe the smoke produced during cold starting. It is mainly a mixture of water vapor and unburnt or partially burnt fuel which is emitted when starting from cold after a lengthy or overnight shutdown, and before all the cylinders have warmed sufficiently for complete combustion.

This type of emission, which is usually accompanied by a high level of combustion noise, is not controlled by legislation. However, it can be a serious nuisance both to the vehicle operator and the public, particularly if a large fleet of buses or trucks is garaged adjacent to residential properties. In some countries, indirect control can be enforced through local authority by-laws which prohibit increased nuisance and allow for withdrawal of the operating license of any bus or truck company not in compliance.

Engine manufacturers can influence cold smoke emissions through the design of cold start features in the fuel injection equipment and devices to give rapid warm-up. Good maintenance by the vehicle operator is also an important factor. Cetane number, volatility, viscosity, and aromatics content are fuel characteristics which have been shown to have an impact on cold smoking.[14,15] The cranking time needed to start a cold engine will influence the amount of white smoke produced, and Figure 14.17[16] shows the influence of cetane number and cetane improver on the cold start characteristics of a DI engine.

The time to start, time to achieve a stable idle and time to reach 50% of the initial smoke level all increase progressively with decreasing cetane number. Increasing the cetane number by means of an additive (e.g., isooctyl nitrate) gives benefits, although performance does not always match results obtained with natural cetane values.

Black smoke occurs at full engine load due to the air-fuel ratio being too rich. It can be caused by many factors within the control of the operator and is often the result of overloading, overfueling or poor maintenance of fuel injectors. Fuel quality will also affect black

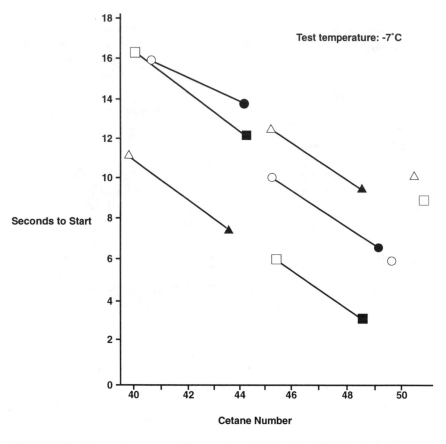

	AVERAGE VISCOSITY cSt at 40°C	AVERAGE VOLATILITY °C at 50% off
○	1.95	245
□	2.55	265
△	3.3	290

● ■ ▲
ADDITIVE TREATED

Fig. 14.17. Cetane improver aids starting of DI engine. [16] *(Reprinted by permission of the Council of the Institution of Mechanical Engineers from Proceedings of an International Conference on Petroleum Based Fuels and Automotive Applications, London, England, 25-26 November 1986.)*

smoke emissions, even from a well-maintained engine, through its influence on fueling rate if its density is higher than that of the reference fuel used for the makers' settings.[16,17,18,19] Cetane number has also been shown to influence black smoke emissions, the level increasing with lower cetane fuels, as shown in Figure 14.18.[16]

Aromatics content, which is not normally specified, can also influence smoke emissions, but the interrelationship between many of the fuel

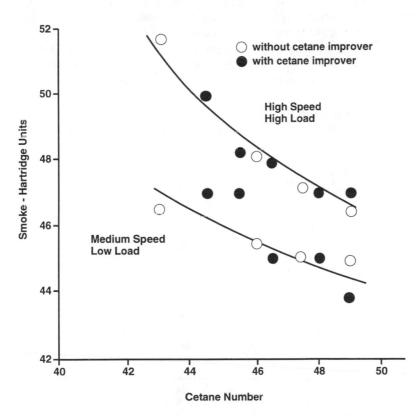

1.6-liter IDI Engine

Fig. 14.18. Influence of cetane number on black smoke.[16] (Reprinted by permission of the Council of the Institution of Mechanical Engineers from Proceedings of an International Conference on Petroleum Based Fuels and Automotive Applications, London, England, 25-26 November 1986.)

365

characteristics makes it difficult to identify the contribution of any individual property. Although they have a higher aromatic content than straight run distillates, catalytically cracked gas oils also have higher densities and lower cetane values, both of which would tend to increase black smoke emissions.

Many independent studies have been carried out to assess the effect of fuel composition, including aromatics, on emissions from a range of heavy-duty diesel engines.[20,21,22,23,24,25] Tests on fuels with 0 up to about 35% total aromatics, typical for currently marketed fuels, showed no influence of total aromatics content on particulates emissions but there were indications of an upturn at higher aromatics contents.[26] However, engines differ in their sensitivity to fuel quality and this is evident from results obtained with two HDD engines (Figure 14.19). Other evaluations suggest that while total aromatics may have no influence on particulates emissions, a more significant role may be played by polycyclic (di- and tri-) aromatics.[27]

It is evident that difficulties in breaking the correlations between aromatics content, density, and cetane number make further studies nec-

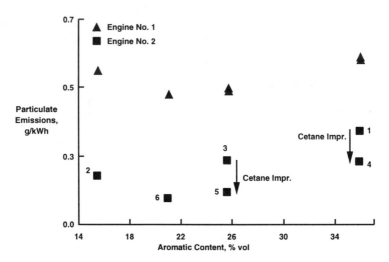

Fig. 14.19. Relationship between particulates emissions and aromatic content.[25] (Reprinted by permission of the Council of the Institution of Mechanical Engineers from Proceedings of the Second Seminar on Fuels for Automotive and Industrial Diesel Engines, London, U.K., 6-7 April 1993.)

essary before definitive conclusions can be drawn on the influence of particular aromatic species.

Limits on aromatics have already been imposed by some authorities. In January 1991, Sweden introduced environmental classifications of diesel fuels for use in urban areas, along with tax incentives for their use. The specifications for the two new fuel classes call for low sulfur levels as well as limiting the aromatics content. Table 14.2 shows specification characteristics for Urban Diesel 1 and 2 and standard Swedish diesel fuel.

Table 14.2. Swedish Urban Diesel Fuels

Fuel Characteristic	Units	Urban Diesel 1	Urban Diesel 2	Typical European
Density	kg/m³	800-820	800-820	820-860
Cetane No. (min)	-	50	47	49
Distillation:				
IBP	°C (min)	180	180	180
95%	°C (max)	285	295	340-370
Sulfur	% m/m	0.001	0.005	0.2-0.3
Aromatics	% v/v (max)	5	20	20-30
P.A.H.	% v/v (max)	0.02	0.1	?
Tax Incentive	$/m³	75	42	-

Since October 1993, U.S. federal regulations have required all highway diesel fuel to meet a maximum sulfur level of 500 ppm (0.05%), and in California a maximum of 10% aromatics is required in addition to the low sulfur limit. European motor companies are also calling for aromatics content to be included in diesel fuel specifications. The ACEA 1994 Fuel Charter recommends limiting the three-ring aromatics content of diesel fuel to 1.0% m/m max.

While these changes are beneficial in lowering particulates emissions, such fuels tend to be low in lubricity. This could result in rapid wear of fuel injection equipment and lead to poorer control of combustion and higher emissions, as discussed in Sections 17.11 and 18.8.

14.5.5 HC, CO and NO$_x$

Current and future legislation is aimed at limiting gaseous and particulate exhaust emissions from diesel engines. Permitted levels vary according to engine or vehicle size and there are also differences from one country or region to another, as indicated in Appendix 3.

The assessment of the influence of some individual diesel fuel properties on emissions summarized in Table 14.1 shows volatility and ignition quality to be relevant properties for both gaseous and particulate emissions, with viscosity affecting only particulates. Fuel sulfur and to a lesser extent aromatics will also have an influence on particulates.

Because of interrelated fuel properties, investigations into the effect of changes in the chemical characteristics of diesel fuels have not enabled clear conclusions to be drawn on the influence of aromatics on engine operation.

A cooperative study undertaken by an engine manufacturer and an oil company in the U.S. showed no major detrimental effects on combustion performance.[28] The tests were carried out on a turbocharged and after-cooled heavy-duty DI engine, using a selection of fuels with aromatics contents ranging between 20 and 50% under different modes of operation. Increasing aromatics had no substantial influence on HC, smoke or particulate emissions during steady-state operation, although NO$_x$ increased.

Emissions all increased with increasing aromatics in the transient tests, as shown in Figures 14.20 and 14.21.[28] These effects may have been due to overfueling because of the test cycle, or due to the density changes mentioned above, or a combination of the two.

Studies carried out in Japan on passenger cars with IDI engines and trucks with DI engines showed that variations in viscosity and cetane number tended to have relatively little influence on passenger car exhaust emissions, but the truck engines were more sensitive to fuel characteristics.[29] Increasing viscosities increased smoke and CO but had little effect on HC and NO$_x$, while cetane levels below 45 gave increased smoke, noise and gaseous emissions. The results are shown in Figures 14.22 and 14.23.[29]

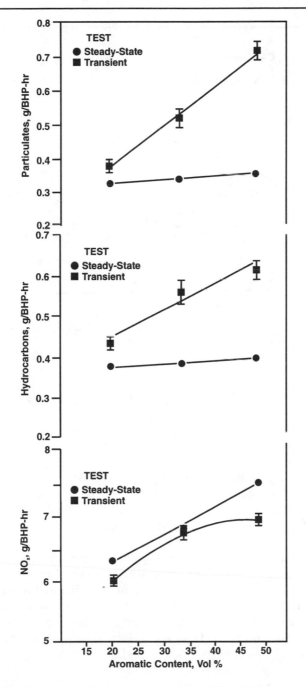

Fig. 14.20. Influence of aromatics content on NO$_x$, HC and particulates.[28]

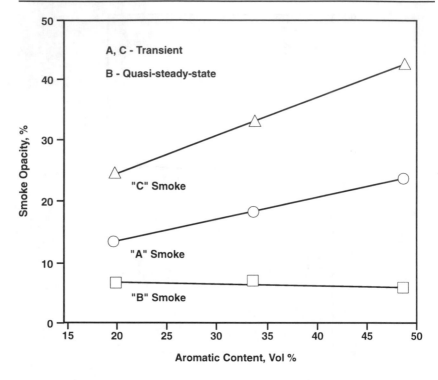

Fig. 14.21. Influence of aromatics on exhaust smoke.[28]

14.5.6 Particulates

Particulates appear in diesel exhaust gases as a result of incomplete combustion of fuel droplets. They consist, essentially, of very small particles of carbonaceous soot onto which molecules of other combustion by-products have been adsorbed or condensed.[30]

Tight particulate emissions standards for heavy-duty diesels came into force in the U.S. in 1991, with a level of 0.25 g/bhp hour for engines in vehicles over 6000 pounds mass and an even lower limit of 0.10 g/bhp hour for urban buses. As mentioned in Section 14.5.3, sulfur in the fuel contributes to particulate emissions, so the proposed reduction of the fuel sulfur limit in the U.S. to 0.05% will help engine manufacturers meet these stringent emissions standards, although it will be a significant cost burden to the oil industry.

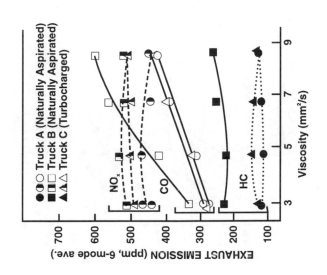

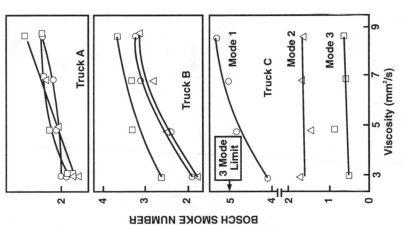

Fig. 14.22. Effect of viscosity on exhaust emissions. [29]

371

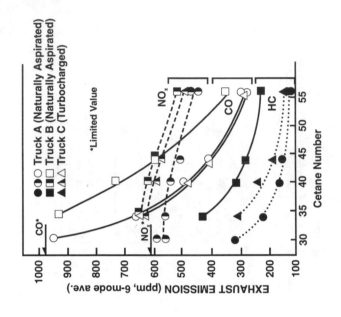

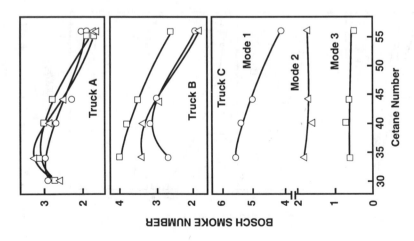

Fig. 14.23 Effect of cetane number on exhaust emissions. [29]

Another way of reducing the level of particulates emissions is to trap them in a filter before they leave the engine exhaust system, although, because of the quantities involved, periodic regeneration of the filter will be required. Various types of filter media have been evaluated to compare retention capabilities and the frequency of regeneration to avoid excessive back-pressure buildup.

Trapping techniques include total filtration of the exhaust gases through monolithic interception filters of porous magnesium silico-aluminate, and partial filtration through impaction and diffusion filters that rely on soot particles being retained on the surfaces of passage-ways through a ceramic foam or a sponge of knitted metal fibers. Interception filters retain a high proportion (80-90%) of the particles but they are blockable and require frequent regeneration to avoid excessive exhaust back-pressure. Impaction and diffusion filters are non-blockable and, consequently, less likely to increase back-pressure, but their particle retention is lower at about 60 to 70%. The three types of particle trap are illustrated in Figure 14.24.[30,31]

Filter regeneration involves burning-off the accumulation of soot. This can be by means of an additional fuel burner or electric heater to raise the exhaust gas temperature to the level needed for combustion, or by the use of a catalyst on the filter medium to lower the ignition temperature of the soot. Further work is in progress on diesel particulate traps to ascertain the viability and endurance of the different systems.

14.5.7 Noise

Noise is also classified as an undesirable emission from diesel engines and much research has been directed into ways of reducing engine noise as well as the other legislated emissions. Low ignition quality of the fuel has already been mentioned as a cause of higher combustion noise levels, when rapid burning of the larger amount of fuel injected before ignition occurs results in higher cylinder peak pressures, producing the characteristic diesel knock. This noise tends to be most evident during cold starting or cool running, as when accelerating after a period of idling or light load operation. Adjustments to injection timing or the amount of fuel injected during the ignition delay period can have some effect in reducing combustion noise with a low cetane number fuel, but retarded timing will result in a fuel consumption penalty. Cetane number is not the only criterion of combus-

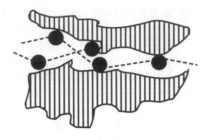

Impaction type: Wire mesh or ceramic fiber; non-blockable

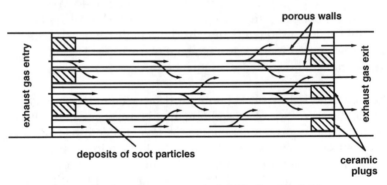

Interception type: Porous ceramic; full filtration; blockable

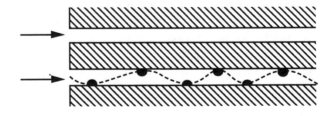

Diffusion type: Porous ceramic; wire mesh and ceramic; non-blockable

Fig. 14.24. The three types of particle trap. [30,31]

tion quality and other fuel characteristics such as volatility and composition can also have an influence on combustion and noise.[32]

In another study,[1] tests were run on different automotive diesel engines: IDI and high-speed DI engines for cars and light commercial
vehicles and heavy-duty DI engines for trucks, at several injection
timings. The results plotted in Figure 14.25[1] show an inverse relationship between combustion noise and fuel consumption for a range
of engine types: four DI engines and two IDI engines, one having
four alternative injection systems.

14.6 Fuel Economy

One of the major advantages the diesel engine has over the gasoline
engine is its better fuel economy. This is brought about mainly by the
higher compression ratio needed to produce autoignition of the fuel,
although other factors such as the absence of an air-throttle and a
higher density fuel also contribute. The inherent design advantages
are supplemented by the combined benefits of a fuel with a lower
basic cost, a higher volumetric heat content and, in many countries, a
significantly lower tax. Where there has been a choice of engine type
as, for example, with passenger cars, the lower fueling costs of the
diesel version usually more than compensate for any initial price difference. However, in Europe this price difference is tending to disappear, making the economics of running a diesel car more attractive,
even without the benefit of a lower fuel tax.

Direct injection diesel engines are widely accepted as the preferred
type of power unit for heavy-duty commercial truck and bus transport
applications, where running and maintenance costs, together with
reliability, are the operator's predominant concerns. Modern trucks
and buses are powered by highly developed DI engines, very often
equipped with turbochargers, intercoolers, and sophisticated electronic management systems for the engine and fuel injection equipment. Such engines have a high power-to-weight ratio and low
specific fuel consumption.

The IDI combustion system is used almost exclusively for the size of
engines required for passenger cars and light commercial vehicles.
Compared with the DI engine, it is relatively insensitive to fuel quality, generally quieter and operates well over a broad speed range. Ad-

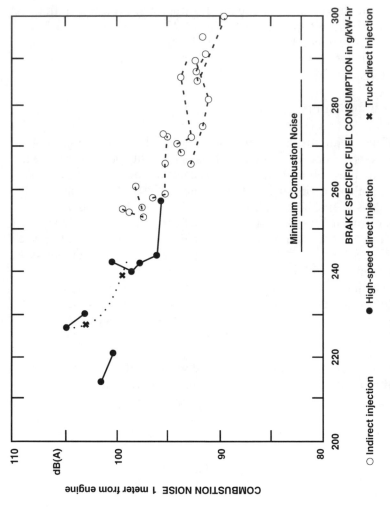

Fig. 14.25. "Trade-off" between combustion noise and fuel consumption for various automotive diesel engines.[1]

ditionally, its structural requirements are not particularly demanding, which means that it is more compact and less expensive to manufacture. However, the IDI engine has poorer fuel economy than the DI engine. Heat losses are proportionately greater with the higher surface-to-volume ratio and there are other losses associated with the rapid transfer of gases into and out of the prechamber, and the higher compression ratio needed to attain autoignition temperatures.[33,34]

Recent years have seen marked improvements in the fuel economy of gasoline engines, which have reduced the efficiency gap between the indirect injection diesel and the latest gasoline engine designs. This has focused attention on the feasibility of developing a small direct injection diesel engine for the passenger car market.[35]

Although the majority of diesel passenger cars still have IDI engines, some models are now available with DI engines. These new-generation small high-speed direct injection engines, ranging in size from 1.9 to 2.6 liters, are turbocharged and equipped with electronic engine management systems and exhaust catalyzers. Some also have exhaust gas recirculation (EGR) for further control of emissions.

Compression ratios of 19.5-22:1 are higher than for large DI engines but important factors in the success of these small DI engines are the very high injection pressures (up to 1000 bar) and very precise control of the timing, volume and injector spray pattern of the minute quantities of fuel required.

Engine design changes to improve fuel economy often provide environmental benefits by reducing one or more of the regulated emissions. Boosting by turbocharging, which is already well-established and applied to a wide range of DI and IDI engines, can give better fuel economy with lower smoke, hydrocarbons and noise as well as increased power. However, measures which are beneficial for some aspects of engine operation may work in the opposite sense for others. For example, changing conditions in the combustion zone to reduce NO_x emissions gives poorer fuel economy and increases particulates. Figure 14.26[34] illustrates the problem of trying to satisfy conflicting emissions requirements. The data, showing the relationship between fuel consumption and NO_x, were obtained on a turbocharged European diesel passenger car. Fuel injection retard and modulated exhaust gas recirculation were evaluated as means of reducing NO_x emissions.

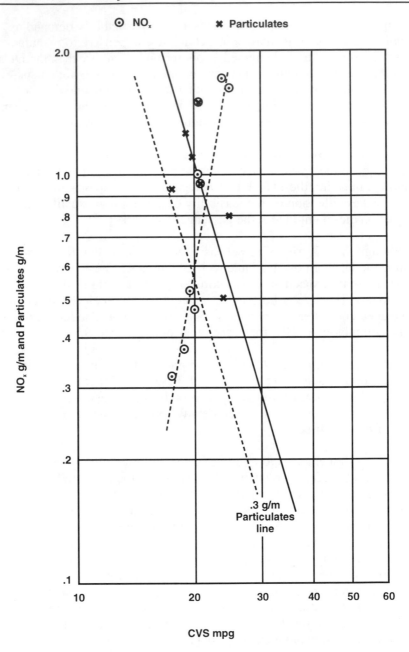

Fig. 14.26. Relationship between NO$_x$ particulates and fuel consumption.[34]

Reducing the NO_x levels worsened fuel consumption and also increased particulates.

A prototype 2.0-liter DI engine,[36] designed for a low emissions vehicle, has a combined pump/injector unit because of advantages in NO_x, smoke and combustion noise levels over the distributor-type pump, used for most small diesel engines.

14.6.1 Looking to the Future

Although diesel fuel quality and the way a vehicle is operated and maintained will have some influence on exhaust emissions and the fuel consumption of individual vehicles, the effects can only be secondary. Prime responsibility for compliance with emissions regulations and improvements in fuel economy rests, inevitably, with designers and manufacturers of engines and fuel injection equipment. To provide a realistic basis for the development of environment-friendly diesel engines, the petroleum industry produces a wide range of reference fuels representing current and future qualities in various countries and regions around the world. Appendix 9 lists some of the reference fuels prepared for this purpose.

References

1. M.F. Russell, "Diesel Engine Noise: Control at Source," SAE Paper No. 820238, 1982.

2. Diesel Engine Reference Book, Ed. L.C.R. Lilly, Butterworths, 1984.

3. Lyle Cummins, Internal Fire, SAE R-100, Society of Automotive Engineers, Inc., 1989.

4. Technical Publication, Lucas Diesel Systems, 1993.

5. Diesel Fuel Injection—An Overview, Bosch Technical Instruction, Robert Bosch GmbH, 1990.

6. Additives Influencing Diesel Combustion, T.J. Russell, Associated Octel Ltd, 1988.

7. "Diesel Fuels," W.H. Kite, Jr., R.E. Pegg, Diesel Engine Principles and Practice, Ed. C.C. Pounder, G. Newnes Ltd 1955.

8. E.N. Cantwell, W.G. Kunz and V.J. Tomsic, "Factors Affecting Demand for Transportation Fuels Through 1995," API Mid-Year Meeting, 1984.

9. *Paramins Post*, Issue 6-1, pp. 8-9, May 1988.

10. I.M. Khan and A.C. Green, "Scheduling Injection Timing for Reduction of Diesel Emissions," SAE Paper No. 750337, 1975.

11. D. Broome, "The Present Status and Future Development of the European Passenger Car Engine," SAE Paper No. 865001, 1986.

12. E.W. Johnson, "The Importance of Diesel and Gasoline Fuel Quality Today and in the Future," Paper No. C309/86, Institution of Mechanical Engineers International Conference on Petroleum Based Fuels and Automotive Applications, London, UK, 1986.

13. "Motor Vehicle Emission Regulations and Fuel Specifications — 1994 Update," Report 4/94, CONCAWE, 1994.

14. E.G. Barry, L.J. McCabe, D.H. Gerke and J.M. Perez, "Heavy-Duty Diesel Engine/Fuels Combustion Performance and Emissions — A Cooperative Research Program," SAE Paper No. 852078, 1985.

15. L.D. Derry and E.B. Evans, "Cold Starting Performance of High Speed Diesel Engines and their Fuels," *Journal of the Institute of Petroleum*, Vol. 36, No. 319, July 1950.

16. R.D. Cole, M.G. Taylor and F. Rossi, "Additive Solutions to Diesel Combustion Problems," Paper No. C310/86, Institution of Mechanical Engineers International Conference on Petroleum Based Fuels and Automotive Applications, London, UK, 1986.

17. F.J. Hills and C.G. Schleyerbach, "Diesel Fuel Properties and Engine Performance," SAE Paper No. 770316, 1977.

18. M. Fortnagel, H.O. Hardenberg and M. Gairing, "Requirements of Diesel Fuel Quality: Effects of Poor-Quality Fuels," API 47th Mid-Year Meeting, May 1982.

19. P. Heinze, "Betriebsverhalten zukunftig moglicher Dieselkraftstoffe in heutigen Motoren,"Technischen Arbeitstagung Hohenheim, 1986.

20. L.C. van Beckhoven, "Effects of Fuel Properties on Diesel Engine Emissions—A Review of Information Available to the EEC-MVEG Group," SAE Paper No. 910608, 1991.

21. W.E. Betts, S.A. Floysand and F. Kvinge, "The Influence of Diesel Fuel Properties on Particulates Emissions in European Cars," SAE Paper No. 922190, 1992.

22. W.W. Lange, "The Effect of Fuel Properties on Particulates Emissions in Heavy-Duty Truck Engines Under Transient Operating Conditions," SAE Paper No. 912425, 1991.

23. L.T. Cowley, A. le Jeune and W.W. Lange, "The Effect of Fuel Composition Including Aromatics Content on Emissions from a Range of Heavy-Duty Diesel Engines," Paper No. CEC/93/ EF03, Fourth International Symposium on the Performance Evaluation of Automotive Fuels and Lubricants, Birmingham, UK, May 1993.

24. M. Booth, J.S. McArragher and J.M. Marriott, "The Influence of Motor Vehicle Emission Control Legislation on Future Fuel Quality," Paper No. CEC/93/EF05, Fourth International Symposium on the Performance Evaluation of Automotive Fuels and Lubricants, Birmingham, UK, May 1993.

25. P. Heinze, "The Influence of Diesel Fuel Properties and Components on Emissions from Diesel Engines," The Institution of Mechanical Engineers Second Seminar on Fuels for Automotive and Industrial Diesel Engines, London, England, April 1993.

26. Y. Asaumi, M. Shintani and Y. Watanabe, "Effects of Fuel Properties on Diesel Engine Emission Characteristics," SAE Paper No. 922214, 1992.

27. W.W. Lange, A. Schafer, A. le Jeune, D. Naber, A.A. Reglitzky and M. Gairing, "The Influence of Fuel Properties on Exhaust Emissions from Advanced Mercedes Benz Diesel Engines," SAE Paper No. 932685, 1993.

28. E.G. Barry, L.J. McCabe, D.H. Gerke and J.M. Perez, "Heavy-Duty Diesel Engine/Fuels Combustion Performance and Emissions — A Cooperative Research Program," SAE Paper No. 852078, 1985.

29. M. Kagami, Y. Akasaka, K. Date and T. Maeda, "The Influence of Fuel Properties on the Performance of Japanese Automotive Diesels," SAE Paper No. 841082, 1984.

30. E. Goldenberg and P. Degobert, "Filtres a Activite Catalytique pour Moteur Diesel," *Revue de l'Institut Francais du Petrole*, Vol. 41, No. 6, Nov.-Dec. 1986.

31. W.R. Wade, J.E. White and J.J. Florek, "Diesel Particulate Trap Regeneration Techniques," SAE Paper No. 810118, 1981.

32. D. Anderson and P.E. Waters, "Effect of Fuel Composition on Diesel Engine Noise and Performance," SAE Paper No. 820235, 1982.

33. R. Chicocki and W. Cartellieri, "The Passenger Car Direct Injection Diesel — A Performance and Emissions Update," SAE Paper No. 810480, 1981.

34. C.C.J. French and D.A. Pike, "Diesel Engined, Light Duty Vehicles for an Emission Controlled Environment," SAE Paper No. 790761, 1979.

35. M.L. Monaghan, "The High Speed Direct Injection Diesel for Passenger Cars," SAE Paper No. 810477, 1981.

36. AVL Focus No. 5/93, AVL List GmbH, A-8020 Graz Austria, 1993.

Further Reading

<u>Bosch Automotive Handbook</u>, 2nd English Edition, 1986.

"Combustion of Diesel Fuel Oils," M.A. Elliott, Diesel Fuel Oils, 19th ASME National Oil and Gas Power Conference, 20 May 1947.

"The Parameters Available for Controlling Diesel Engine Performance and Their Relationship with Performance Parameters," H.C. Grigg, Lucas CAV Ltd., <u>Automotive Microelectronics</u>, Eds: L. Blanco, A. La Bella, Elsevier, 1986.

C.L. Wong and D.E. Steere, "The Effects of Diesel Fuel Properties and Engine Operating Conditions on Ignition Delay," SAE Paper No. 821231, 1982. Ford Motor Company of Europe, Automotive Diesel Fuel — A Specification for Europe, October 1985.

G. Lepperhoff, M. Houben and H. Garthe, "Influences of Future Diesel Fuels on Combustion and Emissions of a DI-Diesel Engine," SAE Paper No. 872244, 1987.

N.J. Beck and O.A. Uyehara, "Factors that Affect BSFC and Emissions for Diesel Engines: Part II Experimental Confirmation of Concepts Presented in Part I," SAE Paper No. 870344, 1987.

G. Greeves and I.M. Partridge, "Experiments and Calculations on Optimizing Fuel Economy and Emissions in Passenger Car Diesel Engines," I.Mech.E Paper, 1979.

G. Greeves and C.H.T. Wang, "Origins of Diesel Particulate Mass Emission," SAE Paper No. 810260, 1981.

G. Greeves, I.M. Khan, C.H.T. Wang and I. Fenne, "Origins of Hydrocarbon Emissions from Diesel Engines," SAE Paper No. 770259, 1977.

H.C. Grigg, "The Role of Fuel Injection Equipment in Reducing 4-stroke Diesel Engine Emissions," SAE Paper No. 760126, 1976.

M.F. Russell, "Recent CAV Research into Noise, Emissions and Fuel Economy of Diesel Engines," SAE Paper No. 770257, 1977.

E.G. Barry, J.C. Axelrod, L.J. McCabe, T. Inoue and N. Tsuboi, "Effects of Fuel Properties and Engine Design Features on the Performance of a Light-Duty Diesel Truck — A Cooperative Study," SAE Paper No. 861526, 1986.

C.J. Potter, J.C. Bailey, C.A. Savage, B. Schmidt, A.C. Simmonds and M.L. Williams, "The Measurement of Gaseous and Particulate Emissions from Light-Duty and Heavy-Duty Motor Vehicles Under Road Driving Conditions," SAE Paper No. 880313, 1988.

J.C. Wall and S.K. Hoekman, "Fuel Composition Effects on Heavy-Duty Diesel Particulate Emissions," SAE Paper No.841364, 1984.

A.K. Smith, M. Dowling, W.J. Fowler and M.G. Taylor, "Additive Approaches to Reduced Diesel Emissions Requirements," SAE Paper No. 912327, 1991.

C.I. McCarthy, W.J. Slodowske, E.J. Sienecki and R.E. Jass, "Diesel Fuel Property Effects on Exhaust Emissions from a Heavy Duty Diesel Engine that Meets 1994 Emissions Requirements," SAE Paper No. 922267, 1992.

T.W. Ryan III and J. Erwin, "Effects on Fuel Properties and Composition on the Temperature Dependent Autoignition of Diesel Fuel Fractions," SAE Paper No. 922229, 1992.

L. Rantanen, S. Mikkonen, L. Nylund, P. Kociba, M. Lappi and N-O Nylund, "Effects of Fuel on the Regulated and Unregulated and Mutagenic Emissions of DI Diesel Engines," SAE Paper No. 932686, 1993.

P. Tritthart, R. Cichocki and W. Cartillieri, "Fuel Effects on Emissions in Various Test Cycles in Advanced Passenger Car Diesel Engines," SAE Paper No. 932684, 1993.

Chapter 15

Diesel Fuel Characteristics
Influencing Combustion

15.1 Formulating Diesel Fuels

Diesel fuels are complex mixtures of hydrocarbon molecules derived
from crude oil by distillation, generally boiling within the temperature
range of 150 to 380°C. They are normally blended from several refinery streams, mostly coming from the primary distillation unit but, in a
conversion refinery, components from other units are often used to
increase diesel production.

The section on manufacture of transport fuels, Chapter 2, described
the various processes available for refiners to use in the preparation of
fuels meeting the appropriate product quality specifications. The
specifications have evolved from the characteristics of fuels originally
supplied for diesel use and the requirements of the engine manufacturer striving for further improvements in performance and reliability.

In the primary refining unit, distillation takes place at atmospheric
pressure, the furnace temperature being set to give maximum distillation without cracking. The quality and quantity of the streams drawn
off will be determined both by their boiling range and by the crude
being used. Table 15.1 gives a general indication of how the type of
crude oil can influence diesel fuel characteristics.[1]

Another important feature of crude oils is that they contain different
proportions of gas, gasoline, middle distillate and residual fuel oil,
which poses problems in meeting the product demand pattern in any
particular market. Figure 15.1 compares the widely varying yields
from atmospheric distillation of different crudes with the demand for

Table 15.1. Influence of Crude Oil Type on
Diesel Fuel Characteristics

DIESEL FUEL CHARACTERISTICS

CRUDE OIL SOURCE	HYDRO-CARBON TYPE	CETANE NUMBER	SULFUR CONTENT	CLOUD POINT	CALOR-IFIC VALUE
UK/Norway	Paraffinic	High	Low/Medium	High	Low
Denmark	Naphthenic	Moderate	Low	Low	Moderate
Mid. East	Paraffinic	High	High	High	Low
Nigeria	Naphthenic	Low	Low	Low	Moderate
Venezuela/ Mexico	Naphthenic/ Aromatic	Very Low	Low/ Medium	Low	High
Australia	Paraffinic	High	Low	High	Low
Indonesia	Paraffinic	High	Low	High	Low

Reprinted by permission of the Council of the Institution of Mechanical Engineers from Proceedings of an International Conference on Petroleum Based Fuels and Automotive Applications, London, England, 25-26 November 1986. On behalf of the Institution of Mechanical Engineers.

the major fuel products in the U.S., Europe and Japan. The only way to balance the refinery production pattern with market demand is by means of downstream conversion processes. Most refineries now have vacuum distillation and at least one type of conversion unit — thermal cracking (including visbreaking and coking), catalytic cracking, and/or hydrocracking — to increase the yield of "white" products by cracking unwanted heavy fractions.

Paraffinic hydrocarbon types are attractive for the manufacture of diesel fuel because of their good ignition quality, although they may pose problems in meeting low-temperature specifications, particularly if they come from crude oils of high wax content. Cracking processes give blend components which have lower wax contents but gas oils from catalytic and thermal cracking also have lower ignition quality, so the refiner has to make a careful choice of blend components to

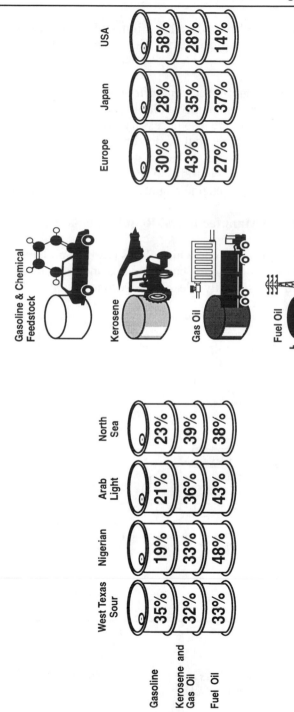

THE REFINER'S CHALLENGE

Fig. 15.1. The refiner's challenge (source: Exxon Chemical International).

387

incorporate in the finished fuel.[2] Figure 15.2[2] summarizes the influence of the different processes on diesel fuel density and ignition quality.

The most important fuel criteria influencing combustion in the diesel engine are its ignition quality, volatility, density and viscosity. Good cold flow properties are also necessary because, if they are not adequate, fuel may not reach the injectors in sufficient quantity to support combustion. However, before going on to discuss individual fuel characteristics, it will be useful to see how a typical diesel fuel specification relates to the engine manufacturers' requirements.

15.1.1 Engine Manufacturers' Requirements

An ideal diesel fuel is one which flows easily at all temperatures, is clean and free from foreign contaminants and separated wax, ignites readily and burns quietly, cleanly and economically.

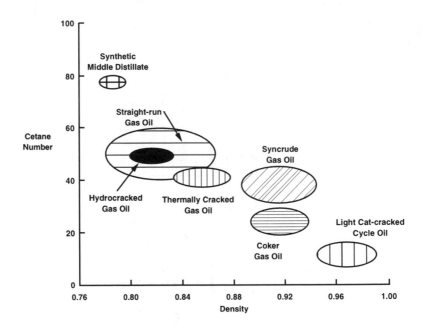

Fig. 15.2. Refinery schemes and gas oil yields (source: Shell International Petroleum Maatschappij).

From the viewpoint of the diesel engine manufacturer, one of the most important fuel properties is density, which relates to the heating value of the fuel and, thus, the energy available to generate power. However, before the fuel can be used effectively in the engine, it has to be drawn from the tank, passed through filters and delivered at a low pressure to the injection pump, where a small volume will be metered precisely, compressed and injected as a finely atomized spray which will ignite and burn in the combustion chamber. To pass through the vehicle fuel system, and also satisfy the performance and reliability needs of the user, involves many other fuel characteristics including cold flow, viscosity, ignition quality, volatility, deposit-forming tendency, cleanliness and corrosivity. Diesel fuel low-temperature characteristics are discussed in Chapter 16.

Viscosity is important for the satisfactory operation of the fuel injection equipment which has to meter, with great accuracy, the small quantity of fuel to be injected.[3] As viscosity varies inversely with temperature, the tolerance band between maximum and minimum viscosity values should be kept as small as practicable. A very high viscosity at low temperature could reduce fuel flow rates and result in incomplete filling of the metering chamber, so that a smaller volume is injected. If the fuel is very viscous, there is a possibility of pump distortion due to heat generated by the shearing action in the small clearances.

Good sealing of the pumping chambers in an injection system depends on the close clearances of the principal elements. While a certain amount of leakage will occur normally, a low viscosity fuel could result in almost total leakage from the pumping elements, particularly at low speeds. In practice, such a situation can arise when attempting a "hot restart" after a brief shutdown, following a period of operation at high load. The temperature of the already-hot fuel injection equipment will be further increased by heat soak-back from the engine, reducing the fuel viscosity so much that leakage may make restarting impossible until the fuel system has cooled down.

Cetane number is normally used to specify the ignition quality of diesel fuel. Running an engine on a lower cetane fuel than it was designed for will make cold starting more difficult. Peak cylinder pressures, combustion noise and hydrocarbon emissions will all increase because more fuel will be injected before ignition, giving less time for combustion to be completed before the exhaust valve opens.

A higher cetane fuel will ignite sooner and will usually reduce combustion noise and emissions, although an extremely high cetane fuel may ignite before adequate fuel-air mixing can take place, possibly causing higher emissions. Power output may be reduced if burning starts too early, causing a big rise in cylinder pressure before the piston reaches top dead center.[4]

Volatility is less directly related to power and economy of the diesel engine but it has an influence on other aspects of engine operation. Starting and warm-up will be helped by a high front-end volatility, while deposits, exhaust smoke and wear may be worsened with very high end point fuels.

The performance characteristics of a diesel engine are determined during its design and development stages. Tests will be carried out to measure power, fuel consumption and exhaust emissions at various speed and loads. To achieve the best compromise between efficiency and emissions, the setting of the fuel injection system for production engines will be finalized using a reference fuel of appropriate quality for the intended sales market. The manufacturers' hope is that the available fuels in that market will be manufactured to tolerances which allow little variation in fuel properties, thus enabling the engine to maintain performance close to the design standard.

A wide range of reference fuels has been agreed on and made available for use by engine manufacturers and other testing laboratories to verify that performance standards and emissions levels will be acceptable, both for home and export markets. Some of the reference fuels available for engine development, homologation and production testing for a wide selection of world markets are listed in Appendix 9.

15.1.2 Typical Diesel Fuel Specifications

Table 15.2 shows typical diesel fuel characteristics and the range of values found in the U.S., Japan and Germany during a worldwide product quality survey carried out in 1994. National standards or guide specifications for automotive diesel fuels marketed in different countries are given in Appendix 8.

Lowest cetane levels are to be found in North America where the specification minimum is 40, while in most other parts of the world the minimum is at least 45 cetane number. This situation has devel-

Table 15.2. Automotive Diesel Fuel: National or Guide Standards and Survey Data, Winter 1994
(source: Exxon Chemical International)

Property	USA East Coast (20 Samples)				Japan Grade 2 (26 Samples)				Germany (20 Samples)			
	Guide Standard ASTM D 975	Values min	mean	max	National Standard JIS K 2204	Values min	mean	max	National Standard DIN 51601	Values min	mean	max
Density, kg/m³ @ 15°C		833.2	846.4	851.9		823.1	835.0	844.5	820-860	824.9	834.3	844.6
Viscosity mm²/s @ 20°C mm²/s @ 30°C mm²/s @ 40°C	1.9-4.1	2.066	2.348	2.711	2.5(min)	3.092	3.859	4.872	2.0-8.0	3.278	3.953	5.120
Sulfur % m/m	0.05(max)	0.013	0.027	0.036	0.2(max)	0.068	0.154	0.194	0.2(max)	0.051	0.124	0.1962
Cetane number	40(min)	41.7	44.9	49.8	45(min)	47.0	54.9	59.0	45(min)	48.0	51.6	53.6
Cetane index 1980 equation 1988 equation		43.5 43.3	45.1 45.2	50.2 50.1		52.8 52.7	55.7 56.5	58.0 59.9		47.8 48.2	50.2 50.2	53.9 53.8
Cloud point, °C		-19	-14	-7		-11	-5	-1		-13	-9	-6
Pour point, °C		-45	-29	-24	-7.5(max)	-33	-16	-8		-41	-32	-26

Table 15.2. Automotive Diesel Fuel: National or Guide Standards and Survey Data, Winter 1994 (Cont.)
(source: Exxon Chemical International)

Property	USA East Coast (20 Samples)				Japan Grade 2 (26 Samples)				Germany (20 Samples)			
	Guide Standard ASTM D 975	Values			National Standard JIS K 2204	Values			National Standard DIN 51601	Values		
		min	mean	max		min	mean	max		min	mean	max
CFPP, °C		-36	-22	-16	-5(max)	-12	-9	-6	-15(max)	-32	-28	-21
Wax content, % m/m, 10°C below Cloud		1.0	1.6	2.1		2.8	4.0	6.1		0.7	1.4	2.8
ASTM D-86 distillation, (°C)									(65% max @ 250°C, 85% min @ 350°C)			
IBP		148	174	206		151	172	209		150	173	186
20%		209	221	229		215	242	260		198	216	235
50%		242	253	264		274	285	298		241	257	280
90%	282-338	305	312	322	330-350	318	336	347		309	328	363
FBP		340	344	352		339	360	375		329	355	388

oped as a result of the large amount of cracking operations needed in North American refineries to meet the high demand for gasoline, leaving relatively low cetane blend components for absorption into diesel fuel. The trend toward higher conversion refineries and the growing need to utilize cracked stocks to increase diesel fuel availability are also evident in other parts of the world. Figure 15.3 illustrates the influence of the different refinery processes on the relative yields of the main fuel products.

Cold property specifications reflect the climate of the country and, because of their influence on cloud point, so, to a certain extent, do distillation characteristics. In Canada, with its extremely cold winter climate, the 90% distillation point is limited to a maximum of 315°C, while in tropical and sub-tropical areas a 90% point as high as 379°C is possible.

Engine manufacturers' requirements are determined by the physical needs of the fuel injection equipment, the performance to be obtained from the engine, and the legislation imposed to control visible and noxious emissions from the engine. These criteria are not always compatible and the end result is usually a compromise.[5]

15.2 Cetane Number

15.2.1 Measurement of Cetane Number

The readiness of a fuel to ignite when injected into a diesel engine is indicated by its cetane number; the higher the number, the easier it is to ignite. As mentioned in Chapter 2, this measure of ignition quality is determined by an engine test, of which the CFR Cetane Engine method, ASTM D 613, is the most widely accepted. The cetane number of a fuel is determined by comparing its ignition quality under standard operating conditions with two reference fuel blends of known cetane number. The reference fuels are prepared by blending normal cetane (n-hexadecane), having by definition a value of 100, with heptamethyl nonane, a highly branched paraffin with an assigned value of 15.

When a fuel has the same ignition quality as a mixture of the two reference fuels, its cetane number is derived from the equation:

Cetane number = % n-cetane + 0.15(% heptamethyl nonane)

393

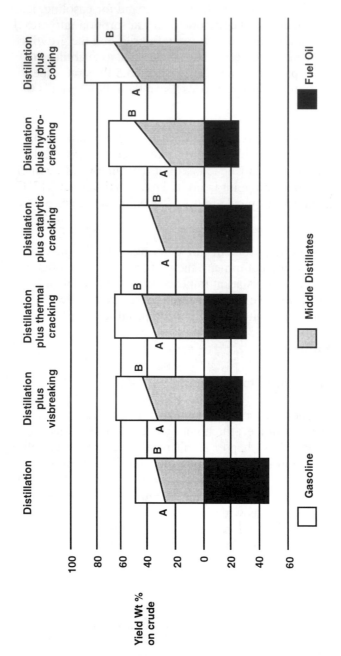

Fig. 15.3. Refinery processing options to meet changing demand (source: Exxon Chemical International).

In practice, the compression ratio of the engine is varied to give the same ignition delay period for the test fuel and two reference blends of higher and lower quality than the test fuel, which differ by less than five cetane numbers. The cetane number of the unknown test fuel is calculated by interpolation between the lowest and highest compression ratios.

15.2.2 Influence of Cetane Number on Engine Performance

Cetane number indicates the readiness of a diesel fuel to ignite spontaneously under the temperature and pressure conditions in the combustion chamber of the engine. The higher the cetane number the shorter the delay between injection and ignition. The graphs in Figure 15.4[6] show the influence of cetane number on ignition delay in different IDI and DI engines.

Most of the engines experienced an increase in ignition delay of about 2° crank angle when the cetane number was reduced from 53 to 41. There were large differences in the levels of delay for the six DI engines but the three IDI engines had similar and relatively short delay periods. Fuel is being injected during the delay period and the amount present when ignition occurs will determine the rate of pressure rise in the initial stage of combustion. This is illustrated by the results plotted in Figure 15.5.[4]

There is some scatter of results, presumably due to compositional differences between the fuels, but a drop in cetane number from 53 to 38 gave a 50% increase in the rate of pressure rise. The use of an additive to improve the cetane number showed a corresponding reduction in the rate of pressure rise. One of the more obvious effects of running on a lower cetane fuel is increased noise, as illustrated by Figure 15.6.[7] Emissions are also increased and the impact of diesel fuel quality on the environment is discussed in Chapter 14.

Specific fuel consumption can be affected slightly by changes in cetane number, as Figure 15.7[7] shows. Three trucks were tested using five fuels varying from 30 to 57 in cetane quality, of which four were blends of straight run fuel with a cracked gas oil.

The general trend shows a reduction in fuel consumption with decreasing cetane number because of the higher heating value of the lower cetane blends. The upturn in consumption below 35 cetane,

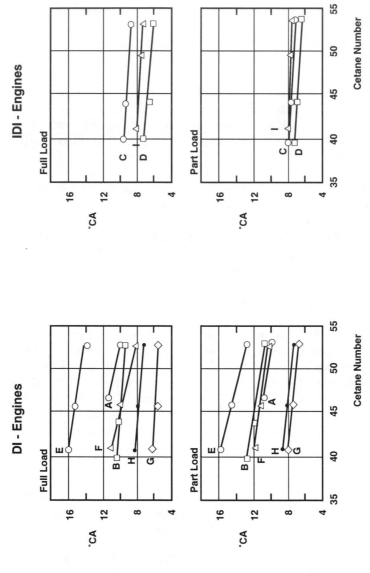

Fig. 15.4. Influence of cetane number on ignition delay.[6] (Reprinted by permission of the Council of the Institution of Mechanical Engineers from Proceedings of an International Conference on Petroleum Based Fuels and Automotive Applications, London, England, 25-26 November 1986.)

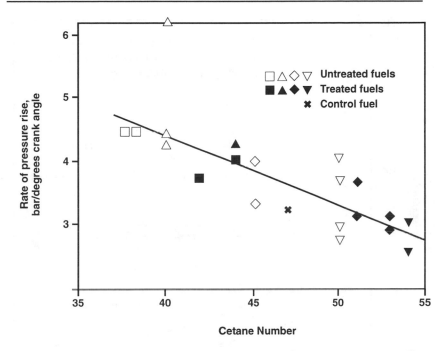

Fig. 15.5. Influence of cetane number on rate of combustion pressure rise. [4]

observed with the one vehicle tested on the 30 cetane fuel, is assumed to be due to poorer efficiency as a result of the longer ignition delay.

15.3 Diesel Fuel Volatility

15.3.1 Measurement of Volatility

The volatility characteristics of a diesel fuel are expressed in terms of the temperature at which successive portions are distilled from a sample of the fuel under controlled heating in a standardized apparatus. Distillation may be carried out by several methods, of which one of the more widely used versions is published as ASTM D 86 in the ASTM Book of Standards.

The distillation or boiling range of the fuel will influence other properties such as viscosity, flash point, autoignition temperature, cetane number and density. As distillation is the way in which the refiner

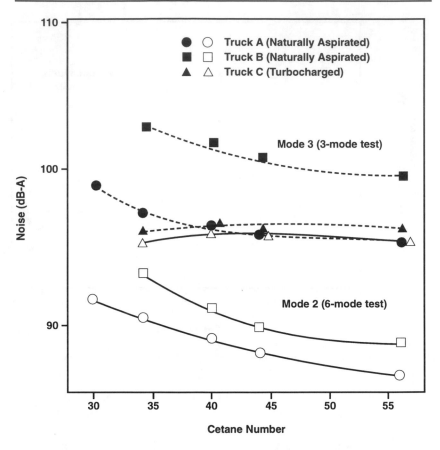

Fig. 15.6. Influence of cetane number on noise. [7]

segregates component streams from which fuels are blended, it is an important factor in the control of fuel quality.

In the ASTM D 86 method, distillation of the fuel is carried out at atmospheric pressure. The vapors, which form as the temperature increases, are condensed and collected in a cylinder graduated in percentages of the initial volume of liquid. Heating is continued until the fuel starts to decompose or until no more condensate can be recovered. Information recorded during the distillation includes:

- initial boiling point (IBP)
- end point (EP) or final boiling point (FBP)
- percent of condensate recovered

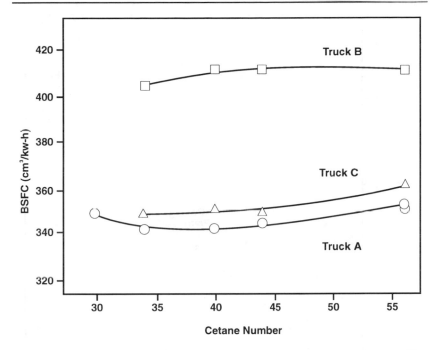

Fig. 15.7. Influence of cetane number on fuel consumption (6-mode test). [7]

- percent residue of non-volatile matter

The temperatures corresponding to the amounts distilled as the test proceeds enable a graph of the distillation curve to be drawn, as illustrated in Figure 15.8.

Automatic apparatus, which duplicates the conditions prescribed for the D 86 manual method, is often used for distillation. A drum carrying a sheet of graph paper rotates as the temperature increases and a photocell detector with an attached pen follows the liquid level in the recovery cylinder, automatically plotting a graph of temperature and volume of condensate as the distillation proceeds.

Above 370°C there is the likelihood of decomposition or cracking of the fuel which effectively terminates the distillation process. However, if no cracking is observed, it is possible to continue to around 400°C, the level to which the thermometer is calibrated, without serious loss of accuracy. The 10% and 90% or 95% recovery temperatures are often used instead of the initial and final boiling points as

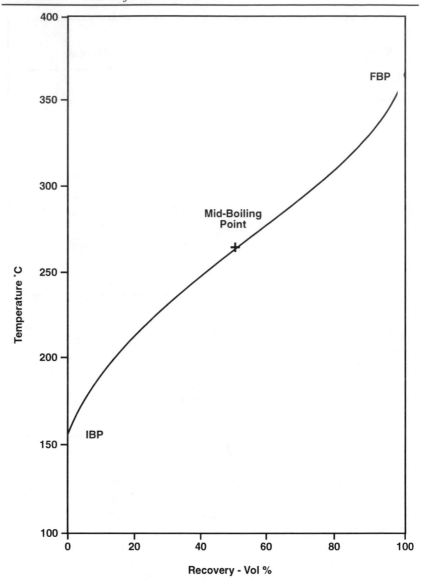

Fig. 15.8. Typical diesel fuel distillation curve (source: Exxon Chemical International).

more precise indicators of the front-end and back-end volatility of the fuel.

15.3.2 Legislative Constraints

Distillation is the means whereby fuel components are produced in the refinery, and some authorities include one or more distillation points in the legal definition of certain product types. This is the situation within the European Community (EC) and, to conform with the stipulated requirements of the common customs tariff, diesel fuels for use in that area must have at least 85% recovery at 350°C and a maximum of 65% recovery at 250°C.

Flash point imposes a constraint on diesel fuel front-end volatility. Established safety practices associated with the classification of storage facilities, transportation and handling of petroleum products, all bring legal implications for flash point. Customs requirements (such as those applying within the European Community) may also be pertinent.

15.3.3 Influence of Volatility on Engine Performance

The distillation range of a diesel fuel does not allow much flexibility to the refiner because of interrelated and interdependent properties and the constraints imposed by other specification items.[8,9] Raising the back-end temperature also raises the cloud point; lowering the temperature at the front-end lowers the flash point and also raises the vapor pressure, which might result in vapor lock in the fuel injection system and cause engine misfiring or failure to restart after a brief shutdown in hot conditions.[10,11]

There are some aspects of the distillation test which can relate to the burnability of the fuel. A low 10% recovery temperature reflects the ease with which the fuel will start to vaporize, while the 90% or 95% temperature gives an indication of the extent to which complete vaporization of the fuel may be expected in the combustion zone of the diesel engine. High boiling components may not burn completely, forming engine deposits and increasing smoke levels.[12] Some diesel fuel specifications impose a limit on the back-end distillation temperature to avoid such problems, although research has shown that, within the 350-400°C range, back-end volatility on its own has a relatively low influence on emissions.[6,13]

Front-end volatility is not usually specified and it will depend on the amount of low boiling material incorporated in the blend. This will be determined by the need to control cloud point by that means, the demand for kerosene for the jet fuel market, and the availability of alternative light fractions. Also, the IBP must be high enough to ensure that the flash point is at or above the specified legal minimum.

The mid-volatility of a diesel fuel has been shown to relate to its smoking tendency, possibly through influence on the injection and mixing of the fuel, but there is also interest in mid-volatility for the calculation of Cetane Index by ASTM D 976. The 50% recovery temperature is specified in some countries and was considered for inclusion in the pan-European diesel fuel specification. This has now been issued as Norme (Standard) Europeene EN 590: 1993. The 50% recovery temperature is not specified, but the test method for Cetane Index is ISO 4264, with the four-variables equation, using density and the 10%, 50% and 90% recovery temperatures (see Section 17.2).

Although engine tests comparing fuels differing widely in boiling range have indicated some differences attributable to volatility,[11] the effects of variations in volatility that are likely with commercial diesel fuels are generally modest and usually far less significant than the influence of individual engine design features.[13,14]

Measurements of exhaust smoke and odor on three DI engines of U.S. origin that were run on typical commercial fuels showed some trends related to volatility. The data are plotted in Figures 15.9 and 15.10.[13] Odor intensity at idle and under load decreased sharply with increasing 95% point, but the effect on smoke level at rated power and at peak torque was small. Engine C, which was supercharged, gave the lowest values of smoke and odor intensity and the data clearly demonstrate that the contribution due to engine design far exceeds that of fuel volatility.[13,15]

The gaseous emissions data shown in Figure 15.11[16] were obtained using the 13-mode test cycle on six other DI engines, of which two were turbocharged. There were no significant trends associated with variation in the 85% recovery temperature of the test fuels but, again, the levels were found to be highly dependent on the engine design.

A similar pattern of results was obtained with three IDI engines at full load and rated speed conditions.[16]

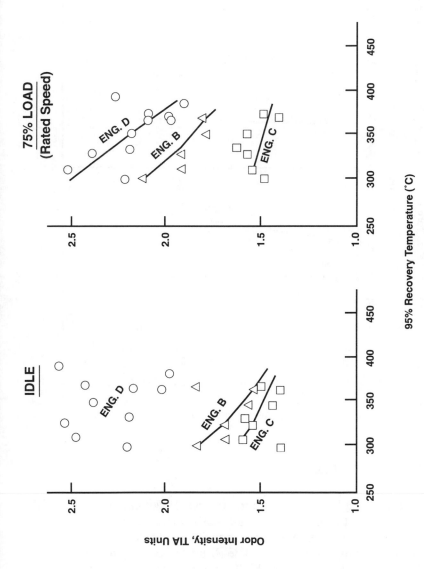

Fig. 15.9. Influence of 95% recovery temperature on exhaust odor intensity.[(13)]

403

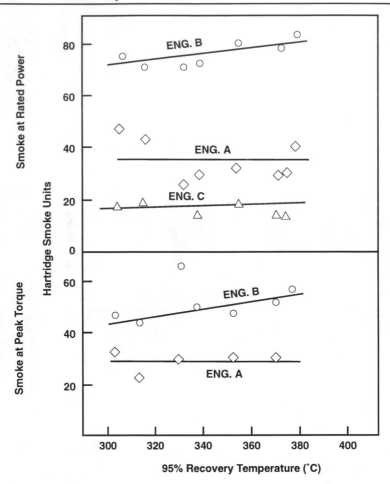

Fig. 15.10. Influence of 95% recovery temperature on exhaust smoke intensity.[13]

15.4 Diesel Fuel Density

Density, the weight of a unit volume of diesel fuel, can provide useful indications about its composition and performance-related characteristics such as ignition quality, power, economy, low-temperature properties and smoking tendency.[10] The SI unit of density is kilograms per cubic meter (kg/m^3), so the higher the density, the heavier the material.

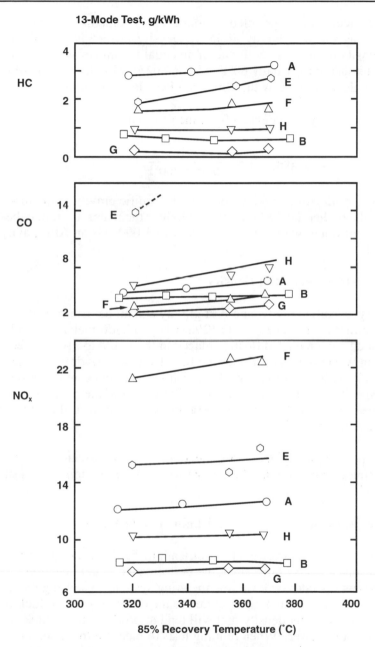

13-Mode Test, g/kWh

Fig. 15.11. Influence of 85% recovery temperature on gaseous exhaust emissions.[16]

This characteristic of petroleum products may sometimes be expressed as Specific Gravity (Relative Density), which is the ratio of the density of the product to that of an equal volume of water at the same temperature, usually 60°F (15.6°C) or as API Gravity, an arbitrary scale developed by the American Petroleum Institute.

The API gravity is calculated from the formula:

$$\text{API gravity (°)} = \frac{141.5}{\text{sp gr } 60/60°F} - 131.5$$

Because of the inverse relationship with specific gravity in the formula, the higher the API gravity, the lighter the material. It may be noted that water, with a specific gravity of 1.000, has an API gravity of 10.0°.

15.4.1 Measurement of Density

Density, specific gravity, and API gravity can all be determined by using either ASTM D 287 or D 1298, where a hydrometer, which is a weighted and graduated float, is placed in the liquid to give a direct reading where the scale crosses the liquid surface. Hydrometers are available with scales that have been calibrated to indicate either density, specific gravity or API gravity. Correction of the reading may be necessary if the fuel sample temperature is not at or near the reference temperature of 15°C.

Both specific gravity and API gravity are now obsolescent terms that are being eliminated with the changeover to the SI system of measurement.[16]

15.4.2 Influence of Density on Engine Performance

Density does not feature in all specifications but it has an importance for various aspects of diesel engine performance. Fuel injection equipment operates on a volume metering system, so a change in density will influence engine output due to the different mass of fuel injected, and a higher density fuel will tend to produce more smoke as well as more power. When running tests to measure the inherent smoking propensity of fuels, the injection pump delivery must be adjusted for equal power to eliminate the density effect.

The relationship between density and power is illustrated by Figure 15.12,[6] which contains bench or vehicle dynamometer data from DI engines for commercial vehicles and IDI-engined passenger cars.[6]

All the engines, which ranged in size from 1.5 to 6.2 liters, were set to manufacturer's specifications to evaluate the influence of density under the standardized full load test conditions of the ECE-R 24 procedure for smoke measurement. Five fuels were tested: a CEC reference fuel, RF-03, two low-cetane fuels and two with higher 85% points. All four had higher densities than the reference fuel. Inspection data on the fuels are given in Table 15.3.[6]

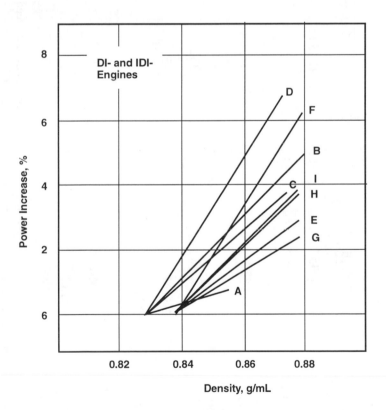

Fig. 15.12. Influence of density on maximum power. [6] *(Reprinted by permission of the Council of the Institution of Mechanical Engineers from Proceedings of an International Conference on Petroleum Based Fuels and Automotive Applications, London, England, 25-26 November 1986.)*

Table 15.3. Test Fuel Properties

Property	Method	Reference CEC RF-03	Test Fuels			
			Lower	Lower Cetane No.	Higher	Higher 85% Pt
Density at 15°C, g/mL	DIN 51757	0.837	0.860	0.874	0.855	0.863
Viscosity at 20°C, mm²/s	DIN 51562	4.00	4.28	4.46	5.74	6.78
Distillation, °C	ASTM D 86					
10% recovery		218	224	227	226	227
50% recovery		262	269	271	283	292
85% recovery		320	322	325	355	367
90% recovery		335	332	336	371	381
Cetane Index	ASTM D 976	51	45	41	49	48
Cetane Number	DIN 51773	53	46	41	49.5	49.5
Aromatic Content FIA, % vol	ASTM D 1319	32	39	46	34	34
Hydrogen, % wt	NMR	13.3	12.6	12.1	13.0	12.9

Reprinted by permission of the Council of the Institution of Mechanical Engineers from Proceedings of an International Conference on Petroleum Based Fuels and Automotive Applications, London, England, 25-26 November 1986.

Improvements in power output for the same injection pump setting ranged from 0.4% to 1.6% per 0.01 g/mL increase in density. Fuel consumption results from the same series of tests are plotted in Figure 15.13[6] and show that volumetric fuel consumption (VFC) generally decreases with increasing density.

The corresponding exhaust smoke readings in Figures 15.14 and 15.15,[6] obtained using a light-obscuration smoke meter, show an upward trend with increasing density. IDI engines were found to be more sensitive than DI engines to the influence of density on smoke levels.

Adjusting the amount of fuel delivered as mentioned above, to achieve the same power output as with the lower density reference fuel, significantly reduced smoke levels, as shown in Figure 15.15. It is important, however, to note that fuel density has a relatively minor effect on levels of smoking compared with the influence of engine design.

15.5 Diesel Fuel Viscosity

The viscosity of a fluid indicates its resistance to flow; the higher the viscosity, the greater the resistance to flow, and it may be expressed either as absolute viscosity or as kinematic viscosity.

15.5.1 Measurement of Viscosity

The unit of absolute viscosity is the poise (P), which is the force in dynes required to move an area of 1 cm^2 at a speed of 1 cm/s past a parallel surface 1 cm away and separated from it by the fluid. The kinematic viscosity unit is the stoke (St), measured in cm^2/s. For numerical convenience, viscosity values are often reported in centipoise (cP) or centistokes (cSt). The term centistoke is being replaced by the preferred SI unit: mm^2/s. The two viscosities are related by the following equation:

$$cP = cSt \times \text{oil density}$$

A widely used laboratory method for determining the kinematic viscosity of diesel fuels is ASTM D 445, which measures the time taken for a fixed volume of the fuel to flow under gravity through a capillary tube viscometer immersed in a thermostatically controlled bath.

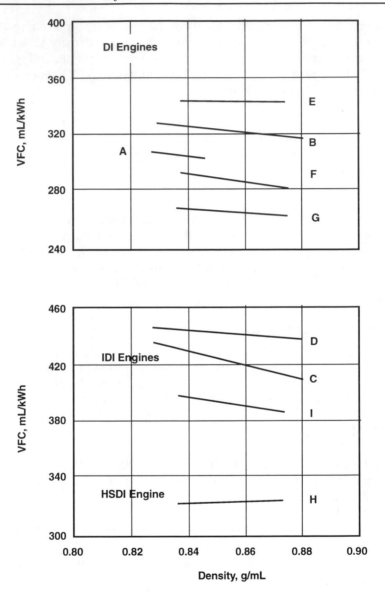

Fig. 15.13. Influence of density on volumetric fuel consumption.[6] (Reprinted by permission of the Council of the Institution of Mechanical Engineers from Proceedings of an International Conference on Petroleum Based Fuels and Automotive Applications, London, England, 25-26 November 1986.)

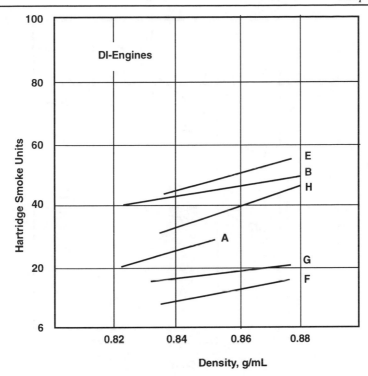

Fig. 15.14. Influence of density on exhaust smoke from DI engines.[6]
(Reprinted by permission of the Council of the Institution of
Mechanical Engineers from Proceedings of an International
Conference on Petroleum Based Fuels and Automotive
Applications, London, England, 25-26 November 1986.)

As viscosity varies inversely with temperature, the relevant temperature at which the viscosity was determined must always be quoted
and, for diesel fuel, it is usually either 20°C or 40°C.

15.5.2 The Influence of Viscosity on Fuel Injector Performance

Detailed observation of the performance of different fuels in selected
injector types has been carried out using a sophisticated test device
designed to simulate the thermodynamic environment encountered in
an engine.[3] High-speed cine photography and laser droplet sizing
were used to facilitate diagnostic analysis. Changes in fuel viscosity

411

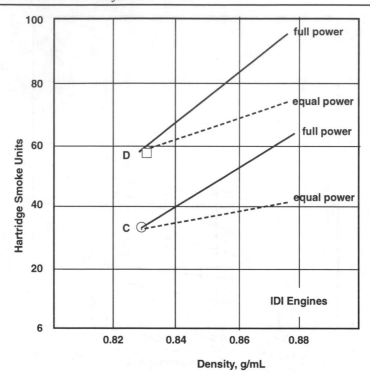

Fig. 15.15. Influence of density on exhaust smoke from IDI engines after adjustment for equal power.[6] (Reprinted by permission of the Council of the Institution of Mechanical Engineers from Proceedings of an International Conference on Petroleum Based Fuels and Automotive Applications, London, England, 25-26 November 1986.)

were found to alter the injector spray penetration rate, cone angle and drop-size distribution.

With a pintle nozzle of the type used for IDI passenger car engines, penetration rate decreased with increasing viscosity while droplet size increased.

The selection of fuel injection equipment and injection timing has to take into account the likely range of viscosities to be encountered as well as typical average values.

15.5.3 The Influence of Viscosity on Engine Performance

Viscosity is an important property of a diesel fuel because of its relevance to the performance of the fuel injection equipment, particularly at low temperatures when the increase in viscosity affects the fluidity of the fuel. However, as mentioned in Section 15.1.1, the viscosity must be high enough to avoid engine-starting difficulties at high temperatures.[3]

In the diesel engine, fuel is injected as a finely atomized spray into the combustion chamber, where it mixes with the hot compressed air and ignites. Fuel is delivered at a relatively low pressure to the metering and pumping chamber of the injection pump, usually a plunger-in-barrel assembly with very fine tolerances, where it is compressed to a pressure of around 200 bar or higher before injection through a single- or multi-hole nozzle. For a given pressure and injector nozzle configuration, the viscosity will obviously influence the quantity of fuel injected.

Diesel specifications usually impose an upper limit on viscosity to ensure that the fuel will flow readily during cold starting, and an additional minimum limit is often also specified to avoid the possibility of serious power loss at high temperatures.

Figure 15.16[3] shows the temperature/viscosity characteristic, plotted on linear/logarithmic paper, for a typical automotive diesel fuel. Also marked on the graph is the viscosity range allowed in the British specification for automotive diesel fuel (BS 2869:1988), together with the cold and hot risk points for the UK climate as defined by Lucas Diesel Systems, a major manufacturer of fuel injection equipment.

On this basis, the maximum level of 5.0 cSt at 40°C just meets the equipment manufacturers' limit for the cold start region, with a viscosity of 47 cSt at -20°C. The lower limit of 2.5 cSt at 40°C would be sufficient to minimize the risk of hot start problems by ensuring a minimum viscosity of 1.6 cSt at 70°C, the estimated pump temperature during operation in UK summer ambient conditions. The British specification was recently revised and the changes included a tightening of the viscosity tolerance band by lowering the upper limit and raising the lower limit, effectively bringing them into line with the two risk-point recommendations.

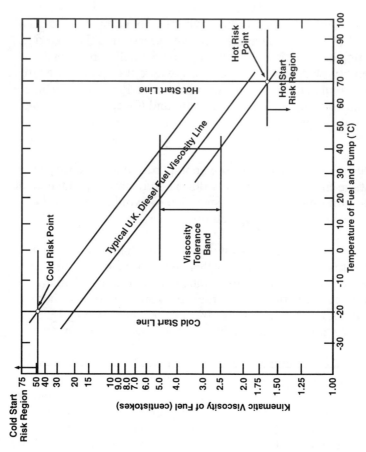

Fig. 15.16. Fuel viscosity requirements for UK climate.[3] (Reprinted by permission of the Council of the Institution of Mechanical Engineers from Proceedings of an International Conference on Petroleum Based Fuels and Automotive Applications, London, England, 25-26 November 1986.)

The interrelationship between many fuel properties makes assessment of individual characteristics difficult, particularly if density variation cannot be avoided. Tests made on three engines of U.S. manufacture, to ascertain the influence of fuel characteristics, have shown that viscosity variation over the range 1.68 to 6.15 cSt at 37.8°C (100°F) had little effect on smoke levels at either rated power or peak torque conditions.[13] There were, however, large differences in smoke emissions between the three models, showing engine design to be the major factor.

15.5.4 The Influence of Aromatics on Engine Performance

A maximum aromatics content of 10% has been imposed by the California Air Resources Board (CARB) and there are calls by the Association of European Automobile Constructors (ACEA) for the aromatics content to be included in diesel fuel specifications.

Aromatic components in the fuel are claimed to contribute to particulates emissions, and the influence of aromatics on combustion is discussed in Section 14.5.4. Opinion on the role of fuel aromatics is varied, with some results showing no influence on the formation of particulates, some showing an influence at levels above 35%, and others indicating that, while mono-aromatics may not have an influence, there could be a contribution from two- and three-ring aromatics. The Fuel Charter published by ACEA in October 1994 sets out guide recommendations for high-quality automotive diesel fuels, which include a maximum limit of 1.0% m/m for the three-ring aromatics content.

References

1. R.E. Williams, CONCAWE Keynote Presentation: "Supply and Economic Considerations," International Conference on Petroleum Based Fuels, Institution of Mechanical Engineers, London, 1986.

2. C.W.C. van Paasen, "Changing Refining Practice to Meet Gasoline and Diesel Demand and Specifications Requirement," Paper No. C316/86, Institution of Mechanical Engineers Symposium on Petroleum Based Fuels, London, 1986.

3. H.C. Grigg, P.S. Renowden and L. Bodo. "The Properties of Diesel Fuel as They Relate to the Operation of Fuel Injection

Equipment," Paper No. C305/86, Institution of Mechanical Engineers Symposium on Petroleum Based Fuels, London, 1986.

4. J.-C. Guibet, <u>Carburants et Moteurs</u>, Tome 1, Editions Technip, Paris, France, 1987.

5. K. Holmes, "The Need for a Global Approach to Diesel Engine Legislation, Coupled to Future Quality and Operation Requirements," Paper No. 317/86, Institution of Mechanical Engineers Symposium on Petroleum Based Fuels, London, 1986.

6. P. Heinze, "Engine Performance and Emissions with Future-Type Diesel Fuels," Paper No. C306/86, Institution of Mechanical Engineers International Conference on Petroleum Based Fuels and Automotive Applications, London, U.K., 1986.

7. M. Kagami, Y. Akasaka, K. Date and T. Maeda, "The Influence of Fuel Properties on the Performance of Japanese Automotive Diesels," SAE Paper No. 841082, 1984.

8. L.D. Derry, E.B. Evans, and C.S. Windebank, "Fuels for High-Speed Diesel Engines," Paper No. 2, Fourth World Power Conference, London, 1950.

9. H.E. Howells and S.T. Walker, "Fuel Limitations on Diesel Engine Development and Application," Paper No. 10, Institution of Mechanical Engineers, London, 1970.

10. P. Eyzat and J.-C. Guibet, "Le Moteur Diesel et son Carburant. Principaux Problemes et Solutions Potentielles," *Revue de l'Institut Francais du Petrole*, Vol. 41, No. 6, Nov.-Dec. 1986.

11. G.M. Barrett and T.D. Freeston, "Fuel Requirements of the Small High Speed Diesel Engine," Proceedings of the Institution of Mechanical Engineers, Papers on Small High-Speed Diesel Engines, No. 8, 1954-55.

12. M. Fortnagel, H.O. Hardenberg and M. Gairing, "Requirements of Diesel Fuel Quality: Effects of Poor-Quality Fuels," API Mid-Year Meeting, 1982.

13. F.J. Hills and C.G. Schleyerbach, "Diesel Fuel Properties and Engine Performance," SAE Paper No. 770316, 1977.

14. C.L. Wong and D.E. Steere, "The Effects of Diesel Fuel Properties and Engine Operating Conditions on Ignition Delay," SAE Paper No. 821231, 1982.

15. C.G. Schleyerbach, J. Drummond and M.J. McNally, "Performance of Future Type Diesel Fuels," Paper No. C114/82, Institution of Mechanical Engineers Conference on Diesel Engines for Passenger Cars and Light Duty Vehicles, London, 1982.

16. P. Heinze, "Betriebsverhalten zukunftig moglicher Dieselkraftstoffe in heutigen Motoren," Technischen Arbeitstagung Hohenheim, Germany, April 1986, Mineraloel Technik 4, March 1987.

Further Reading

"Fuel Injection System Requirements for Different Engine Applications," P. Howes, Symposium Paper No. 9, Institution of Mechanical Engineers, London, 1970.

"Fuel Property Effects on Fuel/Air Mixing in an Experimental Engine," K.R. Browne, I.M. Partridge and G. Greeves, SAE Paper No. 860223, 1986.

"Diesel Fuel Quality and its Relationship with Emissions from Diesel Engines," CONCAWE Report 10/87, 1987.

"The Present Status and Future Development of the European Passenger Car Engine," D. Broome, SAE Paper No. 865001, 1986.

"Heavy-Duty Diesel Engine/Fuels Combustion Performance and Emissions — A Cooperative Research Program," E.G. Barry, L.J. McCabe, D.H. Gerke and J.M. Perez, SAE Paper No. 852078, 1985.

CONCAWE Report No. 86/65, "The Relationship Between Automotive Diesel Fuel Characteristics and Engine Performance," 1986.

E.G. Barry, J.C. Axelrod, L.J. McCabe, T. Inoue and N. Tsuboi, "Effects of Fuel Properties and Engine Design Features on the Performance of a Light-Duty Diesel Truck — A Cooperative Study," SAE Paper No. 861526, 1986.

Ford Motor Company of Europe, "Automotive Diesel Fuel — A Specification for Europe," October 1985.

T.J. Callahan, T.W. Ryan III, L.G. Dodge and J.A. Schwalb, "Effects of Fuel Properties on Diesel Spray Characteristics," SAE Paper No. 870533, 1987.

Chapter 16

Diesel Fuel Low-Temperature Characteristics

16.1 Diesel Fuel Low-Temperature Properties

Diesel fuels are complex mixtures of hydrocarbons, usually prepared from blends of two or more middle distillate streams from the atmospheric crude unit, with the possible inclusion of some vacuum gas oil or cracked material. As much as 20% of the diesel fuel can consist of relatively heavy paraffinic hydrocarbons which have limited solubility in the fuel and, if cooled sufficiently, will come out of solution as wax.

Wax in a vehicle fuel system is a potential source of operating problems, so the low-temperature properties of the fuel are defined by wax-related tests that measure the temperature at which:

- the separation of wax out of solution is first observed.

- the amount of wax is considered sufficient to restrict flow in a vehicle fuel system.

- the amount of wax is sufficient to cause complete gelling of the fuel.

16.1.1 Cloud Point

The cloud point test measures the temperature at which wax first becomes visible when the fuel is cooled, and the method is published by the ASTM with the designation D 2500. In the test, a small sample of the fuel is placed in a glass jar, cooled at a specified rate and examined at intervals of 1°C. A thermometer is immersed in the fuel to measure the temperature at the bottom of the jar, where the first waxes appear. It is a subjective test, depending on the operator's judgment that wax particles are visible.

Interlaboratory correlation tests to establish the precision of the test have shown that duplicate results on identical fuel samples, when tested by the same operator using the same apparatus, should not differ by more than 2°C. This value is referred to as the Repeatability of the test. When comparing results obtained by different operators in different laboratories, the difference should not be more than 4°C. This is known as the Reproducibility of the test. Both values represent the 95% limit of confidence which can be placed on results obtained from the test.

16.1.2 Wax Appearance Point

Another procedure used to determine the temperature at which wax starts to come out of solution is the Wax Appearance Point, ASTM D 3117. In this test, a 25mL portion of the fuel is agitated while being cooled in a double-walled (Dewar-type) jacketed, clear glass tube. The cooling bath is an unsilvered vacuum flask and mixing of the fuel is by means of a twin-helix stirrer operating at 110 strokes/minute through a 50mm amplitude within the test portion. The test is terminated when a distinct swirl of wax crystals in the fuel makes the stirring pattern obvious. The thermometer is read immediately to the nearest 0.2°C.

With agitation there is no temperature gradient, so the wax crystals will be dispersed throughout the fuel and the precise temperature at which this occurs is reported. In the cloud point test the jar is examined at intervals of 1°C to check for wax crystals in the coldest part of the fuel, at the bottom of the jar. The wax appearance point test shows better precision than cloud point, with a repeatability of 0.8°C and a reproducibility of 2.2°C. No data are included in the ASTM Book of Standards to correlate wax appearance point with cloud point. Some studies have shown the two values to be almost equivalent while others found differences of 1 or 2° about the cloud point.

16.1.3 Pour Point

The pour point test is used to measure the temperature at which the amount of wax out of solution is sufficient to gel the fuel. It is carried out in a similar way and under the same controlled cooling conditions as cloud point, except that the thermometer is located with its bulb just below the surface of the fuel, and checks on the condition of the fuel are made at intervals of 3°C.

The pour point test method is published as ASTM D 97. Checks are made by briefly removing the test jar from the cooling bath and tilting it to see whether the fuel flows. This procedure is continued until the fuel fails to move when the jar is held horizontally for 5 seconds. At this point the fuel is not completely solidified; movement is prevented by the formation of an interlocking wax structure. As little as 2% wax out of solution can result in the remaining 98% of liquid fuel being locked in by the wax structure.[1]

The pour point is the lowest temperature at which the fuel was observed to flow. This will be 3° above the temperature of the last check, when the fuel was seen to have gelled. The repeatability of the test is 3°C and its reproducibility, for comparisons between laboratories, is 6°C.

16.1.4 Significance of Cloud and Pour Point

These tests measure two physical properties of diesel fuel. They are the temperatures at which: (1) wax starts to come out of solution as the fuel cools and (2) when there is sufficient wax to prevent the fuel from flowing under the low-shear conditions imposed when the test jar is tilted. Neither test measures the intermediate temperature at which the amount of wax becomes sufficient to restrict flow in a vehicle fuel system.

Both tests have been used for a long time to define the low-temperature properties of diesel fuels, but neither indicates precisely how the fuel will perform in vehicle service. This is particularly true if the fuel contains an additive to improve its low-temperature properties.

In many instances the cloud point underestimates the ability of the fuel to perform at low temperatures, while the pour point tends to be over-optimistic. Data obtained by different researchers have been plotted in Figure 16.1[2] to show how vehicle operability relates to the cloud point of the fuel. The diagonal line on the graph indicates where the points should lie if the laboratory test prediction is correct. A large number of fuels were found to be operating satisfactorily at temperatures well below their cloud point. These included base fuels as well as those containing an additive which improved their low-temperature performance without affecting the cloud point.

A similar plot, showing the relationship between pour point and operability, is also given in Figure 16.1. In this case, most fuels reached their operability limit before cooling to their pour point temperature, irrespective of whether or not they contained a cold flow improver additive.

The low-temperature operability limit is defined as the lowest temperature at which acceptable performance is possible. Interpretation of vehicle operability test results will be covered in more detail later.

To overcome the limitations of cloud and pour point in defining diesel fuel quality, other laboratory tests have been developed to predict fuel performance in service. The new methods aim to simulate realistic operability criteria by assessing the ability of a waxy fuel to flow through a filter screen, as it has to do in a vehicle fuel system.

16.1.5 Cold Filter Plugging Point

A test which has become widely accepted in Europe and other temperate regions of the world to predict low-temperature performance is the Cold Filter Plugging Point of Distillate Fuels (CFPP), IP 309/80.[2,3] The method is also published as a European Standard by CEN, EN116:1981, and as a national standard in various countries around the world. CFPP does not correlate well with the low-temperature performance of North American fuels in North American equipment, and it is not in the ASTM Book of Standards (see Section 16.1.6).

The CFPP test was developed from vehicle operability data generated during a winter field trial on a fleet of European diesel vehicles. Fuels of different origins and base qualities were tested to observe how treatment with a pour depressant affected the low-temperature

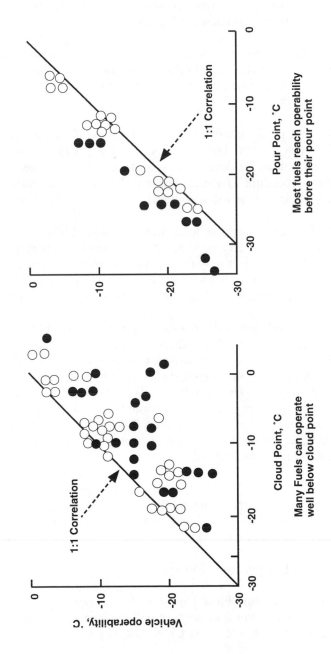

Fig. 16.1. Correlation of cloud point and pour point with diesel vehicle low-temperature operability. [2]

performance of the test vehicles. Additive treating levels were chosen to give good pour point reduction and tests were made when suitable below-cloud temperatures were forecast.

The field trial showed that operation down to lower temperatures was possible when a fuel was treated with the additive. Failures were due to fuel starvation caused by wax accumulation on the filters in the vehicle system.

The field test results confirmed the inadequacy of the pour point as a predictive test for vehicle operability at low temperatures, and provided the stimulus for development of the CFPP as a more realistic laboratory procedure. The new test was found to be relevant for untreated fuels as well as those containing a wax modifying additive.

The CFPP test measures the lowest temperature at which 20 mL of the fuel will pass through a fine wire mesh screen of 45 micrometers nominal aperture in less than 60 seconds. Checks are made at 1°C intervals as the fuel cools under conditions similar to those for cloud and pour point. Figure 16.2 shows the layout of the test apparatus.

The closer correlation with operability given by the CFPP is illustrated in Figure 16.3.[2] While there is some scatter of results around the 1:1 correlation line due to differences between fuels and vehicle systems, the prediction is better than was obtained using either cloud point or pour point. The operability data were obtained from field and chassis dynamometer tests on a wide range of commercial vehicles and passenger cars which were mainly European in origin. The fuels, both with and without flow improver treatment, came from a variety of crude and refining sources.

The precision of the CFPP test, determined in 1988 on a range of European fuels, showed the repeatability to be 1°C, while the reproducibility varied with the CFPP level, ranging from 3 to 6°C over a temperature range of 0 to -35°C.

16.1.6 Low-Temperature Flow Test

The Low-Temperature Flow Test (LTFT) was developed in the U.S. as a test to predict how a diesel fuel will perform at low temperatures. It uses the same concept as the CFPP test by requiring a quantity of

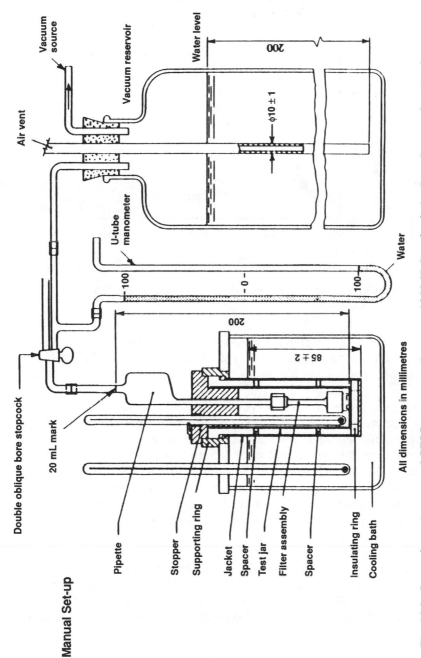

Fig. 16.2. General arrangement of CFPP apparatus (source: 1989 IP Standards for Petroleum and its Products).

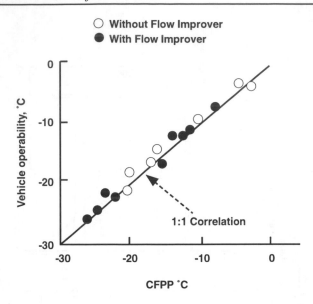

○ Without Flow Improver
● With Flow Improver

Fig. 16.3. Correlation of CFPP with diesel vehicle low-temperature operability. [2]

chilled fuel to pass through a fine mesh screen within a short period of time, but detail features are different. The arrangement of the test equipment is shown in Figure 16.4.[4]

A 200mL portion of test fuel is cooled at 1°C/hour to the chosen test temperature, and then a vacuum of 6 inches of mercury is applied to draw the fuel through a 17 micrometer (17 micron) filter screen. Results are reported as the lowest temperature at which 180 mL passes through the screen in less than 60 seconds. A study by CRC in 1981 showed the LTFT procedure to be a better predictor than CFPP of the low-temperature operability limit of some U.S. diesel vehicles. These findings were confirmed by a later, independent field test and Figure 16.5[5] shows a plot of the data.

A variation of the LTFT procedure uses the same apparatus but employs a faster rate of cooling (2°C/hour) and three different filter screens of 37, 42 and 125 micrometers porosity. The vacuum is applied at 1°C intervals as the fuel is cooling and filtration is made sepa-

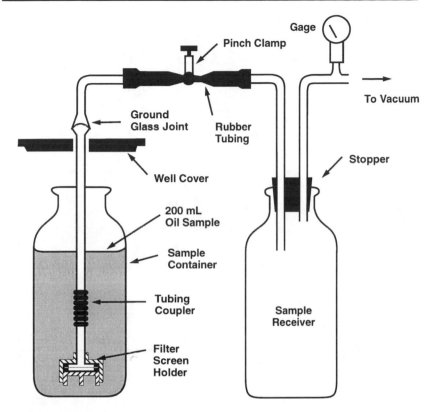

Fig. 16.4. General arrangement of LTFT apparatus. [4]

rately through each screen. This provides a range of test severities to see whether a more satisfactory correlation with vehicle performance can be achieved. New procedures are also being evaluated in the search for a better test.

Another approach to predicting vehicle operability is to use the cloud and pour points of the fuels, because the operability limit generally falls somewhere between those two fuel properties. The Wax Precipitation Index (WPI) is a function of cloud point (CP) and the spread between cloud point and pour point (CP - PP). The empirical equation below correlates with the average estimated minimum operating temperatures of the vehicles in the CRC field test, although there is wide scattering of data points about the average values. [5]

$$WPI = CP - 1.3(CP - PP - 1.1)^{0.5}$$

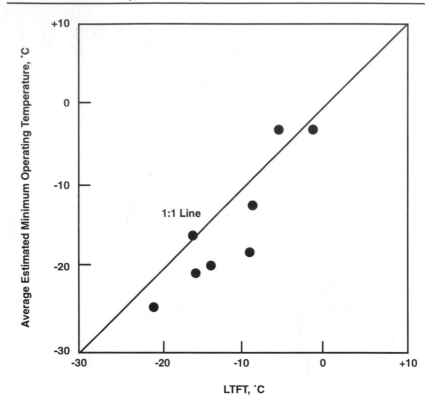

Fig. 16.5. CRC diesel field test 1981 results.[5]

The equation implies that, for a given cloud point, vehicle operability improvements are related to pour point reduction, but this will not necessarily be correct for data from other field tests.

16.1.7 Simulated Filter Plugging Point (SFPP)

The CFPP test has been incorporated into many diesel fuel specifications as the principal criterion of low-temperature quality. Application of the test has enabled refiners to increase production to meet the demand for diesel fuel through the use of flow improver additives.

The late 1980s saw a trend in Europe towards further improvement in the low-temperature quality of diesel fuels. This was stimulated partly by two consecutive severe winters in the mid-1980s and also by the introduction of the concept of product differentiation in the diesel fuel market.

Improved diesel fuel properties were achieved through changes in fuel formulation, increased use of flow improvers, and the application of more advanced flow improver technology. As a result, fuels offering low-temperature operability at temperatures more than ten degrees Celsius below the fuel cloud point became available. For such fuels with a large Cloud-CFPP difference, a new laboratory predictive test was required, since it was found that CFPP was not always a reliable predictor of operability in vehicle service.

The Simulated Filter Plugging Point (SFPP)[6] test was developed by a group of European oil companies in response to that need. Figure 16.6 shows the schematic layout of the test apparatus. The SFPP test was based on a CFPP apparatus, with a number of significant modifications. In particular, these include a controlled (and slower) rate of cooling; a finer mesh filter; a higher vacuum, progressively applied and controlled convection within the test cell. The criterion for a "Pass" result is for the 5 mL pipette to fill within 60 seconds of the vacuum being applied and for the fuel to flow back into the test jar within 40 seconds when the vacuum is switched off.

A special adapter kit is available for quick conversion of an existing standard automatic CFPP apparatus to operate as an SFPP tester. As conversion back to CFPP testing is equally rapid, the need for two separate sets of apparatus is avoided.

The improved protection given by the SFPP is illustrated in Figure 16.7. The correlation with vehicle operability, based on fuel tank temperatures of vehicles critical with regard to cold weather waxing problems, was obtained for 129 fuels from various sources. A few of the fuels did not contain a flow improver additive but the majority had been treated with different types of additive to improve cold weather performance (see Section 16.2).

The precision of the SFPP, as determined in 1993, gave a repeatability of 1°C and a reproducibility of 3.2°C. These values were derived by statistical analysis of results obtained by 13 laboratories evaluating 20 fuels having cold properties covering the full range of temperatures in EN 590, the specification for diesel fuels throughout Europe, which

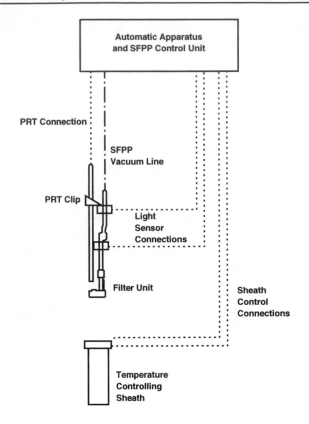

Fig. 16.6. Schematic layout of SFPP apparatus (source: Exxon Chemical International).

was published in 1992 by the European Standards Organization (CEN).

The SFPP has been submitted to CEN for consideration as a new European standard and may at some future stage replace CFPP as the low-temperature quality criterion for certain fuels in EN 590.

16.2 Additives to Improve Cold Weather Performance

Additives to lower the pour points of engine lubricating oils have been in use for many years. The effective application of additives to improve the cold properties of middle distillate fuels is, however, a

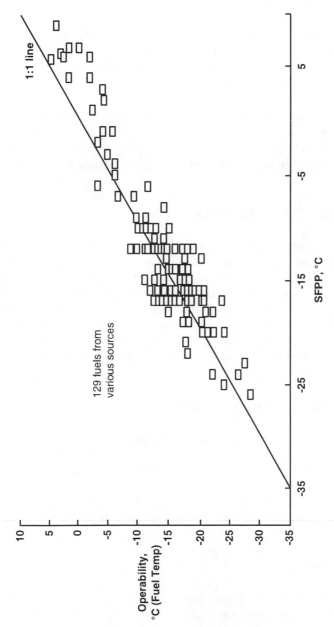

Fig. 16.7. Correlation of SFPP with vehicle operability (source: Exxon Chemical International).

more recent development. Interest in their use for diesel fuels dates from around 1960, after the development of a pour depressant for domestic heating oil.

The selection of additives available at the present time has evolved from pour depressants to include filterability or flow improvers, cloud point depressants, and additives which lessen the tendency for wax crystals to settle in the fuel.

The most commonly used unit for the treat rates of additives for diesel fuels is parts per million, which may be expressed as a percentage, with 0.1% being equal to 1000 ppm. Usually, these treats will be on a weight basis, although some additives may be applied on a volume basis.

16.2.1 Wax Crystal Modifiers (WCM)

Most diesel fuels contain a significant proportion of paraffinic components, which include waxes. Paraffins are important because they have good ignition characteristics but, due to their limited solubility, when the fuel cools, wax crystals form that can restrict flow in the vehicle fuel system. Additives that improve the low-temperature properties of diesel fuels by changing the way the wax crystals grow are known as wax crystal modifiers. The different types of wax modifier additive will be dealt with in the chronological order of their appearance on the scene.

16.2.2 Pour Point Depressants (PPD)

The use of pour point depressants in lubricating oils has been long established, but a different type of additive was needed for distillate fuels and these began to make their appearance in Europe and North America during the 1950s. Pour point depressants work by interacting with the waxes in the fuel to modify their size and shape, making them more compact and less able to form interlocking structures that would prevent the fuel from flowing. The mechanisms of wax crystallization and pour point depression are described in detail by Holder and Winkler.[1]

Some early attempts to lower fuel pour points were unsuccessful because the additives used were lube oil pour depressants. These had little or no effect on the fuel because they had been developed to interact with waxes in lubricating oils, which differ from those normally found in middle distillate fuels. Extensive screening tests were carried out to identify materials capable of modifying the growth habit of the waxes in diesel fuel.

Natural wax crystals that precipitate when an untreated fuel is cooled below its cloud point form as thin plates which overlap and interlock, eventually creating a structure that gels the fuel. When a pour depressant has been blended into the fuel, the resulting crystals are smaller and thicker than those formed naturally, as shown in Figure 16.8. These changes in the size and shape also reduce the tendency for adjacent crystals to interlock, thereby lowering the pour point of the fuel. The depressed pour point will be reached after further cooling to a lower temperature, bringing more wax out of solution. Gelling or solidification of the fuel will occur when all of the additive has been adsorbed by the precipitated wax, allowing the formation of larger, unmodified crystals which readily interlock.

One of the first applications for a fuel pour point depressant was for domestic heating oils in Canada, where the severe winter climate caused problems in the distribution and use of stove oil. Pour point depressants are still used in countries where pour point is the low-temperature specification. Usually it is easy to modify wax crystals and achieve a substantial pour depression as Figure 16.9 shows.

Diesel vehicle road tests in North America and Europe during the 1960s had also confirmed the additive's effectiveness in lowering the minimum temperature at which satisfactory operation was possible.

Unfortunately, the improvement in vehicle operability was smaller than that indicated by the reduction in pour point and, for fuller exploitation of this type of additive, a more satisfactory predictive test method was required. The development of such a test, the Cold Filter Plugging Point (CFPP), is described in Section 16.1.5.

16.2.3 Flow Improvers

Flow improvers are the types of additive used to lower the CFPP, and they are sometimes referred to as middle distillate flow improvers (MDFI) because they can be applied to distillate heating oils as well as to diesel fuels. They function in a similar way to pour point depressants but to a greater extent. They interact with the waxes that

Without pour point depressant

With pour point depressant

Fig. 16.8. Photomicrographs of wax crystals without and with a pour point depressant additive (source: Exxon Chemical International).

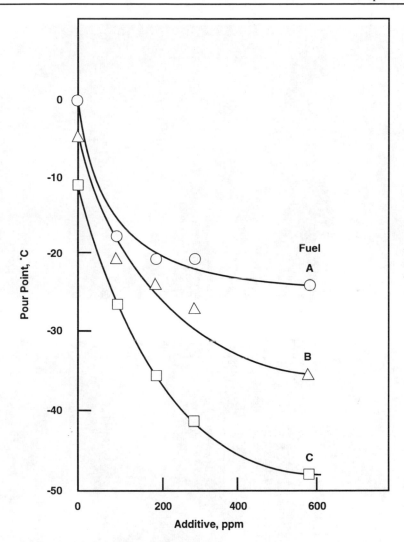

Fig. 16.9. Response of three diesel fuels to pour point depressant (source: Exxon Chemical International).

separate from the fuel as it cools, making small, three-dimensional crystals rather than the larger platelet crystals produced by an untreated fuel.[7] Figure 16.10 shows the greater degree of wax modification given by the flow improver additive.

435

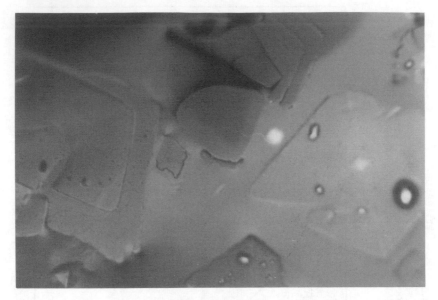

Without MDFI

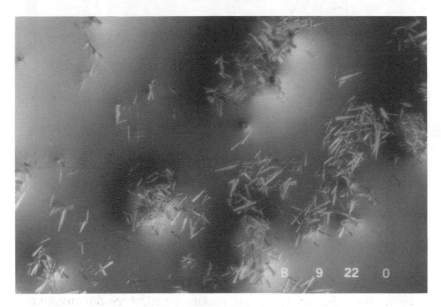

With MDFI

*Fig. 16.10. Photomicrographs of wax crystals without and with a middle
distillate flow improver (MDFI) (source: Exxon Chemical
International).*

436

Due to the presence of the additive molecules, modified wax crystals are also less prone to attach themselves to each other and form agglomerates that could restrict the flow of liquid fuel through the lines and filters in the vehicle system. Modified wax crystals are able to pass the fairly coarse prefilters and strainers that are fitted to hold back large items of foreign material, but they are too large to go through the main filter of paper, felt or cloth protecting the closely machined clearances of the fuel injection equipment. Because of the shape of the modified crystals, the wax layer on the filter is permeable, allowing liquid fuel to pass through, whereas the large unmodified platelet crystals readily interlock, making a structure that will impede flow through the waxy layer on the main filter or even the coarse, woven mesh strainers. The photograph in Figure 16.11 illustrates how the presence of a flow improver additive increases the wax tolerance of the fuel system. The engine gave the same performance as the untreated fuel when operating on the higher cloud point treated fuel, even though there was much more wax out of solution.

A number of different materials have been found to be effective as CFPP improvers, but all the products in current commercial use are ashless copolymers of ethylene and vinyl acetate or other olefin-ester copolymers. These additives have no influence on the fuel other than its low-temperature properties and are compatible with the other types of additive used in automotive diesel fuel. Treating levels can range from 50 to more than 500 ppm, depending on the CFPP improvement required and the responsiveness of the fuel.

The average temperature difference between cloud point and CFPP in a recent survey of European winter diesel fuels was about 12°C, although the spread for individual fuels varied widely, from 5 to as high as 20°. The base CFPP is usually 1-3° below the cloud point so, on average, the additive has to lower the CFPP by about 10°. Typical response curves are shown in Figure 16.12, in which two different additives are compared.

As for most additives, flow improvers are subject to the law of diminishing returns. The CFPP improvement decreases as the treat rate is raised, effectively reducing the cost-effectiveness of additive use. In the example shown, a change in fuel characteristics had made it less responsive to the additive normally used, so that a high treat of around 800 ppm was needed to meet the CFPP target. The second additive enabled the CFPP to be met with about 150 ppm, illustrating the need

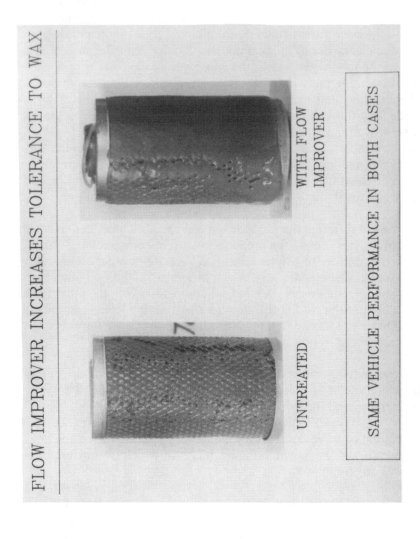

Fig. 16.11.   Photo of waxed fuel filters (source: Exxon Chemical International).

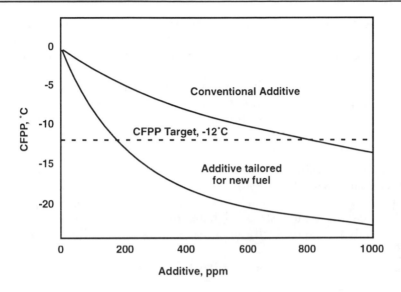

Fig. 16.12. Response of diesel blend to two different flow improver additives (source: Exxon Chemical International).

to ensure that the additive used is the one best suited to the type of fuel being produced.

16.2.4 Cloud Point Depressants (CPD)

Cloud point depressants are another type of wax-modifying additive developed for use in distillate fuels. Producing a diesel fuel with satisfactory properties for winter use is one of the major constraints on the refiner, and before flow improvers were used, winter quality was achieved by blending to a lower cloud point. This improved the low-temperature characteristics by eliminating some of the heavier distillate components but it also reduced the amount of fuel produced. This is the reason why most diesel fuel specifications generally allow seasonal cold property grades.

Cloud point is not always included in diesel fuel specifications but when it is, as well as being a constraint on the refiner (who usually blends to cloud point), it is an important indicator of base fuel quality.

It was formerly considered impossible for a small amount of a chemical additive to influence the wax solubility of a diesel fuel sufficiently to give a reduction in its cloud point. However, close studies into the effect of wax modifiers have shown that some olefin-ester copolymers appear capable of suppressing wax crystallization by a few degrees.[8]

The amount of cloud point depression obtainable, even with high treat rates, is relatively small, rarely exceeding 3 or 4°, whereas flow improver additives provide CFPP improvements two or three times greater at much lower treat levels. Cloud depressant additives also lower the base CFPP by a similar amount, but they tend to be antagonistic to conventional flow improvers. Adding cloud depressant to a fuel treated with flow improver may worsen its CFPP and give little cloud depression. Although some additive combinations can provide dual benefits, treat costs are high and the use of cloud depressants may only be economically attractive in fuels without flow improver.

16.2.5 Wax Antisettling Additives (WASA)

The wax crystals formed in a flow-improved fuel are smaller and more compact than the normal platelet crystals of an untreated fuel, and they have a greater tendency to settle to the bottom of the fuel tank. This is not a new phenomenon; wax settling has been observed ever since flow improvers have been in use.

Observations in storage tanks have confirmed the tendency for additive-modified waxes to settle but they also showed that withdrawals from the tank disturb the waxy layer. This helps redistribution of the waxes in the withdrawn fuel and, unless the tank is completely emptied, minimizes the risk of wax enrichment putting the last portions of fuel out of specification.[9] While not normally likely, this situation might arise if the temperature remains below the fuel cloud point for extended periods or if the fuel has a relatively high wax content. The diagram in Figure 16.13 illustrates the channeling effect that takes place during road tanker refueling at oil company distribution depots.

Changes in refinery operations and fuel blending to meet the growth in demand for diesel fuel, coupled with a spate of unusually cold winters in Europe during the early 1980s, have increased awareness of the potential problems associated with excessive amounts of settled wax. Additives that inhibit the wax settling tendency are being used routinely by some refiners and are also available for secondary treatment of finished fuels. Treating levels are similar to those for flow improvers, in the range of 100 to 500 ppm. Vehicle operability data obtained

Exposed Tankage

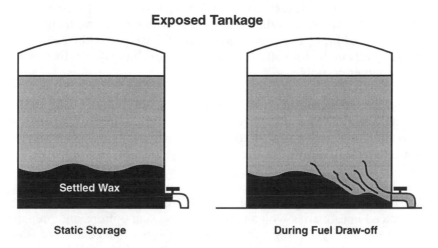

*Fig. 16.13. Wax settling in storage and the channeling effect during fuel
draw-off (source: Exxon Chemical International).*

with fuels treated with a wax antisettling additive are given in Section
16.2.6.

16.2.6 Mechanism of Wax Crystal Modification

Holder and Winkler concluded from their detailed studies on wax
crystallization and the mechanism of pour depressant action[1] that
polymeric additive molecules, by virtue of their chain length and
structure, are able to incorporate themselves at the dislocation step on
the growing crystal and stop the natural spiral growth pattern. Adsorption of the additive onto the wax crystal appeared to hinder its
precipitation from solution but the resulting smaller and more compact wax crystals effectively lowered the pour point of the treated
fuel.

Middle Distillate Flow Improvers (MDFI) are wax modifier additives
developed to give better cold filterability in the CFPP test rather than
only depressing the pour point. Like pour depressants, MDFI have a
polymeric active ingredient, but modification of the crystal growth is
achieved through a dual action of nucleation and growth arresting.
The additive composition is adjusted so that, as the fuel cools to its
cloud point, wax crystal growth sites are provided by polymeric nuclei

from the additive. These wax crystals will develop until further growth is prevented by other additive molecules attaching themselves to the crystal face. The combined effect of these two additive functions results in the formation of many very small crystals rather than fewer larger crystals.

The effect on wax crystal growth is illustrated by Figure 16.14, indicating how the additive molecule attaches itself to the developing wax crystal, inhibiting its normal growth habit and changing its size and form to be less likely to restrict flow in a vehicle fuel system.

The selection of flow improver additives available at the present time is extensive, covering the needs of fuels from a variety of crude types produced in refineries with a range of processing options and prepared for use in widely varying temperature conditions. The choice of flow improver is determined by evaluations of alternative candidates to ascertain the most cost-effective additive for the typical refinery production fuels, as illustrated in Figure 16.12.

Wax Antisettling Additives (WASA) form crystal interactions in the same way as conventional middle distillate flow improver additives but they make even smaller wax crystals, which has the effect of slowing down their rate of settling.[10] Stoke's Law states that the settling of spheres in a fluid is a function of:

$$\frac{(\text{diameter})^2 \times (\text{density difference})}{\text{viscosity}}$$

The possibilities for changing fuel viscosity and density other than by means of a solvent are limited, but an additive can have a marked influence on crystal size. The square law relationship between diameter and settling rate means that, if the wax crystal size is reduced by a factor of ten, they will settle at one-hundredth of the normal rate. This extends the daily rate of settling to a three-month period which is usually well beyond the duration of commercial fuel storage. Photomicrographs of crystals in Figure 16.15 show the extent of size reduction achieved with the antisettling additive.

Full-scale vehicle chassis dynamometer and field tests comparing an additive giving antisettling protection, as well as CFPP improvement with a conventional MDFI, have shown that the new additive provided protection under severe temperature cycling conditions and

Broad Boiling Distillate (BBD)	**Narrow Boiling Distillate (NBD)**

Crystal growth at low rate of wax deposition	Crystal growth at high rate of wax deposition

Time ——▶

Time ——▶

Regular growth of crystal face in BBD	Rapid irregular growth of crystal face in NBD

Time ——▶

Dislocation growth

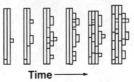

Time ——▶

2-D surface nucleation growth

Effect of conventional additive on growing crystal face in BBD	Effect of conventional additive on growing crystal face in NBD

Time ——▶

Few additive molecules inhibit growth by blocking kink sites

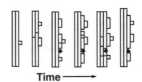

Time ——▶

Absorbed additive molecules rapidly grown over

Fig. 16.14. Mechanism of wax crystal modification in broad and narrow boiling distillates (source: Exxon Chemical International).

maintained the close correlation between vehicle operability and CFPP. Figure 16.16[9] shows results obtained with two portions of the same -1°C cloud point fuel, one treated with conventional filterability improver and the other with an additive that inhibited wax settling and improved filterability. Each had the same CFPP value of -24°C, which was well below the usual level for commercial fuels.

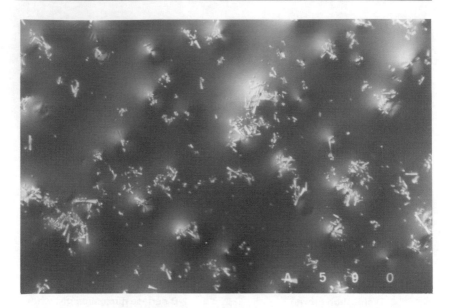

Fig. 16.15. Photomicrograph of wax crystals obtained with a wax antisettling additive (WASA) (source: Exxon Chemical International).

The fuel containing the conventional additive reached its operability limit a few degrees above its CFPP, at about -21°C, while with the antisettling treatment the vehicle was still operating satisfactorily at −26°C.

16.2.7 Factors Influencing Choice of Wax Modifier Additive

European refiners have tended to process mainly low to moderate wax content crudes from the Middle East and the North Sea, with some African crudes from the Sahara and Nigeria that are waxier but have a lower sulfur content. Automotive diesel fuels were prepared from predominantly straight-run gas oil but, with the domestic heating oil market shrinking, more cracked components are being diverted into diesel fuel. Winter CFPP specifications vary from maximum values of -5°C for Greece to -32°C for northern Finland. Flow improver treatment to meet national specifications is routine throughout Europe, the additive used being selected to suit the individual refinery needs.

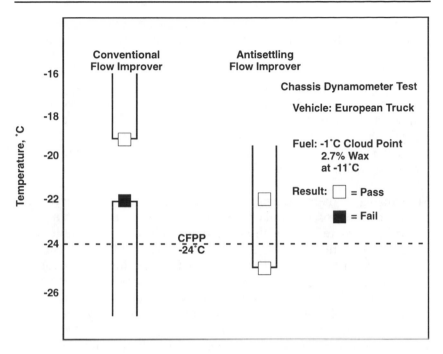

Fig. 16.16. Influence of wax antisettling additive on diesel low-temperature operability.[9]

The need for different additives to give effective wax crystal modification in fuels from different sources is due to the influence of a number of factors on the distribution of wax types in the fuel. The factors include the crude being processed; the availability of blend components and the way they are fractionated; the boiling range of the final blend and the target CFPP temperature to be met. Raising the end point of a fuel will bring in some higher carbon number waxes, which may increase response to the original flow improver or require the use of a different product.

The refiner must also ensure that other diesel fuel properties conform to specification limits. These factors will influence not only the amount of wax to be modified but also the range and distribution of individual normal paraffins (n-alkanes) in the wax and the rate at which they come out of solution as the fuel cools below its cloud point. Because the composition of the flow improver is determined by the type of wax in the fuel to be treated, the additive giving the most cost-effective treatment will vary from one type of fuel to another.

Cracked gas oils tend to be lower in wax content than atmospheric distillates and also more aromatic. This gives them better wax solvency which can reduce the additive requirement. On the other hand, sharper fractionation at the heavy end to increase the yield of distillate will also increase the rate of wax precipitation and, as a result, the amount of additive needed.

Maximizing jet fuel production to meet a higher demand reduces the proportion of kerosene available as a blend component. In a simple hydroskimming refinery, with no alternative blend component to replace kerosene, the only way to keep within the cloud point specification is to lower the end point of the diesel fuel. A similar situation can arise in high conversion refineries, when heavy streams from the atmospheric and vacuum units are preferentially used as cracker feed to reduce fuel oil production or make more gasoline components. The result of lowering the end point is that the boiling range of the diesel fuel becomes narrower, which changes the distribution of paraffins present in the fuel. The variety of fuel types produced in different parts of the world, as a result of the factors referred to above, are illustrated in Figure 16.17, which shows the n-alkane distribution in the waxes which they contain.

The steeper slope of the distribution pattern for the narrow boiling distillate also reflects the sharper fractionation typically used to maximize the yield from a narrower cut. Fuels having a narrow boiling range produce large, irregular-shaped wax crystals that are less easy to treat than the regular crystals formed by broader boiling fuels. The different appearance of the wax crystals from untreated broad boiling and narrow boiling distillates (BBD and NBD) is evident from the photomicrographs in Figure 16.18.

Orthorhombic crystals showing a regular growth pattern are typical of distillates having a broad boiling range and n-alkane distribution, whereas narrow cut fuels under identical cooling conditions produce large, amorphous wax crystals exhibiting no regularity in their growth.[7]

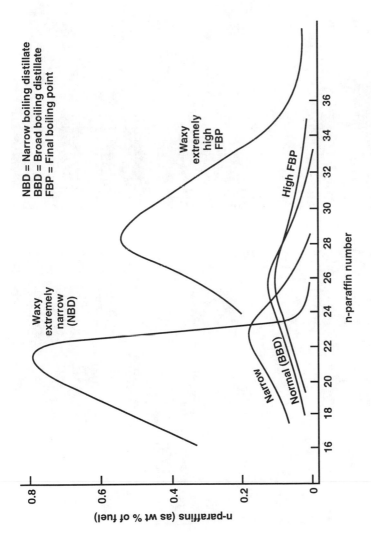

Fig. 16.17. Distribution of n-paraffins separating 10°C below the cloud point (source: Exxon Chemical International).

Broad boiling distillate

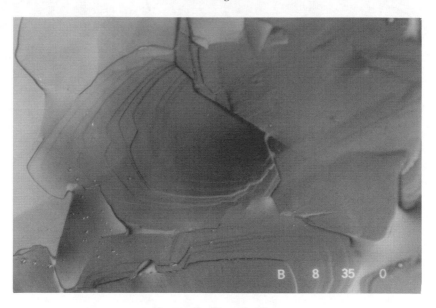

Narrow boiling distillate

Fig. 16.18. Photomicrographs of wax crystals formed by a broad boiling range distillate (BBD) and a narrow boiling range distillate (NBD) (source: Exxon Chemical International).

Figure 16.19 demonstrates the growth-arresting action of the additives developed to suit the particular types of wax crystal.

Narrow boiling distillates have a high initial rate of wax precipitation as the temperature goes below the cloud point. As a result, the formation of attachments on the crystal face is irregular, making it more difficult for a conventional flow improver to arrest growth. New types of additive have been developed for these narrow boiling fuels, but higher treating levels are inevitable because of the large number of growth sites on the crystal.

Fuels produced in different parts of the world vary widely in boiling range and wax content as a result of different crude types, refinery operations and climatic conditions. An indication of the range of variations is illustrated by the chart in Figure 16.20,[10] and a broad selection of flow improvers exists to cover all or most of the distillate categories.

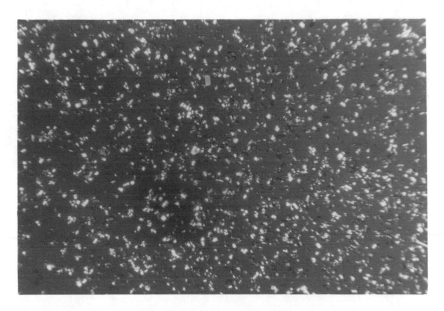

Fig. 16.19. Photomicrograph of additive-modified wax crystals from a narrow boiling distillate (source: Exxon Chemical International).

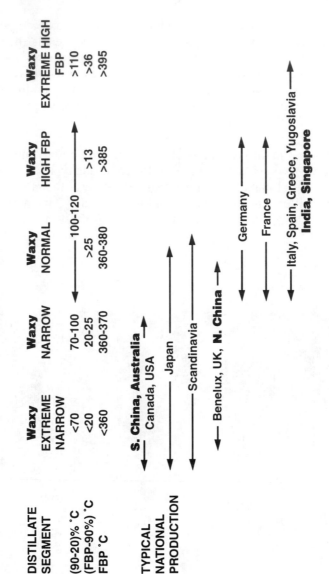

Fig. 16.20. Distillate types around the world (source: Exxon Chemical International).

Descriptions given to the categories are relative. The term "normal" relates to the type of diesel fuel produced in Europe during the early 1970s, before the changed demand pattern for petroleum fuels had necessitated a significant increase in cracking capacity. Final boiling points (FBP) become lower and boiling ranges narrower as the proportion of downstream conversion increases. Fractionation is tending to be sharper, as indicated by the shorter spread between FBP and the 90% distillation temperature, to maximize yield from the narrower cut.

Fuels with higher end points tend to be associated either with warmer climates requiring less stringent CFPP specifications or with refineries able to make low wax fuels from high boiling components. Countries such as Australia, India and China process their own indigenous, waxy crudes, producing distillates of high wax content which require specially developed additives, used at high treat levels, to achieve acceptable CFPP benefits.

16.3 Measurement of Diesel Fuel Low-Temperature Performance

Specifications for diesel fuels define the minimum quality acceptable for low-temperature operation of diesel vehicles in terms of one or more of the rapid laboratory tests described in Section 16.1. Many variations are possible in the composition of diesel fuels and in vehicle fuel systems. In consequence, it is appropriate to consider the procedures to be followed when assessing the validity of specification methods used for defining vehicle performance in service. The justification is to ensure that tests are conducted in a manner that is realistic and acceptable to the manufacturer of the vehicle, the operator and the fuel supplier.

It is standard practice for the motor industry to carry out extensive field testing of prototype and production models to ensure that they will perform acceptably in the different environments in which they are likely to be used. The tests are intended to evaluate the performance of the vehicle under both normal and severe climatic and road conditions. Critical attention has to be given to its various parts, which include the chassis, transmission and braking systems; the bodywork; lighting, heating, ventilation, soundproofing and general comfort of the driver and passenger-carrying areas. All that is in addition to assessing the engine with its ancillary and auxiliary equipment for cranking, fueling, ignition, speed control, cooling, etc. Need for these testing programs applies equally to gasoline and diesel vehicles.

The following sections cover the various aspects that need to be considered when carrying out low-temperature test programs in the field and under simulated conditions in a climate chamber.

16.3.1 Selection of Field Test Site

Siting of the field test is determined by the particular objectives of the tests and can range from tropical to arctic climates, low- and high-altitude areas and a wide variety of road surfaces and grades. For low-temperature testing of diesel vehicles and fuels, the first requirement is a site where there is a high likelihood that temperatures within the required range will be encountered during much of the testing period.

Other important factors to be considered include the accessibility of the site for transportation of test vehicles, fuels and all the other materials and personnel needed to carry out the work; facilities for working on the vehicles; storage for equipment, spare parts and fuels. Suitable routes will be required for realistic driving practices and hotel accommodation arranged for the test team. An additional consideration may be the availability of frequent and accurate forecasting of temperature conditions at the site to facilitate selection of the most appropriate quality fuel for each test run.

16.3.2 Procedure for Low-Temperature Testing

A Code of Practice, M-11-T-89, for low-temperature performance tests on diesel fuels and vehicles has been prepared by the CEC. This is intended to provide a standard procedure which, if adopted by vehicle testing laboratories, will facilitate direct comparison of results obtained by different operators. Many of the recommendations that follow have been taken from this code of practice, which covers cold startability and cold operability testing of vehicles in the field and in cold climate chambers. See Appendix 11 for full details of the code of practice.

16.3.3 Cold Startability

The purpose of cold startability testing is to check the ease with which the engine will start on the test fuel under the prevailing temperature conditions. Factors influencing the result include battery performance, cranking speed and fueling rate, as well as the ignition quality and fluidity of the fuel.

Cold startability at a particular temperature is evaluated on the basis of the total time of starter motor cranking to obtain engine autorotation, the time at which engine speed starts to rise, the number of attempts to start, the number of stalls and the idling speed.

16.3.4 Cold Operability

Cold operability tests are made to ascertain whether the engine will not only start on the test fuel but will also continue to run and perform satisfactorily while the vehicle is being driven under road load conditions. Once the engine has started, the outcome will be determined largely by two factors — the rate at which fuel can be delivered to the engine and the time taken for the fuel system to warm up and melt any wax that might accumulate and starve the engine of fuel. Serious restriction of the flow will reduce the amount of excess fuel recycled back to the tank and slow down the fuel system warm-up.

The objective of some diesel field tests is to ascertain whether satisfactory performance will be obtained from a particular fuel and vehicle at a particular temperature level. In this case, the likely outcome would be a simple "Yes" or "No" result. A more complex objective is to determine the lowest temperature at which satisfactory performance can be achieved. This temperature is often referred to as the operability limit and several tests in the critical temperature range are usually needed for an acceptably precise value to be defined.

16.3.5 Fuel Storage

All diesel fuels contain a significant portion of waxy material which will separate out of solution when the fuel temperature falls below its cloud point. As field test fuels tend to be delivered and stored in barrels at the test site, care must be taken when refueling to avoid the use of a waxy, non-homogeneous product. Such a fuel would not be representative of normal situations, where vehicles are refueled from large underground or overground tanks, in which temperatures are unlikely to go below the fuel cloud point. To minimize the risk of contamination with water or other foreign material, the fuel should be stored in closed containers.

Large quantities of drummed fuel may be left in the open if there is no suitable building, but provision must be made for warming the drums before use to ensure that a homogeneous, wax-free fuel is available for the test vehicles. This may require a heated room in which a few drums of fuel can be stored above the cloud point temperature before they are used. Alternatively, steam coils or electric heaters can be used to warm up individual drums. If the fuel has been stored below its cloud point for several days, the waxes may have settled and sufficient time must be allowed for all the wax to be redissolved. It is advisable to mix the contents thoroughly, by rolling, shaking or stirring the drums, before sampling to check that all the wax is back in solution.

16.3.6 Vehicle Instrumentation

Testing to determine the low-temperature operability limit of diesel fuels and vehicles requires instrumentation to monitor air, fuel system and engine temperatures, pressure and vacuum in the fuel system, and vehicle speed. Figure 16.21 indicates the recommended locations for temperature and pressure measurements to be made in the engine and fuel system. If it is not possible to use a speed-recording instrument, the vehicle speedometer can be used after calibration. Measurement of battery voltage and electrolyte temperature can also provide useful information during cold starting tests.

16.3.7 Preparation of Test Vehicles

Before starting test work, vehicle checks should be carried out to ensure that the correct grades of lubricant are used in the engine, gearbox transmission and auxiliaries, and that the coolant system has sufficient antifreeze for the expected minimum temperature. Drive belt tensions should be adjusted, if necessary, and a check made on the coolant thermostat and fan switch operating temperatures. Checks should also be made on the engine itself, electrical components, starting aids and fuel injection equipment, making corrective adjustments as necessary to fuel leaks, valve clearances and injector opening pressure. Safety considerations make it advisable to ensure that tires and

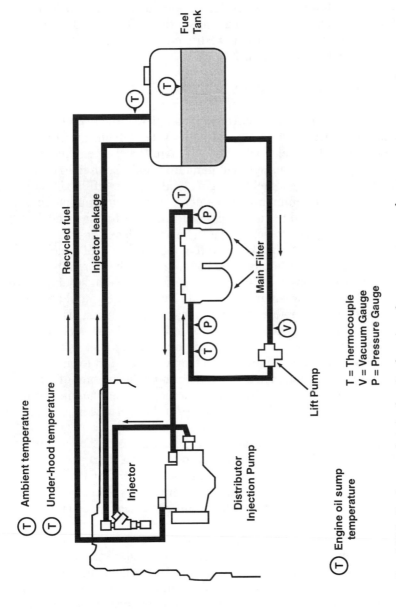

Ambient temperature

Under-hood temperature

Recycled fuel

Injector leakage

Fuel Tank

Main Filter

Lift Pump

Injector

Distributor
Injection Pump

T = Thermocouple
V = Vacuum Gauge
P = Pressure Gauge

Engine oil sump
temperature

*Fig. 16.21. Fuel system layout showing locations for temperature and pressure measurements
(source: Exxon Chemical International).*

455

inflation pressures are appropriate for the driving conditions. When testing for fuel-limited cold operability, normal practice is to fit a new fuel filter for each test. Also, to avoid starting difficulties due to inadequate cranking speed rather than restricted fuel flow, a fully charged battery is recommended for each test.

16.3.8 Operational Procedure

Once the preparatory checks have been completed, the operational procedure is typically based on a 24-hour cycle of events, starting with selection of the fuel to be tested. If there is a choice, a fuel suitable for the expected overnight temperature should be used. A good starting point is to take a fuel having a cloud point or CFPP about 3°C above the expected minimum temperature. After putting in the test fuel, the engine should be run for about 15 minutes to purge the fuel system completely and to verify that there are no leaks or operating faults. The vehicle can then be parked in the open to cool under the prevailing ambient conditions. The battery should be stored in a warm place and, if possible, left on charge to ensure that any failure to start is not due to poor battery condition. The recorder should be left running to monitor ambient and fuel system temperatures. The driving test to see if the engine will perform satisfactorily on waxy fuel is made after the vehicle has been standing overnight, ideally starting early, before ambient and fuel system temperatures start to rise. The overnight minimum temperature will determine the amount of wax separating in the lines and filters between the tank and the injection pump, which are usually the coldest parts of the fuel system.

After refitting the battery, the cold start procedure is carried out as recommended by the vehicle manufacturer. Only a short period of fast idle is recommended before starting the road test, driving at speeds appropriate to road conditions and traffic regulations. For vehicles with air brakes, longer idling will be needed to recharge the system.

A test duration of 30 minutes is generally sufficient to determine whether or not the vehicle will perform satisfactorily. Engine stalling due to wax restriction of the fuel flow generally occurs within the first 15 minutes of driving, before the engine and fuel system have warmed sufficiently to disperse the wax accumulation. Recording fuel system temperature and pressure variations and noting the time of hesitations or power reduction during the test run will provide useful information on the location and severity of any wax accumulation in the fuel system.

When there is sufficient time available and weather conditions permit, several tests should be made in the critical temperature range to provide more data for evaluation of results. Depending on how the vehicle performs on the first test, additional results at lower or higher temperatures may be needed to help define the operability limit of the particular fuel and vehicle combination. If the first test is satisfactory, tests should be made at lower temperatures until performance is affected by wax in the system. Tests at higher temperatures will be needed if wax plugging causes an engine stall on the first test.

16.3.9 Climate Chamber Testing

Cold climate chambers equipped with chassis dynamometers are used to simulate field conditions in a controlled environment. This method of testing overcomes the major problem of unsuitable weather conditions which often afflicts field programs, and it can be used at any time of the year. Also, as the test vehicle is stationary, there are fewer constraints on instrumentation and data gathering. The main limitation is that it is not usually possible to test more than one vehicle at a time.

The vehicle is located with its driving wheels on large rollers coupled to the dynamometer, which allows it to be driven under normal road load conditions. A large fan circulates the air over refrigerator coils to control the room temperature and to provide a cooling air flow at the equivalent road speed of the vehicle. A typical cold climate chassis dynamometer arrangement is shown in Figure 16.22.

Preparation of the vehicle for cold climate chassis dynamometer (CCCD) testing is the same as prescribed for field testing. The starting procedure is also the same but, in the absence of road traffic constraints, a well-defined driving procedure can be followed. This will minimize the test-to-test variability which can arise and possibly affect road test results. The different driving patterns prescribed in the CEC code of practice for passenger cars and for trucks or tractors are shown in Figure 16.23.

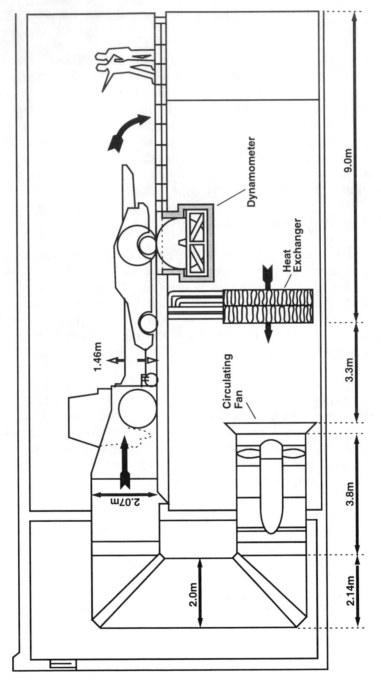

Fig. 16.22. Layout of cold climate chassis dynamometer (source: Esso/Exxon Research Centre, U.K.).

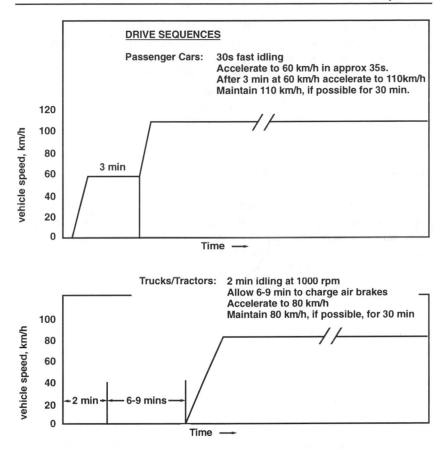

DRIVE SEQUENCES

Passenger Cars: 30s fast idling
Accelerate to 60 km/h in approx 35s.
After 3 min at 60 km/h accelerate to 110km/h
Maintain 110 km/h, if possible for 30 min.

3 min

Trucks/Tractors: 2 min idling at 1000 rpm
Allow 6-9 min to charge air brakes
Accelerate to 80 km/h
Maintain 80 km/h, if possible, for 30 min

←2 min→ ←6-9 mins→

*Fig. 16.23. Drive cycles for diesel vehicle low-temperature operability
(source: Appendix 4).*

When testing in a climate chamber, the temperature programmer is set
to give a controlled rate of cooling to simulate realistic field conditions. This is typically in the range of 1 to 2°C/hr during the overnight period, when vehicles are normally parked. Also the
programmer can be set so that the temperature selected to suit the fuel
being evaluated is reached at a convenient time in the working day.
Taking an approach similar to that for field testing, a realistic starting
point would be at a temperature 3°C below the CFPP, if the fuel contains a flow improver, or 3°C below the cloud point if it is untreated.

Depending on the first result, subsequent tests will be at lower or
higher temperatures. Chassis dynamometer testing is expensive so, as

only one test per day is normally practicable, it is advisable to mini-mize the total number of tests needed to evaluate one fuel/vehicle pair. Steps of 4°C will usually give an early indication of the oper-ability limit zone and then 2°C steps will allow fine-tuning. Between each test the chamber has to be warmed to ambient temperature, or well above the fuel cloud point, for complete draining of the fuel sys-tem and main filter replacement before adding a fresh batch of test fuel.

16.3.10 Interpretation of Results

During cold operability tests, partial or complete power loss may sometimes occur before the engine has warmed sufficiently to redis-solve any wax accumulating in the critical part of the fuel system. For the researcher, whether working for the motor industry, a petroleum company or for an additive supplier, such results are essential to help identify the lowest temperature at which the material under test will operate.

For realistic comparisons to be made of results derived from different test programs, a standardized procedure for interpretation of test data is required. The CEC code of practice, Appendix 11, provides an answer to this need. However, it may be useful to comment on some of the terminology which has been used to describe the low-tempera-ture performance of diesel fuels.

Satisfactory operability, often expressed as a "Pass" result, is reported when the vehicle performs as it would on clear, wax-free fuel, with no hesitations or reduction of power.

Unacceptable operability may be reported as a "Fail" when an accu-mulation of wax in the fuel system restricts the flow of fuel to the engine, so that it is either not possible to start the engine or it stalls during driving and fails to complete the test run.

"Operability limit" is a term which has been used to describe the low-est temperature at which satisfactory performance can be obtained or expected. When testing at temperatures near the operability limit, there may be a partial loss of engine power as the system is warming up, after which the vehicle is able to perform normally and complete the test run. This is sometimes called a "borderline" result and has been used to define the operability limit. When no borderline result is obtained, and the lowest pass and highest fail are within 2 or 3°, the operability limit is sometimes taken as the midway temperature.

To accommodate borderline situations when the result is neither a clear pass nor a clear fail, the CEC code of practice proposes a system of demerit rating. A numerical value is assigned to all misfires, surges, pedal adjustments (to maintain speed) and driving stalls, to provide a quantitative assessment of how the vehicle performed. As there are different opinions on what constitutes acceptable performance, no recommendation is given to indicate a limiting demerit. The choice is left to individual organizations to select a level to suit the particular criteria involved.

16.3.11 Low-Temperature Test Experience

Results obtained by companies and organizations who had carried out low-temperature operability tests were collated and incorporated in a CEC report.[11] Figures 16.24(a), (b), (c) and (d) present data from various sources to illustrate the correlation between the limiting operability temperatures derived from vehicle testing and the predictive values given by the CFPP method.[12]

In the commercial trials of buses and trucks in normal service, reported complaints were associated with periods of operation at temperatures below the CFPP of the fuel under test. No fuel-related problems were experienced when the temperature was above the CFPP of the fuel.

Field trial and climate chamber test results obtained at each temperature were reported as "Pass," "Borderline" or "Fail" and the points on the graphs are identified accordingly. Although the tests had been conducted at different times and places on various models of diesel equipment, including trucks, agricultural tractors and passenger cars, a consistent pattern is discernible. Most of the "Borderline" and "Fail" results occurred when the fuel temperature was at or below its CFPP. The diagonal lines on the graphs show direct 1:1 correlation between the two temperature scales. Relatively few operating problems were experienced at temperatures higher than the CFPP and those deviations from the 1:1 line are mostly within the test precision.

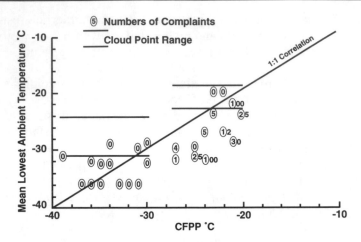

Fig. 16.24. (a) Commercial experience in winters '69, '70, '71, '72.[12]

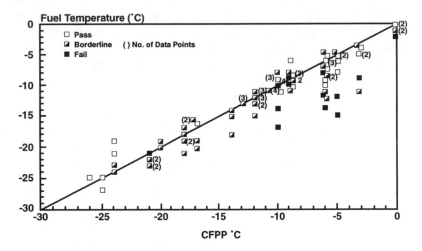

Fig. 16.24. (b) Controlled field trials. [12]

Similar correlation data were obtained from tests on two Japanese diesel vehicles, a light truck and a passenger car tested on a chassis dynamometer.[13]

The different vehicle systems evaluated show significant variability in their sensitivity to wax in the fuel. However, the data indicate that the use of flow improver additives enables satisfactory operability to be

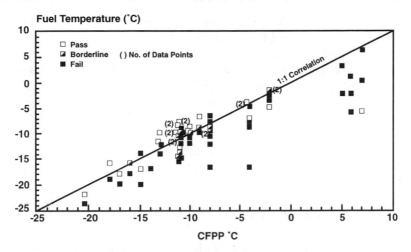

Fig. 16.24. (c) Cold climate chamber tests - 1.[12]

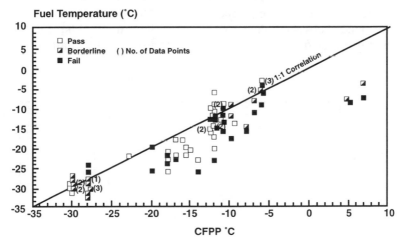

Fig. 16.24. (d) Cold climate chamber tests - 2.[12]

attained to at least 10° below the cloud point for all but the most critical vehicles.

In Canada, a mixed fleet of diesel trucks and passenger cars was field tested in a CRC cooperative program to investigate their low-temperature operability on different fuels and flow improver additives.[5] Significant differences in fuel system design were evident

463

among the test fleet. Minimum operating temperatures derived from the test results indicated that, overall, the LTFT had provided the best available correlation for flow-improved fuels. The CFPP was only partially successful, being unable to give satisfactory predictions of the operability limits of some of the fuel/vehicle combinations. The conclusions on the LTFT were supported by later, independent field tests.[7,14] Data from the three tests are presented in Figure 16.25.[5,7,14]

As mentioned earlier, variant versions of the LTFT procedure are being evaluated to see if a closer correlation with vehicle operability can be obtained.

16.3.12 Reducing Sensitivity to Waxing Problems

A number of the test results collated in the CEC report indicate that satisfactory performance was obtained at temperatures well below the CFPP. It is probable that these may have been due to some vehicles being less sensitive to the presence of fuel wax. Differences in cold performance between vehicles running on the same fuel have been a feature of many field tests, and have led to investigations into the

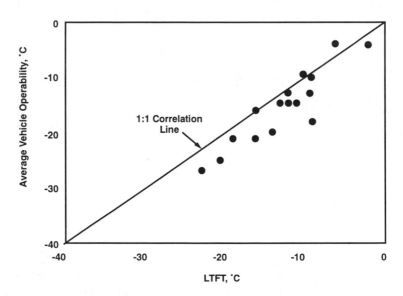

Fig. 16.25. Composite of LTFT correlation with average vehicle operability (1981 CRC trial and 1982 independent field tests).[5,7,14]

causes of sensitivity to wax. Understandably, there have also been studies into ways of improving vehicle operability at low temperatures.

Investigations by the British Technical Council of the Motor and Petroleum Industries into cold weather experiences by diesel vehicle operators have led to recommendations for design and service requirements relating to vehicle fuel systems to minimize the risk of cold weather problems.[15]

Wax in the vehicle tank can be drawn into the fuel lines and restrict fuel flow if it accumulates on fine filters or in sharp bends in the lines, particularly if they are in exposed locations. Water in the fuel system is also a potential hazard at freezing temperatures. Care is needed to prevent water from getting into the tank and the layout of fuel pipes should avoid "U" traps where free water could freeze while the vehicle is parked.

A vehicle fuel system designed to minimize low-temperature problems is illustrated in Figure 16.26,[15] which shows a typical European diesel layout with a separate injection pump mounted on the side of the engine. The recommendations, however, are equally applicable to the typical U.S. diesel fitted with unit injectors, because waxing problems usually occur in the low-pressure system bringing fuel from the vehicle tank to the injection pump.

Water separators should not contain any filter screen, and makers recommend that they be installed close to the tank. Constant bore pipes and connectors with large radius bends fitted in a sheltered location will minimize the likelihood of wax plugs building up. Main filters should be mounted where they can receive engine heat and be of adequate size for the maximum fueling rate. Series filters give better wear protection to the injection pump but offer more resistance to fuel flow than a parallel arrangement.

The fuel recycle line is shown close to the engine supply line, and in some models it goes down to the bottom of the tank. If the recycled fuel is likely to become very hot, a more remote return position may be necessary to avoid vapor lock in the injection pump, but this will slow the rate of fuel system warming.

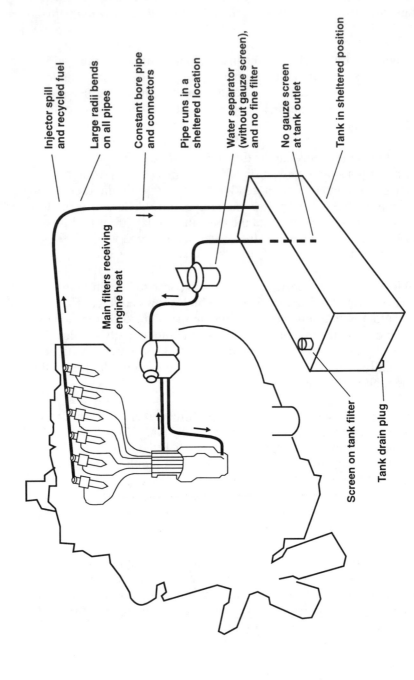

Injector spill
and recycled fuel

Large radii bends
on all pipes

Constant bore pipe
and connectors

Pipe runs in a
sheltered location

Water separator
(without gauze screen),
and no fine filter

No gauze screen
at tank outlet

Tank in sheltered position

Main filters receiving
engine heat

Screen on tank filter

Tank drain plug

Fig. 16.26. Well-designed diesel fuel system to minimize risk of low-temperature problems. [15]

Publications by organizations such as the SAE, BTC and CEC during recent years have directed the attention of diesel vehicle manufacturers to ways of minimizing cold weather problems. These have largely involved relatively low-cost changes to the fuel system layout, of which some are already being applied.

In addition to "rationalizing" the standard fuel system, new devices have been developed that are suitable for fitting to vehicles in service as well as being available either as original equipment or as an option for the purchaser of a new vehicle. These include: heaters for the various parts of the fuel system (such as the tank, lines or main filters) and bypass valves to divert some or all of the recycled fuel back to the injection pump during warm-up, rather than into the tank. Electric filter heaters may be built into a modified replacement filter housing or supplied as a separate unit to be installed close to the main filter inlet of existing vehicle systems.

16.3.13 Experience with Modified Fuel Systems

An assessment of fuel system requirements for satisfactory low-temperature operability identified the use of a tank screen and/or prefilter, the size of the main filter and its location as major factors.[16] A survey of 24 diesel passenger car models from manufacturers in the U.S., Japan and several European countries, revealed a range of differences. Half of the models had no tank or prefilter screen, while main filters were located in various positions around the engine compartment, some close enough to benefit readily from engine heat during the critical warming-up period.

Tank screens and/or prefilters have been found to be prone to blocking in cold weather, particularly when the fuel has not been treated with flow improver to give smaller wax crystals, so.locating the prefilter where it will pick up engine heat can help. Although relocation to a warmer position is not possible for the tank screen, directing the recycled fuel toward it can be beneficial in lowering the rate of wax buildup.

The largest variable was main filter surface area and its influence on the specific fuel flow through the filter, expressed as the rate of flow per unit area. Flows were measured at three engine speeds and averaged. Filter surface areas ranged from 900 to 6400 cm^2, while specific flow rates varied widely between 0.05 and 2.29 mL/s/cm^2.

The main filter tends to be the most sensitive part of the fuel system when flow-improved fuel is used. Most of the modified wax crystals will pass through the mesh screens unless the temperature is well below the CFPP level. Positioning the filter to receive engine heat soon after the engine starts will contribute to melting the wax, but the size and flow rate through the filter are more critical features. When waxy fuel is being drawn from the tank, the filter size influences the amount of fuel that can be filtered before it plugs. A larger surface area will lower the specific flow through the filter and extend the time before it becomes plugged with wax, thereby allowing more time for the fuel system to warm up and melt the wax on the filter.

Chassis dynamometer tests on some of the passenger cars from the survey showed improvements in low-temperature operability as a result of fuel system modifications on the lines referred to above. Figure 16.27[17] indicates the trend in operability improvements brought by the various modifications.

As various fuels were used, the improvements were relative to the CFPP. Several of the vehicles were able to operate to temperatures well below the CFPP with their standard fuel system. Fitting a larger-size filter gave about the same benefit as an electric filter heater, lowering the operability limit of the more critical vehicle models by 8 to 12°. Similar levels of improvement were observed in studies on heavy-duty diesel vehicles.[18]

Electric filter heaters must be used with care to avoid draining the vehicle battery. A recommended practice is to switch on the heater only when the engine has been started. An alternative source of heat for the filter is the engine coolant, which warms up rapidly after engine start-up. Some passenger cars have a water-heated filter connected to the engine coolant system. A thermostatic valve in the coolant line prevents the fuel from overheating.

Where high fuel recirculation rates are needed to avoid overheating in the fuel manifold, the fuel is normally returned to the tank at a point well away from the off-take. This is necessary at normal engine operating temperatures, but getting the engine running properly can be difficult if there is waxy fuel in the tank. The use of a thermostatically controlled bypass between the recycle and feed lines can overcome this problem. The bypass diverts some of the recycled fuel to the main filter inlet, reducing the amount of cold fuel drawn from the

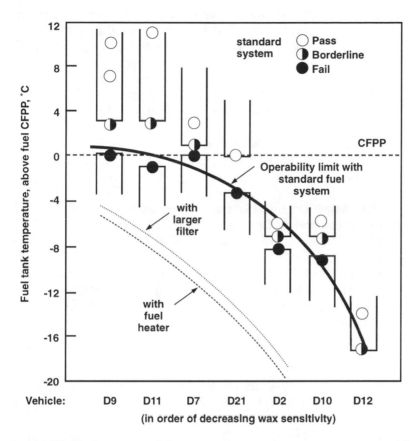

Fig. 16.27. Fuel system modifications to improve low-temperature operability (source: Mobil Oil AG).

tank and allowing the heat in the recycled fuel to melt the wax on the filter. The temperature of fuel coming from the tank controls the opening and closing of the thermostat.

References

1. G.A. Winkler and J. Winkler, "Wax Crystallisation from Distillate Fuels," *Journal of the Institute of Petroleum*, Vol. 51, No. 499, July 1965.

2. T. Coley, L.F. Rutishauser and H.M. Ashton, "New Laboratory Test for Predicting Low-Temperature Operability of Diesel Fuels," *Journal of the Institute of Petroleum*, Vol. 52, No. 510, June 1966.

3. "The Cold Filter Plugging Point of Distillate Fuels," A European Test Method, CEC Report No. P-01-74, 1974.

4. R.C. Tupa and C.J. Dorer, "Gasoline and Diesel Fuel Additives for Performance/Distribution Quality - II," SAE Paper No. 861179, 1986.

5. M.L. McMillan and E.G. Barry, "Fuel and Vehicle Effects on Low-Temperature Operation of Diesel Vehicles - The 1981 CRC Field Test," SAE Paper No. 830594, 1983.

6. P. David, G.I. Brown, E.W. Lehmann, "SFPP—A New Laboratory Test for Assessment of Low Temperature Operability of Modern Diesels Fuels," Paper No. CEC/93/EF15, CEC Fourth International Symposium on the Performance Evaluation of Automotive Fuels and Lubricants, Birmingham, U.K., May 1993.

7. J. Zielinski, F. Rossi and A. Stevens, "Wax and Flow in Diesel Fuels," SAE Paper No. 841352, 1984.

8. B. Damin, A. Faure, J. Denis, B. Sillion, P. Claudy and J.M. Letoffe, "New Additives for Diesel Fuels: Cloud-Point Depressants," SAE Paper No. 861527, 1986.

9. G.I. Brown, R.D. Tack and J.E. Chandler, "An Additive Solution to the Problem of Wax Settling in Diesel Fuels," SAE Paper No. 881652, 1988.

10. T.R. Coley, "Diesel Fuel Additives Influencing Flow and Storage Properties," <u>Gasoline and Diesel Fuel Additives,</u> Edited by K. Owen, J. Wiley and Sons, Chichester; UK, 1989.

11. "Low Temperature Operability of Diesels," a Report by CEC Investigation Group IGF-3, CEC Report No. P-171-82, 1982.

12. T.R. Coley, "Low Temperature Operability of Diesels," SAE Paper No. 830596, 1983.

13. H. Ise, H. Hirano, N. Nozaki, H. Takizawa, M. Tamanouchi and I. Nakajima, "Cold Flow Properties of Automotive Diesel Fuels and Diesel Fuel Systems," FISITA Conference, Dearborn, Michigan, 1988.

14. R.D. Tharby, "Experiences with Diesel Fuel Containing Cold Flow Improver Additives," SAE Paper No. 831753, 1983.

15. "Diesel Fuel Systems for Low Temperature Operation," British Technical Council Report No. BTC/F1/79, 1979.

16. P. Heinze, "Low Temperature Operability of Diesel Powered Passenger Cars," CEC Second International Symposium on Performance Evaluation of Automotive Fuels and Lubricants, Wolfsburg, Germany, 1985.

17. T.R. Coley, F. Rossi, M.G. Taylor and J.E. Chandler, "Diesel Fuel Quality and Performance Additives," SAE Paper No. 861524, 1986.

18. M. Lanoë, "Operabilite a Froid des Poids Lourds Diesel," GFC Comite Technique Carburants Moteurs, Journees d'Etudes Paris, France, October 1982.

Further Reading

Low Temperature Problems with Diesel Vehicle Fuel Systems, BTC/IP Publication, London UK, 1964.

D.E. Steere and J.P. Marino, "Low Temperature Field Performance of Flow Improved Diesel Fuels," SAE Paper No. 810024, 1981.

M. Becker, F.J. Hills and D.P. Osterhout, Jr., "Fuel and Fuel System Factors Influencing Low-Temperature Diesel Operation," API 30th Mid-year Meeting, Montreal, Canada, 1965.

H. Ruf, "Some Methods for Improving the Filterability of Diesel Fuels at Low Temperatures," (original in German and French) *Motorlastwagen/L'Autocamion*, Nos. 19 & 20, Switzerland, 1965.

"Diesel Fuels and Systems for Winter Conditions," Proceedings of BTC Seminar at the Motor Industry Research Association, Nuneaton, UK, 1973.

"Effects of Fuels and Fuel System Designs on Low Temperature Operability of Diesel Vehicles," CRC Project Report No. CD-15-69, 1972.

G.I. Brown, E.W. Lehmann and K. Lewtas, "Evolution of Diesel Cold Flow — The Next Frontier," SAE Paper No. 890031, 1989.

G.I. Brown and G.P. Gaskill, "Additive Developments for Enhanced Diesel Fuel Low Temperature Operability," Paper No. 17F, Third CEC Symposium, Paris, 1989.

K. Mitchell, "The Cold Performance of Diesel Engines," SAE Paper No. 932768, 1993.

J.E. Chandler and G.I. Brown, "Evolution of Faster LTFT and SFPP for Correlation with Low Temperature Operability in North American Heavy Duty Diesel Trucks," SAE Paper No. 932769, 1993.

R.F. Haycock and R.G.F. Thatcher, "Fuel Additives and the Environment," Document 52, published by the Technical Committee of Petroleum Additive Manufacturers in Europe (ATC), 1994.

Chapter 17

Other Diesel Specification and Non-specification Properties

Diesel fuels marketed in many parts of the world have, as a general rule, been prepared from predominantly straight-run components having satisfactory stability characteristics and, usually, good cetane quality. This situation, however, is changing as refiners adjust their operations to produce more clean products and less residual fuel, for which demand is decreasing. Cracking processes are being used to convert heavy streams into lighter fractions suitable for use as blend components for gasoline and diesel fuel.

Distillates from cracking operations are less paraffinic and more olefinic than those from atmospheric distillation and contain more nitrogen compounds. As a result, they have poorer cetane quality and are less stable, being prone to oxidation by free-radical reactions, as discussed in Section 4.3.5.

Studies on an unstable catalytically cracked diesel fuel have led to the identification of two classes of compound that are produced during storage under ambient conditions.[1,2] Their structure is believed to consist of linked indole and phenalene ring systems. Such compounds are soluble in diesel fuel but, if allowed to react with acids, they form an insoluble sediment virtually indistinguishable from the polar, highly colored part of naturally formed sediment.

Generally, cracked gas oils have been blended into heating oil, with the straight-run material being used to make diesel fuel, particularly where cetane numbers of 45 or higher are specified. However, with the heating oil market shrinking, many refiners in Europe are needing to divert some cracked stocks into diesel fuel to meet the growing demand.

Diesel fuels are generally processed with hydrogen to bring their sulfur content down to the specification level. This treatment has an additional benefit of providing improved stability so, unless either plant capacity or hydrogen availability is limited, the stability of fuels containing cracked components should be adequate. There are, however, additives which have been found effective in reducing sediment formation during accelerated oxidation tests. These additives and their mechanism of action will be discussed in Chapter 18.

17.1 Diesel Fuel Stability

The ability of a diesel fuel to remain unchanged during the period between its manufacture and its eventual use in an engine is an important quality requirement. Most commercial deliveries of diesel fuel are consumed within a few weeks of leaving the refinery and the likelihood of degradation is small. However, it is the policy of many governments to lay down stocks of all types of fuel for use in emergencies. These strategic reserves are normally subject to periodic turnover. Quantities will be withdrawn for use and replaced by newer batches at intervals, but it is possible that some fuels may be in extended storage for periods of more than one year before turnover.

Changes that can occur with an unstable fuel are due to oxidative breakdown, resulting in the formation of sediments and gums. These reaction products may degrade the color of the fuel but, more seriously, could also cause engine operating problems due to blocked filters or deposits on the injectors and in the combustion chamber.

17.1.1 Measurement of Diesel Fuel Stability

At the present time there is no universally accepted test to predict the stability characteristics of a diesel fuel which may be kept in normal storage facilities for prolonged periods. The problem is that the rate at which changes are likely to occur under such conditions is usually slow. To obtain a quantitative value for the long-term stability of a diesel fuel, it is necessary to accelerate the degradation process by the use of abnormally severe test conditions. One test which has been used fairly widely for a number of years is ASTM D 2274.[3] It is an accelerated method for oxidation stability which contains the cautionary statement that any correlation between the test and field storage may vary significantly under different field conditions or with distillates from different sources.

A sample of fuel is heated to a temperature of 95°C, which is maintained for 16 hours with continuous bubbling of oxygen. The fuel is then allowed to cool in the dark, prior to filtration and solvent washing to recover sediment and gummy residues. The total amount of insoluble deposits is reported as mg/100 mL of fuel. Typical specification values are usually in the range of 2.0 to 3.0 mg/mL. The precision of the method for amounts greater than 1.0 mg/100 mL has not been determined.

The color of the fuel before and after aging can be determined by ASTM D 1500. A sample is placed in a container and, using a standard light source, is compared with colored glass discs ranging in value from 0.5 to 8.0, in steps of 0.5. The number of the matching glass, or the higher of the two numbers where there is no exact match, is reported as the ASTM Color. Although it does not feature in many diesel fuel specifications, quoted levels tend to range between 2.5 and 5.0 ASTM Color. The precision of the test is not known.

Some of the other tests for predicting diesel fuel stability are variants on the ASTM D 2274 method; the conditions are altered to modify the test severity. The combination of high temperature and continuous oxygen bubbling is considered by some researchers to induce reactions that would not occur under normal conditions. One approach to reduce severity is to replace the oxygen with air, using either continuous bubbling or presaturation, to limit the amount of oxygen available for reaction. Changes to the temperature and duration of exposure to oxidative conditions are other factors influencing severity. Table 17.1 lists some of the alternative test methods to predict the oxidative and thermal stability of diesel fuels.

One of the alternative methods for predicting long-term storage stability, ASTM D 4625, is itself a relatively lengthy test. After presaturation with oxygen, the fuel sample is held at 43°C for a period of 100 days. The results are believed to correlate with storage at ambient conditions for one year. However, even if the 100-day test at 43°C shows close agreement with results obtained after an extended period in normal storage, a rapid test is needed for routine control of product quality at the refinery.

Table 17.1. Some Alternative Oxidation Stability Test Methods

TEST METHOD	ASTM D 2274	ASTM D 4265	ESSO TEST	DU PONT F-21	UOP 413
DURATION	16 hours	100 days	16 hours	1.5 hours	16 hours
TEMPERA-TURE	95°C	43°C	99°C	150°C	100°C
OXIDATION MEDIUM	Bubbling O_2	Dissolved O_2	Presaturation with air	Dissolved O_2	Dissolved O_2

17.2 Cetane Index

Cetane index, as mentioned in Chapter 2, Section 2.2, is widely used for routine monitoring of diesel ignition quality. It is a calculated value, derived from fuel density and volatility, giving a reasonably close approximation to cetane number to avoid the expense and loss of time involved in carrying out a determination in the CFR Cetane engine. The original formula was based on the API gravity and mid-volatility (50% recovery temperature) of the fuel, and the coefficients have been modified from time to time as fuel composition evolved, to improve the correlation with updated data sets. Figure 2.4 in Chapter 2 illustrates the relationship between calculated values using the 1980 revision of the equation:

$$\text{Cetane Index} = 454.74 - 1641.416 \, D + 774.74 \, D^2 - 0.554 \, B$$

$$+ 97.803 \, (\log B)^2$$

where   D = density at 15°C
        B = mid-boiling temperature, °C, corrected to standard barometric pressure

In 1982, an ASTM Task Force was set up to study improved methods for calculating cetane quality. The study led to the development of a new equation that is now included in the ASTM Standards Book as ASTM D 4737, Calculation of Cetane Index by the Four Variables Equation. This new equation is also based on density and volatility parameters but uses three distillation points — 10%, 50% and 90%. An investigation of the four-variables equation was carried out by a panel of the Institute of Petroleum, using 167 European distillate fuels and the results are shown in Figure 17.1.[4]

The new equation for calculating the cetane index, ASTM D 4737, is given below:

$$CI = 45.2 + 0.0892(T_{10} - 215) + 0.131(T_{50} - 260) + 0.0523(T_{90} - 310)$$

$$+ 0.901 \ B(T_{50} - 260) - 0.420 \ B(T_{90} - 310) + 0.0049(T_{10} - 215)^2$$

$$- 0.0049(T_{90} - 310)^2 + 107.0 \ B + 60.0 \ B^2$$

where T_{10}, T_{50} and T_{90} are the 10%, 50% and 90% recovery temperatures in °C, corrected to standard barometric pressure, respectively, D is density at 15°C and

$$B = e^{[-3.5(D-0.85)]} - 1$$

Cetane index has mainly served as an alternative method of determining the ignition quality of diesel fuel, to minimize the need for engine tests. However, the method is now being incorporated into diesel specifications as an additional test to control base fuel quality and to limit the extent of cetane number improvement by means of a cetane improver additive. The UK automotive diesel specification, BS 2869: Part 1: 1988, which came into effect October 1, 1989, had minimum limits of 48 for cetane number and 46 for cetane index. BS 2869 has been superseded by the pan-European specification for automotive diesel fuel, EN 590. This specification, which is included in Appendix 8, was prepared in response to a directive from the European Council of Ministers in Brussels, for adoption by member countries of the EC by the end of 1993, when the Single Europe Act came into effect. The minimum cetane number specified in EN 590:1993 is 49, with 46 as the minimum cetane index. The 1994 Fuel Charter of ACEA calls for a minimum cetane number of 53, with a minimum cetane index of 50.

Harmonizing the various national standards was a logical action, as differences would imply a potential restriction of trade within the European Community. However, the range of geographic and sea-

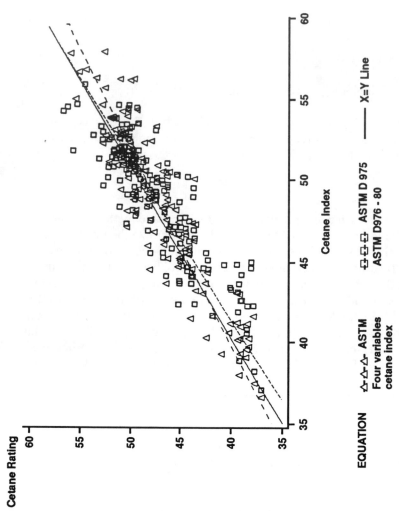

Fig. 17.1. Correlation of ASTM calculated cetane index by four-variable equation with measured cetane number.[4]

sonal climatic differences across Europe made it impractical to apply a single specification. The situation was resolved by establishing a set of generally applicable quality requirements, to which all diesel fuels must comply and a range of climatically related properties, selected to cover the temperature variations across the region.

Table 17.2 shows the general requirements, covering safety, purity, stability, corrosivity and sulfur content. Tables 17.3a and 17.3b present the climatically related properties: density, viscosity, cetane number and index, distillation characteristics and CFPP. National annexes to EN 590 indicate the summer grade, winter grade and (where appropriate) the special grade for more severe needs, such as in mountainous or arctic areas, selected to suit the particular member country.

Table 17.2. Generally Applicable Requirements and Test Methods from European Diesel Fuel Specification EN 590: 1993

Property	Units	Limits Min.	Max.	Test Method
Flash point	°C	above 55		ISO 2719
Carbon residue (on 10% distillation residue)	% m/m		0.30	ISO 10370
Ash content	% m/m		0.01	EN 26245
Water content	mg/kg		200	ASTM D 1744
Particulate matter	mg/kg		24	DIN 51 419
Copper strip corrosion (3 h at 50 °C)	rating		class 1	ISO 2160
Oxidation stability	g/m^3		25	ASTM D 2274
Sulfur content	% m/m		0.20[1]	EN 24260/ISO 8754

1) NOTE: Directive 87/219/EEC, article 2.1 sets a maximum of 0.30.

Table 17.3a. Climatically Related Requirements and Test Methods—Temperate Climates from European Diesel Fuel Specification, EN 590: 1993

Property	Units	Limits Min.	Limits Max.	Test Method
CFPP	°C			EN 116
CFPP grade A			+5	
CFPP grade B			0	
CFPP grade C			–5	
CFPP grade D			–10	
CFPP grade E			–15	
CFPP grade F			–20	
Density at 15 °C	kg/m^3	820	860	ASTM D 4052
Viscosity at 40 °C	mm^2/s	2.00	4.50	ISO 3104
Cetane number		49		ISO 5165
Cetane index		46		ISO 4264
Distillation[1,2]				ISO 3405
Recovered at 250 °C	% V/V		<65	
Recovered at 350 °C	% V/V	85		
Recovered at 370 °C	% V/V	95		

Table 17.3b. Climatically Related Requirements and Test Methods—Arctic Climates from European Diesel Fuel Specification, EN 590: 1993

Property	Units	Limits class 0	class 1	class 2	class 3	class 4	Test method
CFPP	°C max.	–20	–26	–32	–38	–44	EN 116
Cloud point	°C max.	–10	–16	–22	–28	–34	ISO 3015
Density at 15 °C	kg/m^3 min.	800	800	800	800	800	ISO 3675
	kg/m^3 max.	845	845	840	840	840	ASTM D 4052
Viscosity at 40 °C	mm^2/s min.	1.50	1.50	1.50	1.40	1.20	ISO 3104
	mm^2/s max	4.00	4.00	4.00	4.00	4.00	
Cetane number	min.	47	47	46	45	45	ISO 5165
Cetane index	min.	46	46	46	43	43	ISO 4264
Distillation[1,2]							
Recovered at 180 °C	% V/V max.	10	10	10	10	10	ISO 3405
Recovered at 340 °C	% V/V min	95	95	95	95	95	

[1] The limits for distillation at 250 °C and 350 °C included for temperature climate diesel fuel in line with EEC Common Customs Tariff. EEC Common Customs Tariff definition of gas oil does not apply to the grades defined for use in arctic climates.

[2] It should be noted that for the calculation of the cetane index the 10% V/V and 50% V/V recovery points are also needed.

Cetane improver additives are discussed in Section 18.2.

17.3 Flash Point

The flash point is the temperature to which the fuel has to be heated to produce a vapor/air mixture that will ignite when a flame is applied. The most common test method used for diesel fuels is ASTM D 93/ ISO 2719, flash point by Pensky-Martens closed tester. A portion of the fuel is heated slowly at a constant rate in a covered cup, and at regular intervals the cover is opened, briefly, to admit a small flame into the cup. This procedure is continued until the fuel temperature is high enough for flash ignition to occur.

Flash point relates to the front-end volatility of the fuel. Consequently, it will have an influence on the choice of light refinery streams to be included in the fuel blend, as its initial boiling point must be sufficiently high for the flash point to conform with the specification. Assuming that the correct blend components have been used, a low flash point can indicate contamination of the diesel fuel with a more volatile, lower flash point product such as gasoline.

Apart from its relevance to formulation of the fuel, the importance of flash point is primarily related to safe handling of the product. If the flash point is too low, there could be a fire hazard and, for this reason, mandatory minimum limits on flash point have been set and storage criteria established by insurance companies and government agencies. Typical minimum values for automotive diesel fuels range from 38°C in the U.S. (ASTM D 975, No. 1-D fuel) to 56°C in some European countries.

As far as performance in an engine is concerned, the flash point of a diesel fuel has no significance. Variations in flash point will not influence autoignition temperature or other combustion characteristics.

17.4 Electrical Conductivity

Although conductivity is not a specification requirement for diesel fuel, there is a potential safety risk due to the buildup of a static electricity charge during bulk handling of any flammable liquid at high pumping rates. Grounding leads are normally used to conduct away any charge when large quantities of fuel are being transferred into and out of storage tanks, but a volatile product like aviation kerosene is also treated routinely with a static dissipator additive.

The standard test method to determine electrical conductivity of aviation and distillate fuels containing a static dissipator additive is ASTM D 2624. A voltage is applied across two electrodes immersed in the fuel and the current, in picosiemens per meter, is reported as the conductivity value.

The conductivity of diesel fuels influences the likelihood of an excessive charge building up when refueling vehicles at service stations. However, faster filling rates have been suggested to speed up refueling stops for trucks having large-capacity fuel tanks and, as a precaution, some oil companies use an antistatic additive in diesel fuel.

17.5 Water and Sediment Content

Water cannot be completely eliminated from diesel fuel. It can get into the fuel at various stages as it progresses through the distribution network from the oil refinery to the fuel system of the vehicle in which it is finally burned.

The earliest stage at which water can get in is during manufacture, when the hot fuel is in contact with process water. Much of this water will be removed in the stripping units at the refinery and more will separate as the fuel cools. The main risks of water contamination are during transportation and storage in tanks which have not been thoroughly emptied of water or into which water can enter and accumulate. This can be due to rain leaking past the roof seals or breathing-in large volumes of humid air, which can happen during fuel withdrawals and as a result of changes in ambient air temperature. Moisture in the air will condense when the temperature falls and it is recommended practice to maintain tanks as full as possible to minimize water contamination due to breathing. Provision is usually made for regular draining of storage tank bottoms but this may not be completely effective if the tank bottom is distorted. Where natural or man-made caverns are used for intermediate fuel storage, there is no possibility of eliminating contact with water.

The presence of water in storage tanks may encourage growth of fungi or bacteria which feed on the fuel and may block filters if drawn out

482

with the fuel. Also, entrainment of water bottoms when pumping from contaminated tanks can sometimes produce hazy fuel which may be slow to reach the clear and bright appearance required of a merchantable product.

Sediment likely to be found in diesel fuel tends to be mainly inorganic in origin: rust and metal particles from fuel tanks and lines and dirt entering from the atmosphere or because of poor housekeeping practices. Organic deposits may come from degradation of unstable fuel components, bacterial action at the oil-water interface or wax from the fuel. Wax settling may occur if the fuel has cooled below its cloud point as a result of unexpectedly low temperatures or long storage in small exposed tanks. Although settled waxes usually redissolve when the tank is refilled or the ambient temperature rises, tank cleaning is advisable if the amount of accumulated wax present could put the new batch of fuel out of specification.

Water and sediment can contribute to filter plugging in the distribution network or the vehicle, and cause problems due to corrosion and wear in the engine and fuel injection system. A standard test for water and sediment content is a centrifuge method, ASTM D 1796, which is in the ASTM D 975 specification for diesel fuel oils. A 50mL sample of fuel is mixed with an equal volume of water-saturated solvent in each of two cone-shaped graduated tubes, and centrifuged to concentrate the water and sediment in the bottom of the tube. The total volume of contaminant in each tube is noted and the sum is reported as the percentage of water and sediment.

Some diesel specifications impose separate limits for water and sediment. In such cases, water content is determined by distillation (ASTM E 123) or titration with a chemical reagent (ASTM D 1744). Typical maximum limits tend to be 0.05% (500 ppm) for water content and 0.01% for sediment, or 0.05% if combined water and sediment is specified.

17.6 Ash Content

Diesel fuels may contain small amounts of ash-forming material such as suspended solids and soluble organometallic compounds. These can cause damage in the close tolerances of fuel injection equipment and contribute to increased deposit levels and abrasive wear of piston rings and other components in high-temperature zones of the engine.

The standard quantitative method for determining ash from petroleum products is ASTM D 482. A small sample of the fuel is burned in a weighed dish until all combustible material has been consumed. The weight of the unburnt residue is reported as a percentage of the original fuel sample. A typical specification limit for ash content is 0.01% maximum.

17.7 Carbon Residue

A carbon residue test is included in many diesel fuel specifications to indicate the tendency of the fuel to form carbonaceous deposits, although it bears little relationship to coking of injector nozzles in service.[5] Some entrainment of higher boiling material occurs during distillation to separate the diesel fraction, and the test measures the amount of these low volatility components in the finished fuel.

Two tests widely used around the world to measure the carbon residue of automotive diesel fuels are the Conradson method, ASTM D 189, and the Ramsbottom method, ASTM D 524.

As the level of carbon residue in automotive diesel is low, often less than 0.1%, the test is carried out on the 10% residue from the laboratory distillation test in order to increase the accuracy of the result. To avoid erroneous results due to the presence of certain types of additive such as nitrates as cetane improvers, the test is only applicable to base fuels.

In the Ramsbottom procedure, a 4g sample of the 10% bottoms is placed in a tared glass coking bulb and heated at 550°C for 20 minutes. After cooling, the bulb is reweighed to determine the amount of residue, which is reported as percent Ramsbottom carbon residue on 10% distillation residue.

In the Conradson procedure, a larger sample, 10g of distillation bottoms, is put in a tared crucible and subjected to destructive distillation, undergoing cracking and coking reactions during a fixed period of severe heating. After cooling and reweighing to calculate the amount remaining, the result is reported as percent Conradson carbon residue on 10% distillation residue.

The two methods have similar precisions in the region of 0.1%. The 95% confidence level for repeatability is about 0.025% while, for reproducibility, the figures are 0.036% for Ramsbottom and 0.058% for Conradson. There is no exact correlation between results obtained by the two methods but an approximate correlation, which should be used with caution, has been derived by ASTM Committee D-2 (see Figure 17.2).

Permitted maximum levels in specifications range between 0.05% and 0.3%.

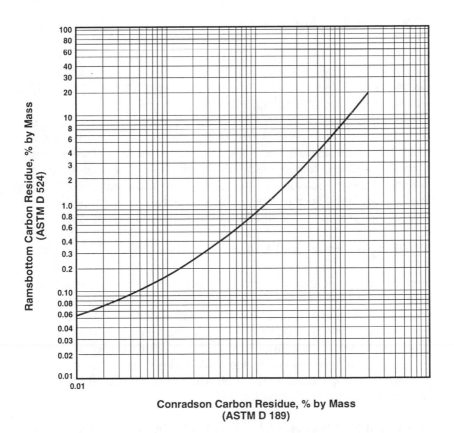

Fig. 17.2. Correlation between carbon residue results by Conradson and Ramsbottom methods (source: ASTM Book of Standards).

17.8 Corrosivity

It is important to ensure that the fuel will not attack metals in the distribution and storage network or in the engine fuel system. Copper and copper alloys are vulnerable to attack by certain sulfur compounds and many specifications, including ASTM D 975, call for a copper corrosion and tarnish test, ASTM D 130. In this test, a polished copper strip is immersed in a portion of the fuel and heated for 3 hours at 50°C, after which it is removed, washed and compared with ASTM standards for copper strip corrosion. The standards are color reproductions of test strips showing increasing degrees of tarnish and corrosion. Typical specification limits allow a measure of tarnishing (ratings of 1 to 3) but no corrosion (rating of 4) is permitted.

17.9 Neutralization Number

Another standard method for indicating whether the fuel has a tendency to corrode metals with which it may come into contact is ASTM D 974, Neutralization Number by Color-Indicator Titration. A sample of the fuel is dissolved in a solvent mixture, a color indicator is added and the mixture is titrated with a standard base or acid solution, potassium hydroxide (KOH) or hydrochloric acid, until a color change indicates that the end point has been reached. Different equations are used to calculate the result (in mg KOH/g) according to whether the specification requirement is total acid number, strong acid number, or strong base number. One or other of these properties may be specified instead of, or in addition to, copper strip corrosion.

17.10 Appearance and Color

A clear and bright appearance is generally a requirement for any diesel fuel, whether or not it is included in the specification, to demonstrate its freedom from water haze, suspended matter and other contaminants which could result in the product being visually unacceptable. Color is sometimes specified to preclude incorporation of materials such as heavy cracked gas oils and residues, which would give a darker appearance to the fuel, while in some specifications a minimum color value is stipulated.

As referred to in Section 17.1.2, ASTM D 1500 is one of the standard methods for measuring color. A sample of the fuel is placed in a cylindrical jar and compared, under a standard light source, with colored glass discs having numerical values ranging in regular steps from 0.5 for the palest to 8.0 for the darkest. The number of the matching glass or the higher of the two numbers where there is no exact match, is reported as the ASTM Color. In specifications where color is included, the limiting values can vary widely — a minimum of 2.5 in Italy, a maximum of 5.0 in France.

17.11 Lubricity

Lubricity, sometimes referred to as film strength, is the ability of a liquid to lubricate. This is very relevant to the satisfactory operation of diesel engines which rely on the fuel to lubricate many of the moving and rubbing metal parts of the fuel injection equipment.

As mentioned in Chapter 15, poor lubricity is often associated with low-viscosity fuel. Lubricity is not provided directly by the fuel viscosity but by other components of the fuel which prevent wear of metal surfaces in contact. These protective components of the fuel are believed to be polycyclic aromatic types with sulfur, oxygen and nitrogen content,[6] which may be rendered ineffective as a result of severe hydrotreatment to desulfurize the fuel. In some areas where the climate necessitates the supply of a kerosene-type fuel as winter-grade diesel, a lubricity agent is sometimes added to the fuel to limit wear in the injection equipment. In 1991 Sweden introduced new environmental classifications for diesel fuels, with tax incentives to encourage their use. The revised specifications, issued in 1992, are given in Table 17.4, together with comparative features of standard grade Swedish automotive diesel fuel.

Although Class 1 and Class 2 fuels can bring benefits of lower emissions levels they may cause premature injection pump wear due to their low lubricity characteristics. Rig endurance tests of a rotary injection pump, running on Swedish Class 1 fuel for 2000 hours of cyclic operation, showed an unacceptable reduction in governed speed and increase in the minimum fuel delivery at high speed. Repeating the tests with a lubricity additive in the Swedish Class 1 fuel maintained governing and fueling within the required tolerances.[6]

At present, lubricity is not included in diesel fuel specifications, but it is considered that some injection equipment may be at risk if operated on fuels of low viscosity or of non-petroleum origin.[7] Bench tests

Table 17.4. Swedish Diesel Fuel Specifications (January 1992)

Property		Urban Diesel Fuel Class 1	Class 2	Standard Grade
Sulfur	%m/m max.	0.001	0.005	0.3
Aromatics	%V/V max.	5.0	20	-
PAH	%V/V max.	0.02	0.1	-
Distillation:				
IBP	ºC min.	180	180	-
10%	ºC min.	-	-	180
95%	ºC max.	285	295	Seasonal
Density	kg/m³	800-820	800-820	Seasonal
Cetane No.	min.	50	47	Seasonal
Tax Rate	$/m³	135	168	210

that measure film strength or load-carrying capacity have been studied as possible methods for assessing the lubricity or antiwear quality of different fuel types.

After initial work by an SAE committee, an ISO working group, WG6, was set up to devise a method of measuring the lubricity of diesel fuel and subsequently setting a standard.[6] A parallel activity in Europe, to assess the merits of alternative lubricity testing machines, was carried out by CEC/PF 006. A subgroup of lubricity experts from ISO WG6 and CEC/PF 006 was set up to analyze the data and recommend a test. Support for the program was provided by oil companies, OEMs, additive companies, and all diesel injection pump manufacturers.

Four test procedures were evaluated: the Ball on Three Seats (BOTS) machine, two versions of the Ball on Cylinder Lubricity Evaluator evolved by Lubrizol (BOCLE-LZ) and SWRI (BOCLE-SWRI), and the High Frequency Reciprocating Rig (HFRR). Figure 17.3 shows schematic arrangements of the three lubricity test machines.

17.11.1 Ball on Three Seats Method (BOTS)

Work on a fuel lubricity test procedure based on the Four Ball lubricant testing machine had been abandoned in favor of the BOTS

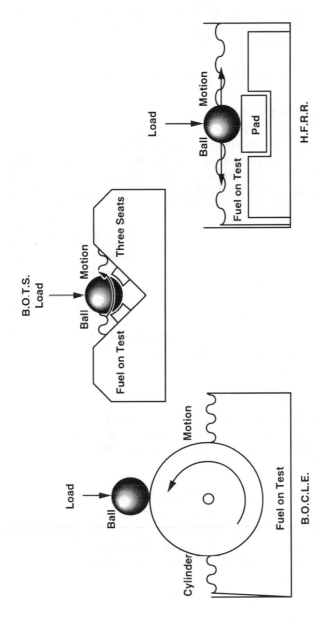

Fig. 17.3. Schematic arrangements of lubricity test machines (source: Lucas Diesel Systems).

method, a variant of the Four Ball machine, in which the three fixed balls are replaced by three seats set at an angle to the axis of the rotating ball. The ball is rotated in loaded contact with the seats, which are immersed in the test fuel and its lubricity is assessed by the weight loss of the three seats due to wear.

17.11.2 Ball on Cylinder Lubricity Evaluator (BOCLE)

Service problems with gas turbine fuel pumps running on hydrofined kerosene led to the development of the BOCLE (ASTM D5001), which became the most widely used test method to measure fuel lubricity. The ball is loaded against a rotating cylinder, partly immersed in the test fuel, which continuously transfers a thin film of the fuel to the ball/cylinder interface. Fuel lubricity is assessed by measurement of the wear scar on the ball.[6,8,9,10]

Initial use of the standard BOCLE procedure to assess diesel fuel lubricity gave poor correlation with field experience. Several variants of the test operation had been recommended to give a better representation of conditions inside the fuel injection pump and improve its suitability for short duration screening tests of diesel fuels and additives. Two methods, intended to represent the scuffing process, one from Lubrizol and the other from SouthWest Research Inc. (SWRI), were studied by the ISO working group.[6]

In the Lubrizol procedure, BOCLE-LZ, the machine is operated at 300 rpm for a conditioning period of 15 min. and then a load of 7 kg is applied for 2 min. The test is carried out with the relative humidity of the air controlled at 50%. The wear scar measured at the end of the test represents the scuffing wear that has occurred when running on the particular test fuel.

The test procedure developed by the SouthWest Research Institute, BOCLE-SWRI, determines the maximum load which can be applied without causing scuffing. The result is attained by running a series of tests at low, non-scuffing loads and high loads which cause scuffing, to bracket closely the highest non-scuffing load which the test fuel can carry.

17.11.3 High Frequency Reciprocating Wear Rig (HFRR)

The HFRR uses a hardened steel ball oscillating transversely in loaded contact with a hardened steel plate immersed in the test fuel. It was designed to give negligible hydrodynamic film formation and low frictional heating. Contact temperature can be controlled as an independent test variable, enabling assessment of temperature-dependent effects as well as the performance of boundary-active additive chemicals.[10] Figure 17.4 shows the general arrangement of the HFRR apparatus and the standard test conditions are given in Table 17.5.

Table 17.5. HFRR Test Conditions

Load	200 g (1.96 N)
Stroke	1 mm (0.5 mm amplitude)
Frequency	50 Hz
Temperature	25°C and 60°C
Metallurgy	Ball: ANSI 52 100 (hardened bearing tool steel 60-65 HRC) (F.A.G. Pt. No. KV6.G28)
	Flat: ANSI 52 100 (bearing tool steel 200 HV 30)
Surface Finish	Ball: <0.05 micron Ra
	Flat: <0.02 micron Ra
Duration	75 minutes

Batch-tested balls and flats can be obtained from:
P.C.S., Sherfield Building, Imperial College, London SW7 2AZ.

Results from the HFRR are based on comparison of the wear scar diameter with that of a reference fuel, usually after test runs at two temperature levels, typically 25°C and 60°C. Measurement of friction and coherent film formation levels can also be used to discriminate between fuels and additives.

17.11.4 Test Machine Evaluations

Data for the program were derived from round-robin tests carried out in 29 laboratories running one or more of the test techniques. Twelve fucls, including three treated with an anti-wear additive, were evaluated in the laboratory test machines, producing 42 data sets from 905 results. All of the fuels were also tested by three fuel injection equip-

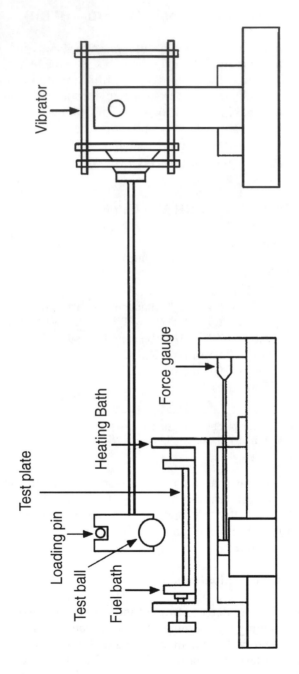

Fig. 17.4. General arrangement of high frequency reciprocating rig. [8]

ment manufacturers in their own pumps, giving over 60 results for comparison with test rig findings.

The test data were statistically analyzed for signal-to-noise repeatability and reproducibility, discrimination between fuels, and the correlation of test rig results with wear measurements from injection pumps. Other factors taken into consideration were ease of use, capital and operational costs, and availability of the equipment.

The HFRR was selected as the test method for use in a draft standard for measuring diesel fuel lubricity, allowing optional use of either of the two temperatures, 25°C and 60°C. The HFRR procedure has been written in conformity with CEC guidelines and a draft meeting the requirements of ISO and ASTM has been prepared. Ratification by national bodies is expected in 1995.

Comparisons of HFRR wear scar results obtained with samples of different grades of diesel fuel are shown in Figure 17.5.[8,11] The fuels tested covered a range of sulfur levels, from 0.3% for a typical pre-1993 European regular grade to the current 0.05% specification level, and also included the special Swedish Class 1 and Class 2, with very low sulfur contents (<0.001 and <0.005 % m/m, respectively).

Results of screening tests on additives used to treat the two Swedish grades are presented in Figure 17.6. Test results shown in Figure 17.7 indicate the good discrimination between fuel grades provided by the HFRR.[8,9]

17.12 Heating Value

Heating value (calorific value, thermal value, heat content, heat of combustion) is an important fuel property. It is a measure of the energy available from a fuel when it is burned, and is the basis for calculating the thermal efficiency of an engine using that fuel.

The procedure for determining the heating value of a diesel fuel is given in ASTM D 240, Heat of Combustion of Liquid Hydrocarbons Fuels by Bomb Calorimeter. The test measures the amount of heat released by burning a known quantity of fuel, and this may be expressed in Joules per kilogram (J/kg) or per liter (J/L). This is reported as the higher or gross heat of combustion, as it includes the

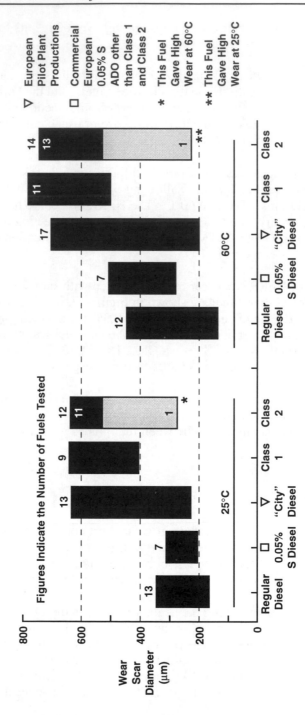

Fig. 17.5. Adhesive wear levels of a range of diesel fuels in the HFRR.[11]

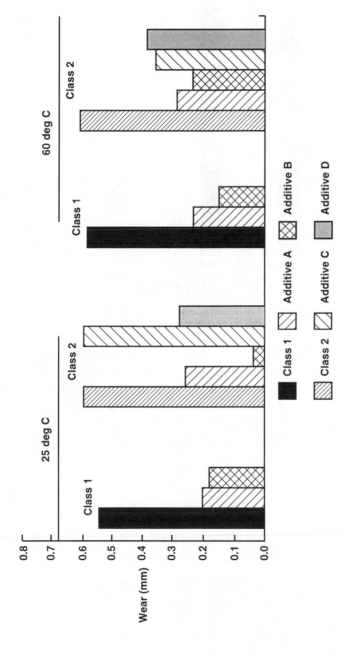

Fig. 17.6. HFRR test results on Swedish city diesel fuels.[11]

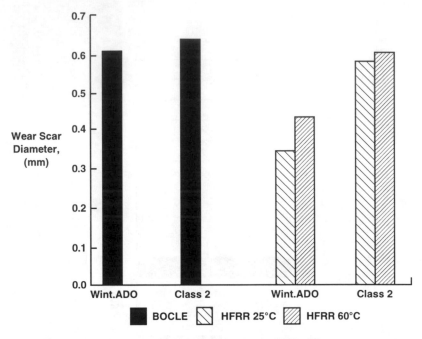

Fig. 17.7. Comparison of HFRR and BOCLE test results.[11]

heat released when water vapor produced during combustion is condensed. However, as water of combustion normally escapes as steam in the engine's exhaust gases, it is usual to subtract the latent heat of condensation of water vapor to give the lower or net heating value, which is used for the calculation of thermal efficiency. The difference between the two values is a function of the hydrogen content of the fuel.

Sulfur content will also have an influence and a number of empirical formulas have been developed for estimating heating values when bomb calorimeter data are not available. Average heating values, tabulated in terms of API gravity and fuel sulfur content, have been published by ASTM.[12]

Some formulas have been used to develop nomographs to facilitate the estimation of heating values using the gravity and one other property such as the aniline point (a measure of aromatics content) or the mid-boiling point of the fuel. Figure 17.8[12] shows a nomograph based on density and mid-boiling point.

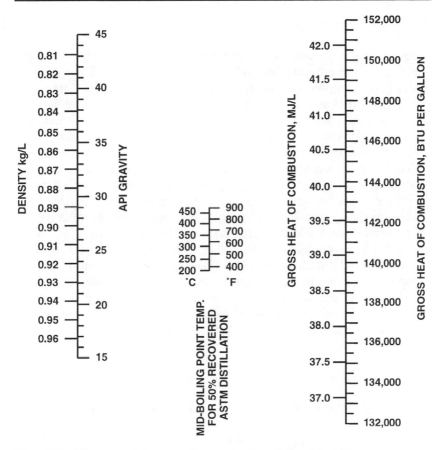

Fig. 17.8. Nomograph for gross heating value of diesel fuel.[12]

Values derived from empirical formulas will generally be of sufficient accuracy for most routine test work, but the bomb calorimeter method should be used when an exact determination of heating value is required.

Heating value is not controlled during manufacture except indirectly through other fuel properties, so it does not normally appear in specifications for automotive diesel fuel. Its application, as mentioned above, relates to the calculation of engine thermal efficiency.

17.13 Aromatics Content

Incorporating or increasing the proportion of cracked stocks in diesel fuel to meet the growing demand has raised concern about the influence of fuel composition on exhaust emissions, with critical attention directed towards aromatics. As a consequence, there is a likelihood that some diesel fuel specifications may impose a limit on aromatics content. At this stage, however, there are difficulties in deciding which analytical measurement techniques could be used effectively for control purposes and also whether the limitation should be on total aromatics or some particular species. An alternative approach proposed for consideration is to calculate the aromatics content from readily available specification parameters. Two equations relating density and viscosity to total aromatics content (TAR, vol%) and to aromatic carbon content (CAR, mole%) have been developed from a study[13] carried out on a total of 52 finished diesel fuels and blend components. From the close correlation between calculated and measured values of the two aromatics parameters, it would appear that the aromaticity chemistry of diesel fuels may be defined with acceptable accuracy by functions of density and viscosity, using the following equations:

$$\%TAR = 0.4358d - 32.67 \log_{10}(KV_{40} + 0.7) - 321.6 \qquad (1)$$

$$\%CAR = 0.3495d - 32.19 \log_{10}(KV_{40} + 0.7) - 259.3 \qquad (2)$$

where d is density in kg/m^3 at 15°C,

KV_{40} is kinematic viscosity in cSt at 40°C.

The influence of aromatics on emissions is discussed in Chapter 14, Section 14.5.5.

References

1. J.F. Pedley and R.W. Hiley, "Investigation of 'Sediment Precursors' Present in Cracked Gas Oil," 3rd International Conference on Stability and Handling of Liquid Fuels, London, 1988.

2. J.F. Pedley, R.W. Hiley and R.A. Hancock, "Storage Stability of Petroleum-Derived Diesel Fuel," *FUEL*, Vol. 66, pp. 1646-1651, December 1987.

3. L.L. Stavinoha, "User Needs for Diesel Stability and Cleanliness Specifications," API 47th Mid-Year Meeting, New York City, NY, 1982.

4. C.J.S. Bartlett, "A New Equation for Calculation of Cetane Index of Distillate Fuels," *Petroleum Review*, Vol. 41, No. 485, p. 48, June 1987.

5. J.-C. Guibet, "Carburants et Moteurs," Tome 1, *Editions Technip*, Paris, France, 1987.

6. H.C. Grigg, "Reformulated Diesel Fuels and Fuel Injection Equipment," Fifth Annual Conference on New Fuels and Vehicles for Cleaner Air, Phoenix, Arizona, January 1994.

7. II.C. Grigg, P.S. Renowden and L.B. Bodo, "The Properties of Diesel Fuel as they Relate to the Operation of Fuel Injection Equipment," Paper No. C305/86, The Institution of Mechanical Engineers International Conference on Petroleum Based Fuels and Automotive Applications, London, UK, 1986.

8. C. Bovington and R. Caprotti, "Latest Diesel Fuel Additive Technology Development (with particular reference to lubricity)," Paper No. CEC/93/EF13, CEC Fourth International Symposium on the Performance Evaluation of Automotive Fuels and Lubricants, Birmingham, U.K., May 1993.

9. S.R. Jenkins, R.G.M. Landells and J.W. Hadley, "Diesel Fuel Lubricity Development of a Constant Load Scuffing Test Using the Ball on Cylinder Lubricity Evaluator (BOCLE)," SAE Paper No. 932691, 1993.

10. R. Caprotti, C. Bovington, W.J. Fowler and M.G. Taylor, "Additive Technology as a Way to Improve Diesel Fuel Quality," SAE Paper No. 922183, 1992.

11. C. Bovington, R. Caprotti, K. Meyer and H. Spikes, "Developing of a Laboratory Test to Predict Lubricity Properties of Diesel Fuels and Its Application to the Development of Highly Refined Diesel Fuels," paper presented at the Esslingen Symposium, Esslingen, Germany, January 1994.

12. ASTM Book of Standards, Part 17, p. 1120: Method for Estimation of Net and Gross Heat of Combustion of Burner and Diesel Fuels.

13. C.J.S. Bartlett, "The Relationship Between Aromatic Content and Specification Parameters in Diesel Fuels," *Petroleum Review*, Vol. 44, No. 517, p. 84, February 1990.

Further Reading

J. Ritchie, "A Study of the Stability of some Distillate Diesel Fuels," *Journal of the Institute of Petroleum*, Vol. 51, No. 501, September 1965.

T.J. Russell, "Petrol and Diesel Additives," *Petroleum Review*, pp. 35-42, October, 1988.

R.C. Tupa and C.J. Dorer, "Gasoline and Diesel Fuel Additives for Performance/Distribution Quality-II," SAE Paper No. 861179, 1986.

R.N. Hazlett, A.J. Power, A.G. Kelso and R.K. Solly, "The Chemistry of Deposit Formation in Distillate Fuels," Materials Research Laboratory Report MRL-R-986, Dept. of Defense, Melbourne, Australia, January 1986.

R.N. Hazlett, D.R. Hardy, E.W. White and L. Jones-Baer, "Assessment of Storage Stability Additives for Naval Distillate Fuel," SAE Paper No. 851231, 1985.

E. Goodger and R.A. Vere, Aviation Fuels, G.T. Foulis & Co. Ltd., 1970.

P.I. Lacey and S.J. Lestz, "Effect of Low-Lubricity Fuels on Diesel Injection Pumps—Part II: Laboratory Evaluation," SAE Paper No. 920824, 1992.

P. Renowden, "The Assessment of Diesel Fuel 'Lubricity' Using the 4–Ball Machine," Paper No. C62/85, The Institution of Mechanical Engineers International Conference, Combustion Engines: Reduction of Friction and Wear, London, UK, 1985.

Chapter 18

Diesel Fuel Additives

Until the 1970s there was little or no use of additives in automotive diesel fuel. The product manufactured at most refineries around the world was generally a blend of straight-run atmospheric distillate components and, apart from sulfur content, the specification points could be met without the need for further processing or the use of additives. In the U.S., where the enormous gasoline market had necessitated a high level of downstream conversion to yield more gasoline components, some cracked gas oils went into diesel fuel.

Routine use of diesel fuel additives effectively started in the late 1960s in Europe, with the introduction of cold flow improvers. With the largest proportion of diesel-powered road vehicles of any world region, the growth in demand for diesel fuel was starting to pose problems for the refining industry. The supply situation was further aggravated by the crude oil price rises during the 1970s. Although total demand for petroleum products went down, refiners had to increase the yield of diesel fuel while reducing crude throughput. The use of flow improvers enabled the refiner to produce more diesel fuel by cutting deeper into the crude oil and using the additive to restore the cold properties of the fuel. Flow improvers and other wax modifier additives are discussed in Chapter 16. A comprehensive report on the various types of diesel (and gasoline) additive, together with background information on their application and the environmental implications, has been published by ATC, the European Additives Technical Committee.[1]

Other additive types are now being used in diesel fuel, as more refineries have been obliged to move towards the typical pattern in the U.S., with downstream conversion units to increase the yield of "clean" products by cracking the fractions used for heavy fuel oil, for which there is a decreasing demand.

503

More low-cetane material is being diverted into automotive diesel fuel because it can no longer be absorbed by the shrinking market for domestic heating oil. This necessitates occasional use of an ignition improver to bring the cetane number on specification.

An additional factor influencing the trends in additive use is a growing awareness of the need for fuel product differentiation in the marketplace. It is common practice in many countries for oil companies to exchange and rebrand products to keep down the costs of fuel transportation, the exchanged product being accepted on the basis of an agreed specification and marketed as such. Nowadays, further additive treatment may be made before an exchanged fuel is sold, in order to support the marketing company's advertising claims for a product of superior quality to those of its competitors. This practice has been widely adopted in Europe and other parts of the world.

As already mentioned in Chapter 16, additive treatment of diesel fuels is usually by weight and expressed either in parts per million (ppm) or as a percentage, where 0.1% is equal to 1000 ppm.

18.1 Stability Improvers/Antioxidants/Stabilizers/MDA

Antioxidants, stabilizers and metal deactivators are types of additive which are sometimes used in diesel fuels considered to be prone to oxidative or thermal instability due to the components used in their preparation. The additives work by terminating free-radical chain reactions that would result in color degradation and the formation of sediment and insoluble gums, as discussed in Chapter 4.[2] If oxidation takes place, engine operation could be affected due to filter blocking or gummy deposits in the injection system and on injector nozzles.[3] In some countries a fuel might be unacceptable for marketing as automotive diesel if the maximum color specification is exceeded.

Cracked gas oils have predominantly been used as blend stocks for distillate and residual heating fuels but, with those markets declining, more cracked gas oil is being diverted into diesel fuel. Distillates from cracking operations are more olefinic than those from atmospheric distillation, and contain more nitrogen compounds. As a result, they are less stable, being prone to oxidation by free-radical reactions. This is the main reason why oxidation stability limits are being introduced into more diesel fuel specifications.

The very complex reactions that cause degradation of the fuel proceed in a series of steps to give rise to the formation of high molecular weight molecules, and have already been discussed in Chapter 4, Section 4.3.5.

Two classes of compound that are produced during storage under ambient conditions have been identified as linked indole and phenalene ring systems. Such compounds are soluble in diesel fuel but, if allowed to react with acids, form an insoluble sediment virtually indistinguishable from the polar, highly colored part of naturally formed sediment.[4,5]

18.1.1 Antioxidants and Stabilizers

Antioxidants used in diesel fuels are usually hindered phenols that prevent high-temperature gum-forming reactions. Stabilizers are amines or other nitrogen-containing basic compounds that prevent sediment formation at ambient temperature by interfering with acid-base reactions. These additive types are not normally used in diesel fuel prepared from straight-run components but, if it contains cracked gas oil, protection may be desirable, especially if the fuel is likely to be in storage for a lengthy period.

The same types of antioxidant are used in diesel fuels and gasolines to prevent high-temperature reactions, but stabilizer additives are more specific in their action and need to be selected to suit the particular fuel to be treated.

Antioxidant/stabilizer additives react with peroxy radicals in unstable fuel, as described in Chapter 4, Section 4.3.5, thereby suppressing the free-radical propagation reaction that would normally occur.

The tests for measuring fuel stability are described in Chapter 17 and the same methods are used to evaluate the various additive types. The effectiveness of additive treatment will depend very much on the dominant fuel characteristics that determine the degradation reactions. The choice of additive is generally decided by trial-and-error to find out which is best for the particular fuel. Treating levels are usually in the range of 25 to 200 ppm.

18.1.2 Metal Deactivators

Metal deactivators are sometimes used in conjunction with stability improvers to prevent oxidation reactions from being catalyzed by heavy metal ions, particularly of copper, which may be present in trace amounts in the fuel. The use of copper is usually avoided in vehicle fuel systems, but the possibility of fuel contamination with copper or its alloys cannot be excluded.

One of the most commonly used metal deactivator additives is N,N[1]-disalicylidene-1,2-propanediamine, which works by chelating the dissolved metal to form a non-catalytically active compound. Treat rates are similar to those for gasolines, typically in the region of 10 ppm.[6]

Additive treatment is not the only means the refiner has to control the oxidation reactions that occur in uninhibited cracked gas oils. Diesel fuels have to conform to a sulfur specification and this is achieved by hydrodesulfurization, which is described in Chapter 3. The process also improves fuel stability by removing nitrogen- and oxygen-containing compounds and saturating the more reactive olefinic compounds which are typically present in catalytically cracked gas oils. As virtually all diesel is hydrodesulfurized, there is normally no need for routine antioxidant treatment. However, when only a mild degree of hydrogen treating is required to meet the sulfur specification, as when running a low-sulfur crude, or if the availability of hydrogen is limited, an antioxidant may be added to ensure that the fuel is adequately stabilized.

18.2 Cetane Improvers

Cetane improvers (ignition improvers) are used to increase the cetane number of a diesel fuel by reducing the delay between injection and ignition when fuel is sprayed into the combustion chamber. Cetane Number is the most widely accepted measure of ignition quality and it is determined using the CFR Cetane engine, ASTM Method D 613. The cetane number of a fuel is determined by comparing its ignition delay, under standard operating conditions, with those of blends of two reference fuels, cetane and heptamethylnonane, having, by definition, cetane numbers of 100 and 15, respectively.

Several types of chemicals — alkyl nitrates, ether nitrates, nitroso compounds and certain peroxides — have been identified as effective cetane improvers. They are all materials that decompose readily and, at elevated temperatures, generate free-radicals that accelerate oxidation of the fuel and initiate combustion. Commercial and safe handling considerations have resulted in most attention being given to primary alkyl nitrates, and various versions have been marketed in recent years. At present, low production costs and good response in a wide range of fuels have identified 2-ethyl hexyl nitrate (isooctyl nitrate) as one of the most cost-effective cetane improvers.[3]

Studies have also looked at the potential of peroxides as non-nitrogen based ignition improvers. Selected peroxy compounds were evaluated in an unmodified single-cylinder four-stroke DI engine to assess the viability of alcohol substitution in diesel-powered generating sets. No cetane improvement data were given but 2% ditertiary butyl peroxide in a blend containing 25% ethanol closely matched the thermal efficiency of diesel fuel. Compared with diesel fuel, CO and HC emissions were higher, NO_x and smoke lower.

The influence of cetane number on ignition delay is shown by Figure 15.4. Although the engines gave different delay periods when tested on the same fuels, they all responded similarly to changes in cetane number. Results from tests on a 1.6-liter passenger car diesel engine to ascertain the effect of a cetane improver additive are given in Figure 18.1,[7] which shows a good correlation between ignition delays observed with untreated and treated fuels.

The data in Figures 15.5 and 18.1 indicate that fully warmed engines do not discriminate between natural and additive-improved cetane values, but an important aspect of cetane quality is its influence on cold starting. To study cold starting performance on a DI engine, a series of fuels was prepared from a straight run diesel fuel of 52.5 cetane number and a cracked gas oil. Blending with 10% and 20% of the cracked stock gave cetane numbers of 47 and 45, respectively, and these qualities were upgraded to a cetane number of about 52 by adding 0.1% and 0.2% of cetane improver. Figure 18.2[3] gives the cold test results, plotted as time to start, time to reach a stable idle, and time for the smoke emission to fall to 50% of the initial level. The cetane-improved fuels gave a performance comparable to that of untreated fuels of the same cetane number.

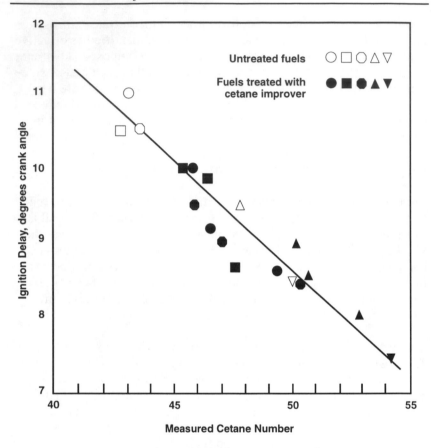

Fig. 18.1. Influence of cetane number on ignition delay.[7] (Reprinted by permission of the Council of the Institution of Mechanical Engineers from Proceedings of an International Conference on Petroleum Based Fuels and Automotive Applications, London, England, 25-26 November 1986.)

Response to cetane improvers is dependent on individual fuel characteristics and on cetane level. Results obtained on a large sample of fuels are presented in Figure 18.3,[8] which shows that, on average, an improvement of about 3 numbers was obtained with a treat level of 500 ppm, and with 1000 ppm the gain was 5 numbers. However, the results show considerable variation about the average line and it was also found that the lower-cetane fuels had the poorest response.

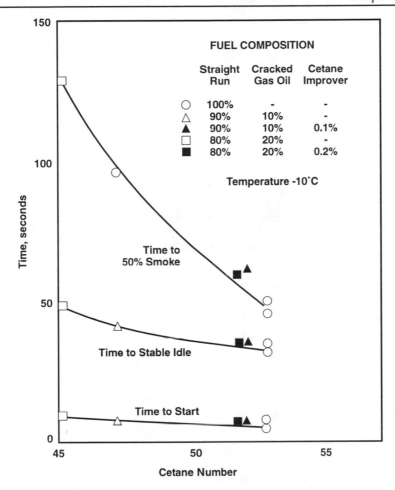

Fig. 18.2. Influence of cetane number and cetane improver on the cold starting time of a DI engine.[3]

In refineries, cetane improvers are used mainly to give fairly modest improvements of 2 or 3 numbers, to bring off-grade fuel blends in specification. This would require additive treat levels in the 500 to 1000 ppm range. This type of additive is used in some multifunctional packages, where the package formulation will give a cetane improver treat level of around 500 ppm.

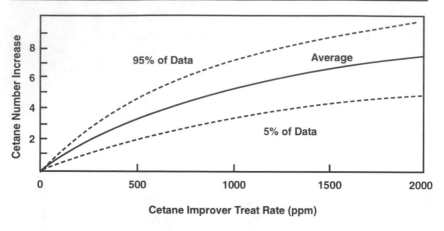

Fig. 18.3. Response to cetane improver additive.[8]

18.3 Combustion Improvers

Combustion improvers are additives that have a catalytic effect on the combustion process in the diesel engine. The most effective additives of this type were organometallic compounds, principally those of barium, calcium, manganese and iron. During the 1960s, they were marketed as antismoke additives. Barium was found to be particularly successful, at 0.2% by volume, in reducing the level of visible black smoke emissions. However, additive use was subsequently discontinued, largely because of concern about the toxicity of barium compounds in the exhaust.

Environmental legislation directed at the diesel engine for the reduction of exhaust emissions has revived interest in the possible use of additives to reduce particulate levels. One concept is the combination of additives with particulate traps. When certain metals such as manganese and copper are deposited with the soot, the autoignition temperature is lowered sufficiently for it to be ignited by the hot exhaust gases. Combustion of the metallic additive will itself produce particulates but the net result is to reduce the total particulates by enabling complete oxidation of the soot and, at the same time, control back-pressure buildup in the particulate trap by burning off trapped deposits.[9]

18.4 Detergents

Detergent additives are considered to be of growing importance in controlling the formation of fuel deposits where they can have a detrimental effect on combustion. Gummy deposits in the fuel injection system can cause sticking of injector needles, resulting in misfires, power loss and increased smoke. The buildup of lacquer and carbonaceous deposits on injector tips can affect the amount of fuel injected and the spray pattern, causing problems of reduced power and higher smoke. Starting may also become more difficult. Figure 18.4 shows how deposits on the needle can spoil the spray atomization of a multihole injector. The finely atomized spray was obtained with a new injector.

The incidence of problems due to nozzle coking with the pintle injectors used in indirect injection engines of diesel passenger cars was reported to be widespread in the U.S. and to a lesser extent in Europe during the early 1980s. The symptoms experienced were misfiring, noisy combustion, exhaust odor and heavy smoking when starting from cold, with irregular and noisy combustion after warm-up if the nozzles were heavily coked.[10] A section through the tip of a pintle nozzle, indicating how the coke deposit restricts the area through which the fuel sprays as needle lift commences, is shown in Figure 18.5.[10] Also shown are two design modifications to make the nozzle less sensitive to restriction by coke. The Central Hole In Pintle (CHIP) nozzle provided an additional passageway for the fuel, but this was also prone to coking. A more effective replacement design is the flatted nozzle which has a segment machined off the pintle tip to increase the area for initial flow through the orifice.

The effect of injector nozzle coking on exhaust emissions and fuel consumption is given in Table 18.1.[8] Two series of tests were run on the same engine with the same fuel, first with clean nozzles and then with coked nozzles. Changing to the coked nozzles markedly increased emissions of hydrocarbons, CO and particulates, while fuel consumption went up by 3.7%.

Detergents for diesel fuels are of the same chemical type as those used in gasolines — amines, amides, imidazolines, etc. These are surfactant additives with a polar group at one end that forms a barrier film on metal or particulate surfaces while a nonpolar, oleophilic group at the other end dissolves in the fuel. Particulates are effectively solubilized

Fig. 18.4. (a) Photo of spray from new nozzle.

Fig. 18.4. (b) Photo of spray from coked nozzle (source: Lucas Diesel Systems).

PINTLE NOZZLE

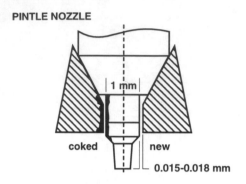

CHIP NOZZLE (CHIP = central hole in pintle)

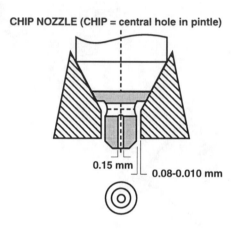

FLATTED NOZZLE

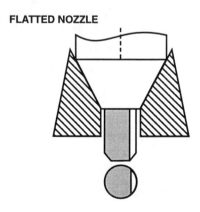

Fig. 18.5. Nozzle types used in prechamber engines.[10]

**Table 18.1. Effect of Nozzle Coking on Exhaust
Emissions and Fuel Consumption
(source: Exxon Chemical International)**

Procedure: ECE 15 Cycle Tests, 1.6L N/A IDI Passenger Car on
Chassis Dynamometer

Program: Two test series on the same fuel: First with clean
injectors, second with coked injectors.

Measurement	Increase with Coked Nozzles
Hydrocarbon, g/test	78%
Carbon Monoxide, g/test	33%
Nitrogen Oxides, g/test	4%
Particulates, g/test	51%
Fuel Consumption, L/test	3.7%

and prevented from agglomerating by the film formed around them.
In the same way, metal surfaces are protected against deposit formation. The function of the detergent additive also gives some antirust
protection. Polymeric dispersants are sometimes used in conjunction
with detergents to help in the dispersion of particulate matter.[6]

The effectiveness of an additive in preventing or controlling deposit
formation is assessed by engine tests, either on the bench or on the
road. A number of test procedures have been developed, mainly by
manufacturers whose engines have been affected by injector coking
problems. The procedures involve running to an engine operating
mode which has been designed to recreate the type of problem experienced in service in a relatively short time.

Although most of these tests originated as "in-house" methods, details
of some procedures have been released to allow oil and additive companies to evaluate their own products and look at ways of overcoming
the problem. Measurements of the coking tendency of different fuels
are illustrated in Figure 18.6.[10] The graph compares the pattern of air
flow with needle lift for new and coked nozzles, with the U.S. fuel

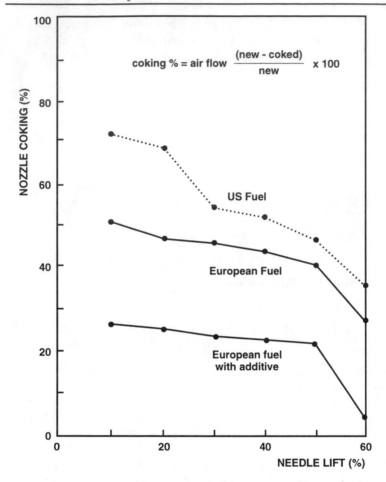

Fig. 18.6. Nozzle coking with different fuels.[10]

causing a higher degree of coking than the European fuel, particularly during the initial opening period. The lowest line shows the benefit in controlling nozzle coke formation by means of a detergent additive in the European fuel.

The choice of additive type and treat rate will be determined by the characteristics of the fuel and also whether the requirement is keep-clean or cleanup performance. A higher treat rate or a more effective additive is usually needed to clean-up dirty injectors. Treat rates to control deposits on new or cleaned injectors are in the 100 to 200 ppm range.[6]

Polymeric dispersants, as mentioned above, are often used to complement the role of the detergent additive. Chemically the dispersants are relatively high-molecular-weight materials, generally either alkenyl succinimides or hydrocarbyl amines. Recommended treat rates are around 200 ppm for the succinimides and up to three times that amount for the hydrocarbyl amines.

18.5 Corrosion Inhibitors

Corrosion inhibitors (or antirust additives) are usually added to diesel fuels being transported by pipeline to protect the pipeline and to avoid rust contaminating the fuel and blocking filters in the distribution system.

These additives are surfactant materials having a polar group at one end and an oleophilic/hydrophobic group at the other. The polar group attaches itself to metal surfaces in the system, while the other group repels water and provides an oily layer to prevent rust formation.

A wide range of chemical types are used as anticorrosion additives. They include esters or amine salts of alkenyl succinic acids, alkyl orthophosphoric acids, alkyl phosphoric acids and aryl sulphonic acids. In many cases, treating rates are controlled to a low level that is just sufficient to protect the distribution systems, leaving only about 1 ppm in the emerging fuel. A typical dose rate for pipeline protection is in the region of 5 ppm, but a higher amount may be added if vehicle system protection is also to be provided. Some multifunctional additive packages contain corrosion inhibitor to protect the vehicle fuel system. Selection of additive type and treat rate is usually determined by using a rusting test, ASTM D 665, which was originally developed for evaluating steam turbine oils. A polished steel rod is immersed in a 9:1 mixture of oil and water which is continuously stirred and kept at a temperature of 60°C for 24 hours. The appearance of the steel rod is compared with a control specimen from a portion of untreated fuel.

18.6 Antistatic Additives

Antistatic or, more correctly, static dispersant additives of the type used in aviation kerosene are occasionally added to diesel fuel to avoid the risk of an explosion due to a charge of static electricity building up during fast rates of pumping, as may occur when loading or unloading large quantities of fuel from road tankers. There is virtually no risk of an excessive charge building up when refueling diesel vehicles at a service station.

The additives work by increasing the conductivity of the fuel, allowing any electrostatic charge generated during pumping to be dissipated. The treatment is usually with a chromium-based additive at the low rate of 2 ppm.

18.7 Dehazers and Demulsifiers

Dehazer treatment may occasionally be needed if the fuel becomes hazy due to the presence of finely dispersed droplets of water. Contamination with water can occur at almost any stage, as the fuel passes from the refinery and through the distribution network until it reaches the vehicle tank. As described in Chapter 17, Section 17.5, it can be the result of dissolved water coming out of solution or condensing from the air when the temperature falls, leakage of rain water into the tank, or entrainment of water accumulated in storage tank bottoms. The situation may be aggravated by the characteristics of the fuel or the type of additives it contains, and by excessive turbulence in the pumping system.

If the haze persists after the normal 1 or 2 days settling time, additive treatment may be necessary to accelerate clearance and meet the usual "clear and bright" requirement. Effective dehazer additives include quaternary ammonium salts, typically used at dose rates between 5 and 20 ppm.

As hazy fuel tends to be a spot problem, the practical approach is for alternative additives to be tested on-site, in cold samples drawn directly from the affected tank. Samples taken away for testing will usually have cleared by the time they reach the laboratory because of a temperature change or contact with the sample container. However, the haze-forming tendency of a diesel fuel can be assessed in the laboratory using a modified version of the Waring Blendor test for gasolines. A portion of diesel fuel, to which is added 10% distilled or synthetic sea water, is stirred for 2 minutes at 10,000 rpm in the standard laboratory blender. The mixture is poured into a glass jar and its appearance is rated at intervals over a period of 24 hours, either visually or using a photometer, to see how long it takes for the haze to clear.

A demulsifier may be included when detergent/dispersant additives are used, to avoid problems due to pick-up of storage tank bottoms. Entrainment of water and debris in pipelines and during product transfer might result in the formation of stable emulsions and suspended matter which could plug filters or otherwise make the fuel unacceptable.

Demulsifiers are highly surface-active chemicals selected for their limited solubility in oil and water. They are usually prepared by reacting a hydrophobic molecule such as a long chain alkylphenol with ethylene or propylene oxide. The effectiveness of different additive types and treat rates can be checked using the 10-cycle multiple contact test in which a small amount of water is successively agitated with ten portions of fuel to represent repeated filling and emptying of a storage tank. The assessment is based on the amount of emulsion and suspended matter at the oil/water interface.[6] Typical treating levels are generally not more than 10 ppm.

18.8 Lubricity Additives

The trend towards cleaner diesel fuels with lower end points and, more specifically, lower sulfur content is causing concern about fuel lubricity. Severe hydrotreatment to reduce the fuel sulfur content in line with stringent future legislation may also lower the fuel's lubricity by removing or inhibiting natural lubricants present in the fuel.[11,12]

The fuel lubricity testing machines described in Section 17.11 are being assessed for their ability to screen lubricity additives as well as evaluating the lubricity of the fuel itself. Tests carried out with the Ball on Cylinder Lubricant Evaluator (BOCLE-LZ) confirmed the poor lubricity of Swedish Class 2 (II) city diesel.[13] Further tests demonstrated that additive treatment can improve its anti-wear performance to match that of conventional winter grade diesel fuels with sulfur levels mainly in the range of 0.1-0.2% m/m, as shown in Figure 18.7.

The High Frequency Reciprocating Rig (HFRR) also demonstrated the effectiveness of additive treatment, at both test temperatures. Figure 18.8 compares wear scar data obtained from a Class 2 diesel fuel

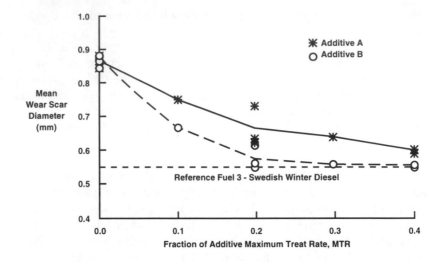

Fig. 18.7. BOCLE test results on Swedish Class 2 fuel treated with lubricity additives.[13]

and two additives with the typical result from a conventional winter diesel fuel.[14] The practical benefits of additive treatment are illustrated by results from field tests on four European passenger car diesels. The cars were run on Swedish Class 1 diesel fuel, with and without a lubricity additive, to see how long it would take for pump failure. With the untreated fuel, catastrophic pump failure occurred in less than 8000 kilometers, while the two additive treated fuels continued running for more than 100,000 kilometers (see Figure 18.9).

Other potential problems associated with severe hydrotreating, mainly because naturally occurring inhibitors can be destroyed, are incompatibility with elastomers in the fuel system and an increased susceptibility of the fuel to oxidize during storage, possibly forming high levels of peroxides.[15] Excessive fuel leakage through the nitrile rubber injection pump seals was experienced soon after the start of a trial on Swedish Class 1 City Diesel by a British bus authority. Nitrile rubber swells when used with conventional diesel fuel, forming a good seal. Although Swedish Class 1 fuel will not swell nitrile rubber, it tends to leach out aromatics and other components of the rubber and will cause previously swollen seals to shrink.[11] The problem can be avoided by replacing swollen seals before changing to Class 1 fuel, using either nitrile or viton rubber, the latter being more resistant to swelling and leaching.

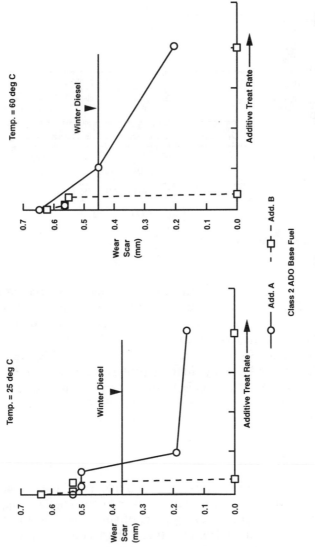

Fig. 18.8. Influence on wear scar diameter of additive treat rate and HFRR test temperature.[14]

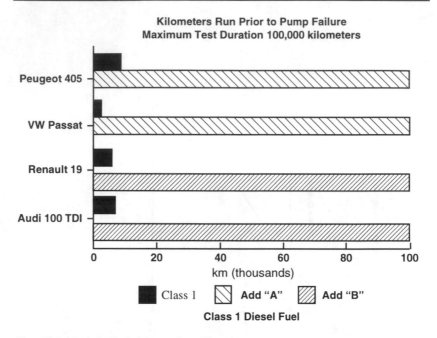

Kilometers Run Prior to Pump Failure
Maximum Test Duration 100,000 kilometers

Fig. 18.9. Lubricity field test data.[14]

Not all low sulfur fuels are unstable but peroxide formation profiles obtained with reactive low sulfur prototype fuel samples indicate the possibility of a future problem. Figure 18.10 shows the amounts of peroxides formed in seven reactive prototype fuels, which had induction periods ranging from 9 to 30 days, at 65°C in the CRC aging test. Commercially available antioxidant additives were found to be effective, either in inhibiting peroxide formation or in minimizing its formation by extending the induction period in the CRC test, as shown in Figure 18.11.

18.9 Anti-icers

In some countries and localities, anti-icers may be added to road tankers making deliveries of diesel fuel. This is to prevent ice plugging of fuel lines by lowering the freezing point of small amounts of free water which may separate from the fuel. These additives are relatively low-molecular-weight alcohols or glycols that are soluble in diesel fuel and have a strong affinity for water, so they are effective at low dose rates in the region of 30 ppm.

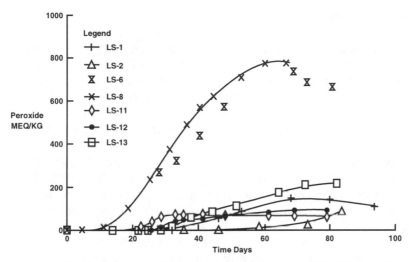

High Peroxide Levels are Possible in Low Sulfur Prototype Fuels (65°C CRC Aging Test)

Fig. 18.10. Peroxide formation in reactive prototype fuels.[15]

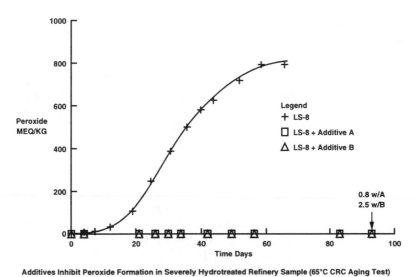

Additives Inhibit Peroxide Formation in Severely Hydrotreated Refinery Sample (65°C CRC Aging Test)

Fig. 18.11. Effectiveness of antioxidants in reactive fuels.[15]

18.10 Biocides

Biocidal treatment of diesel fuel is sometimes carried out to prevent the growth of bacteria and fungi in the bottom of fuel tanks. These aerobic or anaerobic organisms live in the water bottoms and feed at the fuel/water interface. They lead to accumulation of suspended matter that will obstruct flow through filters if drawn out with the fuel. Warm conditions can encourage growth of the organisms.

Regular draining of water bottoms will help minimize the risk of bacterial growths but this is not always possible, especially if natural or man-made caverns are used for long-term storage of strategic reserves. Commercial biocides come from a wide range of chemical types, including boron compounds, amines, imines, etc., which need to be soluble in water and in the fuel to be properly effective. A problem with biocidal treatments is that the bacteria can develop resistance, so the additive type must be changed from time to time.

Evaluation of biocides is by incubation of treated and untreated nutrient media that have been inocculated with the relevant strain of bacteria. A treat level of 200 ppm is used for direct addition to water bottoms but a smaller dose rate would be appropriate for injection into the fuel as it goes into storage.

18.11 Antifoamants

Antifoamants are sometimes added to diesel fuel, often as a component of a multifunctional additive package, to help speed up or to allow more complete filling of vehicle tanks. Their use also minimizes the likelihood of fuel splashing on the ground or onto clothing, avoiding the nuisance of stains and unpleasant odor, and reducing the risk of spills polluting the ground and the atmosphere.

As with lubricating oils, silicon additives are effective in suppressing the foaming tendency of diesel fuels, the choice of silicone and cosolvent depending on the characteristics of the fuel to be treated. Tests used to assess the tendency of a lubricating oil to form a stable foam in an engine are not suitable for diesel fuels. Selection of a diesel antifoamant is generally decided by the speed with which the foam collapses after vigorous manual agitation to simulate the effect of air entrainment during tank filling. Treat levels are normally in the 10 to 20 ppm range.

18.12 Odor Masks and Odorants

As the odor of diesel fuel is considered objectionable by many people, odor masks are occasionally employed to improve the market acceptability of some branded diesel fuels. Because diesel fuel is less volatile than gasoline, the stain and smell of spills will persist, which can be very annoying, particularly if clothing is contaminated.

Attitudes to smells vary widely and are subjective, but odor panel tests suggest that the market preference is for a neutral rather than a positive odor, which puts the emphasis mainly on odor-masking effects. Various products, with a choice of fragrances, are commercially available for use at treating rates of 10 to 20 ppm.

18.13 Dyes and Markers

Dyes are often used in petroleum fuels for identification of particular branded products, or for legal reasons as a means of providing evidence, for example, in cases of theft, tax evasion or fuel adulteration. As it may be possible for the dye to be removed, a colorless chemical marker or tracer is usually added with the dye. Diesel fuels tend to be unidentified because, in Europe, a red dye with furfural as the chemical tracer is mandatory in the lower-taxed domestic heating oil, which might be blended into or substituted for automotive diesel fuel.

18.14 Drag Reducers

Drag reducers are sometimes used in petroleum products when pipeline capacity is limited to increase throughput and postpone the need for investment to construct additional pipelines. They are high-molecular-weight, oil-soluble polymers which shear very rapidly and reduce drag. Treating levels in the region of 50 ppm have been found to give significant reductions in frictional drag.

References

1. R.F. Haycock and R.G.F. Thatcher, "Fuel Additives and the Environment," Document 52, published by the Technical Committee of Petroleum Additive Manufacturers in Europe (ATC), 1994.

2. T.R. Coley, "Diesel Fuel Additives Influencing Flow and Storage; Critical Reports on Applied Chemistry," <u>Gasoline and Diesel Fuel Additives</u>, Ed. by K. Owen, Vol. 25, John Wiley & Sons, 1989.

3. T.J. Russell, "Petrol and Diesel Additives," *Institute of Petroleum Review*, Vol. 42, No. 501, pp. 35-42, October 1988.

4. J.F. Pedley and R.W. Hiley, "Investigation of 'Sediment Precursors' Present in Cracked Gas Oil," 3rd International Conference on Stability and Handling of Liquid Fuels, London, September 1988.

5. J.F. Pedley, R.W. Hiley and R.A. Hancock, "Storage Stability of Petroleum-Derived Diesel Fuel," *FUEL*, Vol. 67, pp. 1124-1130, August 1988.

6. R.C. Tupa and C.J. Dorer, "Gasoline and Diesel Fuel Additives for Performance/Distribution Quality - II," SAE Paper No. 861179, 1986.

7. R.D. Cole, M.G. Taylor and F. Rossi, "Additive Solutions to Diesel Combustion Problems," Paper No. C310/86, Institution of Mechanical Engineers International Conference on Petroleum Based Fuels and Automotive Applications, London, UK, 1986.

8. T.R. Coley, F. Rossi, M.G. Taylor and J.E. Chandler, "Diesel Fuel Quality and Performance Additives," SAE Paper No. 861524, 1986.

9. E. Goldenberg and P. Degobert, "Filtres a Activite Catalytique pour Moteur Diesel," *Revue de l'Institut Francais du Petrole*, Vol. 41, No. 6, Nov.-Dec. 1986.

10. M. Fortnagel, H.O. Hardenberg and M. Gairing, "Requirements of Diesel Fuel Quality: Effects of Poor-Quality Fuels," 47th API Mid-Year Meeting, New York, 1982.

11. H.C. Grigg, "Reformulated Diesel Fuels and Fuel Injection Equipment," Fifth Annual Conference on New Fuels and Vehicles for Cleaner Air, Phoenix, Arizona, January 1994.

12. E. Goodger and R.A. Vere, <u>Aviation Fuels</u>, G.T. Foulis & Co., Ltd., 1970.

13. S.R. Jenkins, R.G.M. Landells and J.W. Hadley, "Diesel Fuel Lubricity Development of a Constant Load Scuffing Test Using the Ball on Cylinder Lubricity Evaluator (BOCLE)," SAE Paper No. 932691, 1993.

14. C. Bovington and R. Caprotti, "Latest Diesel Fuel Additive Technology Development (with particular reference to lubricity)," Paper No. CEC/93/EF13, CEC Fourth International Symposium on the Performance Evaluation of Automotive Fuels and Lubricants, Birmingham, U.K., May 1993.

15. J. Vardi and B.J. Kraus, "Peroxide Formation in Low Sulfur Automotive Diesel Fuels," SAE Paper No. 920826, 1992.

Further Reading

L.L. Stavinoha, "User Needs for Diesel Stability and Cleanliness," API 47th Mid Year Meeting, New York City, NY, 1982.

J. Ritchie, "A Study of the Stability of Some Distillate Diesel Fuels," *Journal of the Institute of Petroleum*, Vol. 51, No. 501, September 1965.

R. Caprotti, C. Bovington, W.J. Fowler and M.G. Taylor, "Additive Technology as a Way to Improve Diesel Fuel Quality," SAE Paper No. 922183, 1992.

P.I. Lacey and S.J. Lestz, "Effect of Low-Lubricity Fuels on Diesel Injection Pumps—Part II: Laboratory Evaluation," SAE Paper No. 920824, 1992.

Chapter 19

Future Trends and Alternative Fuels

In this chapter we discuss the relatively short-term changes (i.e., up to the early part of the next century) that can be envisaged for gasoline and diesel fuel, as well as the present and longer-term use of alternative fuels.

The pressures for change are likely to be:

- Crude oil price and security of supply

- Environmental considerations

- Vehicle technology advances

- Product differentiation

It is not usually possible to make large changes in fuel composition or characteristics because gasolines and diesel fuels need to be suitable for existing and older vehicles as well as for new advanced designs. To market additional types of fuel is difficult and expensive and so fuel quality changes are unlikely to be very dramatic in the short term.

19.1 Influence of Crude Oil Price and Need for Security of Supply

There is a lot of speculation as to what will happen to crude oil prices over the next twenty to thirty years, and it is an important question because they will have an overriding influence on gasoline and diesel fuel composition and quality. It has been suggested[1] that oil prices could remain low throughout this period even if the growth rate of fuel usage were to return to levels of five or more percent per annum. However, over 50% of the world's oil is concentrated in the Middle East and an unstable situation there can make for volatile prices. There is clearly a physical limit to the amount of oil that is in the ground and, as existing reserves become depleted and new finds become less frequent and more difficult to extract, prices must increase. One possibility that could change this situation would be if alternative fuels such as methanol take over well before the oil runs out, perhaps because of advantages they may show in terms of air quality or simply for security of supply. However, this seems unlikely to occur to a large extent before the early part of the next century, so the following comments apply primarily to the period up to that time.

High crude prices might favor the use of heavier crudes because they are somewhat cheaper, but would cause the market for residual fuels to be severely restricted in view of the availability of potentially lower-priced fuels such as coal. It would then be necessary to crack these heavier components for use as automotive fuels, which would give rise to changes in gasoline and diesel fuel composition. Increased cracking could lead to reduced oxidation stability as more and more cracked naphthas and gas oils are used. On the other hand, because of the limitations in terms of octane quality and cetane value of such cracked stocks, a portion of them will be upgraded by reprocessing and this may also improve their stability. Deficiencies in MON would have to be made up by such processes as isomerization and alkylation.

Any increase in the use of cracked stocks will lead to a rise in the use of additives both for gasoline and diesel fuel. These additives will be: antioxidants, particularly for gasoline, to prevent degradation of olefinic compounds; cetane improvers to improve diesel ignition quality; and additives to keep fuel systems free of deposits that will adversely influence performance. The need for cetane improvers had been expected to develop slowly, partly due to downward revision of specification limits during the 1980s and also because of the introduction of minimum cetane index specifications to control base fuel quality; however, the downward trend is likely to be limited. The European Association of Automobile Constructors (ACEA) is pressing for higher cetane number specifications to enable more stringent exhaust emissions targets to be attained. As mentioned in Section 17.2, the ACEA 1994 Fuel Charter recommends a minimum cetane number of

53, with 50 as the cetane index minimum. Engine test results on light-duty diesel vehicles and heavy-duty diesel engines[2] presented in Figure 19.1 and 19.2 show CO and particulates decreasing with increasing cetane number.

Digging deeper into the crude to increase the yield of diesel fuel may mean that more wax crystal modifiers will be required to improve cold flow properties. The heavier paraffinic components incorporated by cutting deeper could have a beneficial effect on cetane number.

Higher crude oil costs would also reemphasize fuel economy considerations and this could favor the greater use of diesel-engined vehicles for personal transport. However, improvements in the fuel economy of the spark ignition engine will probably ensure that it continues to be the dominant engine for personal transport. Some of the methods of improving economy, such as the use of increased compression ratios or of turbo- or superchargers, would increase octane requirements unless compensated for in some way, such as by the wider use of knock sensors to protect the engine from knock damage.

If oil prices remain relatively low over this period, the pressures for change will be mainly as a result of environmental and vehicle technology changes.

Security of supply of energy is of paramount importance to all governments and this can be threatened if countries with no oil reserves have to face high crude oil prices. Alternative fuels based on indigenous material sources such as coal, natural gas or biomass then become very important, as does the need to conserve as much energy as possible.

19.2 Influence of Environmental Pressures

Environmental pressures will continue to be the major factor in influencing fuel quality. For gasoline, in areas where lead is still being used, fiscal incentives will continue to encourage the demise of leaded grades. This, in turn, will put more pressure on meeting octane levels and so cause increased use of oxygenates such as MTBE. More severe reforming will help to meet octane quality, but at the expense of yield and the production of high levels of aromatics which themselves have environmental disadvantages, as discussed in Chapters 11 and

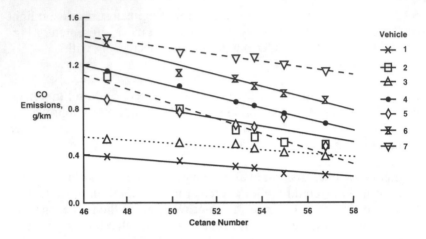

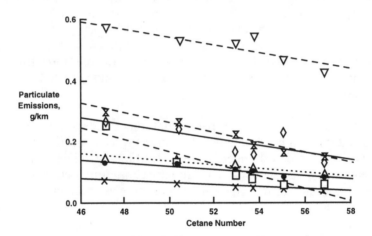

Fig. 19.1. Influence of cetane number on CO and particulates emissions from light-duty diesel engines.[2] (Reprinted by permission of the Council of the Institution of Mechanical Engineers from P. Heinze, "The Influence of Diesel Fuel Properties and Components on Emissions from Engines," Institution of Mechanical Engineers Seminar, 1993 - 2, Fuels for Automotive and Industrial Diesel Engines, 2nd Seminar, Birmingham, U.K., April 1993.)

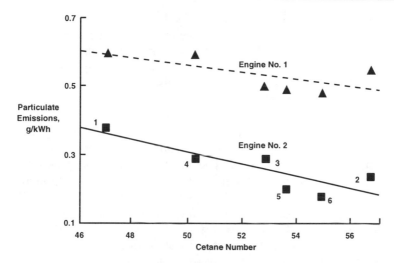

Fig. 19.2. Influence of cetane number on particulates emissions from heavy-duty diesel engines.[2] (Reprinted by permission of the Council of the Institution of Mechanical Engineers from P. Heinze, "The Influence of Diesel Fuel Properties and Components on Emissions from Engines," Institution of Mechanical Engineers Seminar, 1993 - 2, Fuels for Automotive and Industrial Diesel Engines, 2nd Seminar, Birmingham, U.K., April 1993.)

12. With diesel fuels, sulfur specifications are being tightened to reduce acidic and particulate emissions and, because hydrotreating will be used for this purpose, oxidation stability should improve.

Both exhaust and evaporative emissions will certainly continue to be closely controlled with restrictions becoming tighter on a worldwide scale. The use of reformulated gasolines is likely to increase and spread to other countries that have air quality problems.

Emissions of compounds that are presently unscheduled will become closely controlled—materials such as 1,3 butadiene and aldehydes come into this category.

Gasoline evaporative emissions both from the vehicle and during refueling will come increasingly under scrutiny so that the vehicle control systems already in use in such places as the U.S. will spread everywhere. This is because underhood and fuel tank temperatures seem to be increasing as a result of a number of vehicle design factors (see below), and also because such extensive efforts have been extended to reduce exhaust emissions that the relative importance of evaporative emissions has increased dramatically. Vehicle control systems are more effective at reducing evaporative emissions than are changes in gasoline volatility; nevertheless, because in any given car population there are always a large number of uncontrolled cars at the start of any attempts to reduce these pollutants, lower RVP levels are likely in those countries where levels are high during the summer months. Any reduction in RVP will mean less butane in the gasoline and will reduce the likelihood of methanol being used as a blend component.

The increased use of additives, as mentioned with respect to the possible increased levels of olefinic components, is also likely to be important in keeping engine fuel systems free from deposits that could upset air-fuel ratios, cause valves to stick, and block orifices.

Global warming is another environmental threat that, although it has not yet been definitely proven, is causing concern. A major cause is believed to be CO_2 and, although highway transport is considered to contribute only a small percentage of total greenhouse gases worldwide, it could influence the future design of vehicles and the types of fuel they use. Other so-called "greenhouse gases" are CFCs, methane and nitrogen oxides, but CO_2 is the most important and difficult gas to reduce from an automotive viewpoint. The steps that can be taken are: to encourage better fuel economy, and this could mean increased use of diesel-engined vehicles if the other emissions are satisfactory; and a move toward fuels with a lower carbon-hydrogen ratio such as methanol, LPG, CNG, etc., although studies have indicated[3] that CNG and LPG are comparable to gasoline in total greenhouse gas emissions, while methanol from natural gas generally yields more greenhouse gases than gasoline. Hydrogen would be expected to be a clean-burning fuel but combusting it in a vehicle does produce some nitrogen oxides and the production of hydrogen involves very significant CO_2 emissions.

19.3 Influence of Vehicle Technology Changes

With regard to the spark ignition engine, there are a number of design features which may change the susceptibility of vehicles to gasoline quality and which may, therefore, influence future gasoline specifications. These include:

- The wider use of continuously variable transmission systems (CVTs) which, because the engine is likely to be operating at a higher, more constant speed, could increase the importance of MON levels.

- Lower drag coefficients mean that it is more difficult to direct air streams into the engine compartment for cooling purposes. This would raise underhood temperatures and make gasoline front-end volatility more critical.

- Although improved fuel injector designs are less susceptible to fouling problems, there will continue to be a need for additives to ensure they are kept free from deposit formation.

- Engine breathing can be improved by more, or larger, valves which themselves can be susceptible to problems caused by deposits. Again, the need for the wider use of suitable additives is likely.

- Engine management systems aided by sensors will be able to ensure that the optimum conditions of air-fuel ratio, spark advance, etc., are always met and this could make some aspects of fuel quality such as octane and oxygen content less important. On the other hand, the differences in acceleration performance between a high- and a low-octane fuel in such vehicles may become a marketing feature that could make octane even more critical and increase the sales of super-unleaded fuel.

- The progressive introduction by California of so-called Transitional Low Emissions Vehicles (TLEV), Low Emission Vehicles (LEV), Ultra-Low Emission Vehicles (ULEV) and Zero Emission Vehicles (ZEV) will certainly influence fuel type and quality. Each of these types of vehicle has a set of emission standards which are progressively more severe, as shown in Appendix 3, with ZEVs having no exhaust or evaporative emissions at all of any pollutant—being electrically powered. The Act dictates that by 1998 ZEVs must account for 2% of manufacturing sales, and by 2003, for 10% of sales.

Many of the factors in the above list are conflicting and it will be necessary to follow closely the design trends to see which will have the greatest influence on gasoline quality.

Regarding compression ignition engine advances, design changes to improve noxious emissions include charge-air boosting, modifications to injection equipment and timing, exhaust gas recirculation (EGR), particulate traps, and the electronic control of the combustion process:

- Turbocharging gives increased efficiency and specific power output as well as improving the combustion process by reducing ignition delay and the intensity of diesel knock. Although advantages over turbocharging in low-speed torque characteristics can be achieved by boosting with a mechanical supercharger or by Comprex pressure wave techniques, turbochargers are expected to continue as the most common form of charge boosting for diesel engines.[4,5,6]

- A considerable amount of work has been reported on the influence of injector timing on diesel combustion and exhaust emissions. Retarding the start of injection shortens the ignition delay period and lowers the amount of CO, NO_x and unburnt hydrocarbons (UBHC). In the other direction, benefits in specific fuel consumption, smoke and particulates as well as CO and UBHC, are obtained by advancing the end of injection, as shown in Figure 19.3.[7]

  Reducing the duration of injection to approach the optimum situation indicated by these results would necessitate marked changes in injection characteristics to avoid the negative effects of high rates of pressure rise, high peak pressure and NO_x. Possible solutions include higher injection pressures and modified injector nozzles to control the size and distribution of fuel droplets and give controlled burning, as well as limiting the penetration of the fuel spray to avoid wetting of the combustion chamber walls.

- Exhaust gas recirculation can be applied to achieve a low level of NO_x emissions but, understandably, there is a debit with respect to performance, fuel economy and other emissions.[8]

- Electronic engine control systems have become well established for spark ignition engines, but their application to diesel engines is more difficult because of the requirements of the high-pressure injection system. However, the difficulties have been over-

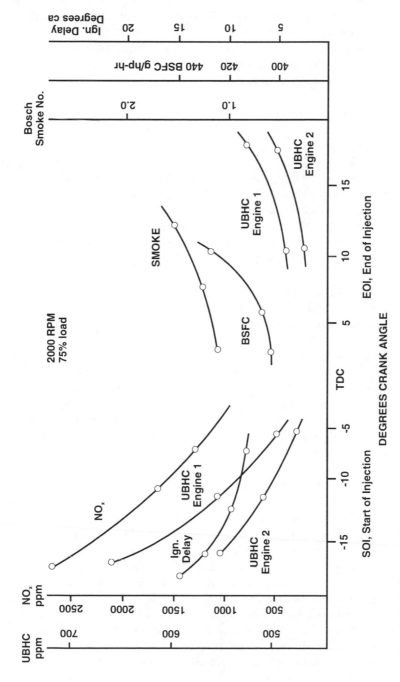

Fig. 19.3. Influence of injection timing on exhaust emissions.

come and electronic controls have been developed by Stanadyne in the U.S. and by the major European manufacturers of fuel injection equipment, Bosch and Lucas (see 14.3.4). Electronic control units are now available for light-duty and heavy-duty diesels, including engines equipped with combined injection pump and injector units.

Bosch electronic controls, which are available for inline and rotary fuel injection pumps, are used by the Swedish company Scania for some of their large truck engines and by BMW for their 2.5-liter, 6-cylinder turbocharged direct injection passenger car engines. The Lucas EPIC control for distributor injection pumps is used by the Ford Motor Company Ltd. for the 2.5-liter direct injection diesel engine in their light commercial vans. In these electronic control systems a computer regulates the operation of the auxiliary injection pump to deliver the required amount of fuel to the injectors at the appropriate timing.

The inability to vary the timing to suit the needs of individual cylinders and also the effects of fuel compressibility in the high-pressure fuel pipes of engines with auxiliary injection pumps, are constraints avoided by the use of combined injection pump and injector units. Caterpillar and Volvo 12-liter truck engines are equipped with Lucas Electronic Unit Injectors (EUI), and Bosch also has developed electronic unit injectors. Computer input signals giving the positions of the accelerator pedal, crankshaft and camshaft, engine speed and temperatures at various locations are used to determine the optimum fuel injection timing and quantity for each cylinder.

19.4 Product Differentiation

It is advantageous when marketing a fuel to be able to claim that your product is better because it contains features that other competitive fuels do not possess. It is difficult to do this by modifying some of the base fuel properties such as volatility, because this might restrict the possibility of exchange agreements with other companies. It would destroy the claims if you supplied them with the same quality. The use of additives to differentiate one brand from another is, however, possible and is widely used. The additives are usually added to the fuel immediately prior to it being delivered to the service station, the appropriate additive being injected as part of an exchange agreement. It seems likely that the use of additives for this purpose will continue and may even expand.

19.5 Alternative Fuels — General Considerations

The use of alternative fuels to gasoline and diesel is already practiced to some extent, as for example, the use of propane and butane (LPG) in some European countries and the use of ethanol in Brazil. Although any substitutes for gasoline must be either gaseous or volatile liquids, the compression ignition diesel engine is potentially capable of burning a wide variety of fuels, from pulverized coal through petroleum residues and middle distillates to gases. However, although some large, slow-speed engines operate routinely on two or more types of fuel, the performance and flexibility demands of the diesel engine for road transport applications have effectively defined and limited the quality requirements to those of the closely controlled middle distillate based hydrocarbon fuel in common use today. Such fuels can be derived from alternative sources, as discussed below. However, partial or complete replacement of petroleum-based fuel for diesel engines has been seriously studied in various parts of the world because of concern about the environment. Alternative types of fuel evaluated include vegetable oils from rape seed oil and soya, methanol and liquefied natural gas (LNG).

Because of the relative importance of LPG and natural gas and also of alcohol fuels for spark ignition engines, these are covered in separate chapters (Chapters 20 and 21).

The main reasons for the interest in alternative fuels are:

- To ensure that when the shortfall in crude oil occurs, there can be a smooth transition to other fuels.

- To provide long-term security of supply because well over half of the world's crude oil is in the Middle East.

- To improve air quality because the alternative fuel may give cleaner exhaust gases as, for example, is claimed for methanol[9,10] as a replacement for gasoline. However, the improvements in exhaust emissions resulting from the use of reformulated gasolines will delay the general introduction of alternative fuels such as methanol.

- To overcome the absence of an indiginous crude oil supply together with an adverse balance of payments situation. An example has been the use of ethanol as an automotive fuel in Brazil, where expensive crude oil had to be imported but ethanol could be manufactured relatively cheaply (when world sugar prices were low) by fermenting sugar cane.

One other type of alternative fuel should be mentioned, already well studied from a military strategic viewpoint, and that is the very broad boiling range petroleum fuel which could be used in wartime for a variety of engine types. Unfortunately, a wide cut fuel usually has poor octane and cetane qualities, and other difficulties such as its deposit-forming tendency make it not particularly suitable for either spark ignition or compression ignition engines. However, the ability to operate on such fuels is often a prerequisite for engines being developed for use in military equipment. The diesel engine can be modified more easily than the spark ignition engine to run on such fuels.

19.6 Alternatives to Crude Oil

In some areas, and particularly North America, tar sands and shale oils are present in massive amounts and many investigations into the use of fuels derived from them have been carried out.[11,12] It is difficult and expensive to obtain satisfactory fuels from these sources because the yield is low and their behavior is apparently not always the same as that of a conventional gasoline[13] or diesel fuel.[14,15] However, the large volumes potentially available make it likely that they will eventually become an important source of hydrocarbons for transportation fuels.

19.7 Synthetic Hydrocarbon Fuels

The Fischer-Tropsch Process for synthesizing hydrocarbons from carbon monoxide and hydrogen was discovered in 1925. The synthesis gas used can be obtained from a number of different sources including coke or natural gas by steam reforming or partial oxidation, and must contain hydrogen and carbon monoxide in the ratio of approximately two to one. The reactions are:

$$nCO + 2nH_2 \rightarrow C_nH_{2n} + nH_2O$$

$$nCO + (2n+1)H_2 \rightarrow C_nH_{2n+2} + nH_2O$$

The products can be varied considerably by the reaction conditions and the catalyst. The hydrocarbon liquids can then be refined using conventional techniques to produce gasoline, diesel fuel and other products.

In South Africa the SASOL process has been used commercially since 1955. Coal, steam and oxygen are reacted together at 30 atmospheres in Lurgi gasifiers to produce synthesis gas which is then made to undergo Fischer-Tropsch reactions in the presence of a suitable catalyst. The process has a low efficiency and is only used where supplies of crude oil are limited.

A new application of the Fischer-Tropsch process, using specifically developed selective catalysts, is to produce very high-quality liquid hydrocarbons from natural gas. This is the Shell Middle Distillate Synthesis (SMDS) process, which is described in 3.5.7.

The SMDS is a two-stage process which converts a mixture of carbon monoxide and hydrogen synthesis gas into products which are completely paraffinic and free from sulfur and nitrogen. The yield is mainly middle distillates, kerosene and gas oil, with some light tops/naphtha fraction which is suitable for use as a gasoline blend component after its octane number has been increased by catalytic conversion. Typical product breakdown on gas oil mode is 60% gas oil, 25% kerosene and 15% tops/naphtha.

The characteristics of the two main products are given in Table 19.1.[16]

Liquid Solvent Extraction (LSE) of coal to produce highly aromatic streams is another process which could become important as a source of fuel hydrocarbons.

A process for producing gasoline from methanol[17] has been investigated in New Zealand and provides an alternative to the Fischer-Tropsch process.

**Table 19.1. Characteristics of Kerosene and Gas Oil
from Shell Middle Distillate Synthesis Process**

SMDS		Typical Property Ranges
Kerosene component		
Boiling range	°C	150 to 250
Density	kg/m³	740 to 760
Freezing point	°C	-47 to -30
Smoke point	mm	g. th. 50
Gas oil component		
Boiling range	°C	250 to 360
Density	kg/m³	770 to 790
Pour point	°C	+10 to -10
Blending cetane number	-	70 to 80

19.8 Hydrogen as a Fuel for Spark Ignition Engines

Hydrogen can be be produced by electrolysis of water or from natural gas or coal. It has a very high flame speed and a wide ignitability range so that it can be used at extremely lean air-fuel ratios. There are no emissions of hydrocarbons or CO from the exhaust and, because of the ability to run at lean ratios, nitrogen oxides are also low.

Apart from the high production cost and the emissions resulting from the manufacturing process, there are a number of severe technical problems to overcome. First of all the weight of the storage tank on the vehicle would be very high if the fuel is to be used either in the liquid or compressed gas form. Safety is also a problem since hydrogen is highly explosive when mixed with air. A dedicated distribution system would be necessary and extremely expensive.

A possible way of overcoming some of these drawbacks that has been investigated is the use of hydrides of which iron-titanium is one of the contenders. Hydrogen is adsorbed by the hydride and can be released by the application of heat which can be obtained from the vehicle's exhaust. Although this system would overcome many of the safety problems, the range of the vehicle would be severely restricted, filling would be slow and the cost would be high.

19.9 Vegetable Oils as Fuel for Diesel Engines

Vegetable oils represent an alternative source of diesel fuel and they were used as emergency diesel fuels in some African and Asian countries during the 1939-45 war period.[20] More recent considerations have included their use in developing countries to enable farmers to become self-sufficient in diesel fuel, thereby reducing dependency (and expenditure) on imported crude oil.

Interest in vegetable oils is not only in developing countries. The European Parliament has supported massive tax concessions on biofuels such as ethanol, methanol, and diesters from agricultural products to offset initial production costs and encourage their use by keeping the price down. However, a proposed Directive, scheduled to take effect January 1, 1995, was amended to allow member states to decide whether to apply lower levels of excise duty on biofuels than on unleaded gasoline.

Evaluations of the potential of biofuels from vegetable crops have covered a number of alternative applications in engine and vehicle tests, as indicated below:

- 100% vegetable oil
- blends with conventional diesel fuel
- blends of different vegetable oils
- methyl esters of vegetable oils
- water:vegetable oil emulsions

In Europe, interest is currently focused mainly on rape seed oil as a biofuel source. Although there are no clear indications of economic or environmental benefits,[21] a number of field trials have been carried out or are underway. Fuels containing various levels of rape seed methyl ester (RME or RSME) are being tested in France, with several bus fleets operating on fuel containing 30% RME. The use of 100% RME has been encouraged in Austria because it is biodegradable.

Vehicle tests comparing RSME with diesel fuel, following European and U.S. legislated driving procedures, indicated HC and CO to be lower and NO_x higher with RSME. Smoke and particulates emissions were mainly lower with RSME than with diesel fuel but some tests showed a big increase and smoke levels also tended to be higher during transient cycles.[22]

A five-month trial on RSME was carried out on three unmodified buses by a U.K. city bus operator. Comparative emissions tests were made on RSME and standard diesel fuel at the start and end of the trial by the Road Research Laboratory of the British Department of Transport. The results showed a significant reduction in exhaust smoke with RSME although, surprisingly, particulates emissions were higher. Average differences in the levels of CO, CO_2, HC and NO_x were relatively small and of the same order as test repeatability.[23]

The cost of RSME is approximately twice that of diesel fuel. In some European countries the fuel duty imposed on RSME has been lowered or removed entirely to make it commercially attractive, but this is not yet the situation in the U.K.

In the U.S., soybeans and peanuts are receiving attention as possible sources of alternative fuels for diesel vehicles. Tests have been run on the peanut and soybean oil and on soybean methyl ester (SOME).

The autoignition properties of vegetable oils are similar to those of diesel fuel but their viscosities are generally higher by as much as one order of magnitude. Tests in a single-cylinder DI engine indicated that soybean oil and peanut oil performed satisfactorily using the standard injection equipment.[24] There was a slight loss of power with the peanut oil compared with soybean oil and ASTM No. 2 diesel. Smoke levels were in line with the 10% lower energy content of the vegetable oils. NO_x levels were similar with the three fuels but reductions were obtained with smaller and larger injection nozzles. Different nozzles also indicated different NO_x-forming tendencies for the fuels, the effect being more pronounced with the larger nozzle. Similar effects were observed with HC formation.

During this test series, No. 2 diesel fuel was used when starting and shutting down the engine, precluding systematic study of the deposit-forming tendencies of the two vegetable oils. However, signs of a gummy coating were seen in the combustion chamber area and there was carbon buildup on the tip of the injector, suggesting the need for more frequent servicing or the use of antioxidants and possibly other types of additive to inhibit deposit formation.

Evaluation of the methyl ester of soybean oil (SOME) in a four-stroke DI engine gave similar combustion performance to that of 51 cetane number diesel fuel,[25] although the methyl ester was more sensitive to changes in injection timing and nozzle hole size. The small differences in thermal efficiency were attributed to the longer ignition delay of the 46 CN SOME. Emissions of CO, HC and smoke were lower for the methyl ester but NO_x levels were similar.

Concerns have been expressed about the possible formation of gummy deposits during extended running on vegetable oils and their high cloud and pour points, which could cause difficulties during cold weather. The smell of RSME exhaust emissions has also been criticized.

An alternative approach to reducing dependency on imported crude oil or finished products is to use vegetable oil as a diesel fuel supplement. In a study carried out as part of a university research program, an unmodified bus engine was used to compare the performance of various blends of diesel fuel and waste soybean oil with 100% diesel fuel.[26]

The soybean oil had been purchased for food frying and contained an antioxidant, tertiary butyl hydroquinone (TBHQ), and an antifoaming agent, dimethylpolysiloxane. A blend of 20% soybean oil and 80% No. 1 diesel oil gave a fuel with a density and viscosity slightly higher than those of No. 1 diesel fuel and a heat content 2.7% lower. The bus, which was powered by a 6V71 Detroit Diesel 424 cu. in. two-stroke engine, averaged 80 km per day running a continuous daily shuttle service between two university campus sites. Smoke readings, using a Bosch smoke meter, were taken during operation on the regular route, with the bus usually loaded to full capacity.

Satisfactory performance was obtained with the blend containing 20% soybean oil. There was no degradation of fuel consumption or excessive deposit formation in the combustion zone. However, smoke readings were slightly higher than for straight No. 1 diesel fuel and some problems of fuel flow restriction were experienced as a result of the higher cloud point of the blend.

Tests on conventional diesel fuel, a 50:50 soybean:rape seed oil blend, and emulsions of each with 10% water have been carried out using a single- and a multi-cylinder engine. The vegetable oil had a cetane number of only 37, while that for the diesel fuel was 52, but similar ignition delays were observed for all the test fuels. In the four-cylinder high-speed engine, smoke and NO_x emissions were lower with the vegetable oil than with diesel fuel, and lower still with the emulsified vegetable oil. The diesel fuel emulsion gave the highest smoke numbers but lowest NO_x levels.[27]

References

1. P.R. Odell and K.E. Rosing, "The future of oil, a reevaluation." *OPEC Review*, Summer 1984.

2. P. Heinze, "The Influence of Diesel Fuel Properties and Components on Emissions from Engines," Institution of Mechanical Engineers Seminar, 1993 - 2, Fuels for Automotive and Industrial Diesel Engines, 2nd Seminar, Birmingham, U.K., April 1993.

3. S.P. Ho and T.A. Renner, "Global Warming Impact of Gasoline vs. Alternative Transportation Fuels," SAE Paper No. 901489, 1990.

4. M.F. Russell, "Recent CAV Research into Noise, Emissions and Fuel Economy of Diesel Engines," SAE Paper No. 770257, 1977.

5. M.L. Monaghan, "The High Speed Direct Injection Diesel for Passenger Cars," SAE Paper No. 810477, 1981.

6. D. Broome, "The Present Status and Future Development of the European Passenger Car Engine," SAE Paper No. 865001, 1986.

7. N.J. Beck and O.A. Uyehara, "Factors that Affect BSFC and Emissions for Diesel Engines: Part II, Experimental Confirmation of Concepts Presented in Part I," SAE Paper No. 870344, 1987.

8. C.C.J. French and D.A. Pike, "Diesel Engined, Light Duty Vehicles for an Emission Controlled Environment," SAE Paper No. 790761, 1979.

9. A. Koenig, H. Menrad and W. Bernhardt, "Alcohol Fuels in Automobiles," Alcohol Fuels Conference, Inst. Chem. Eng., Sydney, 9-11 August 1978.

10. J.C. Ingamells and R.H. Lindquist, "Methanol as a Motor Fuel or a Gasoline Blending Component," SAE Paper No. 750123, 1975.

11. R.H. Thring, "Alternative Fuels for Spark Ignition Engines," SAE Paper No. 831685, 1983.

12. N.R. Sefer, "Regional Refining Models for Transportation Fuels from Shale Oil and Coal Syncrudes," SAE Paper No. 810442, 1981.

13. M.R. Swain, *et al.*, "The Effect of Alternative Gasolines on Knock and Intake Valve Sticking," SAE Paper No. 872040, 1987.

14. D.E. Steere and T.J. Nunn, "Diesel Fuel Quality Trends in Canada," SAE Paper No. 790922, 1979.

15. "Solvent Refined Coal — A Possible Automotive Fuel," *Automotive Engineering*, pp. 53-59, July 1980.

16. M. van der Burgt, J. van Klinken and Tjong Sie, "The Shell Middle Distillate Synthesis Process," paper presented at 5th Synfuels Worldwide Symposium, Washington, D.C., 1989.

17. J.E. Penink, *et al.*, "Mobil Process for Conversion of Coal and Natural Gas to Gasoline," Alcohol Fuels Conference, Inst. Chem. Eng., Sydney, August 9-11, 1978.

18. A.F. Williams and W.L. Lom, Liquefied Petroleum Gases, p. 292, 2nd Edition, John Wiley and Sons, 1981.

19. 1988 Annual Book of ASTM Standards, Vol. 05.02. ASTM, 1916 Race Street, Philadelphia, 1988.

20. G. Onion and L.B. Bodo, "Oxygenated Fuels for Diesel Engines: A Survey of Worldwide Activities," Biomass, Applied Science Publishers Ltd., England, 1983.

21. D.J. Rickeard and N.D. Thompson, "A Review of the Potential for Bio-Fuels as Transportation Fuels," SAE Paper No. 932778, 1993.

22. S. Alfuso, M. Auriemma, G. Police and M.V. Prati, "The Effect of Methyl-Ester of Rapeseed Oil on Combustion and Emissions of DI Diesel Engines," SAE Paper No. 932801, 1993.

23. P. Shepherd, "Bio-Diesel Fuel Trial," Report by Reading Buses, England, 1993.

24. R. Forgiel and K.S. Varde, "Experimental Investigation of Vegetable Oil Utilization in a Direct Injection Diesel Engine," SAE Paper No. 811214, 1981.

25. K.W. Scholl and S.C. Sorenson, "Combustion of Soybean Oil Methyl Ester in a Direct Injection Diesel Engine," SAE Paper No. 930934, 1993.

26. M.K.C. Fishinger, H.W. Engelman and D.A. Guenther, "Service Trial of Waste Vegetable Oil as a Diesel Fuel Supplement," SAE Paper No. 811215, 1981.

27. R.J. Crookes, M.A.A. Nazha and F. Kiannejad, "Single and Multi Cylinder Diesel Engine Tests with Vegetable Oil Emulsions," SAE Paper No. 922230, 1992.

Further Reading

Don Knowles, <u>Alternate Automotive Fuels</u>, Reston Publishing Company, Inc., Virginia., 1984.

<u>Alternative Fuels</u>, SAE Publication SP-480, February 1981.

W.M. Scott, "Alternative Fuels for Automotive Diesel Engines," pp. 263-290, <u>Future Automotive Fuels</u>, Plenum Press, New York, 1977.

E.G. Barry, *et al.*, "Effects of Fuel Properties and Engine Design Features on the Performance of a Light Duty Diesel Truck—A Cooperative Study," SAE Paper No. 861526, 1986.

G. Greeves, *et al.*, "Origins of Hydrocarbon Emissions from Diesel Engines," SAE Paper No. 770259, 1977.

H.C. Grigg, "The Parameters Available for Controlling Diesel Engine Performance and Their Relationship with Performance Parameters," Automotive Microelectronics, Editors: L. Bianco and A. La Bella, Elsevier Science Publishers B.V., North Holland, 1986.

H.C. Grigg, "Fuel Injection Systems and the Techniques Available for Their Operation and Interface with Electronic Control," Automotive Microelectronics, Elsevier Science Publishers B.V., North Holland, 1986.

B. Agnetun, B.-I. Bertilsson and A. Röj, "A Life-Cycle Evaluation of Fuels for Passenger Cars," Paper No. CEC/93/EF09, Fourth International Symposium on the Performance Evaluation of Automotive Fuels and Lubricants, Birmingham, U.K., 1993.

M. Booth, J.M. Marriott and K.J. Rivers, "Diesel Fuel Quality in an Environmentally Conscious World," The Institution of Mechanical Engineers, London, U.K., April 1993.

S. Alfuso, M. Auremma, G. Police and M.V. Prati, "Regulated Emissions of DI Diesel Engines Fueled with Methyl Ester of Rape Seed Oil," Second Seminar on Fuels for Automotive and Industrial Diesel Engines, Institution of Mechanical Engineers, London, U.K., April 1993.

L.A. Perkins, C.L. Peterson and D.L. Auld, "Durability Testing of Transesterified Winter Rape Oil (Brassica Napus L.) as Fuel in Small Bore, Multi-Cylinder DI, CI Engines," SAE Paper No. 911764, 1991.

Chapter 20

Gaseous Fuels for Engines: Natural Gas and Liquefied Petroleum Gas

Christopher S. Weaver, P.E.
Engine, Fuel, and Emissions Engineering, Inc.
Sacramento, CA

Natural gas and liquefied petroleum gas (LPG) have been used on a limited scale as vehicle fuels for more than 50 years. Clean-burning, inexpensive, and abundant in many parts of the world, these fuels have played a significant vehicular role in Russia, Argentina, Italy, the Netherlands, Canada, New Zealand, and the U.S., among other countries. Until recently, the major motivation for using these fuels was economic: the low cost of natural gas and LPG compared to gasoline or diesel made their use attractive in certain applications such as taxicabs, where the fuel savings were sufficient to offset the higher cost of on-board storage and compression/dispensing systems. In recent years, attention has focused increasingly on the environmental as well as the economic benefits of gaseous fuels. Recent advances in the technology for gaseous-fuel vehicles and engines, new technologies and international standardization for CNG storage cylinders, and the production of new, factory-manufactured gaseous-fuel vehicles in a number of countries have all combined to boost the visibility and market potential of these fuels. The stage may now be set for a major expansion in their use.

20.1 Gaseous Fuel Supply and Costs

20.1.1 Natural Gas

After coal, natural gas is the most abundant fossil fuel. Worldwide, the ratio of proven gas reserves to annual production is double that of petroleum. The estimated total U.S. resource base of natural gas, recoverable with existing technology, was estimated in 1988 at 1188 trillion cubic feet (TCF).[1] This is enough to sustain present U.S. rates of consumption for approximately 70 years. Large conventional gas reserves also exist in Canada and Mexico, Russia, Kazakhstan, Iran, and Indonesia. Large quantities of additional gas are available in unconventional resources such as tight sands, coal seams, and geopressurized brines. Many of these would become economically recoverable with a modest increase in wellhead gas prices. Adding these unconventional resources to conventional reserves results in a resource base capable of supplying world consumption for many years to come.

Today, most major urban centers and many minor ones in industrial countries are served by a large network of high-pressure natural gas pipelines. These are connected through "city gate" valves to urban gas distribution networks operating at moderate to low pressures, which transport gas directly to the point of use. Other technologies for natural gas transportation and distribution include liquefaction and shipment in liquid form (LNG), and short-distance transport of compressed natural gas (CNG) in large banks of cylinders. Japan, Taiwan, Korea, and many countries of Western Europe now import significant quantities of natural gas in the form of LNG.

Owing to the difficulty of transportation, the costs of natural gas vary greatly from country to country, and even within countries. Where gas is available by pipeline from the field, its price is normally set by competition with residual fuel oil or coal as a burner fuel. The market-clearing price of gas to industrial customers under these conditions (which pertain in North America, much of Europe, and those parts of Latin America and Asia served by gas pipelines) has typically been about $2.00 to $4.00 per million Btu (higher heating value) in the 1990s. This is equal to about $0.25 to $0.50 per gasoline-equivalent (US) gallon. For typical gas composition, one million Btu is equivalent to about 960 standard cubic feet (scf), or 35 standard cubic meters.

To be used in vehicles, natural gas must be compressed or liquefied for on-board storage. Capital and operating costs of the compression system can add another $0.50 to $2.00 per million Btu ($0.06 to $0.25 per gasoline-equivalent gallon) to the cost of CNG for vehicular use, depending on the size of the facility and the natural gas supply pressure. Other costs to be taken into account include road taxes, if applicable, and the dealer's margin.

The cost of LNG varies considerably, depending on specific contract terms (there is no effective "spot" market for LNG). The cost of small-scale liquefaction of natural gas is about $2.00 per million Btu, making it uneconomic in comparison to CNG in most cases. Where low-cost remote gas is available, however, LNG production can be quite economic. Typical 1993 costs for LNG delivered to Japan were about $3.50 to $4.00 per million Btu.

20.1.2 Liquefied Petroleum Gas (LPG)

The term "liquefied petroleum gas," or LPG, refers to mixtures of three- and four-carbon hydrocarbons such as propane (C_3H_8), propene (C_3H_6), n-butane (C_4H_{10}), isobutane (methyl-propane), and various butenes (C_4H_8). Small amounts of ethane (C_2H_6) may also be included. The major sources of commercial LPG are natural gas processing and petroleum refining. As found in the earth, natural gas often contains excess propane and butanes which must be removed to prevent their condensing in high-pressure pipelines, and to control variation in gas properties. LPG from refineries includes light hydrocarbons originally dissolved in the crude oil and separated during the distillation process, as well as those produced in the process of "cracking" heavy hydrocarbons to lighter products. Refinery LPG often contains significant quantities of olefinic compounds (propenes and butenes) produced in the cracking process.

Whether produced by natural gas processing or petroleum refining, LPG is essentially a by-product. Uses for LPG, in addition to automotive fuel, include petrochemical production, home cooking and heating fuel, and fuel for industry. Presently, LPG supply exceeds the demand in most gas-producing and petroleum-refining countries, so the price is low compared to other hydrocarbons. Wholesale prices for butane and propane in the U.S. have typically been around 30% less than the cost of diesel on an energy basis. Depending on the locale, however, the additional costs of storing and transporting LPG

553

may more than offset this advantage. Because the supply of LPG is limited, and small in relation to other hydrocarbon fuels, large-scale conversion of vehicles to LPG use would likely absorb the existing glut, causing prices to rise.

20.2 Gaseous Fuel Composition and Properties

20.2.1 Gaseous Fuel Components

The properties of the main hydrocarbon constituents of natural gas and LPG are summarized in Table 20.1.

Table 20.1. Properties of the Main Hydrocarbon Fuel Gases

	Methane	Ethane	Propane	Propene	n-Butane	iso-Butane	Butenes
Energy Content (LHV) (MJ/kg)	50.01	47.48	46.35	45.78	45.74	45.59	45.32
Liquid Density (kg/L)	0.466	0.572	0.501	0.519	0.601	0.549	0.607
Liquid Energy Density (MJ/L)	23.30	27.16	23.22	23.76	27.49	25.03	27.51
Gas Energy Density (MJ/m^3)	32.6	58.4	84.4	79.4	111.4	110.4	113.0
Gas Specific Gravity (@ 25°C)	0.55	1.05	1.55	1.47	2.07	2.06	1.93
Boiling Point, °C	-164	-89	-42	-47	-0.5	-12	-6.3 to 3.7
Research Octane No.	>127	-	109	-	-	-	-
Motor Octane No.	122	101	96	84	89	97	77
Wobbe Index (MJ/m^3)	50.66	65.11	74.54	71.97	85.46	84.71	81.27

Natural gas—Natural gas, as it is found in the earth, contains varying amounts of non-methane hydrocarbons, H_2S, CO_2, water vapor, nitrogen, helium, argon, and other trace gases. In most cases, it is necessary to upgrade the gas to pipeline specifications in a gas processing plant before injecting it into the transportation and distribution network. Water and H_2S must be removed to prevent corrosion damage to the pipeline network, and excess amounts of higher hydrocarbons must be removed to prevent them from condensing under the high pressures in the gas transmission network. There is also an economic benefit to recovering these hydrocarbons, since the "natural gas liquids" are more valuable as gasoline feedstock, petrochemicals, or LPG than as components of natural gas. Helium, where found at significant concentrations, is also a valuable by-product. Excess amounts of inert gases such as CO_2, Ar, and N_2 are also removed in processing.

Pipeline-quality natural gas is a mixture of several different gases. The primary constituent is methane (CH_4), which typically makes up 80 to 99% of the total. The remainder is primarily ethane and inert gases such as N_2 and CO_2, with smaller amounts of propane, butanes, and higher hydrocarbons. The mix of minor constituents varies considerably from place to place and from time to time, depending on the source and processing of the gas. In order to ensure consistent combustion behavior, major natural gas pipelines generally impose specifications on the composition of the gas they will accept for transport. These specifications typically limit the percentage of propane, butane, and higher hydrocarbons, the volumetric heating value, and the Wobbe Index (see Section 20.2.2).

Although pipeline gas generally exhibits a limited range of composition and properties, natural gas found in distribution systems may exhibit greater variability. In some cases, distribution systems in gas-producing areas receive gas directly from the well, with minimal processing. The resulting gas may be rich in non-methane hydrocarbons, inert gases, or both. Another factor affecting gas composition in distribution systems is the occasional supplementation of natural gas supplies with propane-air mixtures to meet peak winter demand. Many gas distribution utilities in the Eastern U.S. have such propane-air peak shaving facilities, which are typically used only for a few days per year. The high propane levels resulting from such use may pose problems for CNG fuel systems, since the propane may liquefy at pressures typical of CNG storage.

LPG—The composition of commercial LPG varies greatly from one country to another. In the U.S., automotive LPG is generally more than 80% propane, with small amounts of ethane and butanes, and up to 10% propene. In Europe, countries having relatively cold climates tend to use a high percentage of propane and propene in order to provide adequate vapor pressure in winter, while warmer countries such as Italy use mostly butane and butenes. LPG composition may also vary between summer and winter, with a higher percentage of propane and propenes in the winter months. Table 20.2 summarizes the proportion of C_3 hydrocarbons in commercial LPG for a number of countries.

Table 20.2. Range of LPG Composition for Different Countries

Country	Propane and Propene (%)
Belgium	40-60
Chile:	
commercial propane	70-100
commercial butane	0-30
Finland	100
France	20-50
Germany	100
Netherlands	30-70
United Kingdom	50-100
United States	98-100

20.2.2 Wobbe Index and Fuel Metering

The *Wobbe index*, also referred to as the *Wobbe number*, is an important parameter for gaseous fuels. The Wobbe index of a gaseous fuel is determined by its composition. The value of the Wobbe index, W, is calculated as:

$$W = H/\sqrt{\rho}$$

where H is the volumetric heating value of the gas, and ρ is the specific gravity. Since specific gravity is dimensionless, the Wobbe index has the same units as H: MJ per standard cubic meter in SI and Btu per standard cubic foot in the English units commonly used in the international gas industry. Wobbe indices can be calculated from gas composition data or heating value and density measurements.

The Wobbe index of a gas is proportional to the heating value of the quantity of gas that will flow subsonically through an orifice in response to a given pressure drop. Since virtually all gaseous fuel metering systems are based on orifices, a change in the Wobbe index of the fuel (other things being equal) will result in a nearly proportional change in the rate of energy flow, and thus in the air-fuel ratio. Departures from strict proportionality may occur in fuel systems using choked (sonic) flow, or because of changes in the H:C ratio of the fuel. Even in these cases, however, the Wobbe index provides a good indicator of the change in air-fuel ratio. Figure 20.1 shows the effect of Wobbe index variations on equivalence ratio in a natural gas engine[2] using several different types of fuel metering.

The effect of variations in the Wobbe index for gaseous-fuel vehicles is similar to the effect of varying the fuel's volumetric energy content in gasoline vehicles. A lower Wobbe index results in a leaner air-fuel ratio, while a higher Wobbe index gives a richer mixture. Depending on the fuel metering technology, variations in the Wobbe index may affect engine performance and emissions. Modern, stoichiometric spark-ignition engines with closed-loop control of the air-fuel ratio are able to compensate for reasonable variations in the Wobbe index, just as they compensate for variations in gasoline energy content due to refining differences or use of alcohol blends. For engine control systems without air-fuel ratio feedback, such as those used in heavy-duty lean-burn engines, variations in fuel composition can present a significant problem—possibly resulting either in poor engine performance (due to too lean a mixture) or engine damage due to overheating (with the mixture too rich).

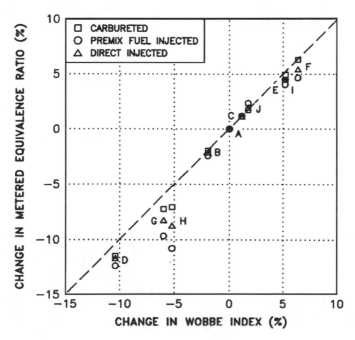

Fig. 20.1. Effect of variation in Wobbe number on equivalence ratio in engines using different fuel metering technologies.[2]

Natural gas—Because changes in air-fuel ratio affect combustion and efficiency in many gas-burning appliances and well as engines, natural gas pipelines and distribution utilities have long striven to maintain close control of the Wobbe index of the gas they deliver. Pure methane has a Wobbe index of 1361 Btu/scf (50.66 MJ/m^3). Increasing concentrations of higher hydrocarbons such as ethane and propane increase the Wobbe index, while increasing concentrations of inert gases lower it. In practice, these two effects are used to cancel each other out, so as to maintain the Wobbe index of natural gas in the pipeline close to the nominal specification.

Figure 20.2 shows the distribution of Wobbe indices for 7000 natural gas samples collected from distribution systems in major cities of the U.S. For gas distributed in most of the U.S., the Wobbe index is maintained close to that of pure methane, i.e., typically in a range of 1320 to 1360 Btu/scf. For historical reasons, gas distributed in the vicinity of Denver and Colorado Springs, Colorado, maintains a dif-

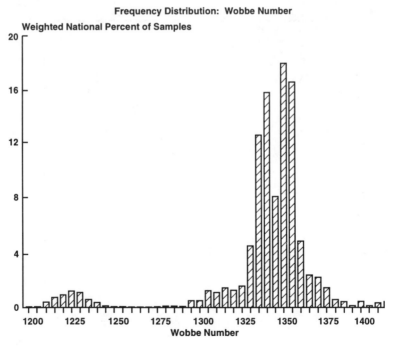

Fig. 20.2. Distribution of Wobbe indices for natural gas in major U.S. cities.[4]

ferent nominal specification from that in the rest of the U.S., typically between 1210 and 1250 Btu/scf. This lower ("leaner") Wobbe index was originally associated with a higher level of inert species in gas produced in that area. Such variation in nominal gas specifications from one geographic area to another is also found in some other countries such as New Zealand.

A heavy-duty, lean-burn engine calibrated for gas with a Wobbe index in the normal range of 1320 to 1360 Btu/scf will experience driveability problems and power loss if fueled with gas at the 1220 Btu/scf typical of Denver. This can be corrected by recalibrating the fuel metering system. A more serious problem occurs when the gas is too rich, resulting in excess power output, overheating, and possibly knock. The author is aware of several cases in which heavy-duty natural gas engines have been destroyed due to receiving too-rich fuel.

LPG—The three-carbon and four-carbon species in LPG differ in volumetric energy content, so that a change in LPG composition can affect the air-fuel ratio in engines and other combustion devices operating on LPG. Although seldom used in reference to LPG, the Wobbe index is equally applicable to assessing the effect of varying fuel composition on air-fuel ratio. The Wobbe indices and other properties of the main constituent gases of LPG were shown in Table 20.1. As this table shows, the indices for propane and propene are nearly identical, so that, from an air-fuel ratio standpoint, these gases are virtually interchangeable. The same is true for normal- and iso-butane. The index for butenes reflects a mixture of several butene isomers, and lies between those for propane and butanes.

The effect of variations in the Wobbe index for LPG vehicles is similar to the effect of variations in gasoline liquid density for gasoline vehicles. The levels of variation in gasoline density that are considered acceptable can therefore serve as a guide to acceptable levels of variation in the Wobbe index. The specific gravity of gasoline ranges from about 0.72 to 0.75—a range of variation of about 4%. Using this criterion, therefore, an acceptable range of variation in the Wobbe index of LPG would also be about 4%, or ±2% around some nominal value. Figure 20.3 plots the Wobbe index versus fuel composition for a binary mixture of propane and equal parts normal- and iso-butanes. Also shown in the figure are boxes corresponding to the permissible ranges of variation in Wobbe index for three nominal LPG compositions: 100% propane, 100% butane, and a 50% propane/butane mixture. For a nominal specification based on 100% propane, the acceptable range in composition would be as low as 75% propane and 25% butane, using the criterion of 4% variation in Wobbe index. For a nominal 50/50 mixture, the acceptable variation in butane content would be between 35% and 65%, if the rest were propane. For a nominal 100% butane specification, the acceptable range would be as low as 65% butane and 35% propane. Thus, this criterion provides for reasonable control of air-fuel ratio variations, while still allowing significant flexibility in LPG composition.

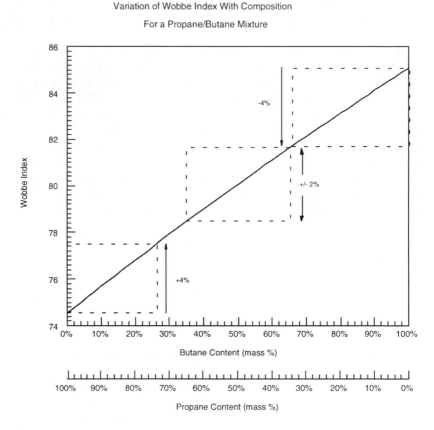

Fig. 20.3. Effect of variation in LPG composition on Wobbe index.

560

20.2.3 Propensity to Knock: Octane and Methane Numbers

As is the case with liquid gasoline, the degree of resistance to engine knock is an important property of gaseous fuels. This tendency is measured in several different ways. Often, the knock resistance of gaseous fuels is reported in terms of the familiar research and motor octane numbers (RON and MON) used with gasoline. However, the present RON and MON methods are intended for liquid fuels and are not well adapted for measuring the knock resistance of natural gas. ASTM has defined a standard (D 2623) for measuring the MON of LPG mixtures, but no RON method has yet been adopted for gaseous fuels.

Natural gas—Natural gas generally has excellent antiknock properties; the knock resistance of most natural gas blends exceeds the maximum range of the ASTM octane scale (120.34). This is one reason that no standard octane testing methods exist for natural gas. In order to better measure the knock resistance of natural gas blends, a separate *methane number* scale has been created.[3] In this scale, the reference fuels are mixtures of methane and hydrogen. Pure methane has a methane number of 100, and pure hydrogen has a methane number of 0. To define the relationship between MON and methane number, workers at Southwest Research[4] extended the ASTM MON method for LPG to a number of typical natural gas blends, as well as samples of pure methane, ethane, and propane, and methane-propane blends. It was found that MON and methane number are closely correlated. The best-fit relationships were found to be

$$MON = 0.679 \times MN + 72.32$$
$$MN = 1.445 \times MON - 103.42$$

with R^2 in each case greater than 0.95.

Because of the excellent knock resistance of natural gas, engines designed specifically for natural gas fuel can use higher compression ratios than gasoline engines, with a consequent improvement in efficiency and power output. Typical compression ratios for natural gas engines range from 10:1 (for large engines) to 13:1. The knock-resistance of natural gas also permits supercharging with much higher boost pressures than gasoline engines, enabling these engines to attain BMEP levels comparable to those of modern heavy-duty diesel engines. The antiknock performance of natural gas is best for pure methane or methane/inert gas mixtures, and declines somewhat with increasing concentrations of non-methane hydrocarbons. This effect is not usually significant for the typical range of pipeline gas composition, but may become important in high-compression engines burning unprocessed gas or propane-air mixtures.

LPG—Of the hydrocarbons commonly included in LPG, propane has good antiknock properties compared to gasoline. The antiknock performance of the other LPG constituents is markedly inferior to that of propane, raising the possibility that an engine optimized for use on high-octane propane might suffer damage from knock if operated on LPG containing significant quantities of propylene, butanes, or butenes. This is the main reason that the U.S. HD-5 standard (ASTM D 1835) for automotive LPG specifies nearly pure propane. This standard was developed in order to accommodate specialized, high-compression propane engines used in heavy-duty applications such as tractors.

With the exception of special-purpose heavy-duty engines, nearly all LPG engines are converted from engines designed for gasoline, and they retain the relatively low compression ratios that gasoline use imposes. Thus, for most engines, the extra octane quality of pure propane far exceeds the octane level required, and provides no benefit. Since pure propane may not be the most economic fuel to supply, a more practical approach could be to specify a minimum octane requirement for LPG, based on considerations both of supply economics and of technical efficiency. At a minimum, this octane requirement should be set high enough to ensure trouble-free operation with gasoline engines. This is the approach that has been taken in Europe with the development of norms for automotive LPG by the European Committee for Standardization (CEN).

European experience indicates that the critical value in establishing antiknock requirements for LPG is the motor octane number (MON). If the MON is satisfactory, this also guarantees that the research octane number (RON) will be high enough. From various studies, the CEN working group concluded that a minimum MON of 89 is required to ensure satisfactory operation on LPG of engines designed for European premium leaded gasoline having MON of 87. It was also concluded that 87 MON LPG would suffice for engines designed for regular or premium unleaded fuel.[5] The actual European standard specifies 89 MON, thus accommodating even engines designed for premium leaded gasoline.

Table 20.3 shows the blending MON values ascribed to the common constituents of LPG by the CEN working group. As this table shows, the MON values of propane and isobutane significantly exceed the required MON level, while n-butane just meets it, and the olefins (propene and especially butenes) fall short. Under the European standard, a binary propane-propene mixture could have as much as 53% propene, compared to a maximum of 5% under the U.S. HD-5 standard.

Table 20.3. Blending MON of LPG Components

Gas Component	Blending MON (mass basis)
Propane	95.9
Propene	82.9
n-Butane	88.9
i-Butane	97.1
Butenes	76.8

Source: CEN document 89/57419

20.2.4 Effect of Gaseous Fuel Composition on Emissions

As further discussed in Section 20.5, emissions from natural gas and LPG vehicles are generally low compared to those from gasoline vehicles at the same level of engine technology. Nonetheless, variations in gaseous fuel composition can affect the level of pollutant emissions. The primary effect is due to variations in the Wobbe index. As discussed above, this can directly affect the air-fuel ratio, and thus pollutant emissions. Reasonable variations in Wobbe index have little effect on emissions from light-duty vehicles using modern engine technology—stoichiometric engines with three-way catalysts and closed-loop feedback control by means of an O_2 sensor. This is because the feedback control makes it possible to compensate for variations in air-fuel ratio.

In addition to their effects on the Wobbe index, differences in the concentration of different hydrocarbons in the fuel can affect the species composition and reactivity of the HC emissions in the exhaust. In the case of natural gas vehicles, this effect is of considerable regulatory importance, since U.S. emissions standards limit emissions only of non-methane hydrocarbons (NMHC). The proportion of non-methane hydrocarbons in the fuel gas directly affects the level of NMHC emissions in the exhaust. Figure 20.4 shows the results of a study sponsored by the Natural Gas Vehicle Coalition.[6] This study tested a Chrysler van and a Ford Crown Victoria using three different test fuels with varying NMHC contents. As the figure shows, NMHC emissions increase with fuel NMHC content.

In order to limit the possible increase in pollutant emissions due to variation in natural gas properties, the California Air Resources Board has established specification limits for natural gas sold commercially as vehicle fuel. Both the Air Resources Board and the U.S. EPA have also established limits for natural gas used in emissions certification testing. These limits are summarized in Table 20.4.

The effects of varying LPG composition on the exhaust hydrocarbon species and reactivity have not been documented. According to the Carter reactivity scale used by the California Air Resources Board, however, olefins such as propene and butenes are much more reactive

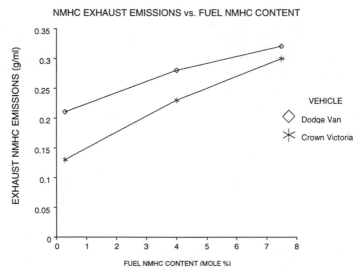

Fig. 20.4. NMHC exhaust emissions as a function of fuel NMHC content.[6]

**Table 20.4. Specifications of Natural Gas Used for Emissions
Certification and for General Vehicular Use**

Constituent (mole %)	CARB Cert Fuel	CARB In-Use Fuel	EPA Cert Fuel
Methane	90.0 ± 1.0	88.0 (min)	89.0 (min)
Ethane	4.0 ± 0.5	6.0 (max)	4.5 (max)
C_3 and higher	2.0 ± 0.3	3.0 (max)	2.3 (max)
C_6 and higher	0.2 (max)	0.2 (max)	0.2 (max)
Hydrogen	0.1 (max)	0.1 (max)	-
Carbon Monoxide	0.1 (max)	0.1 (max)	-
Oxygen	0.6 (max)	1.0 (max)	0.6 (max)
Inert Gases ($CO_2 + N_2$)	3.5 ± 0.5	1.5 - 4.5	4.0 (max)

in contributing to ozone formation than paraffins such as propane and
the butanes. Thus, it is reasonable to suspect that increasing the olefin
content of LPG will result in increased ozone-forming potential in the
exhaust. In order to reduce the possibility of emissions increases due
to variation in LPG composition, the California Air Resources Board
requires that LPG sold for automotive use in California comply with
the HD-5 standard (Table 20.5). Due to concerns about supply availability, however, the maximum 5% propene content required by the
HD-5 specification has been delayed until January 1, 1997. In the interim, LPG having up to 10% propene is permitted.

**Table 20.5. Composition Requirements for HD-5 Propane
for Use as Motor Vehicle Fuel**

Property	Spec.
Propane (vol%)	85 min.
Propene (vol%)	5 max
Butane and heavier (vol%)	2.5 max
Vapor pressure at 100°F	208 psig
Sulfur (ppm mass)	120 max

20.3 Gaseous Fuels Dispensing and Storage

Natural gas may be stored on-board a vehicle either as a compressed
gas in high-pressure cylinders or as a cryogenic liquid. From the

engine's standpoint, CNG and LNG are essentially interchangeable—
it is only the on-board storage medium that is different. LPG is stored
and dispensed at ambient temperatures, as a liquid under pressure.

20.3.1 Compressed Natural Gas (CNG)

Handling of CNG is similar to that of any high-pressure gas. Piping
and connections must be strong and gas-tight. Refueling is accom-
plished by connecting a manifold on the vehicle to a high-pressure gas
line using a positive-lock connection, and then admitting the gas to
the vehicle tanks.

Compressed natural gas refueling systems can be divided into slow-
fill and fast-fill designs. The main components of a typical "fast fill"
station are a compressor to boost the gas pressure from the pressure in
the distribution pipe to about 5000 psi; a bank of storage vessels (often called a "cascade") to store the high-pressure gas; and a dispenser,
which often resembles a common gasoline pump from the outside but
is very different within. The compressor is driven by an electric motor or natural gas engine, and requires considerable power. The cost
of the compressor and motor constitute a large part of the overall costs
of a CNG refueling system. These costs are typically much higher
than those of refueling systems for liquid fuels. The complete package also includes controls and safety devices, and filters to eliminate
oil and particulate matter from the compressed gas. Generally, the
compressor, motor, controls, and auxiliaries are packaged and sold as
a single pre-engineered unit. The package may also include the cascade storage and the dispenser, or these may be sold separately.

"Fast fill" CNG systems are designed to refuel one or two vehicles at
a time, but to refuel each vehicle very quickly—within about 5 minutes. This is achieved by using the compressor to pump high-pressure
gas into the cascade storage, which serves as a "buffer." The cascade
storage vessels are divided into several groups which can be connected independently to the refueling connector. To refill a vehicle,
the cascade storage units are connected in sequence to the vehicle's
fuel intake, and allowed to equalize pressure. The group having the
lowest pressure is connected first, then the next lowest, and so forth,
until the storage pressure on-board the vehicle reaches the desired
level. This arrangement makes it possible to use a smaller compressor
than would be required to achieve the same refueling time by pumping directly from the compressor to the vehicle.

Because of the rapid filling characteristic of fast-fill systems, the gas in the cylinders has no time to lose heat to the environment. At the end of the fill, the gas in the vehicle cylinders will be considerably warmer than the ambient temperature. This makes it difficult to obtain a complete refueling by "fast fill": although the tank pressure may be at the nominal "full" level after refueling, the pressure will drop as the gas in the cylinders cools to ambient temperature.

A "slow fill" CNG fueling system is designed to refuel vehicles such as buses that can be parked overnight. All of the vehicles are connected in parallel to the compressor, eliminating the need for cascade storage, but requiring a large number of high-pressure hoses and connectors. This arrangement makes it possible to achieve good compressor utilization, since the compressor can run continuously all night. For even more effective compressor utilization, it is also possible to design a "hybrid" system that can be used for fast fill during the day and slow fill at night.

The current standard working pressure for compressed natural gas cylinders is 3000 pounds per square inch gauge pressure (psig), or 20 MPa. There has been a recent push to move to 3600 psig (25 MPa) working pressures for CNG tanks to provide better range (using equivalent fuel storage system volume). The Institute of Gas Technology (IGT) is currently studying potential methods of safely increasing current design 3000 psig compressor output up to 3600 psig. Because most current systems operate at 3000 psig, a complete move to 3600 psig systems would be a disadvantage to those people who have already installed the lower pressure systems.

The size and weight of CNG cylinders are often cited as major drawbacks of natural gas use in vehicles. As Figure 20.5 shows, however, recent developments in high-strength composite materials have made it possible to reduce the weight of CNG cylinders substantially. About 122 scf of natural gas, weighing 5.4 lb (national average),[7] are required to equal the energy content (lower heating value) of one U.S. gallon of gasoline. With conventional steel cylinders, the weight of the cylinder required to contain 122 scf of gas is about 38 lb, for a weight of fuel and storage of 43 pounds per (U.S.) gallon of gasoline equivalent. For comparison, the weight of the fuel and tank for gasoline would be about 7.8 lb/gal; for diesel fuel it would be about 8.5 lb/gal.

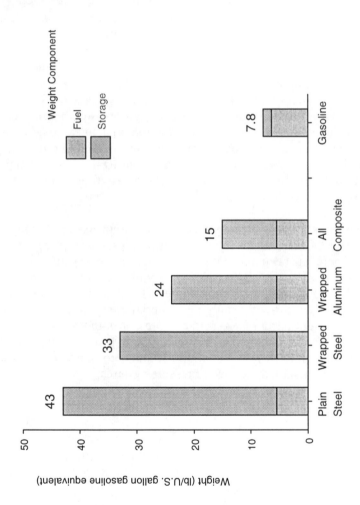

Fig. 20.5. Comparison of fuel and storage system weights for CNG and gasoline.

With present fiber-wrapped steel tanks, the tank weight is reduced by about 30%, so that the total weight of gas and tank would be about 33 lb/gal of gasoline equivalent. Fiber-wrapped aluminum cylinders weigh about 24 lb/gal of gasoline equivalent. Newly released, all-composite cylinders weigh even less—about 15 lb/gal of gasoline equivalent. This is still higher than gasoline (or diesel) fuel storage, but these new tanks are significantly lighter than the plain steel tanks of the past.

20.3.2 Recommended Practice for CNG: SAE J1616

The SAE Technical Committee on Fuels has adopted Recommended Practice J1616 which establishes acceptable compositional limits for natural gas intended for use in CNG vehicles. This recommended practice is intended to protect the interior surfaces of the fuel container, and other vehicle fuel system components such as fuel injector elements, from the onset of corrosion, the deposition of liquids or large dust particles, and the formation of water, ice particles, frost, or hydrates. Among the recommendations are the following:

- Pressure water dew point temperature (at CNG storage pressure) at least 5.6°C below the lowest monthly dry-bulb temperature expected in the area. This generally requires the use of a separate gas dryer attached to the compression station;

- No methanol to be added to the gas at the CNG fueling station;

- Fuel filtration to remove particulate material over 5 μm in size, and use of coalescing filters to remove oil mist carryover from oil-lubricated compressors;

- Wobbe index between 1300 and 1420 Btu/scf (48.5 to 52.9 MJ/m^3);

- Propane and higher hydrocarbon content limited to ensure no more than 1% condensation of the fuel at the lowest expected temperature (see Figure 20.6); and

- Odorization to ensure that gas leaks are clearly detectable at concentrations not over 1/5 of the normal flammability limit in air (this is approximately 1% gas in air by volume).

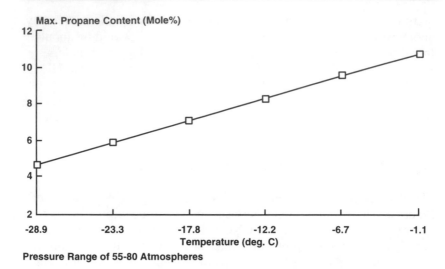

Max. Propane Content (Mole%)

Pressure Range of 55-80 Atmospheres

Fig. 20.6.   Maximum propane content of CNG to avoid excessive condensation (SAE J1616).

20.3.3 LNG Dispensing and Storage

LNG is stored in double-wall, vacuum-insulated cryogenic (−260°F or less) containers (commonly known as "dewars") and then vaporized, usually by engine heat, to produce engine-usable natural gas. A sophisticated fuel tank storage/delivery system is required to allow pumping of LNG to and from the fuel system, while maintaining cryogenic temperatures over long periods of time. Current design LNG tanks are of double-walled construction: an inner vessel made of stainless steel, surrounded by an insulating material and an outer casing made of high-strength carbon steel. The space between the inner and outer shells is vacuum evacuated to 1.9×10^{-6} psig. For applications requiring multiple LNG tanks, vacuum-jacketed piping (double-walled pipe with vacuum separation) is used to interconnect the tanks.

The quality of LNG can vary greatly, and the quality needed by the user may also vary. The liquefaction process generally removes all of the minor natural gas constituents except ethane and nitrogen. Typical merchant LNG is about 87-92% methane, with most of the remainder being liquid ethane. With additional processing, the ethane and nitrogen components can be removed as well, yielding a product that is 99+% pure methane. Since methane has substantially better antiknock properties than ethane, the use of pure liquid methane may be desirable for certain high-performance engines, such as those used in locomotives.

An important concern with the use of liquid methane/ethane mixtures is the possibility of changes in fuel composition during handling and processing. This is known as "aging," "weathering," or "enrichment." The liquid ethane has a higher boiling point than methane. Therefore, at every stage in processing where evaporation can take place—central storage, trucking, transfer, refueling station storage, and final fueling—methane boils off, leaving the ethane behind. At the end of processing, a higher concentration of ethane exists than did at the beginning. The Wobbe index of the fuel thus becomes progressively higher, while the knock-resistance becomes lower. This may be a significant problem for high-compression engines or those with very high BMEP levels, due to the increased potential for knock.

A number of LNG refueling system designs exist. They can be divided into two categories: those with and those without vapor return. It appears that the industry is leaning toward single hose, no-vapor-return refueling systems. Tank depressurizing is automatic, with integrated hoses and nozzles. Both fueling hoses and dispensing units must be designed according to the following needs: accurate flow measurement, accurate pressure regulation, sufficient hose support (hoses are heavier than gasoline or diesel hoses), leak detection, fire detection, frost and ice avoidance (especially in high-humidity areas), and protection of refueling personnel. The lack of a standard LNG nozzle/receptacle design is presently a significant barrier to the commercialization of LNG technology.

The newest design LNG tanks are filled from the top, providing a vapor-only interface at the nozzle connection. This practice is one of design convenience and safety. LNG entering the tank at the vapor interface cools the vapor present in the tank, condensing some of it back into liquid form. This eliminates the need for a vapor return line. If the fuel is pumped into the liquid interface at the bottom of the tank, it will not cool the overhead vapor, but rather will compress it and increase the internal tank pressure. Older LNG tank designs using bottom fill require a vapor return system to maintain proper internal tank pressure during refueling.

Nearly all LNG engines use the fuel in gaseous form. To accomplish this, a "vaporizer" borrows engine heat (usually through heat exchange with the engine coolant) to help expand the cold liquid into its gaseous form. In some applications, the LNG is vaporized to a gas at close to ambient conditions and then pumped to the required pressure with a compressor. In other applications, the fuel is pumped as a liquid to high pressure and then vaporized. The latter requires much less pumping power because liquids are much less compressible than gases. On some designs, a "CNG buffer," or expansion tank, gives additional space for sudden expansion of cold LNG when the fuel demand from the engine is suddenly reduced or interrupted. This buffer is also used to store natural gas for start-up of the engine. When the engine is restarted, there is enough gas in the buffer system to start the engine running and get the LNG vaporizer system working and producing gas for normal operation.

20.3.4 LP Gas Dispensing and Storage

LPG is stored on the vehicle as a liquid under pressure. LPG tanks must be designed to contain an internal pressure of 240-250 psig (1.6-1.7 MPa). They are generally cylindrical, with rounded ends, and are much stronger than tanks used for storing gasoline or diesel fuel, albeit much less so than those used for CNG. LPG can be pumped from one tank to another like any liquid, but the need to maintain pressure requires a gas-tight seal. Except for the need for a standardized, gas-tight connection, LPG used as vehicle fuel can be dispensed in much the same way as gasoline or diesel fuel. To ensure that some vapor space is always available for expansion, LPG tanks used in automotive service must never be filled more than 80% full. Automatic fill limiters are incorporated in the tanks to ensure that this does not occur.

20.4 Gaseous-Fuel Engine Technology

Technology for natural gas and LPG engines resembles that for conventional gasoline spark-ignition engines in many respects. This is especially the case for light-duty vehicles. Modern light-duty natural gas and LPG vehicles and conversion kits commonly employ three-way catalytic converters and stoichiometric air-fuel ratio control systems with feedback control via an oxygen sensor. Except for differences in the fuel metering hardware and the absence of cold starting aids, these systems closely resemble those used in modern light-duty gasoline vehicles.

Heavy-duty engines for gaseous fuels may resemble heavy-duty gasoline engines, with spark ignition and rich or stoichiometric air-fuel ratio. More commonly, however, these engines are based on heavy-duty diesel engine block and "bottom end" designs, and use a high-compression, lean-burn, spark-ignition combustion system. Natural gas and LPG can also be burned in dual-fuel diesel engines, in which the gaseous-fuel charge is ignited by injecting a small amount of diesel fuel.

20.4.1 Fuel Metering Systems

Natural gas occupies considerably more volume than the amount of gasoline having the same energy. As a result, the volumetric energy content of a stoichiometric natural gas/air mixture is less than that of gasoline-air mixture. In addition, natural gas does not benefit from the practice of "power enrichment"—the best power output from natural gas engines occurs at essentially the stoichiometric air-fuel ratio. When a gasoline engine is converted to natural gas, the combination of these two effects typically results in a loss in maximum BMEP and power output of approximately 10%. For LPG, these effects are smaller, and the power loss is typically only a few percent. In dedicated engines, the reduction in power output with gaseous fuel can be compensated by increasing the compression ratio, thus increasing the amount of useful work extracted from a given amount of fuel input. Further increases in BMEP and power output can be achieved by turbocharging. Turbocharging is extremely common for lean-burn, heavy-duty natural gas engines, which can attain BMEP levels exceeding 200 psi (1.4 MPa).

Gaseous-fuel vehicles require precise control of the air-fuel ratio to minimize emissions while maintaining good performance and fuel economy. Until about 1990, nearly all gaseous-fuel metering systems relied on mechanical principles, analogous to the mechanical carburetors used in gasoline engines until the early '80s. Klimstra[8] has summarized the operating features and characteristics of these devices. Although these mechanical systems can be designed to give good engine performance and efficiency, they are susceptible to fuel metering errors due to wear, drift, changes in elastomer properties, changes in fuel and air temperature, changes in fuel properties, and other causes.

These mechanical systems are thus unable to meet the requirement of modern three-way catalytic converter systems for very precise control of the air-fuel ratio.

Over the last decade, air-fuel ratio control systems for light-duty gasoline vehicles have evolved from mechanical systems with electronic trim to full-authority digital electronic fuel injection. Gaseous fuel metering systems have recently undergone a similar evolution. The fuel metering and engine control systems installed on new NGVs produced by Chrysler[9] and Ford are essentially identical to the multipoint sequential fuel injection systems installed on production gasoline vehicles, except for details of the fuel rail and injectors. Several manufacturers of gaseous-fuel retrofit kits now also offer systems using fuel injection under digital electronic control.

20.4.2 Spark-Ignition Engines

Gaseous-fuel engine technology has been reviewed extensively elsewhere.[10] Most of the natural gas and LPG vehicles now in operation have stoichiometric engines which have been converted from engines originally designed for gasoline. Such engines may be either bi-fuel (able to operate on either gaseous fuel or gasoline) or dedicated to the gaseous fuel. In the latter case, the engine can be optimized to take advantage of the knock-resistance of the gaseous fuel by increasing the compression ratio and making other changes. This is not usually done in retrofit situations because of the cost. Nearly all present light-duty natural gas and LPG vehicles use stoichiometric engines, with or without three-way catalysts, as do a minority of heavy-duty natural gas vehicles and most heavy-duty LPG vehicles.

Lean-burn engines use an air-fuel mixture with much more air than is required to burn all of the fuel. The extra air dilutes the mixture and reduces the flame temperature, thus reducing engine-out NO_x emissions, as well as exhaust temperatures. Because of reduced heat losses and various thermodynamic advantages, lean-burn engines are generally 10-20% more efficient than stoichiometric engines. Without turbocharging, however, the power output of a lean-burn engine is less than that of a stoichiometric engine. With turbocharging, the situation is reversed. Because lean mixtures knock less readily, lean-burn engines can be designed for higher levels of turbocharger boost than stoichiometric engines, and can thus achieve higher BMEP and power output. The lower temperatures experienced in these engines also contribute to engine life and reliability. For these reasons, the great majority of heavy-duty natural gas engines are of the lean-burn design.

Large, heavy-duty natural gas engines have been used for many years in stationary applications such as electric generation, irrigation pumping, and driving compressors on natural gas pipelines. More recently, a number of heavy-duty, lean-burn natural gas engines have been developed and marketed specifically for vehicular use, where they are being substituted successfully for diesel engines. Some lean-burn LPG engines have also been developed. Vehicular applications of heavy-duty, lean-burn engines include low-emission buses, trucks, airport service equipment, industrial tractors, and railway locomotives. These engines are able to achieve NO_x emissions less than half those of present diesel engines. Because of the strong regulatory focus on reducing mobile-source NO_x and particulate emissions in the U.S., some observers expect to see greatly increased use of heavy-duty, lean-burn natural gas engines during the next decade.

20.4.3 Dual-Fuel (Compression Ignition) Engines

Dual-fuel diesel engines are a special type of lean-burn engine in which the air-gas mixture in the cylinder is ignited not by a spark plug but by injection of a small amount of diesel fuel, which self-ignites. These have been discussed extensively elsewhere.[11] Most diesel engines can readily be converted to dual-fuel operation using natural gas, while retaining the option to run on 100% diesel fuel if gas is not available. Because of the flexibility this allows, the dual-fuel approach has been popular for retrofit applications, in which diesel engines are modified to use natural gas. Such retrofits have been applied to diesel engines used in trucks, buses, locomotives, and marine vessels, as well as engines used in stationary electric generation and pumping applications. In stationary generation, the dual-fuel capability allows an engine to operate most of the time on low-cost natural gas or biogas, while retaining the ability to switch to 100% diesel fuel in the event that gas supply is disrupted. This makes them useful for emergency standby, as well as in sewage treatment plants and other applications where the gas supply varies over time.

Because of its greater tendency to knock, the use of LPG in dual-fuel applications is subject to severe limitations, and is much less common than natural gas. Where LPG is used in dual-fuel engines, it is generally limited to substituting for 30-50% of the diesel fuel, as compared to 70-99% substitution with natural gas.

The emissions performance of current dual-fuel engine systems is mixed. These engines can be designed to have very low NO_x emissions, but tend to have high HC and CO emissions and poor efficiency at light load. This is because they operate unthrottled, so that the air-fuel mixture becomes leaner as the load is reduced. As the mixture becomes leaner, combustion eventually degrades, leaving large amounts of partial reaction products in the exhaust. Possible solutions to this problem include throttling the intake air at light loads, use of electronically controllable turbochargers to reduce light-load airflow, or the use of skip-firing.

20.5 Gaseous Fuels and the Environment

A major reason for the increased interest in gaseous fuels for vehicles in the last few years has been the promise of environmental benefits. Compared to conventional gasoline and diesel fuels, gaseous fuels for vehicles tend to exhibit lower emissions of pollutants contributing to urban air pollution, lower emissions of greenhouse warming gases, and greatly reduced capacity for water and soil pollution and generation of hazardous waste.

20.5.1 Urban Air Pollution

Light-duty vehicles—NGVs and LPG vehicles require less complex emission control systems than vehicles using gasoline or other liquid fuels in order to achieve extremely low levels of pollutant emissions. Emissions data from a number of gaseous-fuel vehicles using modern electronic emission control systems clearly show the ability to meet California ULEV emissions standards. Two such vehicles, the Chrysler B350 natural gas van and the Chrysler natural gas minivan, have been certified to meet the California LEV and ULEV emission standards, respectively—the first vehicles so certified using *any* fuel (Table 20.6). The actual measured emission levels, in each case, were well below the applicable emissions standard.

Table 20.6. Emissions Certification Data for Light- and Medium-Duty Natural Gas Vehicles

	Mileage	NMOG	Emissions (g/mi) CO	NO_x	HCHO
LEV Std[1]	50,000	0.195	5.0	1.1	0.022
	120,000	0.280	7.3	1.5	0.032
ULEV Std[2]	50,000	0.050	2.2	0.4	0.009
	100,000	0.070	2.8	0.5	0.013
Chrysler B350 Ramvan (5751-8500 lb)					
Gasoline	50,000	0.19	3.4	0.51	NA[3]
CNG	50,000	0.031	2.3	0.05	0.002
(LEV)	120,000	0.040	3.1	0.05	0.003
Chrysler Minivan (3751-5750 lb)					
Gasoline	50,000	0.20	1.2	0.19	NA[3]
CNG	50,000	0.021	0.4	0.04	0.0002
(ULEV)	100,000	0.035	0.4	0.05	0.0002

[1] LEV standard for medium-duty vehicles: 5751-8500 lb GVW
[2] ULEV standard for light-duty trucks: 3751-5750 lb
[3] Not Available

The benefit of NGVs and LPG vehicles in reducing "real world" pollutant emissions from vehicles in consumer use is expected to be even greater. Among the reasons for the superior emissions performance of NGVs are the following:[12]

- Inherently low emissions of non-methane hydrocarbons, since the fuel is 85-99% methane;

- Low ozone-forming reactivity of the residual NMHC, which are primarily ethane;

- Low emissions of toxic air contaminants such as benzene and 1,3 butadiene;

- Low "off-cycle" emissions—unlike gasoline vehicles, NGVs are not calibrated for power enrichment at high loads, so their HC and CO emissions do not skyrocket under "off-FTP" driving conditions;

- Low cold-start emissions—unlike gasoline vehicles, NGVs do not require mixture enrichment for cold starting, so that HC and CO emissions from NGVs are unaffected by low temperatures;

- Likely better emissions durability, due to the reduced complexity of NGV emission control systems and the reduced chance of catalyst damage from overheating due to mixture enrichment;

- Zero evaporative and running losses, due to the sealed fuel system, and negligible refueling emissions;

- Low "fuel cycle" emissions from fuel processing, distribution, and marketing, due to pipeline transport;

- Reduced emissions of greenhouse gases, especially CO and CO_2.

Because of these differences, use of NGVs would result in significant environmental benefits, even compared to the use of gasoline or other alternative-fuel vehicles meeting LEV or ULEV emissions standards. With the exception of inherently low NMHC emissions, the same comments are applicable to LPG vehicles. Table 20.7 shows some of the limited emissions data available for modern LPG vehicles.

Heavy-duty vehicles—Emissions from heavy-duty natural gas engines can be much less than those of heavy-duty diesel or gasoline vehicles. In the last few years, a number of heavy-duty engine manufacturers have developed diesel-derived lean-burn natural gas engines for use in

Table 20.7. Emissions Data for Light- and Heavy-Duty LPG Vehicles

	Emissions		
	NO_x	NMHC	CO
Passenger Car[a] (g/mi)	0.2	0.15	1.0
Heavy-Duty Engine[b] (g/bhp-hr)	2.8	0.5	23.2

[a] CARB, 1991. "Technical Support Document for Proposed Reactivity Adjustment Factors for Transitional Low-Emission Vehicles," Mobile Source Division, Research Division, CARB, El Monte, CA.
[b] CARB certification data for Ford PFM07 LPG engine

emissions-critical applications such as urban transit buses and delivery trucks. These engines incorporate low-NO_x technology used in stationary natural gas engines, and often an oxidation catalyst as well. They are capable of achieving very low levels of NO_x, particulate, and other emissions (less than 2.0 g/bhp-hr NO_x and 0.03 g/bhp-hr particulate) with high efficiency, high power output, and (it is anticipated) long life. Four such engines—the Cummins L10, Detroit Diesel Series 50, and Hercules GTA 5.6 and 3.7 engines—have been certified in the U.S., and all are now in production. The Cummins and Detroit Diesel engines are widely used in new transit buses and some other urban fleet applications such as garbage collection, while the smaller Hercules engines are used in schoolbuses, shuttlebuses, and delivery trucks, among other applications. Table 20.8 summarizes the emissions certification data for these engines.

Table 20.8. Emissions Certification Data for Heavy-Duty Natural Gas Engines in the U.S.

Manufacturer and Model	hp	Emissions (g/bhp-hr)			
		NMHC	CO	NO_x	PM
Cummins					
L10	240	0.2	0.2	1.4	0.02
L10	260	0.2	0.4	1.7	0.02
Detroit Diesel					
Series 50G	275	0.9	2.8	2.6	0.06
Hercules					
GTA 3.7	130	0.6	2.7	3.1	0.08
GTA 5.6	190	0.9	2.8	2.0	0.10

Natural gas engines have also been developed in Europe. For example, a number of CNG-fueled Volvo buses are being used for public transport in Gothenburg, Sweden. They were developed by the Volvo Bus Corporation in response to an order from the City of Gothenburg to provide a low-emission engine to replace the diesel engine normally used in city buses.

The CNG engine was developed from a 9.6-liter diesel engine. The valve timing, combustion chamber, fuel control system and turbocompressor were redesigned and the ignition timing and fuel injection functions controlled by an electronic engine management system. To satisfy the high ignition energy requirement of the lean-burn CNG engine, an inductive spark ignition system, with multi-coil power units mounted directly on the spark plugs, was chosen.

A reasonable operating range is provided by equipping each bus with fifteen 80-liter roof-mounted gas storage tanks. Rated output of the engine is 185 kW at 2200 rpm, and public reaction to the acceleration and smooth ride of the CNG buses has been favorable. Figure 20.7 shows particulate and NO_x emissions well below the European 1995/96 standard for diesel engines.

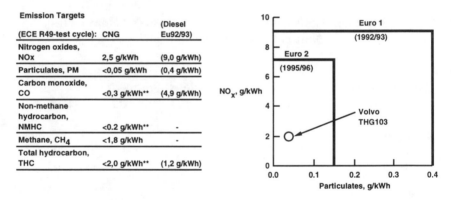

Fig. 20.7. PM and NO_x emissions from Volvo CNG city bus engine.[13]

20.5.2 Global Climate Change

Concern has been growing over the possibility of a global temperature increase due to increased concentrations of "greenhouse" gases in the atmosphere. Two of the most important such gases are CO_2 and methane, which are projected to account for roughly 46% and 9% of the total anticipated warming, respectively (most of the remaining warming effect is due to chlorofluorocarbons [29%], increased tropospheric ozone [4%] and N_2O [6%]).[14] One pound of methane in the atmosphere is estimated by some scientists to contribute about 65-70 times as much to global warming as one pound of CO_2. Unlike CO_2, however, methane in the atmosphere slowly oxidizes through photochemical processes, with an estimated lifetime of 5-10 years. Thus, the relative effects of methane emissions are less than would be anticipated from the effects of methane concentrations alone. The U.S. Office of Technology Assessment, in a recent study, has estimated the global warming potential of a kilogram of methane *emissions* at about 4.6 to 12 times that of a kilogram of CO_2.[15]

One concern sometimes expressed about natural gas vehicles is that their methane emissions might exacerbate global warming. Total global methane emissions are estimated at 400 to 800 million tons per year, mostly from naturally occurring sources such as swamps, wetlands, termites, and ruminant animals.[16] Fossil fuels account for less than 10% of global methane emissions (mostly from venting of associated gas from oil wells). In the most extreme conceivable case, the whole world vehicle population (cars and trucks) of 600 million vehicles, each traveling 15,000 km per year and emitting 0.67 gram of methane per km (typical for current technology NGVs) would result in about 6.6 million additional tons of methane per year, or 1 to 2 percent of total global methane emissions.

Any increase in methane emissions from NGVs would be more than offset by a substantial reduction in CO_2 emissions compared to other fuels. The carbon-hydrogen ratio of natural gas and LPG is less than that of gasoline or diesel fuel, which means that their combustion produces lower CO_2 emissions, even if the energy efficiency is the same. If the engine is optimized for the gaseous fuel by increasing the compression ratio, then the energy efficiency of natural gas and LPG vehicles would also be superior to that of a gasoline vehicle. A further reduction in global warming impacts would result from the lower CO emissions typical of gaseous-fuel vehicles. Although CO is not itself a greenhouse gas, it plays an important role in the atmospheric chemistry that determines concentrations of methane, ozone, and other greenhouse gases. In the atmosphere, CO undergoes photochemical oxidation, producing tropospheric ozone (which is a greenhouse gas) and competes with methane and other species for OH radicals. These radicals play a key role in breaking down methane and some other greenhouse gases. The reduction in CO emissions due to use of gaseous-fuel vehicles would increase the global rate of methane decomposition, and thus reduce methane concentration in the atmosphere.

20.5.3 Spills, Water Pollution, and Hazardous Waste

Spills and leaks of gasoline and diesel fuel from underground tanks, pipelines, trucks, and ships are important sources of water and soil pollution, with cleanup costs at a single site sometimes running into the millions of dollars. The possible environmental liabilities from these leaks are a major concern to users and distributors of transportation fuels. Contamination of various materials with gasoline and diesel fuel is also an important source of hazardous waste. Safe disposal of these wastes is a topic of increasing concern in most industrial and industrializing countries. Because of their non-toxic chemical composition and gaseous form, natural gas and LPG are much less likely to contribute to hazardous waste generation, water pollution, or soil contamination (and the attendant environmental liabilities) than are gasoline or diesel fuel.

20.6 Gaseous Fuel Safety

20.6.1 Compressed Natural Gas

Overall, CNG vehicles are expected to be significantly safer than gasoline vehicles. Compressed natural gas storage cylinders are built to rigorous quality standards. Studies conducted in the late 1970s were unable to identify a single instance of a compressed natural gas cylinder failure due to collision damage, either in the U.S. or in Italy (where more than 300,000 natural gas vehicles are in use). This stands in sharp contrast to the record of collision-caused fuel leakage and fires in gasoline vehicles.

As natural gas is lighter than air, the sudden release of a CNG tank's contents, if it were to occur, would not form a spreading pool or vapor cloud on the ground (such as in a gasoline spill). Instead, the gas cloud would rise into the air and (unless it contacted some nearby ignition source) rapidly disperse. Except for trace contaminants such as hydrogen sulfide (H_2S), natural gas is non-toxic and biologically inert. Due to the high ignition temperature of natural gas (some 600°F higher than that of gasoline), simple exposure to a hot surface (such as an exhaust manifold) is unlikely to lead to a fire. Overall, the risk of fire and hazard to vehicle occupants as a result of uncontrolled release of CNG is much lower than that from gasoline, and comparable to that which might be expected with diesel fuel (which is also difficult to ignite).

In a 1992 survey[17] of the U.S. major natural gas-powered fleet vehicle operators covering more than 8000 vehicles that cumulatively traveled approximately 278.3 million miles from 1987-1990, the injury rate for NGVs per vehicle mile traveled (VMT) was 37 percent lower than the rate for gasoline fleet vehicles and 34 percent lower than the rate for the entire population of registered gasoline vehicles. This study also showed that, in addition to the lower injury rate, no deaths were recorded for NGVs in the 278.3 million miles driven by the sampled fleet. The rate for gasoline fleet vehicles in 1990 was 1.28 deaths per 100 million VMT and the national average rate for all U.S. vehicles registered in 1990 was 2.2 deaths per 100 million VMT. NGVs showed a fire incidence rate of 2.9 incidents per 100 million VMT. Of these 2.9 fires, only one per 100 million VMT was directly attributable to the failure of the compressed natural gas system.

20.6.2 Liquefied Natural Gas

The safety record of LNG vehicles is not as well established as that of CNG, due to the much smaller number of LNG vehicles in use. As with compressed natural gas, vaporized LNG is non-toxic and biologically inert. Being a cryogenic liquid, LNG boils at any temperature above –260°F, and therefore tends to vaporize quickly when spilled. Experience has shown that the vapor cloud above a pool of spilled LNG is very difficult to ignite, due to the narrow range of flammability of natural gas vapor.

One of the major concerns with the use of LNG in vehicles is the possibility that excess vapor pressure might be vented in an enclosed area such as a parking garage, possibly causing an explosion. Although LNG tanks are very well insulated, heat gradually leaks into the tank and raises its temperature and pressure. LNG tanks are equipped with pressure relief valves, so that if the internal pressure rises above a certain level (generally 150 psig in vehicle applications) the excess vapor will be vented. New technology fuel tanks regularly guarantee fuel storage for up to eight to ten days without pressure relief valves being activated. Even with these safety provisions, this could pose a safety problem for inactive vehicles left enclosed for long periods of time.

LNG storage in cryogenic vessels also introduces the risk of a Boiling Liquid Expanding Vapor Explosion (BLEVE). This could occur if the storage vessel were heated rapidly from the outside, causing an equally rapid temperature and pressure rise inside. This could be a consequence of the tank being immersed in a fire of some other origin, for example, after a collision with a diesel-carrying truck. The likelihood of BLEVE incidents is reduced by using materials for the inner vessel that fail predictably and non-catastrophically, so that the fuel is not ejected suddenly. Although LNG storage tanks are heavier than the thin sheet steel gasoline or diesel tanks they would replace, they are also much stronger due to their construction. This added strength means LNG tanks are much more resistant to damage from an impact sustained in a vehicle collision.

Another danger associated with the use of LNG is the possibility of cryogenic burns due to contact with spilled LNG during refueling or as the result of an accident. Because of the level of training required, it is unlikely that members of the general public would be allowed to dispense their own LNG. LNG nozzles, hoses, and dispensers are equipped with valves to prevent excessive release of LNG in the event of an accident. The fact that LNG tanks are tough and designed to fail predictably reduces the likelihood of contact with spilled fuel during an accident.

Various agencies are working on codes and regulations for LNG transportation and use in vehicles. Such regulations are already in place for LNG production and storage. The National Fire Protection Association (NFPA) is working on such standards, as well as AGA Laboratories, ANSI, Society of Automotive Engineers (SAE), and Canadian Gas Association (CGA).[18] LNG tank designs are already covered with American Society of Mechanical Engineers (ASME), CGA, Code of Federal Regulations (CFR), and API (American Petroleum Institute) specifications. Updated codes and regulations for LNG use in ferry boats, fishing boats, and locomotives are being studied by the U.S. Coast Guard and the Federal Railroad Administration.

The National Fire Protection Association is currently in the process of establishing recommended safety practices for NGV liquefied natural gas vehicular fuel systems (NFPA 57/59).[19] As with CNG fuel systems, NFPA 57/59 apply to the design and installation of LNG engine fuel systems on vehicles of all types including aftermarket and OEMs and to their associated fueling (dispensing) systems.

20.6.3 LPG Safety

LPG poses a somewhat greater safety risk than CNG, but the risk is still considerably less than with gasoline. Unlike natural gas, LPG vapors are heavier than air, so that leaks from the fuel system tend to "pool" at ground level, where they are likely to contact ignition sources. The flammability limits of LPG vapor in air are also broader than those for natural gas. A number of serious explosions have been caused by LPG leaks in buildings. Rules for safe handling and use of LPG are compiled in NFPA 58.

Like natural gas, LPG is non-toxic. Also like natural gas, LPG is stored on the vehicle in sealed pressure vessels which are much stronger than typical gasoline fuel tanks. The probability of a tank rupturing and releasing fuel is thus less than for gasoline. The sealed pressure vessel does raise the possibility of a BLEVE explosion, but standard precautions such as pressure relief valves make this a very unlikely occurrence.

References

1. U.S. Dept. of Energy, *An Assessment of the Natural Gas Resource Base of the United States*, Report No. DOE/W/31109-H1, Washington, DC, May, 1988.

2. King, S.R., "The Impact of Natural Gas Composition on Fuel Metering and Engine Operational Characteristics," SAE Paper No. 920593, 1992.

3. T.J. Callahan, T.W. Ryan III, and S.R. King, "Engine Knock Rating of Natural Gases—Methane Number," ASME Paper No. 93-ICE-18.

4. J. Kubesh, S.R. King, and W.E. Liss, "Effect of Gas Composition on Octane Number of Natural Gas Fuels," SAE Paper No. 922359, 1992.

5. Minutes of European Committee for Standardization Working Group 23, second meeting, held November 8, 1989.

6. Jayaraman, V., Consolidated Natural Gas, personal communication, 1992.

7. W.E. Liss, W.H. Thrasher, G.F. Steinmetz, P. Chowdiah, and A. Attari, *Variability of Natural Gas Composition in Select Major Metropolitan Areas of the United States*, GRI-92/0123, Gas Research Institute, Chicago, Illinois, March 1992.

8. Jacob Klimstra, "Carburetors for Gaseous Fuels: On Air-to-Fuel Ratio, Homogeneity, and Flow Restriction," SAE Paper No. 892141, 1989.

9. R.O. Geiss, "Technical Highlights of the Dodge Compressed Natural Gas Ram Van/Wagon," SAE Paper No. 921551, 1992.

10. C.S. Weaver, "Natural Gas Vehicles: A Review of the State of the Art," SAE Paper No. 892133, 1989.

11. C.S. Weaver and S.H. Turner, "Dual Fuel Natural Gas/Diesel Engines: Technology, Performance, and Emissions," SAE Paper No. 940548, 1994.

12. C.S. Weaver and S.H. Turner, *Prospects for Meeting Low Emission Vehicle Standards: A Comparison of Natural Gas and Reformulated Gasoline*, American Gas Association, April 1993.

13. H. Larsson and Lars Karlsson, "The Volvo THG103 Low-Emission CNG Engine," Volvo Technical Report No. 1, 1992.

14. V. Ramanathan, R.J. Cicerone, H.B. Singh, and J.T. Kiehl, "Trace Gas Trends and Their Potential Role in Climate Change," *J. Geophys. Res.*, V. 90, No. D3, pp 5547-5566, June 20, 1985.

15. U.S. Congress, Office of Technology Assessment, *Changing by Degrees: Steps to Reduce Greenhouse Gases*, OTA-O-482, Washington, DC: U.S. Government Printing Office, February 1991.

16. D. Golomb and J.A. Fay, "The Role of Methane in Tropospheric Chemistry," Energy Laboratory, Cambridge, MA, 1989.

17. American Gas Assocation Policy & Analysis Group, "Natural Gas Vehicle Safety Survey: An Update," Issue Brief 1992-3, American Gas Association, Arlington, VA, 1992.

18. Acurex Corporation, *A White Paper: Preliminary Assessment of LNG Vehicle Technology, Economics, and Safety Issues (Revision 1)*, for Gas Research Institute, January 1991.

19. NFPA 52, *Compressed Natural Gas (CNG) Vehicular Fuel Systems*, Effective August 14, 1992, National Fire Protection Association, Quincy, MA, 1992.

Chapter 21

Neat or Near-neat Alcohols as Alternative Fuels

Two alcohols, methanol and ethanol, are suitable as alternative fuels for automotive use. Their major advantage over hydrocarbon fuels derived from crude oil is that they can be readily made from resources that are fairly evenly distributed over the globe, unlike crude oil where over fifty percent is located in the Middle East. In addition, the raw materials are either renewable or are much more extensive than crude oil since methanol can be made from natural gas or coal, and ethanol from the fermentation of agricultural products (see Chapter 3, Section 3.9.1).

Another advantage of these fuels is that it is possible to achieve lower exhaust emissions of hydrocarbons and this, together with their reduced photochemical "smog-forming" activity, means that they can be attractive as alternative fuels in areas where there is severe atmospheric pollution. This is discussed in more detail in Section 21.1.4.

The main disadvantage is that they have a much lower energy content than gasoline or diesel fuel. However, the fuel consumption debit associated with this can be reduced to some extent by suitable engine design. Another difficulty is that alcohols are not volatile enough for easy cold starting in spark ignition engines, unless some form of heating is employed, but near-neat alcohols, containing small amounts of gasoline or other materials, can be used to avoid this problem.

In Brazil, ethanol containing about 5% water has been used successfully in vehicles designed to run on this fuel. The ethanol has been produced by fermentation of sugar cane and, although it is relatively expensive to manufacture, it has been attractive to this country because it avoided costly imports of crude oil. However, when the world sugar price rose and an indigenous source of crude oil became available, its use became less economical and required heavy subsidies. The case for using biofuels has been studied extensively[1,2] and it is generally agreed that the economic balance is so poor that bioethanol must depend heavily on subsidies, the CO_2 balance is negative and there are virtually no energy gains. Although bioethanol can give lower HC and CO emissions, there are increases in NO_x and aldehydes. In the United States, West Germany, Japan, and New Zealand, the interest in neat or near-neat alcohol fuels is primarily directed at methanol because of the manufacturing cost and supply considerations.

21.1 Alcohols as Fuels for Spark Ignition Engines

Both methanol and ethanol have excellent octane qualities, they boil in the right range and they burn cleanly, and so have many desirable properties for use in spark ignition engines. They are well known as fuels for racing engines, as discussed in Chapter 13, where their high-octane qualities allow the use of high compression ratios, and their ability to burn at rich mixture strengths coupled with their high latent heat of vaporization enables them to produce more power than is possible with gasoline.

Some physical and chemical characteristics of methanol and ethanol are summarized and compared with gasoline in Table 21.1.

21.1.1 Fuel Economy and Power Using Alcohol Fuels

In order to make the best use of the special properties of alcohols, engines must be designed specifically for them. Because of the lower energy content, which for methanol is less than half that of gasoline, volumetric fuel economy will always be lower than that achieved with gasoline. However, improvements in thermal efficiency are possible with alcohols because of the following differences in combustion characteristics from gasoline:[3]

- Alcohols burn with lower flame temperatures and luminosity, so that less heat is lost by conduction and radiation to the engine cooling system.

**Table 21.1. Characteristics of Methanol and
Ethanol as Alternative Fuels**

	Methanol	**Ethanol**	**Gasoline**
Oxygen content, wt%	50.0	34.8	0
Boiling Point, °C	65	78	35-210
Lower Heating Value, MJ/kg	19.9	26.8	approx 42.7
Heat of Vaporization, MJ/kg	1.17	0.93	approx 0.18
Stoichiometric air-fuel ratio	6.45:1	9.0:1	approx 14.6:1
Specific Energy, MJ/kg per air-fuel ratio	3.08	3.00	approx 2.92
RON	109	109	90-100
MON	89	90	80-90

- Alcohols burn faster and allow more efficient torque development.

- Alcohol combustion generates a larger volume of combustion products and develops higher cylinder pressures.

In addition, the high-octane quality of alcohols enables higher compression ratios to be used than are possible with gasoline-fueled engines, and this improves both fuel economy and power.[4,5] A 16% improvement in efficiency has been reported[6] by increasing the compression ratio of a single-cylinder engine from 8:1 to 18:1 when operating on methanol. Table 21.2 summarizes some performance data that have been obtained[7] and are consistent with many other studies.[8,9,10,11] These show that, although power output can be maintained for a given vehicle, fuel consumption is always higher. This means that a larger tank is required if the range of the vehicle on a single filling is to be maintained.

The use of turbochargers or superchargers will also allow an improvement in fuel economy by reducing engine weight for the same power output.

Table 21.2. Comparison of Alcohols with Gasoline in Optimized Vehicles[12]

	Methanol optimized	Ethanol optimized	Gasoline	
Performance				
0-100 km/h (sec)	9.0	10.1*	10.0	10.5*
V_{max} (km/h)	214	207*	210	205*
Fuel consumption (L/100 km)				
City cycle	27.0	22.8*	17.3	16.8*
at 90 km/h	15.8	13.9*	9.1	9.4*
at 120 km/h	18.9	17.0	11.3	11.7*

*Automatic transmission version

21.1.2 Octane Quality of Alcohol Fuels

Although Table 21.1 shows that both ethanol and methanol have similar antiknock characteristics, with very high RON levels but with quite high sensitivities (differences between RON and MON), these values cannot be related precisely to the way in which the fuels perform in a vehicle on the road. This is because the standard CFR test methods for determining RON and MON cannot be used satisfactorily for these fuels due to their high latent heats of vaporization. This is particularly true for the MON method, and modified test equipment has been suggested[13] to achieve the required test temperature of 149°C for the fuel-air mixture. This modified octane rating procedure nullifies one of the important benefits of alcohol fuels — that of increased charge density — because of the latent heat effect, and so interpretation of the meaning of octane numbers obtained in this way must be made with caution.

However, as mentioned in Section 21.1.1, these alcohols do show excellent antiknock properties which enable high compression ratios to be used so that improved thermal efficiencies can be achieved over those obtained using gasoline.

21.1.3 Volatility of Neat and Near-neat Alcohols and Their Influence on Vehicle Driveability

Neat alcohols have low vapor pressures (4.6 psi for methanol and 2.3 psi for ethanol as compared with a range of about 8 to 15 for gasoline) and high latent heats of vaporization, and consequently show poor cold weather starting and warm-up performance. This is compounded by the fact that methanol is conductive so that if any liquid fuel bridges the spark plug gap or forms a film over the ceramic, the ignition current passes through the fuel and no spark occurs.[14] Ethanol, on the other hand, is not conductive unless contaminated with ionic material. The minimum cold start ambient temperature for pure methanol-fueled vehicles without starting aids has been reported as being between 0 and 16°C,[15,16,17,18,19] and for ethanol, about 43°C,[20] although lower temperatures of 15 to 20°C have also been claimed.

In Brazil, an on-board auxiliary supply of volatile fuel such as a gasoline/ethanol blend has been used to facilitate cold starting when ethanol is the main fuel. A similar system was used in a field test involving both methanol- and ethanol-fueled vehicles.[12] Propane has also been used to help cold starting of methanol-fueled cars.[21] Improved atomization of the alcohol helps and this can be achieved in many ways, including fuel injection[3] and heating.[22] Dissociation of some of the methanol into carbon monoxide and hydrogen has also been investigated as a cold starting aid.[23]

Near-neat alcohols have been investigated to overcome this cold starting problem. These are blends of alcohol with a minor amount (up to 15% by volume) of a volatile primer which increases the vapor pressure. Gasoline has been used for this purpose as well as dimethyl ether and various hydrocarbons such as isopentane.[3,15,16,24,25,26]

The use of volatile primers[27] to assist cold starting has another benefit in that they give luminosity to an alcohol flame. Methanol, in particular, burns with an almost invisible flame and so presents a special safety hazard. Primers can also enrich the vapor above the fuel in the vehicle's tank and put it outside the flammability limit,[28] thereby reducing a potential explosion hazard. However, primers can cause a number of difficulties such as those described in Chapter 11 for gasoline/alcohol blends, including water sensitivity, hot weather driveability malfunctions, increased evaporative losses, etc.[29]

21.1.4 Exhaust and Evaporative Emissions

There is no large reservoir of knowledge based on practical experience as to the effects of alcohol fuels on exhaust and evaporative emissions since, although Brazil has had a large number of vehicles operating on ethanol, for much of the time they have had no legislated emissions standards. Most of the data available are from experimental programs carried out on modified or prototype vehicles.

Near-neat methanol containing about 15% primer, and often designated as M85, would ensure satisfactory cold starting and warm-up and would help minimize exhaust emissions during this period. For non-optimized vehicles, the lower peak combustion temperatures and the ability to run lean without misfire reduces oxides of nitrogen emissions, but as compression ratios are increased to take advantage of the high antiknock quality of methanol, this benefit is reduced.[15,30,31]

One of the difficulties in assessing the potential benefits of alcohol fuels versus more conventional gasolines in terms of emissions is that ideally the engine needs to be designed for each fuel so as to achieve the maximum benefits. Thus it is not possible to compare data from each fuel using the same vehicle, unless that vehicle is in a non-optimized state on one or other of the fuels. However, quite a lot of work has been carried out using such vehicles which are capable of running on both types of fuel—so-called Flexible or Variable Fueled Vehicles (FFV/VFV)—and the results of this work are discussed in Section 21.1.7.

In one set of tests using vehicles optimized for use with methanol it was shown that the main advantage of methanol is in the emissions of hydrocarbons and that emissions of CO and NO_x are similar to those obtained from vehicles optimized on conventional gasoline.[32] The results of FFVs also indicate that significant reductions in CO are possible but that NO_x emissions may be worse for M85, as discussed later.

Methanol has a lower photochemical reactivity than hydrocarbons in terms of "smog" formation, but unburnt methanol and aldehyde emissions in exhaust gases are of some concern.[33] They can be reduced satisfactorily by the use of oxidation or three-way catalysts, and all exhaust pollutants can be controlled by the use of three-way catalysts, closed-loop control and exhaust gas recirculation. The lower exhaust gas temperature obtained with methanol is reported to reduce the efficiency of three-way catalytic converters, particularly with respect to NO_x reduction.

Table 21.3 summarizes some exhaust emission data from a European car operating on neat alcohol fuels versus the California limits,[9] although generally higher values might be expected in more typical U.S. vehicles.

Table 21.3. Emissions from VW California Alcohol Test Vehicles — Test Method CVS 75

	HC	CO	NO_x
VW Rabbit - methanol (g/mile)	0.14	0.8	0.32
VW Rabbit - ethanol (g/mile)	0.30	1.5	0.29
California limit	0.41	3.4	0.40

Evaporative emissions are lower with neat methanol and increase as the gasoline content increases. They can be be controlled using charcoal canisters as for gasoline, although there are some questions regarding the capacity and life of canisters fitted to vehicles operating on methanol.[34] The toxicity of methanol vapor makes it important to control such emissions apart from any effect on smog formation.

21.1.5 Compatibility with Materials of Construction of Vehicles

Methanol and ethanol are corrosive to many of the metals that are used in conventional fuel systems such as terne plating used in fuel tanks (a lead/tin coating on steel), aluminum, copper, brass, magnesium and die cast zinc.

Elastomers and plastic components can also be attacked. Carburetor and fuel gage floats are often made of plastic and have been known to swell and stick. Fiber gaskets can be softened and fuel hoses and pump diaphragms can swell or harden and crack, particularly if gasoline is used as a primer. Fluorocarbons such as Viton can also be attacked.

The correct choice of fuel system materials is important and several test programs have been carried out to identify suitable and unsuitable materials.[29,35,36]

21.1.6 Engine Wear

Cylinder and piston ring wear have been found to occur when using neat methanol or ethanol.[37] It has been suggested that the oil film coating these metal parts can be washed away by liquid alcohol during cold starting, giving metal-to-metal contact, although this theory has been contended. According to some authors,[38,39,40] wear using methanol could be due to the formation of formic or performic acids formed during combustion directly attacking the iron, and thereby causing corrosive wear. The use of modified engine components such as chrome-plated rings have been found to reduce the problem.[41,42,43] Fuel additives to reduce wear and corrosion have also been investigated.[27,44]

21.1.7 Introduction of Alcohol Fuels into the Market—Flexible and Variable Fuel Vehicles

A considerable amount of work has been carried out to assess the problems and economics of introducing an alcohol fuel into a market that already has a gasoline distribution system that must continue in operation until existing vehicles can either be converted or are no longer important. Studies into introducing methanol indicate that the price must be competitive with gasoline while both fuels are marketed together.[45,46,47] Upgrading existing fuel distribution systems will be expensive[48] and preparations need to be started well before the planned date for general availability.[46]

In Brazil[49] the government initially provided incentives for using ethanol rather than gasoline and actively discouraged the use of gasoline by prohibiting its sale during weekends. This program successfully enabled Brazil to meet its objective of reducing imports of crude oil.

There has been considerable interest in the use of flexible or variable fueled vehicles (FFVs or VFVs) which allow the use of either alcohol or gasoline or of any mixture of the two, since these would clearly help customer acceptance of methanol as a fuel. One version uses a sensor in the fuel system to detect the ratio of gasoline to methanol and, hence, to control the fuel flow to the injectors so that the correct air-fuel is always achieved.[50] Such a vehicle would eliminate the risk of refueling with the wrong fuel and would help the development of a methanol distribution network. However, there are a number of disadvantages associated with such vehicles, such as:

- The potential environmental and self-sufficiency benefits of methanol would be compromised because an engine cannot be optimized for both fuels.

- The initial cost is likely to be high because of the increased complexity, and the vehicle would have a shorter driving range and poorer fuel economy than a pure methanol vehicle.

- It is likely to be a transitional stage only, while a methanol network is being established.

An alternative concept[51] incorporates a high-compression-ratio engine in order to be able to take advantage of the benefits of neat methanol, but has a separate tank for gasoline as an emergency fuel. There is a dual-fuel system and operation on gasoline is achieved only with a significant loss of power and efficiency.

21.1.8 Emissions Performance of FFVs and VFVs

In tests carried out as part of the AQIRP test program[52] (see Chapter 12), the emissions from a series of prototype flexible/variable fueled vehicles (FFV/VFV) running on M85 were compared with those using an industry average gasoline containing no oxygenates and also with a blend containing 10% methanol. The results are summarized in Table 21.4 and compared with test results from a further test program[53] using a 1992 production Chevrolet Lumina variable fuel vehicle.

It can be seen from the table that the data from both studies were reasonably consistent with each other and showed that in general most emissions were lower with M85 with the notable exceptions of formaldehyde and, of course, methanol itself.

21.2 Alcohol Fuels for Compression Ignition Engines

The physical and thermodynamic characteristics of alcohols do not make them particularly suitable fuels for compression ignition en-

Table 21.4. Comparison of FTP Composite Tailpipe Emissions from FFV and VFV Vehicles Using Industry Average 1992 Gasoline and M85[53]

	1992 Lumina VFV[53]			AQIRP Data[52]		
	Gasoline	M85	%Change	Gasoline	M85	%Change
Emissions (g/mile)						
OMHCE*	0.189	0.117	−38	0.29	0.18	−37
NMOG**	0.157	0.199	+27	0.24	0.28	+14
CO	3.33	1.27	−62	2.81	1.93	−31
NO$_x$	0.283	0.278	−1.5	0.40	0.49	+23
Methanol	0	0.132	—	0	0.20	—
Toxic Emissions (mg/mile)						
Benzene	13.16	1.35	−90	15	2.4	−84
1,3 Butadiene	1.14	0.203	−82	1.2	0.1	−93
Formaldehyde	1.37	17.03	+1143	2.8	15	+436
Acetaldehyde	0.685	0.330	−52	0.8	0.2	−70
Total Toxics	16.35	18.91	+16	19.4	17.4	−10
Reactivity of Emissions						
Specific Reactivity (g O$_3$/gNMOG)	3.87	1.77	−54	2.4	1.0	−57
Ozone Forming Potential (OFP) (g O$_3$/mile)	0.615	0.352	−43	0.62	0.32	−52

* OMHCE = Organic Matter Hydrocarbon Equivalent
** NMOG = Non-methane Organic Gases

gines, but they offer a means of reducing exhaust emissions of sulfur compounds, smoke, particulates and NO_x.

Methyl and ethyl alcohols have much higher autoignition temperatures than conventional diesel fuel, in the region of 1000 K compared with 550 K, so the use of alcohol as a direct replacement primary fuel would not be straightforward. Higher compression ratios and other aids such as glow plugs and ignition improvers would be required for starting from cold, even at normal ambient conditions, although operation on alcohol fuel would be feasible on a warmed-up engine.[54]

Ignition improver treatment of methanol has been shown to be effective but a large amount, as much as 16% by volume, is needed to achieve reasonably short ignition delay periods.[55] Power output will be reduced unless the volume of fuel injected is increased to compensate for the lower specific heat content of alcohols. Other factors requiring consideration are the lower viscosity and lubricity of alcohols, which would cause excessive wear in conventional fuel injection equipment, and their higher volatility, which might increase the risk of vapor lock and cavitation.

The State of California Energy Commission is sponsoring projects to confirm the use of neat methanol as a cost-effective means of controlling exhaust emissions from compression ignition engines.[56] Dedicated fuel systems are needed for methanol to avoid materials compatibility problems and suitable engines have been developed by Komatsu in Japan, M.A.N. in Germany, and Detroit Diesel Corporation in the U.S., to study operation on methanol. The Komatsu IDI engine and the M.A.N. DI engine both have spark assistance. The D.D.C. two-stroke DI engine incorporates glow plugs to help ignition at light loads and an electronically regulated intake air heating system for high load operation.[57,58,59] The Komatsu engine has been field tested in tractors and the other two in transit bus service.

Comparisons of emissions from the two methanol-fueled DI engines running on a chassis dynamometer under city driving conditions, as defined by the Central Business District (CBD) Cycle, are shown in Table 21.5.

Simple blends of alcohols with diesel fuel are impractical because of poor solubility, water pick-up and the likelihood of phase separation, but other techniques to employ alcohols as auxiliary fuels have been

**Table 21.5. Comparison of Emissions from Methanol
Diesels and Conventional Diesels
(Source: API publication 4261, 2nd Edition, July 1988)**

Exhaust Emissions	Methanol Diesels		Typical Diesels
	MAN	DDC	
HC, g/km	0.53	75.0	1.7-2.5
CO, g/km	0.48	55.0	10-18
NO_x, g/km	8.80	4.9	14-19
Particulates, g/km	0.06	0.96	1.7-3.9
Aldehydes, g/km	0.10	1.20	not detected
Methanol, g/km	0.35	62.0	not detected

tried. "Diesohol" (diesel/alcohol) blends with additives to improve phase stability when water is present have been evaluated in an unmodified Ducati DM 175 single-cylinder, air-cooled DI diesel engine. Performance tests were carried out to assess the effectiveness of various nitrogen-free compounds as compression ignition improvers and to see whether the diesohol blends showed benefits in exhaust emissions over conventional deisel fuel.[60]

When more than 10% alcohol was present in the blend, brake thermal efficiencies approaching those obtained with 100% diesel fuel could only be achieved with the addition of large amounts of an ignition improver and at high loads. The peroxide compounds used as ignition improvers varied in their effectiveness, with 0,0-t-Amyl-0 (2-Ethyl hexyl) monoperoxycarbonate being rated best of the five tested, while synergistic benefits were seen with some combinations of the additives. Emissions of NO_x and soot from the diesohol blends were slightly lower than for diesel fuel but CO and unburned hydrocarbon levels were generally higher.

Aspiration (or fumigation) of methyl and ethyl alcohol is an effective way to increase smoke-limited power output. Carburetors, manifold injectors, heated vaporizers and mist generators have been used to introduce alcohol into the intake airstream. This approach has shown small benefits in thermal efficiency with increasing alcohol substitution (up to 30% by energy content), possibly due to nearly constant volume combustion. Alcohol fumigation lengthens the ignition delay, the effect being greater with methanol and also with increasing water content. NO_x and particulate emissions decrease as alcohol substitution increases due to the lower combustion temperature, but CO tends to increase, particularly at low fuel rack settings.[61]

In other dual-fuel studies, a separate injection system has been used to inject alcohol into the combustion chambers of DI and IDI engines. As with fumigation, the benefits of dual injection are improved thermal efficiency and reduced emissions of NO_x and particulates.[62]

21.2.1 Acceptance of Alternative Diesel Fuels

A considerable amount of testing has been carried out to assess the feasibility of using various alternative types of fuel as auxiliary or replacement fuels for compression ignition engines and data from these tests have been discussed in earlier chapters. Most alternative fuels would involve some worsening of performance or exhaust emissions, while certain fuel types would necessitate different degrees of engine modification to enable complete substitution, as may be required in California. Also, the costs of modifying vehicles and distribution systems to accommodate a radically different fuel such as methanol would be substantial.[63] Additionally, environmental considerations are now being directed at the global impact of switching to alternative fuels. The "Well to Wheels," or "Cradle to Grave" concept[64] takes into account the energy consumption and the associated CO_2 and other pollutants released into the atmosphere during the production, transportation, processing and distribution, and burning of a fuel, instead of being concerned only about emissions from the vehicle exhaust. Excluding the extreme possibilities, unless local or national situations provide a realistic economic incentive, demand for conventional petroleum-sourced fuels is expected to continue more or less as predicted through the year 2000. Petroleum reserves are currently estimated as sufficient for the next 40 years and a general switch to a different type of fuel is only likely when these reserves are seriously depleted or if environmental or other factors present an inevitable motive.

21.3 Distribution, Storage and Safety of Alcohol Fuels

It will probably require about 1.8 times as much neat methanol to provide the same driving range in a vehicle as can be obtained with gasoline, and this means that increased storage and distribution capability would be required for such a fuel. In addition, there are a number of other factors which must be taken into account when considering this type of fuel, as follows:

- Distribution and storage equipment must be compatible with neat or near-neat alcohol to ensure no corrosive attack or adverse effect on elastomers.

- Contamination of other petroleum products must be avoided, particularly in pipelines.

- Water contamination must be avoided, particularly with near-neat alcohols where phase separation could be a problem. Storage should be in fixed-roof tanks with internal floating covers.[65]

- Fires involving neat or near-neat alcohols require the use of alcohol-resistant foam concentrates or dry chemicals.[66] Neat methanol burns with an almost invisible flame and the addition of hydrocarbons or other additives is effective in increasing flame luminosity.[27,67,68]

- The vapor space above neat methanol will normally be in the flammable range at bulk liquid temperatures between about 45 and 110°F but, when blended with hydrocarbon fuels, the vapor space flammability limits approach those of gasoline.[54,69]

- Methanol can cause blindness and acidosis in man and continuous exposure to the vapor must be avoided.

References

1. K. Taschner, "Who Needs Biofuels?," European Environment Bureau, Brussels, June 1993.

2. D.J. Rickeard and N.D. Thompson, "A Review of the Potential for Bio-Fuels as Transportation Fuels," SAE Paper No. 932778, 1993.

3. API, "Alcohols and Ethers, A Technical Assessment of Their Application as Fuels and Fuel Components," <u>API Publication 4261</u>, Second Edition, July 1988.

4. Swedish Motor Fuel Technology Co., <u>Alcohols and Alcohol Blends as Motor Fuels</u>, Vol. IIB, p. 8:39, STU Information No. 580, 1986.

5. H. Menrad and M. Haselhorst, <u>Alcohol Fuels</u>, Monograph. Springer, New York, ISBN 3211816968, 1981.

6. N.D. Brinkman, "Effect of Compression Ratio on Exhaust Emissions and Performance of a Methanol-Fueled Single Cylinder Engine," SAE Paper No. 770791, 1977.

7. D.H. Huttbraucker, "Optimization of Vehicle Concepts for Alcohol Fuels," *Proc. Fifth Int. Fuel Technol. Symposium*, Aukland, May, 1982.

8. R.D. Fleming and T.W. Chamberlain, "Methanol as an Automotive Fuel, Part 1—Straight Methanol," SAE Paper No. 750121, 1975.

9. H. Menrad, *et al.*, "Alcohol Fuel Vehicles of Volkswagen," SAE Paper No. 820968, 1982.

10. R.J. Nichols, "Further Development of the Methanol-Fueled Escort," *Proc. Second Int. Conf. Autom. Eng.*, Tokyo, Nov. 1983.

11. R.A. Potter, "Neat Methanol Injection Fleet," Paper at Fourth Washington Conference on Alcohol, November 1984.

12. H. Menrad, G. Decker and K. Weidman, "Alcohol Fuel Vehicles of Volkswagen," SAE Paper No. 820968, 1982.

13. G. Heilmann, *et al.*, "Determination of Antiknock Performance (RON and MON) of Alcohols and Alcohol-Gasoline Blends," German Soc. Petr. Sci. and Coal Chem., Hamburg, Report No. 260.01, 1981.

14. C.J. Dasch, N.D. Brinkman and D.H. Hopper, "Cold Starts Using M85: Coping with Low Fuel Volatility and Spark Plug Wetting," SAE Paper No. 910865, 1991.

15. R.H. Thring, "Alternative Fuels for Spark Ignition Engines," SAE Paper No. 831685, 1983.

16. A.A. Reglitzky, R.W. Hooks and W.E. Bernhardt, "Alternative Fuels for Existing Engines, Their Potential and Drawbacks," Tenth World Petroleum Congress, September, 1979.

17. S. Ito, S. Sato and T. Ichimaya, "Performance of a Methanol Fueled Car," SAE Paper No. 811383, 1981

18. D.P. Gardiner and M.F. Bardon, "Cold Starting Tests on a Methanol Fueled Spark Ignition Engine," SAE Paper No. 831175, 1983.

19. D.P. Gardiner, M.F. Bardon, *et al.*, "Review of Cold Starting Performance of Methanol and High Methanol Blends in Spark Ignition Engines," SAE Paper Nos. 902154 and 902181, 1990.

20. M. Kirik, "Alcohol as an Alternative Fuel," *The Ontario Digest and Engineering Digest,* pp. 29-31, September 1977.

21. R.A. Potter, "Neat Methanol Fuel Injection Fleet Alternative Fuels Study," Fourth Washington Conference on Alcohol, November 1984.

22. L.G. Dodge, D.W. Naegeli, *et al.*, "Improved Atomization of Methanol for Low Temperature Starting in Spark Ignition Engines," SAE Paper No. 920592, 1992.

23. M.S. Pike, T.J. Guglielmello and J.W. Hodgson, "Development of a Cold-Start Device for Methanol-Fueled Engines," SAE Paper No. 932772, 1993.

24. A. Koenig, H. Menrad and W. Bernhardt, "Alcohol Fuels in Automobiles," Alcohol Fuels Conference, Inst. Chem. Eng., Sydney, 9-11 August 1978.

25. J.L. Keller, G.M. Nakaguchi, J.C. Ware and E.L. Wiseman, "Methanol Fuel Modification for Highway Vehicle Use, Final Report," HCP/W 368318, July 1978.

26. R.M. Tillman, "Blending, Distribution and Marketing Aspects of Alcohols as Alternative Fuels," Symposium on Alcohols as Alternative Fuels, Ontario, November 19, 1976.

27. E.R. Fanick, L.R. Smith, *et al.*, "Laboratory Evaluation of Safety-Related Additives for Neat Methanol Fuel," SAE Paper No. 902156, 1990.

28. P.A. Machiele, "Flammability and Toxicity Tradeoffs with Methanol Fuels," SAE Paper No. 872064, 1987.

29. R. Nichols, "Field Experience with U.S. Methanol Vehicles: Future Design Considerations," CEC Second International Symposium on the Performance Evaluation of Automotive Fuels and Lubricants, Wolfsburg, 1985.

30. A. Koenig, H. Menrad and W. Bernhardt, "Alcohol Fuels in Automobiles," Alcohol Fuels Conference, Inst Chem. Eng., Sydney, 9-11 August 1978.

31. N.D. Brinkman, "Effect of Compression Ratio on Exhaust Emissions and Performance of a Methanol-Fueled Single Cylinder Engine," SAE Paper No. 770791, 1977.

32. R.J. Nichols, "Update on Ford's Dedicated and FFV Methanol Vehicles," Presentation to the Light-duty Methanol Vehicle Workshop, California Air Resources Board, Los Angeles, April 1987.

33. C.A. Harvey, *et al.*, "Toxicologically Acceptable Levels of Methanol and Formaldehyde Emissions from Methanol-Fueled Vehicles," SAE Paper No. 841357, 1984.

34. K.R. Stamper, "Evaporative Emissions from Vehicles Operating on Methanol/Gasoline Blends," SAE Paper No. 801360, 1980.

35. G. Decker and H. Menrad, "Field Experience with German Methanol Vehicles," Paper EF 10, CEC Second International Symposium on the Performance Evaluation of Automotive Fuels and Lubricants, Wolfsburg, 1985.

36. N.E. Gallopoulos, "Alcohols for Use as Motor Fuels and Motor Fuel Blends," Paper EF 11, CEC Second International Symposium on the Performance Evaluation of Automotive Fuels and Lubricants, Wolfsburg, 1985.

37. E.J. Owens, *et al.*, "Effects of Alcohol Fuels on Engine Wear," SAE Paper No. 800857, 1980.

38. T.W. Ryan III, T.J. Bond and R.D. Schieman, "Understanding the Mechanism of Cylinder Bore and Ring Wear in Methanol Fueled SI Engines," SAE Paper No. 861591, 1986.

39. K. Otto, *et al.*, "Steel Corrosion by Methanol Combustion Products: Enhancement and Inhibition, " SAE Paper No. 861590, 1986.

40. T. Ichimiya, K. Takahashi and Y. Fuwa, "Wear Mechanisms of Methanol Fueled Engines," SAE Paper No. 852199, 1985.

41. G.K. Chui and R.J. Nichols, "Development of Experimental Vehicles for Methanol Fuel," American Chemical Society, Division of Petroleum Chemistry, presentation, Kansas City, September 1982.

42. E.T. King and G.K. Chui, "Hardware Effects on the Wear of Methanol-Fueled Engines," SAE Paper No. 841377, 1984.

43. R.J. Ernst, R.K. Pefley and F.J. Wiens, "Methanol Engine Durability," SAE Paper No. 831704, 1983.

44. R.M. Estefan and J.G. Brown, "Evaluation of Possible Methanol Fuel Additives for Reducing Engine Wear and Corrosion," SAE Paper No. 902153, 1990.

45. T.O. Wagner and D.F. Taterson, "Comparative Economics of Methanol and Gasoline," SAE Paper No. 872061, 1987.

46. B.D. McNutt and E.E. Eckland, "Is There a Government Role in Methanol Market Development?" SAE Paper No. 861571, 1986.

47. H.L. Walters, "Necessary Incentives for a Methanol Fuel Market," SAE Paper No. 861572, 1986.

48. J.G. Holmes, "Distribution Systems and Costs for a Regional Fuel Methanol Market," SAE Paper No. 861573, 1986.

49. U.E. Stumpf, "Brazilian Research on Ethyl Alcohol as an Automotive Fuel," Alcohols Fuels Conference, Inst. Chem. Eng., Sydney, 1978.

50. R.J. Wineland, "The Ford Flexible Fuel Vehicle," *Proc. Sixth Int. Symp. Alcohol Fuels Technology*, Ottawa, 1984.

51. W. Bernhardt, "Status Report on Volkswagen's Research and Development Efforts on Light-Duty Methanol Vehicles," Light-duty Methanol Workshop, California Air Resources Board, Los Angeles, April 1977.

52. R.A. Gorse, J.D. Benson, *et al.*, "The Effects of Methanol/ Gasoline Blends on Automobile Emissions," SAE Paper No. 920327, 1992.

53. J.E. Kirwan, "Effects of Methanol and Fuel Sulfur Concentration on Tailpipe Emissions in a Production Variable Fuel Vehicle," SAE Paper No. 932774, 1993.

54. D.L. Siebers and C.F. Edwards, "Autoignition of Methanol and Ethanol Sprays Under Diesel Engine Conditions," SAE Paper No. 870588, 1987.

55. H.O. Hardenburg, "Comparative Study of Heavy-Duty Engine Operation with Diesel Fuel and Ignition Improved Methanol," SAE Paper No. 872093, 1987.

56. Acurex Corporation, "California's Methanol Program — Evaluation Report," Energy Commission, State of California, Sacramento, Executive Summary P50086012, 1986.

57. K. Kamiyama and I. Hashimoto, "Spark Assisted Diesel for Multifuel Capability," SAE Paper No. 810072, 1981.

58. R.R. Toepel, J.E. Bennethum and R.E. Heruth, "Development of Detroit Diesel Allison 6V92Ta Methanol Fueled Coach Engine," SAE Paper No. 831744, 1983.

59. T.L. Ullman, C.T. Hare and T.M. Baines, "Emissions from Two Methanol Powered Buses," SAE Paper No. 860305, 1986.

60. T.G. Holland, M.N. Swain and M.R. Swain, "Using Ethanol/Diesel Mixtures in a Compression Ignition Engine with Ignition Improver Additives," SAE Paper No. 922191, 1992.

61. J.B. Heisey and S.S. Lestz, "Aqueous Alcohol Fumigation of a Single Cylinder DI Diesel Engine," SAE Paper No. 811208, 1981.

62. P.S. Berg, E. Holmer and B.I. Bertilsson, "The Utilization of Different Fuels in a Diesel Engine with Two Different Injection Systems," Paper delivered at Third International Symposium on Alcohol Fuels Technology, Asilomar, California, 1979.

63. American Petroleum Institute, "Storing and Handling Ethanol and Gasoline Ethanol Blends at Distribution Terminals and Service Stations," API Recommended Practice 1626, 1st ed., April 1985.

64. B. Agnetum, B.-I. Bertilsson and A. Roj, "A Life-Cycle Evaluation of Fuels for Passenger Cars," Paper No. CEC/93/EF09, Fourth International Symposium on the Performance Evaluation of Automotive Fuels and Lubricants, Birmingham, U.K., 1993.

65. American Petroleum Institute, "Background Paper on the Cost of Mandates: Transportation, Storage and Distribution," 1988.

66. J.L. Keller, *et al.*, "Methanol Fuel Modification for Highway Vehicle Use," Contract No. EY76C043683, U.S. Department of Energy, July 1978.

67. J. Panzer, "Characteristics of Primed Methanol Fuels for Passenger Cars," SAE Paper No. 831175, 1983.

68. M. Singh, "A Comparative Analysis of Alternative Fuel Infrastructure Requirements," SAE Paper No. 892065, 1989.

69. R.H. Vaivads, V.K. Rao, *et al.*, "Volatility and Flammability of Variable Fuel Tank Contents," SAE Paper No. 932776, 1993.

Appendix 1

Introduction to Fuel Chemistry

This appendix gives an introduction to those aspects of organic chemistry (i.e., the chemistry of carbon compounds) that are important to the understanding of automotive fuels.

1.1 Hydrocarbons

As the name indicates, these compounds contain carbon and hydrogen only, but there are many thousands of different possibilities depending on how the individual atoms are arranged. Carbon is quadrivalent, which means that in almost all cases it has four chemical bonds that allow it to combine with other atoms. Hydrogen has one chemical bond and so has a valency of one.

The simplest hydrocarbon is methane (CH_4), which can be written as:

Although the formula above is shown in two dimensions, in fact the carbon atom behaves as if it were in the center of a tetrahedron with carbon bonds going out to each point, so that the angle between the bonds is about 109°.

The valency or combining power of an atom is determined by the number of electrons in its outer shells. When the outer shells are full, either by sharing electrons with other atoms (giving a covalent bond, as in most organic compounds) or by having electrons donated by another atom (ionic bond), a relatively stable chemical compound results. The stability will depend on the strength of the chemical bonds and this, in turn, depends on the nature and structure of the various groupings present.

Carbon can combine with itself with single, double or triple bonds, as shown below:

$$\begin{array}{ccc}
\text{H H} & \text{H H} & \\
| \; | & | \; | & \\
\text{H-C-C-H} & \text{H-C=C-H} & \text{H-C}\equiv\text{C-H} \\
| \; | & & \\
\text{H H} & & \\
\text{ethane} & \text{ethene or ethylene} & \text{ethyne or acetylene}
\end{array}$$

and these are more normally written:

$$CH_3 \cdot CH_3 \text{ or } C_2H_6 \qquad CH_2{=}CH_2 \text{ or } C_2H_4 \qquad CH{\equiv}CH \text{ or } C_2H_2$$

Hydrocarbons with only single bonds are described as saturated whereas those with double or triple bonds are unsaturated. The unsaturated hydrocarbons, which contain a lower proportion of hydrogen, are less stable than the saturated compounds since other compounds such as oxygen can react quite readily with them across the double bond to form new materials. Although unsaturated compounds with double bonds are commonly present in automotive fuels, compounds with triple bonds are not. Hydrocarbons can also occur as ring compounds in which several carbon atoms are joined together to form a ring.

1.2 The Alkanes or Paraffins, C_nH_{2n+2}

This class of compound consists of a homologous series of saturated hydrocarbons. Methane is the simplest member, then ethane having two carbon atoms, propane with three carbon atoms, and so on. The carbon atoms can be arranged as a straight chain or as branch chain compounds, and the alternative forms of compounds having the same basic formula are known as *isomers*. Thus, there are two forms of butane, which is an alkane with four carbon atoms, as follows:

$$CH_3.CH_2.CH_2.CH_3 \qquad\qquad \begin{array}{c} CH_3.CH.CH_3 \\ | \\ CH_3 \end{array}$$

$$\text{n-butane} \qquad\qquad \text{Isobutane or 2-methylpropane}$$

The "n" in n-butane stands for normal and this means that all the carbon atoms are arranged in a straight line. Although both butanes have the same simple formula of C_4H_{10}, they are two entirely separate compounds which differ in boiling point, octane quality, and many other characteristics. The more carbon atoms there are in a compound, the more isomers are possible.

The group $CH_3\cdot$, derived from methane, is called a methyl group or radical so that it is easy to see why isobutane, which is propane ($CH_3\cdot CH_2\cdot CH_3$) with one of its hydrogen atoms on the second carbon atom (counting from the left) replaced by a methyl group, can also be called 2-methylpropane. There are no less than 18 different isomers of octane (having eight carbon atoms), two of which are n-octane and isooctane:

$$CH_3\cdot CH_2\cdot CH_2\cdot CH_2\cdot CH_2\cdot CH_2\cdot CH_2\cdot CH_3$$

$$
\begin{array}{c}
\quad CH_3 \quad\quad CH_3 \\
\quad | \quad\quad\quad | \\
CH_3\cdot C\cdot CH_2\cdot CH\cdot CH_3 \\
\quad | \\
\quad CH_3
\end{array}
$$

n-octane   isooctane or 2,2,4 trimethylpentane

The boiling point of n-octane is 126°C and its octane quality is very low — it blends as if it had an octane number of about -17. The highly branched compound isooctane, on the other hand, has a boiling point of 99°C and an octane value of 100.

Other groups or radicals that can be attached to carbon atoms are: the ethyl group, C_2H_5; the propyl group, C_3H_7; the butyl group, C_4H_9; and so on. In the case of groups with more than two carbon atoms, different arrangements of the atoms means that there are: two propyl groups, isopropyl and n-propyl; three butyl groups, n-butyl, isobutyl and tertiary butyl; and so on through the series. Such groups derived from alkanes are called alkyl groups.

1.3 The Cycloparaffins or Naphthenes, C_nH_{2n}

As the name suggests, these are cyclic compounds consisting, in their simplest form, of CH_2 groups arranged in a circle. The most stable structures, which give the minimum distortion of the carbon bond angles, have either five or six carbon atoms, i.e., cyclopentane and cyclohexane, and these form more readily than any other ring structures. The hydrogens attached to each carbon atom can be substituted by methyl or other groups.

cyclopropane   cyclobutane   cyclopentane

613

1.4 The Alkenes or Olefins, C_nH_{2n}

Although the alkenes have the same carbon-to-hydrogen ratio and the same general formula as the cycloparaffins, their behavior and characteristics are entirely different. They are straight or branch chain compounds with one or more double bond. The position of the double bond is indicated by the number of the first carbon atom to which it is attached, i.e., the compound having the formula:

$$CH_2=CH.CH_2.CH_2.CH_3$$

can be called pentene-1 or 1-pentene or even pent-1-ene.

Similarly the compound:

$$CH_3.CH=CH.CH_3$$

is called either butene-2, 2-butene or but-2-ene.

The double bond is a very reactive group so that the alkenes, or olefins as they are frequently called, can react readily with many other compounds. For this reason the olefinic compounds are easily oxidized and so tend to have a poor oxidation stability. Again, since the double bond can occur in different parts of the molecule, different compounds having the same general formula are possible.

More than one double bond can occur in a compound as, for example, butadiene:

$$CH_2=CH.CH=CH_2$$

Dienes are particularly undesirable in fuel due to their high reactivity and consequent poor resistance to oxidation. When the double bonds are arranged alternatively with single bonds, they are known as conjugated double bonds, and compounds having such structures are even more prone to oxidation and other reactions.

Olefins can be present as groups or radicals in the same way as alkyl groups, and examples are the vinyl group, $CH_2=CH-$, and the allyl group, $CH_2=CH-CH_2-$.

1.5 The Aromatic Hydrocarbons, C_nH_{2n-6}

Aromatic hydrocarbons were so called because some of these compounds have a pleasant "aromatic" odor. However, the term is now applied to a class of hydrocarbons based on a six-membered ring having three apparently conjugated double bonds. The simplest member is benzene, C_6H_6, as follows:

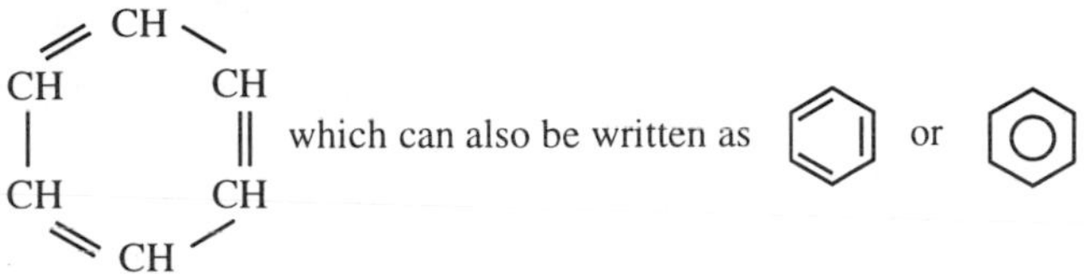

CH
CH   CH
CH   CH   which can also be written as ⬡ or ⬡
CH

The double bonds do not behave in the way that would be expected if they were a non-cyclic molecule in that the aromatic ring itself is not very reactive. This is because the double bonds behave as if they were not in fixed positions but "resonate" between the two possible structures:

⬡ ⟷ ⬡

The hydrogen atoms can be substituted by other groups, the position of which are denoted by numbering the carbon atoms or by the use of names for the relative positions, as follows:

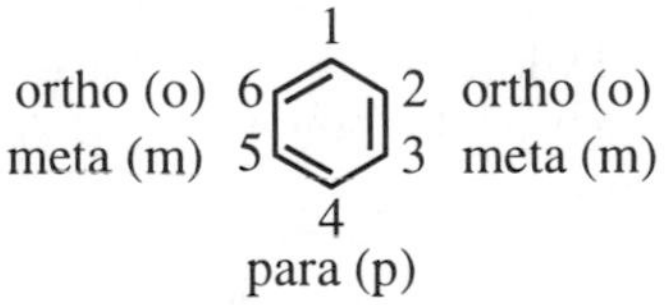

1
ortho (o) 6 ⬡ 2 ortho (o)
meta (m) 5 ⬡ 3 meta (m)
4
para (p)

The following are examples of aromatic compounds:

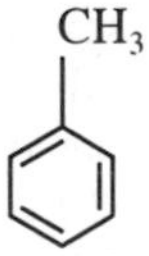

CH_3
⬡
Toluene or methylbenzene

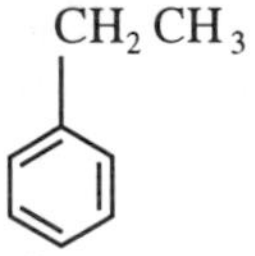

CH_2CH_3
⬡
Ethylbenzene

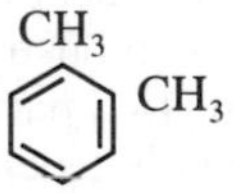

CH_3
⬡ CH_3
o-xylene or
1, 2-dimethylbenzene

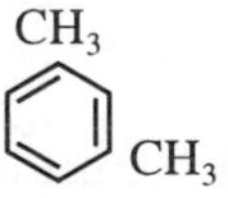

CH_3
⬡ CH_3
m-xylene or
1, 3-dimethylbenzene

CH_3
⬡
CH_3
p-xylene or
1, 4-dimethylbenzene

The aromatic rings can be fused together to give polynuclear aromatics such as naphthalene and anthracene:

naphthalene

anthracene

or much more complex compounds:

2, 3, 6,-trimethylnaphthalene

benzo (α) pyrene

Benzo(α)pyrene is of concern because it can be present in small amounts in exhaust gases and it is highly carcinogenic. These polynuclear aromatics (PNAs) are also sometimes called polycyclic aromatic hydrocarbons (PCAs or PAHs).

1.6 Combustion of Hydrocarbons

Under ideal conditions all hydrocarbons, no matter what type, combust when ignited in the presence of oxygen to form water and carbon dioxide, as illustrated below for n-heptane:

$$C_7H_{16} + 11O_2 \rightarrow 7CO_2 + 8H_2O$$

This equation says that n-heptane reacts with oxygen in the proportion of one molecule of heptane to eleven molecules of oxygen to form seven molecules of carbon dioxide and 8 molecules of water. Carbon has a molecular weight of 12, hydrogen of 1 and oxygen of 16 so that it can be calculated that 100 grams of n-heptane reacts with 352 grams of oxygen to give 308 grams of carbon dioxide and 144 grams of water. This ratio of oxygen to hydrocarbon to give theoretically complete combustion is known as the stoichiometric ratio.

In practice, air is used instead of oxygen so that an oxygen-fuel ratio of 3.52:1 translates to 15:1 for air on a weight basis for the combustion of n-heptane. For gasoline, one part by weight requires about 14.5 parts of air by weight, although the exact stoichiometric amount of air will depend on the precise composition of the fuel.

Using air instead of pure oxygen allows some of the oxygen to combine with nitrogen at the temperatures reached during fuel combustion, so that small amounts of oxides of nitrogen are formed. There are three common oxides of nitrogen and a mixture of them is usually formed so that they are not given a precise formula, but simply NO_x since the relative proportions of each will depend on the combustion conditions. Sometimes there is less than the stoichiometric amount of air present and, in this case, a mixture of carbon monoxide (CO) and carbon dioxide is formed as well as water. If there is an excess of air, the amount of carbon monoxide formed will be very low but the oxides of nitrogen may be high (see Chapter 6). In practice, because the air-fuel mixture is not entirely homogeneous and because the metal of the cylinder head quenches the flame near it, combustion does not proceed exactly as indicated by the theoretical equation. In addition, there are various impurities present in the fuel in small amounts and some of them, such as the polynuclear aromatics, are quite difficult to combust, and a percentage of them will survive the combustion process. Finally, partially combusted materials can undergo reactions after the combustion chamber to form compounds that were not present in or before the combustion chamber.

2.0 Monohydric Alcohols, $C_nH_{2n+1}OH$

These compounds are used as automotive fuels and fuel components, and include methanol (methyl alcohol), ethanol (ethyl alcohol), propanol (propyl alcohol) and butanol (butyl alcohol), as the compounds of most interest in automotive fuels. The OH group, which replaces one of the hydrogen atoms in an alkane, gives these compounds their characteristic properties such as solubility in water, which gets progressively less as the number of carbon atoms increases. Their use as automotive fuels is described in detail in Chapters 11, 12 and 20.

Methanol (CH_3OH, sometimes written as MeOH) is the simplest of the monohydric (i.e., having one OH group) alcohols.

Ethanol (C_2H_5OH, sometimes written as EtOH) is the next in the series, followed by propyl alcohol (C_3H_7OH) of which there are two isomers:

$$CH_3.CH_2.CH_2.OH \qquad \text{and} \qquad$$

$$\begin{array}{l} CH_3 \\ | \\ CH.OH \\ | \\ CH_3 \end{array}$$

n-propyl alcohol	isopropyl alcohol
or propan-1-ol	or propan-2-ol
or 1-propanol	or 2-propanol

There are three different butyl alcohols:

$$CH_3.CH_2.CH_2.CH_2.OH$$

n-butyl
alcohol
or 1-butanol

$$CH_3.CH_2.CH.OH$$
$$|$$
$$CH_3$$

secondary butyl
alcohol
or 2-butanol

$$CH_3$$
$$|$$
$$CH_3.C.OH$$
$$|$$
$$CH_3$$

tertiary butyl
alcohol
or 2-methyl-2-propanol

3.0 Alkyl Ethers, $(C_nH_{2n+1})_2.O$

The alkyl ethers are isomeric with the monohydric alcohols, but contain oxygen linked to two alkyl groups instead of to an alkyl group and a hydrogen atom. The two alkyl groups can be the same or different, i.e.:

$$CH_3.O.CH_3 \text{ is dimethyl ether}$$

$$CH_3.CH_2.O.CH_3 \text{ is ethyl methyl ether}$$

The above simple ethers are too volatile to be readily used in automotive fuels and the ethers of most interest are:

$$CH_3.O.C(CH_3)_3 \qquad \text{Methyl tertiary butyl ether (MTBE)}$$

$$(CH_3)_2C.O.CH_3$$
$$|$$
$$C_2H_5$$
$$\qquad \text{Tertiary amyl methyl ether (TAME)}$$

$$C_2H_5.O.C(CH_3)_3 \qquad \text{Ethyl tertiary butyl ether (ETBE)}$$

4.0 Combustion of Oxygenates

Oxygenates require proportionately less oxygen for complete combustion as illustrated by the following equations for the combustion of methanol and MTBE:

$$2CH_3OH + 3O_2 \rightarrow 2CO_2 + 4H_2O$$

i.e., 64g of methanol require 96g of oxygen for complete combustion of methanol and so 100g requires 150g of oxygen.

$$2CH_3OC(CH_3)_3 + 15O_2 \rightarrow 10CO_2 + 12H_2O$$

i.e., 176g MTBE requires 480g of oxygen and so 100g would require 273g oxygen.

For the hydrocarbon heptane, which is similar to gasoline in terms of stoichiometric air-fuel ratio, 100g requires 352g oxygen, as outlined in Section 1.6 above, which is 2.3 times as much as for methanol and 1.3 times as much as for MTBE. Most other oxygenates used in gasoline lie between these two figures. It can be seen that if methanol or any other oxygenate is mixed with gasoline and used in an engine calibrated for hydrocarbon fuels, there will be an excess of oxygen present which will reduce the formation of CO and unburnt hydrocarbons in the exhaust gases provided that the mixture is within the flammable range.

Appendix 2

Physical Properties of Hydrocarbons

Physical Constants of Hydrocarbons

	Formula	Molec. Wt	Boiling Point °F	Melting Point °F	°API	Sp Gr 60°/60°	lb/gal	Gross	Net
NORMAL PARAFFINS C_nH_{2n+2}									
Methane	CH_4	16.0	-258.9	-296.5	340	0.30	2.50	23,860[1]	21,500[1]
Ethane	C_2H_6	30.1	-128.0	-297.8	247	.374	3.11	22,300[1]	20,420[1]
Propane	C_3H_8	44.1	-43.8	-305.7	147	.508	4.23	21,650[1]	19,930[1]
Butane	C_4H_{10}	58.1	+31.1	-216.9	111	.584	4.86	21,290[1]	19,670[1]
Pentane	C_5H_{12}	72.1	96.9	-201.5	92.7	.631	5.25	21,070[1]	19,500[1]
Hexane	C_6H_{14}	86.2	155.7	-139.5	81.6	.664	5.53	20,780	19,240
Heptane	C_7H_{16}	100.2	209.2	-131.1	74.2	.688	5.73	20,670	19,160
Octane	C_8H_{18}	114.2	258 2	-70.3	68.6	.707	5.89	20,590	19,100
Nonane	C_9H_{20}	128.2	303.4	-64.5	64.5	.722	6.01	20,530	19,050
Decane	$C_{10}H_{22}$	142.3	345.2	-21.5	61.3	.734	6.11	20,480	19,020
Undecane	$C_{11}H_{24}$	156.3	384.4	-14.1	58.7	.744	6.19	20,450	19,000
Dodecane	$C_{12}H_{26}$	170.3	421.3	+14.7	56.4	.753	6.27	20,420	18,980
ISO-PARAFFINS C_nH_{2n+2}									
Isobutane	C_4H_{10}	58.1	10.9	-255.0	120	.563	4.69	21,240[1]	19,610[1]
2-Methylbutane (Isopentane)	C_5H_{12}	72.1	82.2	-255.5	94.9	.625	5.20	21,030[1]	19,450[1]
2,2-Dimethylpropane (Neopentane)	C_5H_{12}	72.1	49.0	+2.1	105	.597	4.97	20,960[1]	19,330[1]
2-Methylpentane (Isohexane)	C_6H_{14}	86.2	140.5	-245	83.5	.658	5.48	20,750	19,210
3-Methylpentane	C_6H_{14}	86.2	145.9	-180	80.0	.669	5.57	20,760	19,220
2,2-Dimethylbutane (Neohexane)	C_6H_{14}	86.2	121.5	-147.6	84.9	.654	5.44	20,700	19,160
2,3-Dimethylbutane (Diisopropyl)	C_6H_{14}	86.2	136.4	-198.8	81.0	.666	5.54	20,740	19,200
2-Methylhexane (Isoheptane)	C_7H_{16}	100.2	194.1	-180.8	75.7	.683	5.68	20,650	19,140
3-Methylhexane	C_7H_{16}	100.2	197.5	-182.9	73.0	.692	5.76	20,660	19,150
3-Ethylpentane	C_7H_{16}	100.2	200.2	-181.5	69.8	.703	5.85	20,670	19,160
2,2-Dimethylpentane	C_7H_{16}	100.2	174.6	-190.8	77.2	.678	5.64	20,600	19,090
2,3-Dimethylpentane	C_7H_{16}	100.2	193.6		70.6	.700	5.83	20,640	19,130
2,4-Dimethylpentane	C_7H_{16}	100.2	176.9	-183.1	77.2	.678	5.64	20,620	19,110
3,3-Dimethylpentane	C_7H_{16}	100.2	186.9	-211.0	71.2	.698	5.81	20,620	19,110
2,2,3-Trimethylbutane (Triptane)	C_7H_{16}	100.2	177.6	-13.0	72.1	0.695	5.78	20,620	19,110
2-Methylheptane (Isooctane)	C_8H_{18}	114.2	243.8	-165.1	70.1	.702	5.84	20,570	19,080
3-Ethylhexane	C_8H_{18}	114.2	245.4	-	65.6	.718	5.98	20,570	19,080
2,5-Dimethylhexane (Diisobutyl)	C_8H_{18}	114.2	228.4	-130	71.2	.698	5.81	20,550	19,060
2,2,4-Trimethylpentane ("Isooctane")	C_8H_{18}	114.2	210.6	-161.2	71.8	.696	5.79	20,540	19,050
OLEFINS C_nH_{2n}									
Ethylene	C_2H_4	28.0	-154.7	-272.5	273	.35	2.91	21,640[1]	20,290[1]
Propylene	C_3H_6	42.1	-53.9	-301.4	140	.522	4.35	21,040[1]	19,690[1]
Butene-1	C_4H_8	56.1	20.7	-	104	.601	5.00	20,840[1]	19,490[1]
Cis-Butene-2	C_4H_8	56.1	38.6	-218.0	94.2	.627	5.22	20,780[1]	19,430[1]
Trans-Butene-2	C_4H_8	56.1	33.6	-157.7	100	.610	5.08	20,750[1]	19,400[1]
Isobutene (Isobutylene)	C_4H_8	56.1	19.6	-220.5	104	.600	4.99	20,720[1]	19,370[1]
Pentene-1 (Amylene)	C_5H_{10}	70.1	86.2	-216.4	87.2	.647	5.38	20,710[1]	19,360[1]
Cis-Pentene-2	C_5H_{10}	70.1	86.2	-290.2	87.2	.661	5.50	20,660[1]	19,310[1]
Trans-Pentene-2	C_5H_{10}	70.1	98.6	-211.0	84.9	.654	5.44	20,640[1]	19,290[1]
2-Methylbutene-1	C_5H_{10}	70.1	88.0	-	84.5	.655	5.45	20,610[1]	19,260[1]
3-Methylbutene-1 (Isoamylene)	C_5H_{10}	70.1	68.4	-292.0	92.0	.633	5.27	20,660[1]	19,310[1]
2-Methylbutene-2	C_5H_{10}	70.1	101.2	-207.0	80.6	.667	5.55	20,570[1]	19,220[1]
Hexene-1	C_6H_{12}	84.2	146.4	-218.0	77.2	.678	5.64	20,450	19,100
Cis-Hexene-2	C_6H_{12}	84.2	155.4	-231.0	73.9	.689	5.73	20,420	19,070
Trans-Hexene-2	C_6H_{12}	84.2	154.2	-207.0	75.7	.683	5.68	20,400	19,050
Cis-Hexene-3	C_6H_{12}	84.2	153.7	-211.0	75.4	.684	5.69	20,420	19,070
Trans-Hexene-3	C_5H_{12}	84.2	154.6	-171	76.0	.682	5.68	20,400	19,050
DIOLEFINS C_nH_{2n-2}									
Propadiene	C_3H_4	40.1	-30.1	-213.0	106	.595	4.95	20,880[1]	19,930
Butadiene-1,2	C_4H_6	54.1	+50.5	-	83.5	.658	5.48	-	-
Butadiene-1,3	C_4H_6	54.1	24.1	-164.0	94.2	.627	5.22	20,230[1]	19,180[1]
Pentadiene-1,2	C_5H_8	68.1	112.8	-85.0	71.5	.697	5.80	-	-
Cis-Pentadiene-1,3	C_5H_8	68.1	111.6	-	71.8	.696	5.79	20,150[1]	19,040[1]
Trans-Pentadiene-1,3	C_5H_8	68.1	108.1	-	76.0	.682	5.68	20,150[1]	19,040[1]
Pentadiene-1,4	C_5H_8	68.1	78.9	-234.0	81.3	.665	5.53	20,320[1]	19,210[1]
3-Methylbutadiene-1,2	C_5H_8	68.1	104	-184.0	82.9	.685	5.70	-	-
2-Methylbutadiene-1,3 (Isoprene)	C_5H_8	68.1	93.3	-231.0	74.8	.686	5.71	20,060[1]	18,950[1]

[1] Heat of combustion as a gas—otherwise as a liquid. *Mixture of cis- and trans-isomers. **Sublimes

Physical Constants of Hydrocarbons ... continued

	Formula	Molec. wt	Boiling Point °F	Melting Point °F	°API	Density Sp Gr 60°/60°	lb/gal	Heat of Combustion at 60°F – Btu/lb Gross	Net
DIOLEFINS *(Cont.)*									
Hexadiene-1,2	C₆H₁₀	82.1	172	–	64.5	0.722	6.01	–	–
Hexadiene-1,3*	C₆H₁₀	82.1	163	–	67.8	.710	5.91	–	–
Hexadiene-1,4*	C₆H₁₀	82.1	149	–	70.6	.700	5.83	–	–
Hexadiene-1,5	C₆H₁₀	82.1	139.3	–221.4	71.8	.696	5.79	20,130	18,980
Hexadiene-2,3	C₆H₁₀	82.1	154.4	–	75.1	.685	5.70	–	–
Hexadiene-2,4*	C₆H₁₀	82.1	176	–	63.7	.725	6.03	–	–
3-Methylpentadiene-1,2	C₆H₁₀	82.1	158	–	65.0	.720	5.99	–	–
4-Methylpentadiene-1,2	C₆H₁₀	82.1	158.0	–	67.0	.713	5.93	–	–
2-Methylpentadiene-1,3*	C₆H₁₀	82.1	169	–	63.9	.724	6.03	–	–
3-Methylpentadiene-1,3*	C₆H₁₀	82.1	171	–	59.7	.740	6.16	–	–
4-Methylpentadiene-1,3	C₆H₁₀	82.1	169.3	– 94.0	63.9	.724	6.03	–	–
2-Methylpentadiene-1,4	C₆H₁₀	82.1	133	–	70.9	.699	5.82	–	–
2-Methylpentadiene-2,3	C₆H₁₀	82.1	162.0	–	66.1	.716	5.96	–	–
2,3-Dimethylbutadiene-1,3	C₆H₁₀	82.1	155.7	–105	62.1	.731	6.08	19,880	18,730
2-Ethylbutadiene-1,3	C₆H₁₀	82.1	167	–	61.0	.735	6.12	–	–
ACETYLENES CnH₂ₙ₋₂									
Acetylene	C₂H₂	26.0	–119**	–114	209	.416	3.46	21,470¹	20,740¹
Methylacetylene	C₃H₄	40.1	– 9.8	–153	94.9	.625	5.20	20,810¹	19,860¹
Butyne-1 (Ethylacetylene)	C₄H₆	54.1	+ 47.7	–188.5	86.2	.650	5.41	20,650¹	19,600¹
Butyne-2 (Dimethylacetylene)	C₄H₆	54.1	80.4	– 26.0	71.2	.698	5.81	20,510¹	19,460¹
Pentyne-1 (Propylacetylene)	C₅H₈	68.1	104.4	–159	71.8	.696	5.79	20,550¹	19,440¹
Pentyne-2	C₅H₈	68.1	132.8	–148	66.1	.716	5.96	20,450¹	19,340¹
3-Methylbutyne-1 (Isopropylacetylene)	C₅H₈	68.1	82	–	79.7	.670	5.58	20,500¹	19,390¹
Hexyne-1 (Butylacetylene)	C₆H₁₀	82.1	160.9	–205.6	65.0	.720	5.99	–	–
Hexyne-2	C₆H₁₀	82.1	184.1	–126.4	60.8	.736	6.13	–	–
Hexyne-3	C₆H₁₀	82.1	179.2	–149.8	63.1	.727	6.05	–	–
4-Methylpentyne-1	C₆H₁₀	82.1	142.1	–157.1	67.5	.711	5.92	–	–
4-Methylpentyne-2	C₆H₁₀	82.1	162	–	65.3	.719	5.98	–	–
3,3-Dimethylbutyne-1	C₆H₁₀	82.1	100.0	–114.2	78.7	.673	5.60	–	–
OLEFINS-ACETYLENES CnH₂ₙ₋₄									
Buten-3-yne-1 (Vinylacetylene)	C₄H₄	52.1	42	–	73.9	.689	5.73	–	–
Penten-1-yne-3	C₅H₆	66.1	138.6	–	58.7	0.744	6.19	–	–
Penten-1-yne-4 (Allylacetylene)	C₅H₆	66.1	107	–	49.4	.782	6.51	–	–
2-Methylbuten-1-yne-3	C₅H₆	66.1	90	–	–	–	–	–	–
Hexen-1-yne-3	C₆H₈	80.1	185	–	56.4	.753	6.27	–	–
Hexen-1-yne-5	C₆H₈	80.1	158	–	32.8	.861	7.17	–	–
2-Methylpenten-1-yne-3	C₆H₈	80.1	169	–	–	–	–	–	–
3-Methylpenten-3-yne-1*	C₆H₈	80.1	156	–	–	–	–	–	–
AROMATICS CnH₂ₙ₋₆									
Benzene	C₆H₆	78.1	176.2	41.9	28.6	.884	7.36	17,990	17,270
Toluene	C₇H₈	92.1	231.1	–139.0	30.8	.872	7.26	18,270	17,450
o-Xylene	C₈H₁₀	106.2	292.0	– 13.3	28.4	.885	7.37	18,500	17,610
m-Xylene	C₈H₁₀	106.2	282.4	– 54.2	31.3	.869	7.24	18,500	17,610
p-Xylene	C₈H₁₀	106.2	281.0	+ 55.9	31.9	.866	7.21	18,430	17,540
Ethylbenzene	C₈H₁₀	106.2	277.1	–138.9	30.8	.872	7.26	18,490	17,600
1,2,3-Trimethylbenzene	C₉H₁₂	120.2	349.0	– 13.8	25.7	.900	7.49	–	–
1,2,4-Trimethylbenzene (Pseudocumene)	C₉H₁₂	120.2	336.5	– 47.3	29.1	.881	7.34	18,570	17,620
1,3,5-Trimethylbenzene (Mesitylene)	C₉H₁₂	120.2	328.3	– 48.6	31.1	.870	7.24	18,620	17,670
Propylbenzene	C₉H₁₂	120.2	318.6	–147.1	31.9	.866	7.21	18,660	17,710
Isopropylbenzene (Cumene)	C₉H₁₂	120.2	306.3	–140.8	31.9	.866	7.21	18,670	17,720
1-Methyl-2-Ethylbenzene	C₉H₁₂	120.2	329.2	–126.6	28.7	.883	7.35	–	–
1-Methyl-3-Ethylbenzene	C₉H₁₂	120.2	322.7	–	31.1	.870	7.24	–	–
1-Methyl-4-Ethylbenzene	C₉H₁₂	120.2	324.5	– 82.7	31.5	.868	7.23	–	–
CYCLOPARAFFINS CnH₂ₙ									
Cyclopropane	C₃H₆	42.1	– 27.0	–196.6	98.6	.615	5.12	–	–
Cyclobutane	C₄H₈	56.1	+ 54.7	– 58.0	74.8	.686	5.71	–	–
Cyclopentane	C₅H₁₀	70.1	120.7	–136.7	56.9	.751	6.25	20,350¹	19,000
Methylcyclopentane	C₆H₁₂	84.2	161.3	–224.4	56.2	.754	6.28	20,110	18,760
1,1-Dimethylcyclopentane	C₇H₁₄	98.2	189.5	–105	54.7	.760	6.33	–	–
1,2-Dimethylcyclopentane-cis.	C₇H₁₄	98.2	210.7	– 62	50.4	.778	6.48	20,020	18,670
1,2-Dimethylcyclopentane-trans.	C₇H₁₄	98.2	197.4	–182	55.4	.757	6.30	20,020	18,670
1,3-Dimethylcyclopentane-trans.	C₇H₁₄	98.2	195.4	–213	57.2	.750	6.24	–	–
Ethylcyclopentane	C₇H₁₄	98.2	218.2	–217	52.0	.771	6.42	20,110	18,760
Cyclohexane	C₆H₁₂	84.2	177.3	+ 44	49.0	.784	6.53	20,030	18,680
Methylcyclohexane	C₇H₁₄	98.2	213.6	–195.6	51.3	.774	6.44	20,000	18,650

¹Heat of combustion as a gas – otherwise as a liquid.   *Mixture of cis- and trans- isomers.   **Sublimes.

Source: Exxon

Appendix 3

Worldwide Survey of Emissions Legislation

Reproduced from Report 4/94, Section 3, by kind permission of:

CONCAWE, Brussels, Belgium

3 Worldwide Survey of Emissions Legislation

3.1 ECE/EU Vehicle Emission Regulations

Emissions regulations in Europe have in the past been formulated primarily by the United Nations Economic Commission for Europe (UN-ECE) through its technical advisory body GRPE.

The ECE is supported by most European nations, EU and non-EU, plus some Eastern European countries. Its role is to produce model standards which may be adopted by member nations, but it has no power to enforce compliance. Generally, standards promulgated by the EU are reissued by ECE and vice-versa. The main ECE emissions regulations are:

- ECE 15.04   emissions from light duty vehicles up to Directive 83/351/EEC
- ECE 83   equivalent to EU Directive for light duty vehicles (88/76/EEC)
- ECE 49.01   heavy duty vehicle emissions
- ECE 40.01   motorcycle emissions
- ECE 47   moped emissions

In the past, the European Union generally tended to adopt regulations which were technically identical with the ECE requirements. This position has changed over the years, with the European Community, now the European Union, gradually assuming a major role in formulating automotive emissions standards. GRPE is now unlikely to adopt any proposal which has not been previously discussed by MVEG and agreed within the EU.

European Union regulations, published as Directives, have the force of law within EU Member States under the provisions of the Treaty of Rome. EU countries may not prohibit marketing of vehicles which comply with the provisions of the directive, but may prohibit vehicles which do not comply. See also the notes in **Section 2.1.1** regarding the mandatory nature of the *"Consolidated Emissions Directive."*

Present membership of the EU is:

Belgium	Greece	Netherlands
Denmark	Ireland	Portugal
France	Italy	Spain
Germany	Luxembourg	United Kingdom

Austria, Finland, Norway and Sweden have applied to join the EU. A four year transitional period has been agreed for environmental legislation, after which limits must either be harmonised to EU or further renegotiated.

Up to now, a number of other countries, including Finland, Hong Kong, Israel and Singapore, have accepted vehicles which meet the ECE regulations, or require mandatory compliance up to ECE 15. However Austria, Sweden, Norway and Switzerland revoked their agreement to accord with ECE regulations and adopted US standards for passenger cars. Denmark, a member state of the EU, decided to do so in 1990 (but see the notes following **Table 3.3**). Finland adopted 1983 US standards for new models from 01.01.90 and for all new cars from 01.01.1992.

3.1.1 Light Duty Vehicle Emissions

Light duty vehicles (passenger cars) were the first to be regulated under the ECE process, and their limit values have subsequently been amended four times. Equivalent EU directives are:

		Implementation Date
Original ECE 15	70/220/EEC	1970
ECE 15/01	74/290/EEC	1974
ECE 15/02	77/102/EEC	1977
ECE 15/03	78/665/EEC	October 1979
ECE 15/04	83/351/EEC	October 1984/86
New ECE 83	88/76/EEC	See **Table 3.2**

ECE 15 exhaust emission regulations up to the 04 amendment are summarized in **Table 3.1**:

Table 3.1 European Exhaust Emission Limits - ECE Regulation 15 (g/test)

Vehicle Type	CO g/test			Unburnt HC g/test		NOx g/test		HC+NOx g/test
ECE 15 Level	**02**	**03**	**04**	**02**	**03**	**02**	**03**	**04**
Type I Test Reference Weight (kg)								
ò750	80	65	58	6.8	6.0	10.0	8.5	19.0
751 - 850	87	71	58	7.1	6.3	10.0	8.5	19.0
851 - 1020	94	76	58	7.4	6.5	10.0	8.5	19.0
1021 - 1250	107	87	67	8.0	7.1	12.0	10.2	20.5
1251 - 1470	122	99	76	8.6	7.6	14.0	11.9	22.0
1471 - 1700	135	110	86	9.2	8.1	14.5	12.3	23.5
1701 - 1930	149	121	93	9.7	8.6	15.0	12.8	25.0
1931 - 2150	162	132	101	10.3	9.1	15.5	13.2	26.5
ò2150	176	143	110	10.9	9.6	16.0	13.6	28.0

All vehicles (Type II Test)	Maximum concentration of CO at end of last urban cycle: 02 levels - 4.5%; 03 and 04 levels - 3.5%

All vehicles (Type III Test)	No crankcase emissions permitted

*NO$_2$ Equivalent

Notes:
1. Regulation 15 applies to vehicles up to 3.5t GVW. Only gasoline-fuelled vehicles are covered by 01/02/03 Amendments, but the 04 Amendment also applies to diesel-powered vehicles.
2. The constant volume sampling (CVS) measurement technique was introduced with the 04 Amendment. Fuel consumption and power measurement procedures are detailed in the Regulation, but do not include any limits.
3. The 03 Amendment came into force on 1 October 1979 and the 04 Amendment on 1.10.84 for new models, 1.10.86 for existing models. Mandated introduction dates in individual countries vary and may be later than these dates.
4. The limits quoted are those for Type Approval. Production vehicles are permitted to exceed these figures by up to 30% for HC, and up to 20% for CO and NOx. The tolerance for HC+NOx in the 04 Amendment is 25%.

The EU agreed further reductions from the 04 levels in gaseous exhaust emissions limits for vehicles less than 3.5 t GVW. Particulate limits for diesel vehicles were also agreed. This Directive, 88/76/EEC, was then adopted by ECE as Regulation ECE 83. The new Directive 88/76/EEC, which amends Directive 70/220/EEC, allows the certification of cars with an engine displacement above 1.4 litres on the basis of the 1983 US procedure and limits.

In practical terms, this regulation was not implemented by any European country in anticipation of the adoption by EU of the Consolidated Emissions Directive (see overleaf).

Table 3.2 European Exhaust Emission Limits - ECE Regulation 83

Cubic capacity, (cm³)	Emission (g/test)				Effective Date	
	CO[e]	HC+NOx[e]	NOx[e]	Pm[e]	New Models	All Production
Gasoline vehicles						
>2000	25	6.5	3.5		1.10.88	1.10.89
1400 to 2000	30	8	-		1.10.91	1.10.93
<1400	45[a]	15[a]	6[a]		1.10.90	1.10.91
	30[b]	8[b]	-		1.10.92	1.10.93
Diesel vehicles						
>2000	30	8	-	1.1[d]	1.10.88	1.10.89
1400 to 2000	30	8	-	1.1[d]	1.10.91[c]	1.10.93[c]
<1400	45[a]	15[a]	6[a]	1.1[d]	1.10.90[c]	1.10.91[c]
	30[b]	8[b]	-	1.1[d]	1.01.91	

a) These limits are interim values.
b) Limits adopted by the EU Council in November 1988. A further reduction to 19 g/test CO and 5 g/test HC + NOx was adopted by the EU Council in June 1989 as 89/458/EEC. However, it was superseded by the Consolidated Directive and is not included in ECE 83.
c) Proposed implementation dates for direct injection-engined vehicles are 1.10.94 for new models and 1.10.96 for all production.
d) A separate EU Directive, 88/436/EEC, specifies the limits for particulates (Pm) as 1.1 g/test for Type Approval and 1.4 g/test for conformity of production, with implementation dates of 1.10.1989 for new models and 1.10.1990 for all production.
e) Measurements are made over the ECE 15 driving cycle. Member States must accept on the market all vehicles meeting the above limits.

On 26th June 1991 the Council of Ministers of the European Community adopted the "*Consolidated Emissions Directive*", 91/441/EEC. This covers not only exhaust emission standards (including durability testing) but also limits for vehicle evaporative emissions (see **Section 4.4**).

According to the Directive, exhaust emission standards have to be certified on the basis of the new combined ECE-15 (urban) cycle and EUDC (extra-urban) test cycle (see **Section 4.1**). In contrast to previous Directives a common set of gaseous emission standards will apply to all private passenger cars (both gasoline and diesel-engined) irrespective of engine capacity. Limit values are shown in **Table 3.3**.

Subsequently in December 1993 the Environment Council agreed more stringent limits for 1996 onwards and these were adopted as Directive 94/12/EC in March 1994. Compared with current standards, separate limits

Table 3.3 Limits for the Consolidated Emissions Directive (91/441/EEC) and (94/12/EC) - Light Duty Vehicles less than 3.5 tonnes GVW. (Test Method ECE 15+EUDC)

Vehicle Type	Effective Date	Type Approval				Conformity of Production			
		CO g/km	HC+NOx g/km	Pm g/km	Evap. g/test	CO g/km	HC+NOx g/km	Pm g/km	Evap. g/test
all LDV except DI diesels	01.07.92 (new models) 31.12.92 (all models)	2.72	0.97	0.14	2.0	3.16	1.13	0.18	2.0
DI diesels	current 01.07.94	2.72 2.72	.36 .97	0.19 0.14		3.16	1.13	0.18	
gasoline	01.01.96 (new models)	2.2	0.5	-	2.0				
IDI diesels	01.01.97 (all models)	1.0	0.7	0.08		all production must meet Type Approval limits			
DI diesels		1.0	0.9	0.1					
DI diesels	1.10.99	1.0	0.7	0.08					

Note:
1. From 01.07.92, Member States can no longer grant Type Approval against previous Directives.
2. For vehicles fitted with pollution control devices an additional durability test is required. This represents an ageing test of 80 000 km.

are given for gasoline- and diesel-fuelled vehicles. These represent respectively reductions of 30% CO, 55% HC+NOx for gasoline cars and 68% CO, 38% HC+NOx and 55% particulate emissions for diesel vehicles. Implementation dates are 1 January 1996 for new models and 1 January 1997 for existing models. Slightly less stringent limits apply to DI diesels initially but they have to comply with the full standard by 30 September 1999. Contrary to the earlier standards, production vehicles must comply with the Type Approval Limits. The revised limits are shown in **Table 3.3**.

The Commission is due to submit proposals by the end of 1994 for further reductions in exhaust emissions to apply from the year 2000. A "Tri-Partite" group was set up comprising representatives of the European Commission, motor industry (ACEA) and oil industry (EUROPIA), to develop a rational basis to set future emission limits and fuel specifications. This includes study of current air quality and modelling trends to identify future requirements.

A review of existing data on the effects of fuel properties on emissions has led to the setting up of a test programme (EPEFE) to provide further information on future European vehicles and fuels. This programme should be completed by the end of 1994, to allow proposals for future legislation to be developed in 1995.

It should be noted that the implementation of the limit values by EU Member States is mandatory and no longer (as in previous exhaust emissions Directives) left for the decision of individual national governments. GRPE is considering these limit values with a view to recommending that they are adopted by UN-ECE.

Denmark

Denmark introduced emissions standards equivalent to US 1987 limits from 1 October 1990. However, all vehicles meeting the limit values of the EU *"Consolidated Directive"* are permitted in Denmark. The Danish EPA also drafted an action plan calling for the future adoption of Californian standards and suggested that the EU adopt these norms from 1996.

3.1.2 Light Commercial Vehicles

Light Commercial Vehicles (Maximum Mass 3500 kg) and Heavier Cars Designed to transport more than 6 People.

These vehicles (Classes M and N; M_1 and N_1) are the subject of Directive 93/59/EEC, dated 28 June 1993. In this proposal the vehicles have been further classified according to their mass, to reflect the differences in their power train layouts and body shapes:

Class I: Reference mass equal or less than 1250 kg
Class II: Reference mass more than 1250 kg but less than 1700 kg
Class III: Reference mass more than 1700 kg

The Directive is intended to ensure the following performance:

- *Class I* can comply with the limit values established by Directive 91/441/EEC for passenger cars, i.e. considered to be as least as severe as current US limits.

- *Classes II* and *III* should be considered to be of equivalent stringency to the present US *"Light Duty Truck"* standards.

Vehicles will be tested over the ECE15 + EUDC test cycle. However, as many vehicles in these classes have low power-to-mass ratios and low maximum speeds, the following modification to the procedure is proposed:

- For vehicles with a power-to-mass ratio of not more than 40 kW/t and a maximum speed of less than 130 km/h, the EUDC maximum speed will be reduced to 90 km/h.

The requirements relating to evaporative emissions and durability of anti-pollution devices specified in Directive 91/441/EEC will also apply.

Table 3.4 EU Emission Limits for Light Commercial Vehicles (93/59/EEC)

Vehicle	Type Approval			Conformity of Production		
	CO g/km	HC+NOx g/km	Pm* g/km	CO g/km	HC+NOx g/km	Pm*
I RM ≤ 1250 kg	2.72	0.97	0.14	3.16	1.13	0.18
II RM > 1250 kg ≤ 1700 kg	5.17	1.40	0.19	6.0	1.6	0.22
III RM ≥ 1700 kg	6.9	1.7	0.25	8.0	2.0	0.29

*Applies to diesels only.

Effective dates: 01.10.93   -   New models
                          01.10.94   -   All models.

3.1.3 Heavy Duty Vehicle Emissions

ECE Regulation 49 Heavy Duty Engine Gaseous Emission Limits

ECE Regulation 49 applies to gaseous emissions from diesel engines used in vehicles with GVW over 3.5 t. Limits are in g/kWh determined by using an engine test procedure based on the former US 13 mode test (see **Section 4.7**). **Table 3.5** shows both the original ECE 49 limits, and also those adopted in September 1989 as amendment 01 (these are identical with EU Directive 88/77 - see below).

Table 3.5 ECE Regulation 49/49.01 Emission Limits

	CO (g/kWh)	HC (g/kWh)	NOx (g/kWh)
ECE 49	14	3.5	18
ECE 49.01	11.2	2.4	14.4

With the introduction of ECE 15/04, diesel engined vehicles under 3.5t GVW must comply with ECE 15 limits.

ECE Regulation 24.03 Diesel Black Smoke Emissions

The current ECE Regulation 24.03 governing black smoke emissions from diesel engines is given in **Table 3.6**. These were adopted in EU Directive 72/306/EEC but will be superseded by the particulate emissions requirements in the EU "Clean Lorry" Directive 91/542/EEC (see **Table 3.7**).

EU Regulations for Heavy Duty Diesels

The European Council of Ministers, through Directive 88/77/EEC, adopted Type Approval limits for gaseous emissions from vehicles over 3.5 tonnes based on ECE regulation 24.03. This was permissive but not mandatory and the suggested application dates were from 1 April 1988 for new models and from 1 October 1990 for all production.

Table 3.6 Smoke Limits Specified in ECE Regulation 24.03 and EU Directive 72/306/EEC

(a)   Smoke emission limits under steady state conditions.

Nominal Flow (litres/second)	Absorption Coefficient (m^{-1})
42	2.26
100	1.495
200	1.065

Intermediate values are also specified

(b)   Opacity under free acceleration should not exceed the approved level by more than 0.5 m^{-1}.

Note: Although the free acceleration test was intended as a means of checking vehicles in service it has not proved entirely successful. A number of different methods have been proposed by various countries, but there is no generally accepted alternative method of in-service checking.

On 1 October 1991 the European Council adopted the "Clean Lorry"
Directive which reduces in two phases the limit values for gaseous and
particulate (Pm) emissions for diesel engines and other heavy utility vehicles.
This sets norms, which will be compulsory throughout the EU in two stages,
as shown in **Table 3.7**.

Table 3.7 EU Limits for Heavy Duty Vehicles
Directives 88/77/EEC and 91/542/EEC

Effective	Type Approval (g/kWh)				Conformity of Production (g/kWh)			
Date***	CO	HC	NOx	Pm	CO	HC	NOx	Pm
01.04.88	11.2	2.4	14.4	**	13.2	2.64	15.8	**
01.07.92	4.5	1.1	8.0	0.36*	4.9	1.23	9.0	0.4*
01.10.95	4.0	1.1	7.0	0.15*	4.0	1.1	7.0	0.15*

*   In the case of engines of 85 kW or less, the limit value for particulate emissions is
        increased by multiplying the quoted limit by a coefficient of 1.7
**   Smoke according to ECE Regulation 24.03, EU Directive 72/306/EEC (see **Table 3.6**)
*** Dates for new models, all production has to comply according to the following
        schedule:

New Models	All Production
01.04.88	01.10.90
01.07.92	01.10.93
01.10.95	01.10.96

3.1.4   Motor Cycle and Moped Emissions Regulations

ECE Regulation 40 was adopted in September 1979 and applies to
two-wheeled and three-wheeled vehicles with an unladen weight of less than
400 kg. and having a maximum design speed exceeding 50 km/h. and/or a
cylinder capacity exceeding 50 cubic centimetres. The vehicle is required to
meet emissions limits over a driving cycle (Type I Test) and also at idle
(Type II Test). The emission limit for the Type II test is 4.5% vol CO.
Separate limits are specified in the Type I test for certification and production
vehicles, as shown in **Table 3.8.1** and **Table 3.8.2**.

Regulation 40 was amended on May 31 1988 to become ECE 40.01, with
lower limits for both CO and HC emissions. The EU has yet to introduce
legislation with respect to emissions from motor-cycles and mopeds. However, draft Directives, based on ECE Regulations, are under preparation.

The following countries accept vehicles meeting ECE 40.01, or require mandatory compliance:

Germany	Hungary	France	Italy
Netherlands	Norway	Belgium	Finland
United Kingdom	Romania	CIS	Czech Republic and Slovakia

Table 3.8.1 ECE Regulation 40/40.01 for Exhaust Emission Limits for Motorcycles with 4-stroke Engines

Reference Weight R [1] (kg)	CO (g/km)		HC (g/km)	
	ECE 40 [2]	ECE 40.01 [2]	ECE 40 [2]	ECE 40.01 [2]
< 100	25 {30}	17.5 {21}	7 {10}	4.2 {6}
100 - 300	$25 + 25(\frac{R-100}{200})$	$17.5 + 17.5(\frac{R-100}{200})$	$7 + 3(\frac{R-100}{200})$	$4.2 + 1.8(\frac{R-100}{200})$
{100 - 300}	$\{30 + 30(\frac{R-100}{200})\}$	$\{21 + 21(\frac{R-100}{200})\}$	$\{10 + 4(\frac{R-100}{200})\}$	$\{6 + 2.4(\frac{R-100}{200})\}$
> 300	50 {60}	35 {42}	10 {14}	6 {8.4}

Table 3.8.2 ECE Regulation 40/40.01 for Exhaust Emission Limits for Motorcycles with 2-stroke Engines

Reference Weight R [1] (kg)	CO (g/km)		HC (g/km)	
	ECE 40 [2]	ECE 40.01 [2]	ECE 40 [2]	ECE 40.01 [2]
< 100	16 {20}	12.8 {16}	10 {13}	8 (10.4)
100 - 300	$16 + 24(\frac{R-100}{200})$	$12.8 + 19.2(\frac{R-100}{200})$	$10 + 5(\frac{R-100}{200})$	$8 + 4(\frac{R-100}{200})$
{100 - 300}	$\{20 + 30(\frac{R-100}{200})\}$	$\{16 + 24(\frac{R-100}{200})\}$	$\{13 + 8(\frac{R-100}{200})\}$	$\{10.4 + 6.4(\frac{R-100}{200})\}$
> 300	40 {50}	32 {40}	15 {21}	12 {16.8}

Notes:
1) Reference weight (R) = Motorcycle weight + 75 kg.
2) Limits are for Type Approval. Limits given in parenthesis { } apply to Conformity of Production.

ECE Regulation 47 was issued in August 1981 and applies to vehicles of less than 400 kg equipped with an engine having a cylinder capacity of less than 50 cubic centimetres, namely mopeds. Emission limits are given in **Table 3.9.**

3.2 Europe - Non-EU/ECE Vehicle Emissions Regulations

Table 3.9 ECE Regulation 47 for Exhaust Emission Limits for Mopeds

Vehicle type	2-Wheeled		3-Wheeled	
Pollutant (g/km)	CO	HC	CO	HC
Licensing	8.0	5.0	15.0	10.0
Production	9.6	6.5	18.0	13.0

At a meeting in Sweden in July 1985, a number of countries agreed in principle to adopt US 1983 standards. The signatories to the "*Stockholm Agreement*" were Austria, Canada, Denmark, Finland, Norway, Sweden and Switzerland.

Austria

US 1983 exhaust emission standards were introduced from 01.01.1987 for gasoline vehicles over 1500 cc displacement (Amendment 18 to KDV) and from 28.07.1987 for all gasoline and diesel vehicles up to 3500 kg weight (Amendment 22 to KDV). The US Standard for evaporative emissions became effective from 01.01.1989 for gasoline vehicles.

Slightly modified standards are applied for manufacturers who do not conduct an 80 000 km durability run.

For heavy duty vehicles, the following standards, based on the ECE 49 test procedure, were adopted with effect from 1 January 1991: CO 4.9 g/kWh, HC 1.23 g/kWh, NOx 9.0 g/kWh and particulates 0.7 g/kWh. The particulate limit was reduced to 0.4 g/kWh from 1 January 1993.

The current Austrian standards are summarized in **Table 3.10**. It is interesting to note that the limits for mopeds are very severe, requiring the use of catalysts.

Table 3.10 Austrian Emission Limits

Vehicle Category		Limits (g/km)				
	Test Proc.	CO	HC	NOx	Pm	Impl. Date
Passenger cars						
diesel	FTP 75	2.1	0.25	0.62	0.37	25.05.86
	FTP 75	2.1	0.25	0.62	0.12	01.10.93
otto cycle	FTP 75	2.1	0.25	0.62	-	01.01.87
Commercial vehicles <3.5 tonnes GVW	FTP 75	6.2	0.5	1.43	0.37	from 01.01.89
		g/kWh	g/kWh	g/kWh	g/kWh	
Commercial vehicles >3.5 tonnes GVW (limits in g/kWh)	ECE 49	11.2	2.8	14.4	-	01.01.88
		4.9	1.23	9.0	0.7	01.01.91
		4.9	1.23	9.0	0.7	01.01.93
Diesel black smoke	ECE 24				AS ECE 24	
		g/km	g/km	g/km		
Motorcycles (<50 cc>40 km/h)						
2 stroke		13	6.5	2		before 01.10.91
		8	7.5	0.1		01.10.91
4 stroke		18	6.5	1		before 01.10.91
		13	3.0	0.3		01.10.91
Motorcycles (<50cc)						
2 stroke		12-32	8-12	1		before 01.10.90
		8	7.5	0.1		01.10.90
4 stroke		17.5-35	4.2-6	0.8		before 01.10.90
		13	3.0	0.3		01.10.90
Mopeds (<50cc<40km/h)		1.2	1.0	0.2		01.10.88

Note: For passenger cars, deterioration factors of 1.3 after 80,000 km or as determined by manufacturer's tests.

Finland

Table 3.11 Finland - Current and Planned Emissions Regulations

Vehicle Category	Approval	Effective date	CO g/km	HC g/km	NOx g/km	Pm g/km	Evap g/test	Equiv. Reg.
Cars	all new	01.06.92	2.1	0.25	0.62	0.373	2.0	US 83
		01.01.93	2.1	0.25	0.62	0.124	2.0	US 87
		1995 planned			California or future EU			
LDV<3500 kg Jeeps	all new	01.01.93 01.07.93	6.2	0.5	1.1	0.162	2.0	US 90
			g/kWh	g/kWh	g/kWh	g/kWh		
HDV >85 kW <85 kW	type approval	01.10.93	4.5 4.5	1.1 1.1	8.0 8.0	0.36 0 61		91/542 /EEC
>85 kW <85 kW	conform prodn.	01.10.93	4.9 4.9	1.23 1.23	9.0 9.0	0.4 0.68		
	type approval*	01.10.95	4.0	1.1	7.0	0.15		ECE49
Motorcycles Mopeds		01.01.93	Refer to **Tables 3.81** and **3.82**					ECE40.01 ECE47

Durability 80 000 km for HDV
* Conformity of Production: 01.10.96.

Norway

Table 3.12 Norway - Current and Planned Emissions Legislation

Vehicle Category	L1	L2	L3
Gross Weight Net Weight	<3.5 >760	<3.5 >760	>3.5

Vehicle Category	Effective Date	Regulation	Limits
Passenger Cars Gasoline L1	01.01.89	US 83	See **Table 3.16**
Lights Commercial & Light "Combined"	01.10.90	US 83	See **Table 3.16**
All Diesel L1	01.10.90	US 87	See **Table 3.16**
All LDV L2	01.10.92	US 90	See **Table 3.17.1**
All Vehicles L1 & L2	01.10.95	91/441/EEC and 93/59/EEC	See **Table 3.3** See **Table 3.4**
All HDV L3	01.10.93 01.10.96	91/542/EEC(A) 91/542/EEC(B)	See **Table 3.7** See **Table 3.7**

Note: There are no regulations for motor-cycles and mopeds.

Sweden

Sweden has introduced more stringent limits for all vehicle classes as given in **Table 3.13**. Vehicle manufacturers must also meet conformity guarantees, as in US legislation.

In addition, limits have been published in the A14 Regulation of 18 March 1992 for Low Emitting Vehicles (LEVs). Voluntary adoption will be encouraged within the framework of taxes according to the two environmental categories - Classes C.1 and C.2, with C.1 having the more stringent levels. Tests are carried out according to the US FTP 75 procedure. The limit values are based on US Clean Air Act Limits and are also given in **Table 3.13.**

The limits apply to spark-ignition engines (gasoline-, gaseous- and alcohol-fuelled or hybrid electric vehicles) or compression-ignition engines (diesel- and alcohol-fuelled or hybrid electric vehicles). Excluded from this legislation are motorcycles, vehicles with maximum speeds not exceeding 50 km/h and vehicles with GVWs exceeding 3500 kg.

Table 3.13 Swedish Emission Limits by Vehicle Category

Vehicle Category A-14 Regulation	GVW tons	GVW-service Wt kg	maximum speed km/h	Test Method	Useful life (1) Yrs	Km
L1 Passenger Car		<690	>50	US FTP 75	10	160 000
L2 Light Duty Vehicle	<2.7	>690		US FTP 75	10	160 000
L3 Heavy Duty Vehicle	>3.5		>31	ECE R49	11	200 000

Emissions Standards	Effective (model yr.)	CO g/km	HC g/km	NOx g/km	Nox-hwy g/km	Pm (2) g/km	Evap. g/test
L1	1989	2.1	0.25	0.62	0.75	0.124	2
L2	1992	6.2	0.5	1.1	1.4	0.162	2
		g/kWh	g/kWh	g/kWh		g/kWh	
L3	1993	4.9	1.2	9.0		0.4	

Table 3.13   Swedish Emission Limits by Vehicle Category (Cont.)

Category and Model Year	Low Emitting Vehicle (LEV) Emissions Standards (A-14)	Dura-bility	CO	CO @-7°C	HC	NMHC (3)	NOx	NOx hwy	Pm (2)	Evap	HCHO (4)
		km	g/km	g/km	g/km	g/km	g/km	g/km	g/km	g/test	mg/km
L1 (1993)	L1 : C1 (effective 1993 model yr.)	80k	2.1	6.2	0.25	0.078	0.25	0.33	0.05	2.0	9
		160k	2.6			0.097	0.37		0.06	2.0	11
	C2	80k	2.1	7.5	0.25	0.16	0.25	0.33	0.05	2.0	9
		160k	2.6			0.19	0.37		0.06	2.0	-
L2 (1993)	L2 : C1 (effective 1993 model yr.)	80k	2.7	7.5	0.50	0.10	0.43	1.2	0.05	2.0	11
		160k	3.4			0.124	0.61		0.06	2.0	14
	C2	80k	2.7		0.50	0.20	0.43	1.2	0.05	2.0	11
		160k	3.4			0.25	0.61		0.06	2.0	-
L3 (1994)	L3 : C1 (effective 1994 model yr.)	80k	2.7	7.5	0.50	0.10	0.43	1.2	0.06	2.0	14
		200k	4.0			0.143	0.61		0.06	2.0	17
	C2	80k	2.7		0.50	0.20	0.43	1.2	0.06	2.0	14
		200k	4.0			0.29	0.61		0.06	2.0	-

(1)   Whichever is the shorter. Intermediate useful life 5 years or 80 000 km.
(2)   Applies to diesel vehicles only, which also have smoke limits of 3.5 Bosch/45 Hartridge.
(3)   Expressed as NMOG for Class 1, NMHC for Class 2.
(4)   For methanol fuelled vehicles (those designed to operate on more than 50% methanol) only.   Also HC/NMOG/NMHC limits refer to "organic equivalents"
*   hwy = highway

639

Switzerland

Switzerland formally adopted the US 1977 standards using the FTP 75 procedure as a compulsory requirement from 1986. It has also implemented US 1983 exhaust emission standards, including particulates and evaporative emissions from October 1987 for cars and October 1988 for light commercial vehicles. Heavy trucks have been regulated against the ECE 49 procedure since October 1987. There are also stringent limits for motor cycles and mopeds, the latter requiring the use of catalysts. **Table 3.14** summarizes current limits.

Table 3.14   Swiss Emission Limits by Vehicle Category

Vehicle Category	Effective Date	CO g/km	HC g/km	NOx g/km	Pm g/km	Evap. g/test	Cycle
Cars<760	01.10.86	9.3	0.9	1.2	-	-	FTP 75
payload	01.10.87	2.1	0.25	0.62	0.37	2.0	
	01.10.88	2.1	0.25	0.62	0.12	2.0	
Light Trucks	01.10.88	8.0	0.65	1.8	0.48	2.0	FTP 75
>760 kg payload <3500 kg GVW	01.10.90	6.2	0.5	1.1	0.16	2.0	
		g/kWh	g/kWh	g/kWh	g/kWh		
Heavy Trucks	01.10.87	8.4	2.1	14.4	-		ECE 49
>3500 kg GVW	01.10.91	4.9	1.23	9.0	0.7		
		g/km	g/km	g/km			
Mopeds	01.10.88	0.50	0.50	0.10			ECE 47
Motorcycles: {	01.10.87	8.0	7.5	0.10			ECE 40
2-stroke	01.10.90	8.0	3.0	0.10			
Motorcycles: 4-stroke	01.10.87	13.0	3.0	0.30			ECE 40

Other Current Limits

Vehicle Category	Effective Date	Idle (mogas only) CO %v	HC ppm	Cross Country Cycle NOx g/km
Cars <760 kg payload	01.10.88	0.5	100	0.76
Light Trucks >760 kg payload <3500 kg GVW	01.10.90	1.0	200	1.8
Motorcycles 2 and 4-strokes		2.5		
		g/min		
Mopeds		0.1		

Eastern European Countries

Most East European countries apply some combination of ECE and EU regulations, as shown in **Table 3.15**.

Table 3.15 Summary of Vehicle Emissions Legislation in Eastern Europe

Country	Vehicle Type	Implementation Date	Regulation	Comments
Hungary	Passenger Cars	July 1992	ECE R83	As above, for imported vehicles
Czech Republic and Slovakia	Passenger Cars	Type Approval 01.10.92 All vehicles 01.10.93	89/458/EEC	See **Table 3.2**, note b, for limit values
	Light Duty Vehicles	As above	83/351/EEC	See **Table 3.1** for limit values
	Heavy Duty Vehicles	As above	91/542/EEC	See **Table 3.7** for limit values
Commonwealth of Independent States	See comments	See comments	See comments	The Commonwealth adopted 'all relevant' ECE regulations into national legislation. Details remain unclear.
Poland	Passenger Cars	1995	ECE R83.01	See **Table 3.2**
	Heavy Duty Vehicles	1995 1988	ECE R49.02 ECE R24.03	See **Table 3.7** See **Table 3.6**
	Motor-Cycles	1988	ECE R40.01	See **Tables 3.8.1** and **3.8.2**
	Mopeds	1988	ECE R47	See **Table 3.9**
Hungary	Heavy Duty Vehicles	1990	ECE R49	Steady State CO 14, HC 3.5, NOx 18 g/kWh
			ECE R24	Full load smoke (See **Table 3.6**)
			Ordinance 6/1990	Free acceleration smoke

3.3 Europe - Evaporative, Distribution and Refuelling Emissions Legislation

3.3.1 Evaporative Emissions Legislation

The European Commission stated its intention to control evaporative emissions from motor vehicles by requiring all cars to be fitted with small carbon canisters and *"Stage I"* controls applied to the distribution system. Evaporative emission limits have thus been included in the EU Consolidated Emissions Directive (see **Section 3.1**), with details as follows:

- Limit: 2g/test for all cars.
- Test procedure based on SHED (Sealed Housing for Evaporative Determination) but differs from the US procedure, see **Section 4.4**
- Test includes diurnal emissions (temperature rise 16-30°C) and hot soak emissions (SHED temperature 23-31°C) but not running losses
- Test fuel volatility is 56-64 kPa (same as the exhaust emissions test reference fuel).

3.3.2 Distribution and Refuelling Emissions Legislation

European Union

On the 4 October 1993, the EU Environmental Ministers Council adopted a common position on a draft Directive on emissions of VOC resulting from the storage of gasoline and its distribution to service stations (the "Stage I" draft Directive COM(92)277). Control of refuelling emissions is now being considered by the EU.

The draft Stage I Directive includes the following emissions requirements:

- the annual gasoline loss from the loading and storage at each terminal should not exceed 0.01 %m of the gasoline throughput.
- the annual loss from loading and unloading of gasoline containers at terminals should not exceed 0.005 %m of the throughput.
- the concentration of the vapours exhausted from vapour recovery units must not exceed $35g/Nm^3$ in any one hour.

The implementation schedule will be as follows :

Implementation	**Terminals**	**Service Stations**
Immediate	New	New
Within 3 years	Road/rail >150 kt/yr.	>1000 m^3/yr.
Within 6 years	Road/rail barge/ship >25kt/yr.	500 m^3/yr.
Within 9 years	<25kt/yr.*	>100 m^3/yr.**
Exempt		<100 m^3/yr.

* vapour balancing may be substituted for vapour recovery at terminals with throughputs less than 25 kt/yr.
** service stations with throughputs less than 500 m^3/yr. may be exempted if they do not present a significant environmental or health hazard.

Germany

In parallel with the EU developments, the German government has developed two regulations of its own for the introduction, maintenance and inspection of vapour recovery systems based upon Article 23 of the Bundes-Immmissionsschutzgesetz. The two regulations separately cover terminals (20.BimSchV) and service stations (21.BimSchV) as follows:

20.BimSchV		**21.BimSchV**	
(effective date 08.10.92)		**(effective date 01.01.93)**	
Facilities	**Implementation**	**Service Stations**	**Implementation**
New loading/ despatch	Immediately	New	Immediately
		Existing >5000m^3	3 years
Existing loading/ despatch	2-3 years depending on technical issues	2500-5000 m^3	3-4 years depending on location
Road tankers	3 years	1000-2500m^3	5 years
Rail tank cars	5 years	<1000 m^3	None

Sweden

In January 1991, the Swedish Environmental Protection Agency decided on regulations for vapour recovery systems on service stations. The regulation covers Stage I and Stage II measures. Stage II measures (and Stage I) shall be installed following the schedule shown.

Schedule	Service Stations < 2000m³	All Service Stations
1.1.1992	> 50%	
1.1.1993	> 75%	> 25%
1.1.1994	100%	> 75%
1.1.1995		100%

To be accepted, the Stage II equipment has to have an efficiency of 85% under controlled conditions and 70% for a vehicle stock representative for Sweden and operated by "*normal customers*". The measurement method is the one used by the German Technische Uberwachungsverein. Installation of the system shall be controlled by an authorized company and regular inspection and maintenance will be controlled by spot checks carried out by a national test agency.

Switzerland

In Switzerland, with effect from 1 February 1992, all gasoline retail outlets must be equipped with Stage II vapour recovery systems. The timing of installation must be fixed individually for each site, but all must be converted by the end of 1994. The systems fitted must be certified to have 90% total hydrocarbon control efficiency. However, special 2-stroke pumps are exempted. Vehicle filler nozzles will be required to conform to the US SAE 1140.

3.4 US Vehicle Emissions Regulations

Background

National exhaust emission limits for cars were first set in the Clean Air Act of 1968. However, in 1970 the US Congress passed amendments to this Act which incorporated the so-called "*Muskie*" proposals. These amendments required exhaust emission reductions of 90% from the then current levels to take effect in 1975-76. After lengthy debate between the motor industry and the EPA, implementation of these regulations was delayed until technology was available to meet them. This led to the establishment of interim standards for 1975 and 1976 which were subsequently extended to 1979. Oxidation catalysts were required for most vehicles to meet the 1975 and subsequent limits so unleaded gasoline was made widely available in 1975 to cater for these catalyst cars. In 1977 Congress amended the Clean Air Act and set revised standards to achieve a 90% HC reduction in 1980, and 90% CO and 75% NOx reduction in 1981. This led to the widespread introduction of 3-way catalyst technology. The original 90% NOx reduction (0.4 g/mile) was left as a research goal, although it was adopted in California and now forms part of the new Clean Air Act. **Table 3.16** gives full details of all these and future limits.

Separate emission limits for light-duty trucks were introduced in 1975, but these did not require the use of catalysts. More severe limits in 1984 did, however, require catalysts and the 1988 limits essentially required 3-way catalysts, as for passenger cars - see **Table 3.17.**

Emission limits for heavy duty engines were originally set in 1970 and, in the 1977 Clean Air Act, stringent reductions in HC and CO emissions were proposed to take effect in 1981. These standards were, however, also deferred until technology was available and finally implemented in 1987, requiring catalysts on heavy duty gasoline engines. Further reductions in NOx and diesel particulate limits to be implemented from 1990 to 1995 will require 3-way catalysts for heavy duty gasoline engines and radical changes in diesel engine technology. **Table 3.18** provides details of emission limits for heavy duty vehicles.

The US standards apply over the *"useful life"* of the vehicle, which for cars is defined as 50 000 miles (80 000 km) or 5 years. The durability of the emission control device must be demonstrated over this distance, within allowed deterioration factors, and in some cases over 100 000 miles (160 000 km). The heavy duty truck regulations for 1987 (and later) require compliance over longer periods, representative of the useful life of the vehicle.

Table 3.19 shows emission limits for motorcycles. Federal limits have not been updated since 1980, but Californian limits applied during the 1980s are more stringent.

The State of California has always been a leader in emission control legisla-tion and has generally adopted limits more severe than Federal (Clean Air Act) limits which apply in the rest of the USA. The main reason for this is the atmospheric smog and poor air quality in the Los Angeles area. Although there has been a significant improvement in air quality over the last fifteen years, Los Angeles still has the highest ozone levels of any city in the United States.

As a consequence the California Air Resources Board (CARB) has decided to implement even more stringent emission limits in California over the next fifteen years, culminating in the introduction of the *"Zero Emissions Vehicle"* by the end of the century. This legislation is discussed below and the latest limits are incorporated in **Tables 3.16 to 3.18.**

Table 3.16 Historical Review of US Light Vehicle Emissions Regulations

Model year	Federal					Durability mileage	California				
	CO g/mile	HC g/mile	NOx g/mile	Pm g/mile	Evap. g/test		CO g/mile	HC g/mile	NOx g/mile	Pm g/mile	Evap. g/test
Pre-control	90	15	6.2	-	6.0		15	90	6.2	-	6.0
1970[1]	34	4.1	-	-	-		34	4.1		-	6.0[2]
1971[1]	34	4.1		-	6.0[2]		34	4.1	6.2	-	6.0
1972[1]	28	3.0		-	2.0		28	2.8	3.2	-	2.0
1973-74[1]	28	3.0	3.1	-	2.0		28	2.8	2.0	-	2.0
1975-76	15	1.5	3.1	-	2.0		9.0	0.9	2.0	-	2.0
1977	15	1.5	2.0	-	2.0		9.0	0.41	1.5	-	2.0
1978-79	15	1.5	2.0	-	6.0[3]		9.0	0.41	1.5	-	6.0[3]
1980	7.0	0.41	2.0	-	6.0	50 000 [4]	9.0	0.39[5]	1.0[6]	-	2.0
						100 000 A	9.0	0.39	1.5	-	
						100 000 B	10.6	0.46	1.5	-	
1981	3.4[7]	0.41	1.0[8]	-	2.0	50 000 A	3.4	0.39	1.0	-	2.0
						50 000 B	7.0	0.39	0.7	-	
						100 000 A	3.4	0.39	1.5	-	
						100 000 B	4.0	0.46	1.5	-	
1982-83[9]	NC	NC	NC	0.6	NC	50 000 A	7.0	0.39	0.4	0.6	2.0
						50 000 B	7.0	0.39	0.7[10]	0.6	2.0
						100 000 A	7.0	0.39	1.5	0.6	2.0
						100 000 B	8.3	0.46	1.5	0.6	2.0
1984[11]	NC	NC	NC		NC	50 000 A	7.0	0.39	0.4	0.6	2.0
						50 000 B	7.0	0.39	0.7	0.6	2.0
						100 000 A	7.0	0.39	1.0	0.6	2.0
						100 000 B	8.3	0.46	1.0	0.6	2.0
1985	NC	NC	NC	NC	NC		NC	NC	NC	0.4	NC
1986	NC	NC	NC	NC	NC		NC	NC	NC	0.2	NC
1987-88	NC	NC	NC	0.2	NC		NC	NC	NC	NC	NC
1989-92	NC	NC	NC	NC	NC		NC	NC	NC	0.2	NC
1993-94	NC	NC	NC	NC	NC	50 000	3.4	0.25[18]	0.4	0.08	2.0
						100 000	4.2	0.31	0.7	0.08	2.0
1994-96[12]											
50 000 miles	3.4	0.25	0.4	0.08	2.0	TLEV[16,18]	3.4	0.125	0.4	0.08	2.0[15]
100 000 miles	4.2[13]	0.31	0.6[14]	0.10	2.0[15]	LEV[16,18]	3.4	0.075	0.2	0.08	2.0
1999-2003	NC	NC	NC	NC	NC	ULEV[16,18]	0.04	1.7	0.2	0.04	2.0
2004[17]	1.7	0.125	0.2	0.08	2.0	ZEV	0	0	0	0	0

Notes:

Note: "A" and "B" California emission limits refer to the limits at start and end of the durability test schedule.

(1) Pre 1975 standards are expressed as equivalent 1975 test values.
(2) Ca1rbon canister trap method.
(3) Sealed Housing Evaporative Determination (SHED) technique - 6.0 g/test by SHED method represents approximately 70% less emissions than 2 g/test by the carbon trap method.
(4) Refers to 50 000 mile and 100 000 mile Certification options.
(5) Non-methane HC. Compliance with total HC standard of 0.41 g/mile is an acceptable alternative.
(6) Maximum NOx emissions allowed during highway cycle: 1.33 x standard.
(7) Waivers up to 7.0 g CO/mile for 1981 were granted by EPA for some car models.
(8) Waivers up to 1.5 g NOx/mile were granted for some 1981 and 1982 diesel vehicles.
(9) High altitude standards for 1982 and 1983 - 0.57 HC, 1.0 NOx, 7.8 CO, 2.6 evap.
(10) This option (0.7 g NOx/mile standard for 1983 and later) requires limited recall authority for 7 years/75 000 miles.
(11) All cars must meet standards at all altitudes.
(12) Limits phased in over a 3 year period (see **Table 3.20**).
(13) Additional CO limit of 10 g/mile at 20°F (-7°C).
(14) Diesel NOx limits 1.0/1.25 g/mile.
(15) Revised test procedure with extra diurnal tests.
(16) Different categories to be phased in over a 10 year period from 1994 to meet NMOG emission targets (see **Table 2.7**).
(17) Implementation of these limits at the discretion of the EPA.
(18) Additional separate limits for formaldehyde (HCHO) (see **Table 2.7**).
NC = No Change.

Table 3.17 Historical Review of US light duty truck emissions regulations (less than 8500 lbs GVW)

Table 3.17.1 Federal Exhaust Emissions Standards by 1975 FTP [1]

Model Year	Weight Category (lbs)	Durability Mileage	CO g/mile	HC g/mile	NOx g/mile	Pm g/mile
1970	<6000 GVW[2]		Same as passenger cars			
1975	<6000 GVW		20	2.0	3.1	-
1976-77	<6000 GVW		NC	NC	NC	-
1978	<6000 GVW		NC	NC	NC	-
1979	<8500 GVW[3]		18	1.7	2.3	-
1980	<8500 GVW		NC	NC	NC	-
1981	<8500 GVW		NC	NC	NC	-
1982[7]	<8500 GVW		18	1.7	2.3	0.6
1983	<8500 GVW		NC	NC	NC	-
1984[9]	<8500 GVW		10	0.8	2.3	0.6
1985	<8500 GVW		NC	NC	NC	0.6
1986	<8500 GVW		NC	NC	NC	0.6
1987	<8500 GVW		NC	NC	NC	0.26[12]
1988[11]	<8500 GVW		10	0.8	1.2/1.7	0.26[12]
1989	<8500 GVW		NC	NC	NC	NC
1991	0-3750	120 000	10	0.8	1.2	0.26[12]
	3750-8500	120 000	10	0.8	1.7	0.13
1993	No Change					
1994-6[13]	0-3750	50 000	3.4	0.25[14]	0.4	0.08
		100 000	4.2	0.31	0.6	0.10
	3750-5750	50 000	4.4	0.32	0.7	0.08
		100 000	5.5	0.40	0.97	0.10
	5750-8500	50 000	5.0	0.39	1.1	-
		120 000	7.3	0.56	1.53	0.12
2003[15]	0-3750	100 000	1.7	0.125	0.2	

See notes following **Table 3.17.2.**

647

Table 3.17.2 California Exhaust Emission Standards

Model Year	Weight Category	Durability Mileage	CO g/mile	HC g/mile	NOx g/mile	Pm g/mile
1970	<6000 GVW[2]			Same as passenger cars		
1975	<6000 GVW		20	2.0	2.0	-
1976-77	<6000 GVW		17	0.9	2.0	-
1978	<6000 GVW		17	0.9	2.0	-
	6000-8500GVW		17	0.9	2.3	-
1979	<4000 IW		9.0	0.41	1.5	-
	4000-5999 IW		9.0	0.50	2.0	-
	6000-8500 IW		17	0.9	2.3	-
1980	<4000 IW		9.0	0.39[4]	1.5/2.0[5]	-
	4000-5999 IW		9.0	0.50	2.0	-
	6000-8500 IW		17	0.9	2.3	-
1981	<4000 IW	50 000	9.0	0.39	1.0[6]	-
	4000-5999 IW		9.0	0.50	1.5	-
	6000-8500 IW		9.0	0.6	2.0	-
	<4000 IW	A 100 000	9.0	0.34	1.5	-
	<4000 IW	B 100 000	10.6	0.40	1.5	-
	4000-5999 IW		9.0	0.5	2.0	-
	6000-8500 IW		9.0	0.6	2.3	-
1982[7]	<8500 GVW		NC	NC	NC	0.6
1983	<4000 IW	A 50 000	9.0	0.39	0.4	0.6
	<4000 IW	B 50 000	9.0	0.39	1.0[8]	0.6
	4000-5999 IW		9.0	0.5	1.0	0.6
	6000-8500 IW		9.0	0.6	1.5	0.6
	<4000 IW	A 100 000	9.0	0.39	1.5	0.6
	<4000 IW	B 100 000	10.6	0.46	1.0	0.6
	4000-5999 IW		9.0	0.5	1.5	0.6
	6000-8500 IW		9.0	0.6	2.0	0.6
1984[9]	<8500 GVW		NC	NC	NC	NC
1985	<8500 GVW		NC	NC	NC	0.4
1986	<8500 GVW		NC	NC	NC	0.2
1987	<8500 GVW		NC	NC	NC	NC
1988[11]	<8500 GVW		NC	NC	NC	NC
1989	<8500 GVW		NC	NC	NC	0.08
1993	0-3750	50 000	3.4	0.25[14]	0.4	0.08
	"	100 000	4.2	0.31	"	"
	3750-5750	50 000	4.4	0.32	0.7	0.08
	"	100 000	5.5	0.40	"	"
1994	**TLEV**					
-2003	0-3750	50 000	3.4	0.125	0.4	-
	"	100 000	4.2	0.156	0.6	0.08
	3750-5750	50 000	4.4	0.160	0.7	-
	"	100 000	5.5	0.20	0.9	0.08
	LEV					
	0-3750	50 000	3.4	0.075	0.2	-
	"	100 000	4.2	0.09	0.3	0.08
	3750-5750	50 000	4.4	0.10	0.4	-
	"	100 000	5.5	0.13	0.5	0.08
	5750-8500	50 000	5.0	0.195	1.1	-
		120 000	7.3	0.28	1.5	0.12
	ULEV					
	0-3750	50 000	1.7	0.4	0.2	-
	"	100 000	2.1	0.055	0.3	0.04
	3750-5750	50 000	1.7	0.050	0.2	-
	"	100 000	2.8	0.070	0.5	0.04
	5750-8500	50 000	2.5	0.117	0.6	-
		120 000	3.7	0.167	0.8	0.06

Notes: NC = No Change

"A" and "B" California emission limits refer to the limits at start and end of the durability test schedule.

(1)   Evaporative emission standards same as those for passenger cars.
(2)   GVW - Gross Vehicle Weight   IW - Inertia Weight.
(3)   Prior to 1979 heavy duty standards applied to medium duty (6000 - 8500 lbs GVW) vehicles.
(4)   Non-methane HC. Compliance with a total HC standard of 0.41 g/mile is an acceptable alternative.
(5)   NOx standard of 2.0 g/mile for 4-wheel drive vehicles.
(6)   Maximum NOx emissions allowed during highway cycle: 2.0 x standard.
(7)   High altitude exhaust standard established for 1982 and 1983 - HC 2.0, CO 26, NOx 2.3 g/mile, evap. 2.6 g/test.
(8)   This optional 1.0 g/mile NOx standard for 1983 and later requires limited recall authority for 7 years/ 75 000 miles.
(9)   High altitude exhaust standard established for 1984 and later - HC 1.0, CO 14, NOx 2.3 g/mile, evap. 2.6 g/test. Light duty trucks may also be certified using half life option (1984 only).
(10)   Gasoline fuelled vehicles only. Standard does not apply in high altitude areas.
(11)   Maximum allowed NOx emission 1.2 g/mile for <6000 lbs GVW and 1.7 g/mile for 6000 - 8500 lbs GVW.
(12)   The particulate (Pm) emission standards apply to diesel powered trucks only and were relaxed for vehicles over 3750 lbs GVW. Limits are 0.5 g/mile for 1987 model year and 0.45 g/mile for 1988-90.
(13)   Standards to be phased in over a 3 year period (See **Table 3.22**).
(14)   These are all NMHC (non-methane hydrocarbon) limits. Total HC limits of 0.38/0.46 apply over 50 000 miles.
(15)   The EPA must decide by 1997 whether to apply these limits or set different standards.
(16)   TLEV, LEV and ULEV limits to be phased in over a 10 year period to meet NMHC targets. Higher targets apply for vehicles above 3750 lb. GVW (See **Table 3.22**).

Table 3.18 Historical Review of US heavy duty vehicle emissions regulations (> 8500 lbs GVW)[1]

Year	Federal (g/bhp.h)						California (g/bhp.h)			
	CO	HC	NOx	HC+NOx	Pm	Evap g/test	CO	HC	NOx	HC+NOx
1969	-	-	-	-	-	-	63.6	6.55	-	-
1970-71	63.6	6.55	-	-	-	-	63.6	6.55	-	-
1972	63.6	6.55	-	-	-	-	41.1	4.21	-	-
1973	63.6	6.55	-	-	-	-	40	-	-	16
1974	40	-	-	16	-	-	30	-	-	10
1975	40	-	-	16	-	-	25	-	-	5
or							25	1.0	7.5	-
1979	25	1.5[2]	-	10	-	-	25	1.5[2]	7.5	-
or	25	-	-	5	-	-	25	-	-	5
1980-83	25	1.5	-	10	-	-	25	1.0	-	6
or	25	-	-	5	-	-	25	-	-	5
1984[3]										
Transient	25	1.5	10.7	10	-	-	25	0.5	-	4.5
Idle	0.5%	-	-	-	-	-	-	-	-	-
Diesel Option	15.5	0.5	9.0	-	-	-	15.5	1.3	5.1	-
1985-86[4]										
A	37.1	1.9	10.6	-	-	3.0[5]		No Change		
B	40.0	2.5	10.7	-	-	3.0				
Diesels	15.5	1.3	10.7	-	-	-				
1987										
<14000 GVW	14.4	1.1	10.6	-	-	3.0		No Change		
Idle[6]	0.5%	-	-	-	-	-				
>14000 GVW	37.1	1.9	10.6	-	-	4.0				
Diesels	15.5	1.3	10.7							
Idle[6]	0.%	-	-	-						
1990										
Both classes + Diesels	NC	NC	6.0	-	0.6[7]	4.0				

For model year 1991 forward see **Table 2.10**

Notes: NC= No Change. A and B Federal limits apply at the start and end of the durability test schedule.

(1)   Apply to engines in vehicles over 6000 or 8500 lbs for which no light duty of medium duty standard applies. Standards apply to gasoline vehicles only through 1972 for California and 1973 for Federal, and to gasoline and diesel thereafter. Test procedure is 13 mode cycle up to 1984.

(2)   HC measurement method changed from NDIR to FID for gasoline engines (FID had been previously specified for diesels) resulting in higher readings for equivalent emissions. Optional use of former test procedures allowed for 1979 models.

(3)   The HC and CO standards represent 90% reductions from the uncontrolled baseline. A new transient test procedure (EPA cycle) was introduced from 1984. The NOx standard shown is an interim standard to maintain the 1982 level of control with the revised test procedure.

(4)   Different dynamometer schedules used for options A and B.

(5)   Evaporative standard for 8 500-14 000 lbs. HDTs over 14 000 lbs must meet 4.0 g/test.

(6)   For heavy duty gasoline engines utilizing catalyst technology.

(7)   Diesels only.

**Table 3.19   US Emissions Limits for Motorcycles over 50 cm³ Capacity
(Modified FTP 75)**

Regulation	Model Year	Engine Capacity (cc)	CO g/km	HC g/km
Federal	1978-9	50-170	17.0	5.0
and		170-750	17.0	5.0 + 0.0155(D-170)
California		>750	17.0	14.0
	1980	all	12.0	2.0
California	1982-4	50-279	12.0	1.0
only		>280	12.0	2.5
	1985	50-279	12.0	1.0
		>280	12.0	1.4*
	1988	280-699	12.0	1.0
		>700	12.0	1.4*

Notes   D = Engine displacement in cubic centimetres (ccs)
         *Applied as corporate average

The 1990 US Clean Air Act Amendments (CAAA)

The objective of US legislation has always been to improve air quality, particularly in large cities which were experiencing problems of ozone formation in summer and high ambient CO concentrations in winter. It became apparent, however, that the 1977 Act had not achieved these objectives. Figures indicated that 9 cities have failed to meet the minimum standards for ozone and 41 failed to meet CO standards. This led to calls from a number of quarters for a revision of the 1977 Act.

In July 1989 President Bush proposed a major revision of the Act. Following protracted negotiations a House and Senate conference reached a compromise agreement which was signed by President Bush in November 1990.

The full CAAA is a massive document and contains 7 *"titles"*. However, only Title 2 refers to *"mobile sources"* (i.e. motor vehicles). The other titles cover a wide range of emission sources and air quality issues. Title 2 of the CAAA relates to motor vehicles, fuels, and their emissions, and its major features are as follows:

-   The imposition of tighter tailpipe emission standards.

- The establishment of compliance testing and maintenance programmes related to the above.

- The establishment of a reformulated gasoline programme.

- Legislation relating to clean fuels and clean fuels vehicles which could lead to the introduction of alternative fuels.

- Legislation covering operators of vehicle fleets in areas with specific air quality problems.

- Reaffirmation of the rights of individual states with particular air quality problems to set more severe emission standards, but these must be identical to California limits.

Since the Amendments were approved the EPA has been working hard with the assistance of the oil and motor industries to develop detailed rules to put the legislation into place. The most important of these are the 'Tier I' exhaust emissions limits for light duty vehicles, evaporative emissions procedures and limits and the rules for reformulated gasolines. The requirement for on-board engine diagnostic systems is also close to a final rule.

Tailpipe Emissions Standards

A complete summary of US Federal and Californian current and future legislation will be found in **Table 3.21**. Implementation of these limits is complex and is summarized in **Table 3.22**. The Tier I and Tier II limits are also summarized separately in **Table 3.20**.

Tailpipe Emissions Standards - Light Duty Vehicles (Cars and Trucks below 3750 lbs GVW)

There are two sets of standards defined in the CAAA, Tier I and Tier II (Tier 0 is the current legislation), and they are given in **Table 3.20**. Tier I is now covered by a final regulation, published 5 June, 1991 and will be intro-duced progressively from 1994. Starting in 1996, vehicles must be certified up to 100 000 miles, or to the higher *"useful life"* limits (see **Table 3.21**)

In-use (recall) standards are also specified which must be met under ran-domized testing of in-service cars by the EPA. If the limits are not met an *"emissions recall"* may be triggered where the manufacturer has to recall and rectify any emissions defects.

Tier II emission limits are being considered for 2004 which are 50% lower than the Tier I limits. However, these will only come into effect if the EPA Administrator decides that they are necessary, technically feasible and cost-effective. The Amendments require the Administrator to carry out a study to determine this information and following this he can pursue one of three courses of action:

1) Reaffirm the existing Tier I Standards
2) Impose the Tier II Standards
3) Impose a new set of standards more stringent than Tier II

Table 3.20   US Light Duty Vehicle Emissions Regulations

Effective Date (% Production)	Durability miles	CO g/mile	CO @ 20° F g/mile	NMHC g/mile	NOx g/mile	Pm g/mile	Evap. g/test
Current		3.4	-	0.41[a]	1.0		2.0
US Tier I							
1994 (40%)	50 000	3.4	10.0*	0.25	0.4	0.08	**
1995 (80%)							
1996 (100%)	100 000	4.2	10.0*	0.31	0.6	0.10	**
US Tier II							
2004***		1.7		0.125	0.2	0.08	*

(a)   Total HC.
*   A CO standard of 10 g/mile is specified at 20° F, commencing in 1994. However, if despite the
     "reformulated gasoline" programme six or more cities remain out of compliance with CO air quality
     targets between now and 1996, the more stringent limit of 3.4 g/mile will be phased in over three years,
     starting in 2001.
**   See **Table 2.9**
*** Implementation at the discretion of the EPA

Cold Temperature CO Emissions

The Clean Air Act Amendments specify a CO standard of 10 g/mile at 20°F (-7°C) starting in 1994. However if, despite the oxygenate gasoline programme, six or more cities remain out of compliance with CO air quality targets between now and 1996, the more stringent limit of 3.4 g/mile will be phased in over three years starting in 2001.

The EPA is reviewing its stance regarding whether it should allow the averaging of the results of CO emissions at low temperatures from a given manufacturer's product line. The present regulations require that 40 percent must meet the standard by 1994, 80 per cent by 1995 and all by 1996. Since the technology already exists to meet the limits, retaining the averaging option would allow manufacturers to produce cars easily meeting the limits while allowing heavy trucks, where compliance is more difficult, to exceed the standard. This would lead to higher than average CO emissions in the West where heavy trucks predominate.

Tailpipe Emission Standards - Heavy Duty Vehicles (Trucks and Buses >3750 lbs GVW)

Emission standards will depend on vehicle weight and are as given in **Tables 3.17, 3.18, 3.21 and 3.22.**

The first emissions limits for heavy duty engines were set in 1970. The recent reductions in NOx and diesel particulates, to be implemented from 1990 to 1995 require the use of three-way catalysts for heavy duty gasoline engines and major advances in diesel engine technology. Further reductions are required for 1998.

The EPA put back the implementation of a 0.1 g/bhp.h particulate limit for urban buses from 1991 to 1993. However on 4 March 1993, the EPA published a ruling reducing the limit further to 0.07 g/bhp.h in 1994/95 and 0.05 g/bhp.h in 1996. Coupled with this standard is a requirement for all diesel vehicles to use fuel containing 0.05% mass sulphur. This helps engine manufacturers reduce particulate emissions when using catalytic converters or trap oxidizers.

In April 1993, the EPA issued a final rule establishing a retrofit programme for urban bus engines. It applies to 1993 and earlier model year buses which operate in metropolitan areas with populations of over 750,000 and have their engines rebuilt or replaced after 1 January 1995. The objective is to reduce particulate emissions from older buses by upgrading their particulate emission control systems. Operators must either upgrade engines, the standard required depending on cost and availability of components, or meet a fleet average target level for particulate emissions each year beginning in 1996.

California

The CARB is required by the US Senate to adopt low-emission standards for transit buses. In 1992 the CARB proposed limits for urban buses of 4.0 and 0.50 g/bhp.h for NOx and particulates respectively. An earlier proposal of a 2.5 g/bhp.h limit for NOx is now being considered as an option with credits which could be sold to stationary sources under a mobile source credit programme being developed by several Californian air districts. These limits were approved in June 1993 for adoption in 1996 together with a further reduction in particulates to 0.07 g/bhp.h starting in 1994. Each engine manufacturer would be allowed to apply for an exemption for up to 10 per cent of its output, if it could be demonstrated that not all of its

Table 3.21 Summary of Current and Future US and California Exhaust Emission Standards and Categories

Vehicle Type	Emission Category	Certification[A] Exhaust Emission Standards - gpm																	
		5 Yrs. / 50,000 Miles						10 Yrs. / 100,000 Miles						11 Yrs. / 120,000 Miles					
		THC[B]	NMHC or NMOG[C]	CO	NOx	PM	HCHO	THC[B]	NMHC or NMOG[C]	CO	NOx	PM	HCHO	THC[B]	NMHC or NMOG[C]	CO	NOx	PM	HCHO
PC	Fe Tier 0	0.41	-	3.4	1.0	[1]0.20	-												
	Fe Tier 1	0.41	0.25	3.4	[2]0.4	0.08	-	-	0.31	4.2	[3]0.6	0.10	-						
	Fe CFV 1	-	0.125	3.4	0.4	-	0.015	-	0.156	4.2	0.6	0.08	0.018						
	Fe Tier 2							-	0.125	1.7	0.2	-	-						
	Fe CFV 2	-	0.075	3.4	0.2	-	0.015	-	0.090	4.2	0.3	0.08	0.018						
	CA Tier 0	-	0.39	7.0	0.4	[1]0.08	[4]0.015												
	CA Tier 1	-	0.25	3.4	0.4	[1]0.08	[4]0.015	-	0.31	4.2	-	-	-						
	CA TLEV	-	0.125	3.4	0.4	-	0.015	-	0.156	4.2	0.6	[1]0.08	0.018						
	CA LEV	-	0.075	3.4	0.2	-	0.015	-	0.090	4.2	0.3	[1]0.08	0.018						
	CA ULEV	-	0.040	1.7	0.2	-	0.008	-	0.055	2.1	0.3	[1]0.04	0.011						
	CA ZEV	0.0	0.0	0.0	0.0	0.0	0.0	0.0	0.0	0.0	0.0	0.0	0.0	0.0	0.0	0.0	0.0	0.0	0.0

Table 3.21 Summary of Current and Future US and California Exhaust Emission Standards and Categories (continued)

Certification[A] Exhaust Emission Standards - gpm

Vehicle Type	Emission Category	5 Yrs. / 50,000 Miles						10 Yrs. / 100,000 Miles						11 Yrs. / 120,000 Miles					
		THC[B]	NMHC or NMOG[C]	CO	NOx	PM	HCHO	THC[B]	NMHC or NMOG[C]	CO	NOx	PM	HCHO	THC[B]	NMHC or NMOG[C]	CO	NOx	PM	HCHO
LDT 1 (LVW: 0 - 3750 lb.)	Fe Tier 0	-	0.25	3.4	[2]0.4	0.08	-												
	Fe Tier 1	-	0.125	3.4	0.4	-	0.015	0.80	0.31	4.2	[3]0.6	0.10	-	0.80	-	10	1.2	[1]0.26	-
	Fe CFV 1							-	0.156	4.2	0.6	0.08	0.018						
	Fe Tier 2							-	0.125	1.7	0.2	-	-						
	Fe CFV 2	-	0.075	3.4	0.2	-	0.015	-	0.090	4.2	0.3	0.08	0.018						
	CA Tier 0	-	0.39	9.0	0.4	[1]0.08	[4]0.015												
	CA Tier 1	-	0.25	3.4	0.4	[1]0.08	[4]0.015	-	0.31	4.2	-	-	-						
	CA TLEV	-	0.125	3.4	0.4	-	0.015	-	0.156	4.2	0.6	[1]0.08	0.018						
	CA LEV	-	0.075	3.4	0.2	-	0.015	-	0.090	4.2	0.3	[1]0.08	0.018						
	CA ULEV	-	0.040	1.7	0.2	-	0.008	-	0.055	2.1	0.3	[1]0.04	0.011						
	CA ZEV	0.0	0.0	0.0	0.0	0.0	0.0	0.0	0.0	0.0	0.0	0.0	0.0						
LDT 2 (LVW: 3751 - 5750 lb.)	Fe Tier 0	-	0.32	4.4	[5]0.7	0.08	-												
	Fe Tier 1	-	0.160	4.4	0.7	-	0.018	0.80	0.40	5.5	0.97	0.10	-	0.80	-	10	1.7	[1]0.13	-
	Fe CFV 1							-	0.200	5.5	0.9	0.08	0.023						
	Fe CFV 2	-	0.100	4.4	0.4	-	0.018	-	0.130	5.5	0.5	0.08	0.023						
	CA Tier 0	-	0.50	9.0	1.0	[1]0.08	[4]0.018												
	CA Tier 1	-	0.32	4.4	0.7	0.08	[4]0.018	-	0.40	5.5	-	-	-						
	CA TLEV	-	0.160	4.4	0.7	-	0.018	-	0.200	5.5	0.9	[1]0.08	0.023						
	CA LEV	-	0.100	4.4	0.4	-	0.018	-	0.130	5.5	0.5	[1]0.08	0.023						
	CA ULEV	-	0.050	2.2	0.4	-	0.009	-	0.070	2.8	0.5	[1]0.04	0.013						

Federal & CA GVWR: 0 - 6000 lb.

Table 3.21 Summary of Current and Future US and California Exhaust Emission Standards and Categories (continued)

Certification[A] Exhaust Emission Standards - gpm

Vehicle Type	Emission Category	5 Yrs. / 50,000 Miles						10 Yrs. / 100,000 Miles						11 Yrs. / 120,000 Miles					
		THC[B]	NMHC or NMOG[C]	CO	NOx	PM	HCHO	THC[B]	NMHC or NMOG[C]	CO	NOx	PM	HCHO	THC[B]	NMHC or NMOG[C]	CO	NOx	PM	HCHO
TW. 0 - 3750 lb. / MDV1	Fe CFV	-	0.125	3.4	[5]0.4	-	0.015							-	0.180	5.0	0.6	[1]0.08	0.022
	CA Tier 0[6]	-	0.39	9.0	0.4	[1]0.08	[4]0.015												
	CA Tier 1	-	0.25	3.4	0.4	-	[4]0.015							-	0.36	5.0	0.55	[1]0.08	-
	CA LEV	-	0.125	3.4	0.4	-	[4]0.015							-	0.180	5.0	0.6	[1]0.08	0.022
	CA ULEV	-	0.075	1.7	0.2	-	0.018							-	0.107	2.5	0.3	[1]0.04	0.012
TW. 3751 - 5750 lb. / LDT3	Fe Tier 0[6]	-												0.80	-	10	1.7	[1]0.13	-
	Fe Tier 1	-	0.32	4.4	[6]0.7	-	-							0.80	0.46	6.4	0.98	0.10	-
	Fe CFV	-	0.160	4.4	[6]0.7	-	0.018							-	0.230	6.4	1.0	0.10	0.027
MDV2	CA Tier 0[6]	-	0.50	9.0	1.0	[1]0.08	[4]0.018												
	CA Tier 1	-	0.32	4.4	0.7	-	[4]0.018							-	0.46	6.4	0.98	[1]0.10	-
	CA LEV	-	0.160	4.4	0.7	-	0.018							-	0.230	6.4	1.0	[1]0.10	0.027
	CA ULEV	-	0.100	2.2	0.4	-	0.009							-	0.143	3.2	0.5	[1]0.05	0.013
TW. 5751 - 8500 lb. / LDT4	Fe Tier 0[6]	-												0.80	-	10	1.7	[1]0.13	-
	Fe Tier 1	-	0.39	5.0	[8]1.1	-	-							0.80	0.56	7.3	1.53	0.12	-
	Fe CFV	-	0.195	5.0	[8]1.1	-	0.022							-	0.280	7.3	1.5	0.12	0.032
MDV3	CA Tier 0[6]	-	0.60	9.0	1.5	[1]0.08	[4]0.022												
	CA Tier 1	-	0.39	5.0	1.1	-	[4]0.022							-	0.56	7.3	1.53	[1]0.12	-
	CA LEV	-	0.195	5.0	1.1	-	0.022							-	0.280	7.3	1.5	[1]0.12	0.032
	CA ULEV	-	0.117	2.5	0.6	-	0.011							-	0.167	3.7	0.8	[1]0.06	0.016

Federal & CA Tier 0 GVWR: 6001 - 8500 lb.

CA (not Tier 0) GVWR: 6001 - 14,000 lb.(b)

Table 3.21 Summary of Current and Future US and California Exhaust Emission Standards and Categories (continued)

Vehicle Type	Emission Category	5 Yrs. / 50,000 Miles						10 Yrs. / 100,000 Miles						11 Yrs. / 120,000 Miles					
		THC[B]	NMHC or NMOG[C]	CO	NOx	PM	HCHO	THC[B]	NMHC or NMOG[C]	CO	NOx	PM	HCHO	THC[B]	NMHC or NMOG[C]	CO	NOx	PM	HCHO
MDV4 TW. 8501 - 10,000 lb.	CA Tier 1	-	0.46	5.5	1.3	-	[4]0.028							-	0.66	8.1	1.81	[11]0.15	-
	CA LEV	-	0.230	5.5	1.3	-	0.028							-	0.330	8.1	1.8	[11]0.12	0.040
	CA ULEV	-	0.138	2.8	0.7	-	0.014							-	0.197	4.1	0.9	[11]0.06	0.021
MDV5 TW. 10,001 - 14,000 lb.	CA Tier 1	-	0.60	7.0	2.0	-	[4]0.036							-	0.86	10.3	2.77	[11]0.18	-
	CA LEV	-	0.300	7.0	2.0	-	0.036							-	0.430	10.3	2.8	[11]0.12	0.052
	CA ULEV	-	0.180	3.5	1.0	-	0.018							-	0.257	5.2	1.4	[11]0.06	0.026
MDV (OPT) GVWR 8501 - 14,000 lb. (D).(F).(G)	CA Tier 1													-	[10]3.9	14.4	[10]	[11]0.10	[8]0.10
	CA LEV													-	[10]3.5	14.4	[10]	[11]0.10	0.05
	CA ULEV													-	[10]2.5	7.2	[10]	[11]0.05	0.025
CA (not Tier 0) GVWR: 6001 - 14,000 lb. (a) HD[F] (E) 8501 - 14,000 GVWR	HDGE[G]													1.1	[11]0.9	14.4	[7]5.0	-	[8]0.10
>14,000 GVWR	HDGE[G]													1.9	[11]1.7	37.1	[7]5.0	-	[8]0.10
>8500 GVWR	HDDE[H]													1.3	[11]1.2	15.5	[7]5.0	[9]0.25	[8]0.10
8501 - 20,000 GVWR	Fe CFV[G]													-	[10]3.15	(12)	(10)	[9]0.10	-

Certification[A] Exhaust Emission Standards - gpm

658

Table 3.22   Summary of US and California Exhaust Emission Implementation Schedule

Vehicle Type	Emission Category	Emission Standards Compliance Requirements - - by Model Year														
		1992	1993	1994	1995	1996	1997	1998	1999	2000	2001	2002	2003	2004	2005	2006
PC Federal & CA	Fe Tier 0	100%	100%	60% Max.	20% Max.	0%										
LDT1 Federal & CA GVWR: 0 - 6000 lb.	Fe Tier 1(A)			40% Min.	80% Min.	100%	↑	↑	↑	↑	↑	↑	↑	(1)100%	Undetermined	Undetermined
LVW: 0 - 3750 lb.	Fe Tier 2													(2)100%	Undetermined	Undetermined
	Fe CFV 1(B)					(3)CA Pilot	CA Pilot	CA Pilot (3)30% CFV	(4)CA Pilot 50% CFV	CA Pilot 70% CFV						
	Fe CFV 2(B)										CA Pilot 70% CFV	CA Pilot ↑	CA Pilot ↑	CA Pilot ↑	CA Pilot ↑	CA Pilot ↑
	CA Tier 0	100%	60% Max.	20% Max.	0%											
	CA Tier 1		40% Min.	80% Min. and												
	CA TLEV			0.250 NMOG Fleet Avg.	0.231 NMOG Fleet Avg.	0.225 NMOG Fleet Avg.	0.202 NMOG Fleet Avg.	0.157 NMOG Fleet Avg.	0.113 NMOG Fleet Avg.	0.073 NMOG Fleet Avg.	0.070 NMOG Fleet Avg.	0.068 NMOG Fleet Avg.	0.062 NMOG Fleet Avg.	Undetermined		
	CA LEV															
	CA ULEV															
	CA ZEV							2% Min.	2% Min.	2% Min.	5% Min.	5% Min.	10% Min.	Undetermined		

659

Table 3.22 Summary of US and California Exhaust Emission Implementation Schedule (continued)

Vehicle Type	Emission Category	Emission Standards Compliance Requirements -- by Model Year

Vehicle Type	Emission Category	1992	1993	1994	1995	1996	1997	1998	1999	2000	2001	2002	2003	2004	2005	2006
LDT2 **Federal & CA** GVWR: 0 - 6000 lb. LVW: 3751 - 5750 lb.	Fe Tier 0	100%	100%	60% Max.	20% Max.	0%										
	Fe Tier 1 (A)			40% Min.	80% Min.	100%	↑	↑	↑	↑	↑	↑	↑	(1)100%	↑	↑
	Fe CFV 1 (B)					CA Pilot	CA Pilot	CA Pilot (3)30% CFV	CA Pilot 50% CFV	CA Pilot 70% CFV						
	Fe CFV 2 (B)										CA Pilot 70% CFV	CA Pilot ↑	CA Pilot ↑	CA Pilot ↑	CA Pilot ↑	CA Pilot ↑
	CA Tier 0	100%	60% Max.	20% Max.	0%											
	CA Tier 1		40% Min.	80% Min. and												
	CA TLEV			0.320 NMOG Fleet Avg.	0.295 NMOG Fleet Avg.	0.287 NMOG Fleet Avg.	0.260 NMOG Fleet Avg.	0.205 NMOG Fleet Avg.	0.150 NMOG Fleet Avg.	0.099 NMOG Fleet Avg.	0.098 NMOG Fleet Avg.	0.095 NMOG Fleet Avg.	0.093 NMOG Fleet Avg.			
	CA LEV															
	CA ULEV												Undetermined			

Table 3.22 Summary of US and California Exhaust Emission Implementation Schedule (continued)

Vehicle Type	Emission Category	Emission Standards Compliance Requirements -- by Model Year														
		1992	1993	1994	1995	1996	1997	1998	1999	2000	2001	2002	2003	2004	2005	2006
LDT3 & LDT4 Federal GVWR: 6001 - 8500 lb. TW: 0 - 8500 lb.	Fe Tier 0	100%	100%	100%	100%	50% Max.	0%									
	Fe Tier 1					50% Min.	100%	↑	↑	↑	↑	↑	↑	↑	↑	↑
	Fe CFV							(3)30% CFV	50% CFV	70% CFV	↑	↑	↑	↑	↑	↑
MDV1 - MDV5 MDV (OPT) California GVWR: 6001 - 14,000 lb. TW: 0 - 14,000 lb.	CA Tier 0	100%	100%	100%	50% Max.	0%										
	CA Tier 1				50% Min.	100%	100%	73% Max.	48% Max.	23% Max.	0%					
	CA LEV							25% Min.	50% Min.	75% Min.	95% Max.	90% Max.	85% Max.	Undetermined		
	CA ULEV							2% Min.	2% Min.	2% Min.	5% Min.	10% Min.	15% Min.	Undetermined		
HD Federal & CA Tier 0 GVWR: >8500 lb.	HDGE	100%	↑	↑	↑	↑	↑	100% (NOx)	↑	↑	↑	↑	↑	↑	↑	↑
	HDDE	100%	↑	100% (PM)	↑	↑	↑	100% (NOx)	↑	↑	↑	↑	↑	↑	↑	↑
CA Tier 1 GVWR: >14,000 lb.	Fe CFV							(3)50% CFV						↑	↑	↑

661

Table 3.21   Explanatory Notes.

(A)   Different in-use exhaust standards may apply.
(B)   OMHCE for methanol fuel.
(C)   OMNMHE for methanol fuel.
(D)   Optional certification available for all incomplete & all diesel.
(E)   GVWR -CA: Tier 0 > 8 500, Tier 1 > 14 000, Federal: > 8 500.
(F)   Standards in gm/bhp - hr.
(G)   Useful life 8 yrs/110 000 miles.
(H)   Useful life 8 yrs/110 000; 185 000; 290 000 miles as specified by manufacturer.
(1)   Diesel only.
(2)   Diesel through 2003 MY - 1.0.
(3)   Diesel through 2003 MY - 1.25.
(4)   Methanol fuelled vehicles only.
(5)   Except diesel.
(6)   Use LVW definition for TW.
(7)   4.0 NOx in 1998 MY.
(8)   CA Methanol fueled vehicles only - 0.05 in 1996 MY.
(9)   0.10 PM in 1994 MY.
(10) Combined NMHC & NOx standard in g/bhp-hr.
(11) CA optional.
(12) HDGE or HDDE standard applies.

(Source: General Motors).

Table 3.22:   Explanatory legend and notes.

Legend:

Fe: Federal.
CA: California.
Tier 0: Existing Standards - Pre - Tier 1.
Tier 1: New, General Application Standards; CA 1993 MY, Fe 1994 MY.
CFV: Clean Fuel Vehicle Standards.
         CFV 1: Phase 1 for PC, LDT1, LDT2.
         CFV 2: Phase 2 for PC, LDT1, LDT2.
TLEV: Transitional Low Emission Vehicle (CA).
LEV: Low Emission Vehicle (CA).
ULEV: Ultra Low Emission Vehicle (CA).
ZEV: Zero Emission Vehicle (CA).
HDGE: Heavy Duty Gasoline Engine.
HDDE: Heavy Duty Diesel Engine.

Notes:

(A)   PC & LDT1 combined with LDT2 for Tier 1 phase-in.
(B)   PC & LDT1 combined with LDT2 for CFV requirements.

(1)   NOx standard change for diesel.
(2)   Tier 2 standards pending EPA study by 12/31/99.
(3)   Percent of new vehicle purchases by centrally-fuelled fleets in 22 cities.

(Source: General Motors).

engine models would meet the 1996 standards. Furthermore, low aromatic and low sulphur diesel fuels would be allowed for certification testing for the 1996 and 1997 model years.

CARB projections show that heavy-duty vehicle exhaust will contribute more than 50 per cent of NOx and more than 84 per cent of particulates emitted by all on-road vehicles by 2000. It has been calculated that the proposed Federal heavy-duty vehicle standards for 1998 will reduce NOx by only 20 per cent. This is considered insufficient for California to meet the National Air Quality Standards set by the Clean Air Act. CARB is therefore developing an extension of its LEV programme to apply to heavy-duty vehicles. Such a programme would require the further development of catalytic converters and particulate traps, plus engines capable of using alternative fuels such as methanol and CNG. It is also encouraging local air quality districts to develop mobile source credit programmes to enable the adoption of LEV buses in exchange for emissions offsets.

CARB is also proposing to change the GVW rating of buses from 14,000 to 33,000 pounds which would align its definition with that used by the EPA. This revision will not affect large and medium sized transit companies, most of which generally operate buses over 33,000 pounds. It would, however, minimize the impact of the regulations on small transit districts which operate a greater proportion of smaller buses. The revised classification will not significantly affect control of NOx emissions as larger buses are the main contributors.

CARB is proposing that all transit buses be equipped with positive crankcase ventilation (PCV) systems beginning with the 1996 model year. At present PCV systems are required on all heavy-duty vehicles except for turbocharged diesel engines. The cost of adding PCV systems is estimated at around USD 100 to 240 per vehicle but is claimed to result in a significant reduction in HC emissions.

Refuelling Emissions

A report published by The National Highway Traffic Safety Administrator (NHSTA) in September 1991 concluded that on-board refuelling controls are significantly less safe than the alternative Stage II vapour recovery systems. As a result, the EPA decided not to issue a rule requiring large carbon canisters to be fitted to vehicles. However, this decision was overturned by a Federal Court ruling that the EPA must comply with the Clean Air Act, which explicitly states that the EPA shall promulgate regulations requiring on-board controls.

The EPA issued its rule on the control of refuelling emissions on 24 January 1994. The rule requires on-board refuelling emissions controls for passenger cars and light trucks (e.g. pickups, minivans and most delivery and utility vehicles). It will not require on-board control of refuelling emissions for heavy duty vehicles and trucks over 8500 pounds GVW. The rule covers 97 per cent of new vehicles and 94 per cent of refuelling emissions.

For passenger cars the controls will be phased in over three model years with 40 per cent, 80 per cent and 100 per cent of new car production being required to meet the standard in model years 1998, 1999 and 2000, respectively. Comparable proportions of light trucks will require on-board controls over three-year periods 2001-2003 (GVW <6000 lbs) and 2004-2006 (GVW 6000-8500 lbs).

The rule establishes a refuelling emission standard of 0.20 grams per gallon of dispensed fuel and is expected to yield a 95 per cent reduction over current uncontrolled levels.

Evaporative Emissions - US Federal

The EPA has issued regulations, effective from 23 April 1993, specifying revised procedures and limits for evaporative emissions, with implementation phased-in over the 1996 to 1999 model years. The regulations apply to light- and heavy-duty vehicles and heavy-duty engines fuelled with gasoline, methanol or gasoline/methanol mixtures.

The EPA has also specified that from the 1 January 1996, the dispensing rates from gasoline and methanol pumps may not exceed 10 US gallons (37.9 litres) per minute. Facilities with throughputs below 10,000 gallons per month have been given a further two years to comply. This requirement is consistent with the dispensing rates specified in the new test measuring spillage during refuelling.

The current test procedure, which has changed little since its introduction, measures emissions from fuel evaporation during parking (diurnal emissions) and immediately following a drive (hot soak emissions).

The new procedures, described in detail in **Section 4.4**, consist of vehicle preconditioning (including an initial loading of the carbon canister with fuel vapour), exhaust emission testing, a running loss test and three diurnal emissions cycles. Fuel spillage during refuelling (spitback) is also measured. A supplemental procedure omitting the running loss test but involving two diurnal cycles following the emissions cycles is included. This procedure ensures that all the emissions resulting from purging the evaporative canister are measured during the emission and diurnal cycles and do not escape during the running loss test. Because of its increased severity, the limits specified for this test are more relaxed than those for the three-diurnal sequence. The supplemental procedure can also be used in conjunction with the test procedures devised by the CARB, which are not yet in effect.

The procedures for heavy duty vehicles are similar except that the driving sequence for the running loss test consists of three consecutive UDDS cycles, which reflect the different driving pattern experienced in service. The testing of heavy-duty engines, without the vehicle chassis or body, requires that the test engine be equipped with a loaded evaporative canister and will be expected to demonstrate a sufficient level of purge during engine testing.

The EPA has pointed out that it has powers to deny certification upon determination that a particular control system design constitutes a defeat device, i.e. an auxiliary emission control device that reduces the effectiveness of the system under conditions which may reasonably be expected to be encountered in normal vehicle operations.

The limits are given in **Table 3.23**

Table 3.23 US Federal Evaporative Emissions Requirements for all Vehicles from 1996

Implemen- tation Schedule* % prodn.	Vehicle GVW lb	Durability Mileage	3- Diurnal Hot Soak g/test	Supple- mentary 2- Diurnal g/test	Running Loss g/mile	Spitback g liquid/ test
1996 20%	<6000	**	2.0	2.5	0.05	1.0
1997 40%	>6000 <8500	120 000	2.5	3.0	0.05	1.0
1998 90%	>8500 <14000	120 000	3.0	3.5	0.05	1.0
1999 100%	>14000	120 000	4.0	4.5	0.05	-

*   Implementation for methanol-fuelled vehicles 1998 model year. Manufacturers selling less than 10,000 vehicles per year do not have to comply until the 1999 model year.
**   Durability mileage: LDV 2 years or 24 000 miles if device cost less than $200, 8 years or 80,000 if deemed "specified major emission components", light-duty trucks <3,750lbs, 10 years or 100,000 miles, >3,750lb; 120,000miles.
*** Limits for methanol-fuelled vehicles in g/carbon per test or mile.

Vehicle Maintenance & In-Service Testing **(see also Section 5).**

The CAAA requires the introduction, starting in 1994, of onboard diagnostic systems for light duty vehicles and trucks. These must cover at least the catalytic converter and the oxygen sensor, they must also alert the operator of any possible malfunction or need for repair to emission control parts.

The Amendments make provision for extended compliance testing starting in 1996, this will permit 25% higher CO and NMHC emissions and 50% higher NOx emissions for vehicles having covered between 50,000 and 100,000 miles.

Enhanced inspection and maintenance programmes were planned to be introduced in the most polluted areas from 15.11.92. They consist of a biennial inspection of all trucks and light vehicles from the 1984 model year and later.

The EPA issued a draft detailing the test procedures and related requirements for its controversial IM 240 test on the 5 April 1993 with the intention of promulgating them in the Code of Federal Regulations under Section 207(b) of the Clean Air Act as the official IM test. The EPA's recommended procedure includes three features (see **Section 4.8**).

    a) A pressure test of the evaporative emissions control system.
    b) A purge test of the evaporative emissions control system.
    c) A transient exhaust emissions test.

These procedures have attracted much criticism for two main reasons. Firstly it is suggested that inspection and maintenance schemes can do nothing to prevent subsequent tampering. Conversely roadside remote monitoring can both identify "gross polluters" and monitor the performance of large numbers of vehicles. Secondly the cost of the sophisticated measuring equipment is claimed to be beyond the means of many of the small garages currently conducting inspection and maintenance tests. As a result a number of areas are adopting their own procedures as alternatives to the EPA proposals (see **Section 2.3.2**).

Alternative Fuels

Contrary to President Bush's original proposals, the final version of the Clean Air Act Amendments contain no mandate for the introduction of alternative fuels. Instead it describes performance criteria for "*Clean alternative fuels*" which may include:

"methanol and ethanol (and mixtures thereof), reformulated gasoline, natural gas, LPG, electricity and any other fuel which permits vehicles to attain legislated emission standards."

Since the standards set in the CAAA appear likely to be achievable by future conventional vehicles it is likely that *"conventional"* gasoline and diesel will qualify as clean fuels under certain specific circumstances.

The Amendments do make provision for a Clean Fuels programme which will apply from 1998 to fleets of 10 or more vehicles that are capable of being centrally refuelled (but NOT including vehicles that are garaged at personal residences under normal circumstances) which operate in areas which have problems achieving air quality standards. This programme mandates emission standards for these vehicles which are the same as those specified in California's Low Emission Vehicle (LEV) programme.

This part of the CAAA also specifies a pilot programme for the introduction of lower emitting vehicles in California, beginning in 1996. Under this programme 150 000 clean fuel vehicles must be produced for sale in California in 1996 and this figure will rise to 300 000 in 1999. These vehicles will initially be required to meet Transitional Low Emission Vehicle (TLEV) standards. These limits remain in force until 2000 when the LEV standards outlined above come into operation.

Starting 1994, buses which operate more than 70% of the time in large urban areas will be required to cut particulate emissions by 50% i.e. to 0.05 g/bhp.h, although this may be relaxed to a 30% reduction. EPA will test buses meeting this standard to ensure compliance and if it determines that more than 40% of buses do not comply, they must establish a low pollution fuel requirement. This provision allows the use of exhaust after-treatment devices provided they work in the field, if they fail EPA will mandate alternative fuels.

**Table 3.24 US Emissions Limits for
"Clean Alternative Fuels" Programme**

Programme	Effective Date	Equivalent Standard	CO g/mile	NMHC g/mile	NOx g/mile	HCHO g/mile
Fleet Refuelled	1998	LEV	3.4	0.075	0.2	0.015
California Pilot	1996	TLEV	3.4	0.125	0.4	0.015
Programme	2000	LEV	3.4	0.075	0.2	0.015

California

California has always set more stringent emission limits than the rest of the US, and has now established a plan for the progressive reduction of vehicular emissions which is designed to enable the state to achieve national air quality standards by the year 2010. This plan involves the progressive introduction of so-called Transitional Low Emission Vehicles (TLEV) and Low Emission Vehicles (LEV) referred to earlier. However, it also extends to the introduction of Ultra-Low Emission Vehicles (ULEV) and Zero Emission Vehicles (ZEV).

ZEVs are defined as vehicles which have no exhaust or evaporative emissions of any pollutant, the standards for the other categories are defined in **Table 3.25**.

A fleet average NMOG standard has been established (see **Table 3.22**) and manufacturers will be permitted to manufacture any combination of vehicles as long as sales-weighted emissions do not exceed the fleet average standard. Manufacturers will also be able to accrue marketable credits for complying with, or exceeding, standards.

The introduction of ZEVs is, however, mandated. The Act dictates that by 1998 ZEVs must account for 2% of manufacturers sales, this figure rises to 10% in 2003.

Equivalent standards will be applied to light and medium duty trucks (See **Tables 3.21 and 3.22**). Three new categories of standards (TLEV, LEV and ULEV) have been established of equivalent stringency to the light duty standards.

Table 3.25 Californian Standards for Low Emissions Vehicles

Category	CO (g/mile)	NMOG (g/mile)	NOx (g/mile)	HCHO* (g/mile)	Pm (diesel only)
1993 Standards	3.4	0.25	0.4	0.015	0.08
TLEV	3.4	0.125	0.4	0.015	0.08
LEV	3.4	0.075	0.2	0.015	0.08
ULEV	1.7	0.04	0.2	0.008	0.08
ZEV	0	0	0	0	0

* HCHO = Formaldehyde (methanol fuelled vehicles only)

Intermediate in-use standards, which are up to 30% less stringent than the above certification standards, will be applicable for two years only after a model's introduction. Compliance with the 100,000 mile useful life standards will also be suspended for these two years. However, manufacturers will be required to submit to CARB an *"engineering evaluation"* of the effectiveness of each engine emission control system over the range 20-86°F (-7 to 30°C) and to demonstrate that vehicles can comply with the standards at 45-55°F (7-13°C).

California has also ruled that, from 1994 *"major gasoline suppliers"** will be required to make available alternative fuels at retail outlets. However, the application of this rule will be subject to the availability of a reasonable number of suitable vehicles.

CARB has introduced a procedure which enables the NMOG emissions of a vehicle operating on a fuel producing low reactivity NMOG components to be adjusted:

Adjusted Emissions Test Result = Measured NMOG (g/mile)
Reactivity Factor (RAF)

Each vehicle category/fuel combination will have a unique RAF determined by CARB.

The reactivity adjustment factor (RAF) is given by:

RAF = (g ozone/g NMOG) Alternative Fuel/Vehicle Combination
(g ozone/g NMOG) Conventional Gasoline Vehicle

The reactivity (g ozone/g NMOG) for a passenger car operating on current conventional gasoline has been defined by CARB as 3.42 g ozone/g NMOG. CARB had defined the RAFs of M85, CNG, and LPG as 0.36, 0.18, and 0.15 for passenger cars respectively, but withdrew the figures for LPG and CNG because of a poor statistical basis and modified that for M85 to 0.41.

CARB has defined a 10% correction factor for alcohol and LPG RAFs to account for *"potential modeling and protocol biases"* which increases in the RAF by 10%. The M85 of 0.41 includes this correction.

Note: (*) Those companies having a refinery within California which has a capacity greater than 50 000 barrels/day and who own or lease more than 25 retail stations in the South Coast region.

The RAF for *"reformulated gasoline/passenger cars"* was to be published in the summer of 1992. The RAFs for LPG and CNG will be clarified once a sufficiently large vehicle population has been tested to determine statistically reliable numbers. The RAFs for diesel and reformulated diesel are proving problematic to determine as the heavy hydrocarbons C_{12}-C_{20} are proving difficult to speciate. CARB is currently working on this problem.

A manufacturer can accept the RAF determined by CARB for each vehicle category and fuel combination during the durability vehicle/data vehicle certification process. Thus the manufacturer would apply two deterioration factors to the vehicles certification test results as follows:

Certification Value = Measured Emissions (4 k miles)xRAFxRAFDFxDF

Where RAFDF = reactivity adjustment factor deterioration factor
DF = conventional deterioration factor applied to NMOG

$$RAFDF = \frac{\text{(g ozone/g NMOG) 100 k miles}}{\text{(g ozone/g NMOG) 4 k miles}}$$

$$DF = \frac{\text{g NMOG 100 k miles}}{\text{g NMOG 4 k miles}}$$

CARB has also published Incremental Reactivity Factors for various NMOG compounds to enable development engineers to establish the reactivity of the NMOG exhaust profiles from their vehicles. As for Federal Regulations, Phase II reformulated gasoline will be introduced in 1996 and will be classified as a clean fuel.

State-Autonomy

The Clean Air Act reaffirms the authority of individual states to adopt more stringent emission standards if they wish to do so. However, they will only be permitted to adopt the standards set by California. This restriction was imposed in order to prevent motor manufacturers having to produce individual models for each state, instead they will only need to produce two models, one complying with Federal standards and one complying with Californian standards.

It is probable that the Californian exhaust emissions limits will be adopted by many North-Eastern and Mid-Atlantic States. Governors from Virginia, Maryland, Pennsylvania, New York, Massachusetts, Delaware, New Jersey, New Hampshire, Maine, Vermont, Connecticut and Washington DC have all signed an agreement to pursue adoption of the California limits as is their right under the "State Autonomy" section of the Clean Air Act. Rhode Island is still studying the plan. The area covered by this agreement will contain over a third of the US population. Other states studying the possibility of opting-in include Illinois, Wisconsin and Texas. Cross-border sales between states with different legislative limits could be seriously affected. Sales of Californian cars are currently authorized in Arizona, Connecticut. Massachusetts, Nevada, New Jersey, Oregon, Pennsylvania and Vermont. The position regarding adoption of Californian standards by other states at September 1993 was as follows

Table 3.26 Status of Adoption of Californian Exhaust Emission limits

State	Status
Delaware	Plans under consideration
Maryland	Legislation being considered
Massachusetts	Adopted Californian standards effective 1995 model year
New Jersey	Endorsed Californian rules in principle under review
New York	Under review
Pennsylvania	Endorsed Californian rules in principle
Rhode Island	Needs legislative approval
Vermont	Governor supports rules for DC suburbs
Virginia	Needs legislative approval

There is a vociferous debate as to whether the Californian Emission Limits can be adopted by these other states without mandating the use of Californian "Reformulated" gasolines (see **Section 7.2.2**).

3.5 Canadian Vehicle Emissions Regulations

Canadian emission regulations have always followed US EPA test cycles and procedures, although from 1975-1987 less stringent limits were applied which did not require the use of catalysts. However, many Canadian vehicles were in fact identical to US specification vehicles and fitted with catalysts. In 1987 the Canadian legislation was brought into line with current US limits. Emission limits for heavy duty trucks are the same as the 1988 US limits. An agreement signed in February 1992 between the Federal Transport Minister and the motor manufacturers and importers means that engines being supplied currently are to the 1991 US emissions standards. Furthermore, the oil industry has promised to make 0.05%m/m sulphur diesel fuel available no later than October 1994 so that 1994 US specification engines can be supplied from that date. Another memorandum of understanding means that cars sold in Canada from 1994 to 1996 will be to the same emissions standards as those sold in the US. It is expected that, by 1996, the Canadian government will have passed legislation harmonizing Canadian and US emissions regulations according to the requirements of the US "Clean Air Act". A summary of Canadian emissions regulations is given in **Table 3.27**.

Table 3.27 Canadian Emissions Regulations

Vehicle Class	Year	Test Procedure	CO g/km	HC g/km	NOx g/km	Pm* g/km
Cars and Lt Trucks	1975-87	FTP 75	25.0	2.0	3.1	-
Cars	1988	FTP 75	2.11	0.25	0.62	0.12
Lt Trucks:						
<1700 kg	1988	FTP 75	6.2	0.20	0.75	0.16
>1700 kg	1988		6.2	0.50	1.1	0.16
			g/bhp.h	g/bhp.h	g/bhp.h	g/bhp.h
HD Vehicles Gasoline:						
<6350 kg	1988	US	14.4	1.1	6.0	-
>6350 kg	1988	Transient	37.1	1.9	6.0	-
Diesels	1988		15.5	1.3	6.0	0.6
	1994		15.5	1.3	5.0	0.1

* Diesels only

A number of Province-controlled programmes have also been agreed:-

- Vehicle inspection and maintenance programmes in provinces with ozone problems from 1992.

- A limit on summer gasoline RVP of 10.5 lb/in^2 (72 kPa) from 1990. Lower limits may be set in provinces with ozone problems.

- Regulation to control gasoline vapour emissions from distribution and marketing installations starting in 1991.

3.6 Japanese - Vehicle Emissions Regulations

Emission control in Japan started in 1966 when simple CO limits were introduced, but the first long-term plan was established in 1970 by the Ministry of Transport (MOT). This plan proposed limits for CO, HC and NOx from 1973, with separate sets of limits for a 10-mode hot-start cycle and an 11-mode cold-start test. For heavy duty vehicles a 6-mode steady state test was introduced. In 1971 however, the Central Council for Environmental Pollution Control (CCEPC), an advisory body of Japan's Environment Agency (EA), recommended that legislation should follow the US "Muskie proposal" and submitted recommendations for much more stringent exhaust emission standards. This led to tough limits introduced in 1975 which required the use of catalysts on gasoline cars. These limits (with an NOx reductions in 1978) have not changed since then, but revisions to the test procedures (see below) have effectively made them more severe. Emissions limits for trucks, both gasoline and diesel were also introduced in 1974/5, but these limits have been tightened by varying degrees over the intervening years. A useful summary of Japanese Emission Legislation development is given in SAE paper 922178 "Motor vehicle emission control measures of Japan".

In December 1989 the CCEPC recommended new emission limits with both short-term and long-term targets. Their aim was to set up the most stringent standards which were technologically feasible, and to apply the same standards for both gasoline and diesel fuelled vehicles in the near future. Based on this proposal the MOT revised the emission regulations in May 1991 to incorporate the short-term limits. The longer term targets are being evaluated by the Environment Agency. The major changes are as follows:

- Reductions in NOx emissions from commercial vehicles. Recently the major concern in Japan has been over NOx emissions.

- Introduction of particulate standards and more stringent smoke limits for diesel vehicles. Smoke limits were reduced by 20% in 1993 for light and medium duty diesel vehicles. Heavy duty and passenger vehicle limits follow suit in 1994.

- Revision of test cycles and measurement modes. The 10-mode light duty test cycle has been modified to include a high speed section, and is now called the 10-15 mode cycle and is applied to light duty gasoline and diesel vehicles. The 6-mode gasoline and diesel cycles have been replaced by two new 13-mode cycles with emission measurements changed from ppm to g/kWh. Full details are given in **Section 4.3**.

Current and future emission limits are shown in **Tables 3.28 and 3.29** for gasoline and diesel passenger cars, and in **Tables 3.31 and 3.32** for gasoline and diesel trucks.

More recently a joint MOT/MITI/EA study has proposed legislation to further reduce NOx in urban areas. These proposals had been discussed since spring 1992 and were adopted in December 1992, to take effect December 1993. The objective is to control both the population of older vehicles and impose even tougher emission limits for new vehicles. It is proposed that the regulations will apply to all diesel vehicles in the specified areas. However there are a number of derogations for various classes of older vehicles.

Table 3.28 Current and future emissions standards for gasoline and LPG fuelled passengers cars [1]

Test Method	Exhaust Emission Limits(g/km)						
	CO		HC		NOx		
(2)	Max (5,6)	Mean (4)	Max (5,6)	Mean (4)	Max (5,6)	Mean (4)	Impl. Date
10.15	2.7	2.1	0.39	0.25	0.48	0.25	1991 (2)
		g/test		g/test		g/test	
11 mode	85.0	60.0	9.5	7.0	6.0	4.4	1978 (3)
Evap.				2.0			1978 (3)

Table 3.29 Current and future Japanese emission standards for diesel fuelled passenger cars

Vehicle weight Tonnes	Test Method (2)	Exhaust Emission Limits (g/km)								Impl. Date
		CO		HC		NOx		Particulate		
		Max (5,6)	Mean (4)	Max (5,6)	Mean (4)	Max (5,6)	Mean (4)	Max (5,6)	Mean (4)	
Ó1.265	10.15	2.7	2.1	0.62	0.40	0.98	0.70			1986
" "	" "	" "	" "	" "	" "	" "	0.72	0.50		1990
" "	" "	" "	" "	" "	" "	" "	" "	0.34	0.20	1994
" "	" "	" "	" "	" "	" "	" "	" "		0.40	0.80	Long term
Ò1.265	10.15	2.7	2.1	0.62	0.40	1.26	0.90			1986
" "	" "	" "	" "	" "	" "	" "	0.84	0.60		1992
" "	" "	" "	" "	" "	" "	" "	" "	0.34	0.20	1994
" "	" "	" "	" "	" "	" "	" "	" "		0.40	0.08	Long term
All	Smoke								50%	1972
"	test								40%	1994
"	(3-mode)								25%	Long term

See footnotes following **Table 3.32.**

The proposed regulation, the "Specific Measures for Trucks and Buses" will apply the strictest NOx emission standard for a given weight category to ALL vehicles in service, i.e. gasoline, LPG and diesel. Thus the following limits will apply:

GVW < 1.7 tonnes:	1988 gasoline limits i.e. 0.48 (0.25 mean) g/km
1.7t<GVW<2.5t:	1989 gasoline limits i.e. 0.98 (0.70 mean) g/km
2.5t<GVW<5t:	1994 IDI diesel standards applied after 1994 i.e. 6.8 (5.0 mean) g/kWh
GVW >5t:	1994 DI diesel standards applied after 1994 i.e. 7.80 (6.0 mean) g/kWh

These new limits will be applied in over 150 city districts, covering the main urban areas of greater Tokyo, Yokohama and Osaka. Diesel powered vehicles which exceed the NOx limits will be refused new registration or will not be allowed to renew their old registration. In addition older vehicles will not be eligible for the "Shaken" annual inspection test (see **Section 5.6.1**) and thus cannot be used on the roads. A transitional period till December 1994 will be allowed for such older vehicles before their use is prohibited. The maximum permitted age for various vehicle classes is shown in **Table 3.30**.

The MOT has also established environmental "guidelines" for methanol fuelled vehicles, proposing that NOx should be 40% less than the limits for gasoline fuelled vehicles. CO and HC limits are the same as for gasoline, but formaldehyde limits are also included. The guidelines are only recommendations as yet and it remains to be seen if they will be converted into official standards.

Motorcycles

Japan currently has no limits for motorcycle emissions and the EA set up a committee in 1991 to consider the introduction of limits for motorcycles and "special" vehicles (e.g. off road). The committee is expected to report by the end of 1994.

Table 3.30 Maximum permitted age for commercial vehicles in Urban areas

GVW tonnes	Transition Year		Long Term	
Light trucks/vans	Max. Age - years	Effective Date	Max. Age - years	Effective Date
<3.5	10	Dec. 1994	8	Dec. 1995
3.5-5.0	9	April 1996	8	April 1997
Heavy trucks			9	
Pass car/RV etc.			10	
Buses			12	

Table 3.31 Current and Future Japanese Emission Standards for Gasoline and LPG Fuelled Commercial Vehicles

| Vehicle weight | Test method | | Exhaust Emissions Limits | | | | | | Impl. Date |
| | | | CO | | HC | | NOx | | |
Tonnes	(2)	Units	Max (5,6)	Mean (4)	Max (5,6)	Mean (4)	Max (5,6)	Mean (4)	
61.7	10.15	g/km	2.7	2.1	0.39	0.25	0.48	0.25	1988/9
" "	11-mode	g/test	85.0	60.0	9.5	7.0	6.0	4.4	1988/9
1.7-2.5	10.15	g/km	17.0	13.0	2.70	2.10	0.98	0.70	1989/90
" "	" "	" "	" "	" "	" "	" "	" "	0.40	long term
" "	11-mode	g/test	130	100	17.0	13.0	8.5	6.5	1989/90
ò 2.5 gasoline	6-mode	ppm (%)	1.6%	1.2%	520	410	850	650	1987
ò 2.5 LPG	6-mode	ppm (%)	1.1%	0.8%	440	350	850	650	1987
ò 2.5 gasoline	13-mode	g/kWh	136	102	7.90	6.20	7.20	5.50	1992
								4.50	long term
ò 2.5 LPG	13-mode	g/kWh	105	76	6.80	5.40	7.20	5.50	1992

Table 3.32 Current and Future Japanese Emission Standards for Diesel Fuelled Commercial Vehicles

| Vehicle weight | Test method* | Exhaust Emission Limits | | | | | | | | Impl. Date |
| | | CO | | HC | | NOx | | Particulate | | |
Tonnes	(2)	Max (5,6)	Mean (4)	Max (5,6)	Mean (4)	Max (5,6)	Mean (4)	Max (5,6)	Mean (4)	
61.7	10.15	2.7	2.1	0.62	0.40	1.26	0.90			1988
" "	" "	" "	" "	" "	" "	0.84	0.60	0.34	0.20	1993
" "	" "	" "	" "	" "	" "	" "	0.40		0.08	long term
1.7-2.5	6-mode	980	790	670	510	350(IDI)	260(IDI)			1988
						500(DI)	380(DI)			
	10.15	2.7	2.1	0.62	0.40	1.82	1.30	0.43	0.25	1993
	" "	" "	" "	" "	" "	" "	0.70		0.09	long term
ò 2.5	6-mode	980	790	670	510	520 (DI)	400 (DI)			1988/9
						350 (IDI)	260 (IDI)			
	13-mode	9.20	7.40	3.80	2.90	7.80 (DI)	6.00(DI)	0.96	0.70	1994
						6.80 (IDI)	5.00 (IDI)			
	" "	" "	" "	" "	" "	" "	4.50		0.25	long term

*   Units : See **Table 3.31** above.

Notes:

1. Covers vehicles (no mass limitation) which serve exclusively for the transport of passengers (maximum 10 people).

2. New Hot Start Test (10·15-Mode) superseded the 10-mode test with effect from 1.11.91 for new models; 1.4.93 for importers. The exhaust emission limits remain unchanged.

3. 80 000 km durability run optional; acceptance of US durability run possible. *Advantage:* if standards are met over 80 000 km, the mandatory periodic catalyst change does not apply. Alternatively certification is allowed with a 30 000 km durability run and demonstration of compliance over 45 000 km (by extrapolation).

4. To be met as a Type Approval limit and as a production average (for production control 1% of production has to be tested). If sales exceed 2000 per vehicle model per calendar year, the NOx standards are only applicable if reference mass is > 1 000 kg.

5. To be met as a Type Approval limit if sales are less than 2000 per vehicle model per calendar year and generally as an individual limit in series production. For gasoline and diesel engines (Hot Start Test only) deterioration factors from the durability runs have to be applied.

6. Applicable for simplified certification procedure if sales are less than 1000 per vehicle model per calendar year without durability run. Exhaust emission testing is necessary for every 50th production example per vehicle model.

7. *3-Mode:*   Full load smoke test at three specified engine speeds.

    Free Acceleration: Start from idle; integrated smoke measurement over a 15 second cycle (4 sec. maximum acceleration, followed by 11 sec. coast).

3.7 Other Far Eastern Vehicle Emissions Regulations

China

The following information about current and future emissions legislation in China has come from an unconfirmed source.

Light Duty Vehicles

The limits according to Regulation No. GB-11641-89 which are given in **Table 3.33**, are based on ECE 15.03 limits.

Heavy Duty Vehicles

Current regulations according to Regulation No. GB 3842-83 consist only of CO and HC limits determined at idle and apply both to certification and in-use (see **Table 3.34**).

China is considering legislation for gasoline-engine heavy duty vehicles based on two runs of the US 9-mode cycle. Proposed limits are given in **Table 3.35**.

Table 3.33 Current Chinese Light Duty Vehicle Exhaust Emission limits

Vehicle Ref. Mass kg	Type Approval (g/test)			Conformity of Production (g/test)		
	CO	HC	NOx	CO	HC	NOx
<750	65	10.8	8.5	78	14.0	10.2
750-850	71	11.3	8.5	85	14.8	10.2
850-1020	76	11.3	8.5	91	15.3	10.2
1020-1250	87	12.8	10.2	104	16.6	12.2
1250-1470	99	13.7	11.9	119	17.8	14.3
1470-1700	110	14.6	12.3	132	18.9	14.8
1700-1930	121	15.5	12.8	145	20.2	15.4
1930-2150	132	16.4	13.2	158	21.2	15.8
>2150	143	17.3	13.6	172	22.5	16.3

Table 3.34 Current Chinese Gasoline-Powered Heavy Duty Vehicle Exhaust Emission Limits

Vehicle	Idle CO (%V)	Idle HC(ppm)
New	5	2500
In -use	6	3000
Imports	4.5	1000

**Table 3.35 Future Chinese Gasoline-Powered Heavy Duty Vehicle
Exhaust Emissions Limits**

Year	Vehicle	CO g/kWh	HC+NOx g/kWh
Up to 1997	Certified before 1992	80	32
	Produced after 1992	50	20
	Type approved after 1992	40	16
1997	Certified before 1992	50	20
	Produced after 1992	34	13.6
	Type approved after 1992	28	11

South Korea

Until July 1987, exhaust emission regulations in South Korea were based on
Japanese test procedures. New regulations were introduced for spark ignition
vehicles based on US test procedures as shown in **Table 3.36**. Limits for
diesel vehicles are less severe and continue to be based on Japanese test
procedures.

Legislation has also been introduced for 2-stroke and 4-stroke motorcycles
which will require the use of catalysts. Limits are as follows, but the test
procedure is not known.

Taiwan

Legislation in Taiwan requires vehicles to be certified to current US Limits
for gasoline vehicles from July 1990. In July 1993, light duty diesel vehicles
must be certified to US 1984 light duty truck limits, including particulates.
Heavy duty truck particulate limits, based on the US transient test cycle (see
Section 4.6), have also been proposed. Information regarding parallel NOx
limits are not yet confirmed. Details are given in **Table 3.43**.

Table 3.36 South Korean Passenger and Commercial Vehicle Emission Standards

Type of vehicle		Date	CO	Exhaust HC	Evap. HC	NOx	Smoke (Pm)	Test procedure	Equivalent to
Passenger cars			g/km	g/km	g/test	g/km	(g/km)		
(gasoline	ó 800 cc	July 1987	8.0	2.1	4.0[1]	1.5	-	FTP 75	Unique
and LPG)		Jan 2000	2.11	0.25	2.0	0.62	-		Unique
Passenge cars	ò 800 cc	July 1987	2.11	0.25	2.0	0.62	-	FTP 75	US 1983
(gasoline	but								
and LPG)	<2.7 tons	Jan 2000	2.11	0.25	2.0	0.16	-	FTP 75	Unique
Passenger cars	ò 800 cc but	Jan 1993	2.11	0.25	-	1.25	(0.25)	FTP 75	Unique
(Diesel)	<2.7 Tons	Jan 1996	2.11	0.25	-	0.62	(0.12)	FTP 75	US 1987
		Jan 2000	2.11	0.25	-	0.62	(0.05)	FTP 75	US 1994
Light duty trucks	<2.7 tons	July 1987	6.21	0.50	2.0	1.43	-	FTP 75	US 1984
(gasoline and LPG)		Jan 2000	2.11	0.25	2.0	0.62	-	FTP 75	Unique
Light duty Trucks	ò 2.7 tons		ppm	ppm		ppm			
(Diesel)	DI	July 1987	980	670	-	850	50%[2]	Japanese 6-mode	Japan 1977
	IDI	July 1987	980	670	-	450	50%[2]	Japanese 6-mode	Japan 1977
	DI	Jan 1993	980	670	-	750	40%	Japanese 6-mode	Unique
	IDI	Jan 1993	980	670	-	350	40%	Japanese 6-mode	Japan 1993
			g/km	g/km		g/km	(g/km)		
	DI & IDI	Jan 1996	6.21	0.5	-	1.43	(0.31)	FTP 75	Unique
	DI & IDI	Jan 2000	2.11	0.25	-	0.62	(0.05)	FTP 75	Unique
Heavy duty vehicle	ò 2.7 tons	July 1987	g/bhp.h 15.5	g/bhp.h 1.3	g/test 4.0	g/bhp.h 10.7	-	US Transient Diesel	US 1984
(gasoline and LPG)		Jan 2000	33.5	1.3	4.0	5.5	-	Japanese Gasoline 13-mode	Unique
Heavy duty vehicles	DI	July 1987	ppm 980	ppm 670	-	ppm 850	50%[3]	Japanese 6-mode	Japan 1977
(Diesel)	IDI	July 1987	980	670	-	450	50%[3]	Japanese 6 mode	Japan 1977
			g/bhp.h	g/bhp.h		g/bhp.h	(g/bhp.h)		
	DI&IDI	Jan 2000	4.9	1.2	-	6.0	(0.25) (0.10)[4] 25%	ECE R 49 13 mode	Unique

(1) Reduced to 2.0 g/test WEF 01.01.1996.
(2) Reduced to 40% WEF 01.01.1991.
       (Note: Test Procedure - Japanese 3-mode or free acceleration).
(3) Reduced to 40% WEF 01.01.1992.
(4) 0.10 g/bhp.h Pm limit applies to buses.

Table 3.37 South Korean Motorcycle Emission Standards

Period	2-stroke		4-stroke	
	CO % v	HC ppm	CO % v	HC ppm
1/91 to 12/92	5.5	1100	5.5	450
1/93 to 12/95	4.5	1100	4.5	450
1/96 forward*	3.6	450	3.6	400

*   Proposed

Taiwan has also introduced limits for motorcycles which required catalysts be fitted from 1991 and will probably require 3-way catalysts from 1994. Limits over the ECE 40 test cycle are as follows:

**Table 3.38 Taiwan - Two and Four Stroke Motorcycle
Emissions Regulations**

Effective Dates	Emissions limits		
	CO	HC	NOx
01.88 to 06.91	7.3	4.4*	
07.91 to 06.94	4.5	3.0*	
01.94 forward	1.5	1.2	0.4
01.97 (proposed)	0.75	0.6	0.2

*   Combined HC+NOx limits

Singapore

Singapore has adopted intermediate gasoline emissions limits from January 1992, i.e. European ECE 83 or Japanese 1978 limits. However, from July 1993 the authorities plan to introduce EU *"Consolidated Emissions Directive"* limits (**see Table 3.3**). New diesel vehicles have had to comply with the ECE R24 smoke emission standard (see **Section 3.13**) since 01.01.91. From 01.01.92 this requirement has been extended to include used vehicles imported for registration. Motorcycles have been required to comply with US EPA Standards (see **Table 3.19**) since 1 October 1991.

Hong Kong

Hong Kong has introduced legislation requiring catalyst cars from Jan 1992. Vehicles are required to meet current US limits, but may optionally be certified to meet current Japanese limits (10·15 mode cycle). Certification via EC Directive 88/76 (i.e. ECE 83) will be considered, but the exact status is not clear.

India

India is federal and different states have different motor vehicle regulations. The central government enacted a revised Motor Vehicle Act in 1990, in which emission regulations have become a central subject. Detailed technical and implementation procedures are being developed.

Idle CO and smoke tests for gasoline and diesel vehicles respectively have been in force since 1986 in a number of states. In 1990 they became mandatory throughout India for both new and in-use vehicles. Limits for gasoline mass emissions and diesel full load and free acceleration smoke became effective in 1991. Mass emissions from diesel vehicles will be controlled from 1992. Details of the limit values are not known.

Limits and test procedures for gasoline vehicles have been adopted from ECE R15-04 but modified, using an Indian driving cycle. Diesel smoke and mass emissions have been adopted from ECE R24 and ECE R49 respectively.

Currently no evaporative emissions limits and deterioration factors/endurance tests have been prescribed. Conformity of production tests have been developed.

Malaysia

In Malaysia ECE 15.04 regulations were introduced in September 1992. A further requirement for all new gasoline-fuelled vehicles to be fitted with catalysts has been postponed.

Singapore

Table 3.39.1 Singapore Emission Standards

Vehicle	Effective Date	Limits
Motor Cycles/ Scooters	01.10.91	US Federal Regulation 40 CFR 86.410-80
Gasoline	1986	UN ECE R15.04
	01.07.92	UN ECE R83 or Japanese Standard JIS 78
	01.07.94	91/441/EEC or JIS 78
Diesel	Prior 01.01.91	<50HSU at idle
	01.01.91	UN ECE R24.03
	1997 (proposed)	UN ECE R24.03 EN ECE R49.02 or Phase 1 of 91/542/EEC for HD vehicles with GVW > 3.5t 93/59/EEC for LD vehicles

Thailand

ECE test cycles and limits have been proposed or adopted by Thailand for its emission regulations as follows:

Table 3.39.2 Thai Emission Standards

Vehicle	Effective Date	Cycle	Equivalent Limits
Gasoline	01.01.1995	ECE 83	ECE 91/411/EEC
Light Diesel	01.01.1995	ECE 83	ECE 91/411/EEC
Heavy Diesel	01.01.1993	ECE 49	ECE 91/542 (A) /EEC
	01.01.2000	ECE 49	ECE 91/542 (B) /EEC
Motor Cycles	01.01.1994	ECE 40	as Taiwan (**Table 3.39**)
	01.01.1997	ECE 40	"

A summary is given of current and future emission legislation for Far Eastern countries excluding Japan, together with those for Central and South American countries in **Table 3.43.**

3.8 Central and Southern America Vehicle Emissions Regulations

Argentina Vehicle Emissions Regulations

On 27 June 1994 Argentina passed its vehicle pollution control legislation. It is understood that the legislation is based closely on the Brazilian standards (see below) although implementation is delayed because of the current limited availability of unleaded gasolines.

Brazil Vehicle Emissions Regulations

Light Duty Vehicles

The original Brazilian emissions programme "*PROCONVE*" was published as an official "*resolution*" on May 6, 1989 having been drawn up by CONAMA, the National Environment Board. A month later CONAMA also established separate limits for aldehyde (formaldehyde and acetaldehyde) emissions for both "gasolina" and "alcool" fuelled vehicles (composition of these fuels are described in **Section 7.5.3**). The Brazilian Congress has more recently passed law No. 8723, effective 1 October 1993, setting strict emission standards for passenger vehicles covering the rest of the decade. The limits are measured using the US FTP 75. The limits set do not correspond exactly with US standards but the 1988 standard is roughly equivalent to US 1973 limits and the 1992 standard lies between US 1973 and 1975 limits. The 1997 standard is equivalent to US 1981 standard. More stringent limit values will be introduced by 2000 and will match the US standards.

A smoke emissions standard is under consideration to be confirmed by 31 December 1994, possibly with limits of 30 HSU and 40 HSU for naturally aspirated and turbocharged diesels respectively, under free acceleration conditions. Meanwhile manufacturers must report smoke emissions under wide open throttle acceleration conditions in certification tests.

The law also empowers state environmental agencies with more effective enforcement, including the operation of vehicle inspection stations. Another two regulations have been established by CONAMA. One requires manufacturers to provide the information needed for establishing tune-up and maintenance standards. The other establishes minimum requirements for inspection and maintenance programmes, which involves a two speed test with 30 seconds preconditioning. Sao Paulo will introduce the first inspection programme beginning in 1995. It is planned to require the annual inspection of cars more than two years old. Cars that exceed limits will not be permitted on the road until repaired.

Table 3.40 Light Duty Emission Standards - Brazil
(US FTP - 75 Test Cycle)

Year	CO g/km	CO Idle %v	HC g/km	RCHO g/km	NOx g/km	Pm g/km	Evap g/test
01.01.88	24	3.0	2.1	-	-	-	-
01.01.90	24	3.0	2.1	-	2.0	-	6.0
01.01.92	24	2.5	1.2	0.15	1.4	-	6.0
01.03.94	12	2.5	1.2	0.15	1.4	0.05	6.0
01.01.97	2.0	0.5	0.3	0.03	0.6	0.05	6.0
2000	Limits in line with US limits						

Notes.
- Aldehydes (RCHO) limit for alcohol fuelled vehicles only
- Particulate matter (Pm) for diesel fuelled vehicles only
- Idle CO for alcohol and gasoline vehicles only.
- Evaporative emissions expressed as propane for gasohol and ethanol for alcohol fuelled vehicles.

Heavy Duty Vehicles

New emission standards were established for heavy duty vehicles by the National Environmental Council (CONAMA) on 31 August 1993 (see **Table 3.41**). A revised implementation schedule introduces the limits in four Phases (I- IV), with a more rapid introduction for urban buses and imported vehicles. The limits for Phase IV will be confirmed by the end of 1994. Phase I, which is current, applies to all diesel vehicles and consists of a smoke limit of 2.5 k* only.

Manufacturers of light duty trucks (over 2000 kg GVW) have the option to chose either the LDV or HDV test procedures for certification. Thereby light duty trucks with low speed diesel engines are still technically feasible while high speed diesels have to be tested by the FTP procedure and comply with the 0.05 g/km particulate emission standard. Although diesel engines are not presently allowed for most LDVs, a particulate standard has been adopted in case regulations change.

* See **Table 3.4.1** for a definition of k

Table 3.41 Heavy Duty Emission Standards - Brazil
(ECE R49 test cycle)

Vehicle Class	Date	Phase	% Vehicles	CO g/kWh	HC g/kWh	NOx g/kWh	Pm g/kWh	Smoke k**
All	Jan 1996	II	20	11.2	2.45	14.4	-	2.5
Vehicles		III	80	4.9	2.23	9.0	0.4*	-
	Jan 2000	III	20	4.9	2.23	9.0	0.4*	-
		IV	80	4.0	1.1	7.0	0.15	-
	Jan 2000	IV	100	4.0	1.1	7.0	0.15	-
All	March 1994	III	100	4.9	2.23	9.0	-	2.5
Imports	Jan 1996	III	100	4.9	2.23	9.0	0.4*	-
	Jan 1998	IV	100	4.0	1.1	7.0	0.15	-
Buses	March 1994	II	20	11.2	2.45	14.4	-	2.5
		III	80	4.9	2.23	9.0	-	2.5
	Jan 1996	III	80	4.9	2.23	9.0	0.4*	-
	Jan 1998	III	20	4.9	2.23	9.0	0.4*	-
		IV	80	4.0	1.1	7.0	0.15	-
	Jan 2002	IV	100	4.0	1.1	7.0	0.15	-

Notes
* Particulate emissions (Pm) 0.7 g/kWh for engines up to 85 kWh, 0.4 g/kWh for engines up to 85 kWh.
  Crankcase emissions must be nil, except for some turbocharged diesel engines if there is a technical justification

** k = soot (g/m^3) x gas flow (l/sec)

Emissions must be warranted for 80 000 km for LDVs, 160 000 km for HDVs, or alternatively emissions must be 10 per cent below the limits set.

Mexico Vehicle Emissions Regulations

On June 6, 1988 the Mexican authorities announced a decision to introduce more stringent standards for light duty vehicles, culminating in full US 1981 limits by 1993. Interim standards for 1989 through 1992 are consistent with the proposal made by the automobile manufacturers.

With respect to heavy duty vehicles, discussions are being held between the Mexican Social Development Secretariat (SEDESOL) and a new industry organization (ANPACT), representing diesel engine and vehicle manufacturers, with the view to adopting the US EPA emission standards for 1994, but no confirmation has been received at the time of publication of this report. Until the state oil company, PEMEX, can guarantee meeting a 0.05%w sulphur limit, engines certified to the EPA requirements would not need to be fitted with any after-treatment devices necessary to comply with the regulations.

Table 3.42 Mexican Emissions Standards (US FTP 75 procedure)

Vehicle Type	Effective Date	CO g/mile	HC g/mile	NOx g/mile
Cars	1989	35.2	3.20	3.68
	1990	28.8	2.88	3.20
	1991	11.2	1.12	2.24
	1993	3.4	0.41	1.00
Light Duty Trucks <6012lb GVW	1990	35.2	3.20	3.68
	1994	14.0	1.00	2.30
6013-6614lb GVW	1990	56.0	4.80	5.60
	1992	35.2	3.20	3.68
	1994	14.0	1.00	2.30

Chile - Proposed Vehicle Emissions Regulations

Legislation is being developed in Chile which will require all new cars to meet current US emission limits. This applied from September 1992 to the greater Santiago area (which has a significant smog problem), and from September 1994 to the rest of the country. Such cars will be permitted to use the roads in Santiago at all times, while the existing non-catalyst cars will be subject to a 20% off-the-road restriction on weekdays. 91 RON unleaded gasoline will be made available for the new catalyst equipped cars. Retrofitting some of the older cars with catalysts is also under consideration.

Table 3.43 Summary of Emissions Legislation - Far East and South America

3.43.1 Gasoline Vehicles

Country	Effective Date	Cycle	CO g/km	HC g/km	NOx g/km	Evap g/test	Equiv. Limit
Brazil	Jan 1989	FTP 75	24.0	2.1	2.0		
	Jan 1992	FTP 75	12.0	1.2	1.4	6.0	
	Jan 1997	FTP 75	2.0	0.3	0.6		
Chile	Sept. 1994**	FTP 75	2.11	0.25	0.62	2.0	US 83
	{	FTP 75	2.11	0.25	0.62	2.0	US 83
Hong Kong	Jan {	Jap 10	2.7	0.39	0.48	2.0	Jap 78
	1992 {	ECE 15			See **Table 3.2**		
Malaysia	Sept. 1992	ECE 15	2.72		0.97	2.0	91/441/EEC
Singapore	July 1994	ECE 15 +EUDC	2.72		0.97	2.0	91/441/EEC
S.Korea	July 1987	FTP 75	2.11	0.25	0.62	2.0	US 83
	2000		2.11	0.16	0.25		
Taiwan	July 1990	FTP 75	2.11	0.25	0.62		US 83
Thailand (proposed)	Jan 1995	new	2.72		0.97	2.0	91/441/EEC
China				See **Table 3.33**			

Notes : See overleaf

3.43.2 Light Duty Commercial Vehicles and Diesel Cars

Country	Effective Date	Cycle	CO g/km	HC g/km	NOx g/km	Pm g/km	Equiv. Limit
Brazil	Jan 1992	FTP 75	12.0	1.2	1.4		
	Jan 1997		2.0	0.3	0.6		
Chile	Sept. 1994**	FTP 75	2.11	0.25	0.62	0.12	US 87
Hong Kong	Jan 1992	FTP 75	2.11	0.25	0.62	0.12	US 87
		Jap 10	2.7	0.62	0.72	-	Jap 91
Mexico	1993	FTP 75	2.1	0.25	0.62		
Singapore (proposed)	1997	ECE 15 +EUDC	See **Table 3.4**				91/542/EEC
S.Korea	July 1987	Jap 6	980*	670*	850*		Jap 77/9
Taiwan	July 1993	FTP 75	6.2	0.5	1.4	0.38	US 84
	July 1995		2.1	0.25	0.62	0.12	
Thailand (proposed)	Jan 1995	new	2.72	0.97		2.0	91/441/EEC
China			See **Table 3.33**				

*   ppm
**   from Sept. 1992 for Santiago only

3.43.2 Heavy Duty Emissions Legislation

Country	Effective Date	Cycle	CO	HC	NOx	Pm	Equiv. Limit
				g/b	hp.h		
Brazil			See **Table 3.41**				
Singapore (proposed)	1997	ECE R49	4.5	1.1	8.0	0.36*	91/542/EEC
Taiwan (proposed)	01.07.93	US Trans			6.0	0.7	
					5.0	0.25	
China			See **Tables 3.34 and 3.35**				

* *Note:* In the case of engines of 85kW or less, the Pm limit is increased by multiplying the quoted limit by 1.7.

3.9 Australian Emissions Regulations

Australia operates under a federal system of government, but contrary to the US situation the Australian (federal) government does not have authority over motor vehicle legislation. This power, including the ability to introduce emissions regulations, lies with the various state governments.

The Australian Transport Advisory Council (ATAC) comprises federal and state transport ministers. It meets twice a year and can resolve the adoption of emissions and safety standards which, although not binding on the states, are usually adopted in state legislation. However, states have acted unilaterally when agreement within ATAC is not reached.

ATAC is advised on emissions matters, by a hierarchy of committees: the Motor Transport Groups (comprising senior federal and state public servants), the Advisory Committee on Vehicle Emissions and Noise (ACVEN) which comprises lower-level federal and state public servants, and the ACVEN Emissions Sub-Committee, which includes public servants, representatives from the automotive and petroleum industries as well as consumers. ACVEN also provides advice to the Australian Environment Council, which has some emissions responsibilities.

Petrol engined vehicles emission standards

Prior to 1986, passenger car emissions standards (ADR27) were based on the United States 1973-74 requirements. From January 1986, manufacturers are required to meet the ADR37 standard, which is equivalent to United States 1975 requirements. Current requirements for commercial gasoline-engined vehicles are based on the New South Wales 26(3) and Victorian 9 regulations. Details are given in **Table 3.44**.

Table 3.44 Australian Passenger Car Emission Regulations

Regulation	Effective Date	CO (g/km)	HC (g/km)	NOx (g/km)	Test Procedure	Evaporative Emissions (g/test)
ADR27 A/B/C	July 1976	24.2	2.1	1.9	FTP 75	2.0 (Canister)
	Jan. 1982	22.0	1.91	1.73	FTP 75	6.0 (SHED)
ADR37*	Jan. 1986	9.3	0.93	1.93	FTP 75	2.0 (SHED)
		8.45	*0.85*	*1.75*		*1.9 (SHED)*

Note:
*   The higher figures apply to production vehicles, which must meet the limits from 150 km to 80 000 km or for 5 years, whichever occurs first. The figures in italics apply to certification vehicles.

Table 3.45 Australian Gasoline-Powered Commercial Vehicle Emission Regulations

Regulation	Effective Date	CO (g/km)	HC (g/km)	NOx (g/km)	Test Procedure	Evaporative Emissions (g/test)
NSW (Clean Air Act)	Current	9.3	0.93	1.93	FTP 75	2.0 (SHED)
VIC (Statutory Rules)						

Note: Individual state regulations differ in some respects from these regulations.

Proposals have been made by the FCAI (Federal Chamber of Automotive Industries - representing both Australian vehicle manufacturers and the importers) to reduce emission limits in two stages in 1996 and 2000 as follows:

Passenger cars from 01.01.96

Table 3.46 Australian Proposed Legislation for Passenger Cars

Date	CO g/km	HC g/km	NOx g/km	Pm g/km
1996	4.34	0.26	1.24	2.0
2000	2.11	0.26	0.63	2.0

Commercial vehicles from 01.07.96

Proposed legislation would allow engines to be used which meet current European, US or Japanese emission limits.

Further reductions in emission limits to, say, US 1983 standards could be considered after the year 2000.

Diesel-engined vehicle emissions standards

Diesel exhaust smoke emission limits (ADR30 and ADR55) are set for the opacity of the exhaust smoke when the engine is tested under prescribed conditions. Three alternative test procedures are allowed, equivalent to 1974 British, European or United States standards:

- EPA (US) Federal diesel smoke regulations Part 85 and 86.
- BS AU 141a: 1971.
- ECE Regulation 24.

The above diesel regulations are currently under review but no proposals have yet been published.

3.10 World - Wide Off Highway Emission Limits

The following describes the limit values adopted in Europe and North America. In each case the test cycle adopted is taken from ISO 8178-4, full details of which will be found in **Section 4.9**.

3.10.1 Europe - EU

Draft Directives are under preparation to control emissions from off-highway engines. Initially there will be a Directive for agricultural and forestry tractors and a separate Directive for self-propelled mobile off-highway equipment. Further Directives are expected to cover the other off-highway applications.

Table 3.47 Proposed EU emission limits for Agricultural and Forestry Tractors

Effective Date	Power Band	CO g/kWh	HC g/kWh	NOx g/kWh	Pm g/kWh
Oct 96	>130kW	5.0	1.3	9.2	0.54
Oct 97	75-130kW	5.0	1.3	9.2	0.70
Oct 98	37-75kW	6.5	1.3	9.2	0.85

N.B. The effective dates are the dates at which all new tractor registrations must comply with the limits.

**Table 3.48 Proposed EU emission limits for Self Propelled
Mobile Off-Highway Equipment**

Effective Date	Power Band	CO g/kWh	HC g/kWh	NOx g/kWh	Pm g/kWh
Jan 97	130-560kW	5.0	1.3	9.2	0.54
Jan 98	75-130kW	5.0	1.3	9.2	0.70
Jan 99	37-75kW	6.5	1.3	9.2	0.85

N.B. The effective dates refer to the build date of the engine.

Notes. (Applicable to both tractors and mobile off-highway equipment).

- The Directives are not yet finalized and changes to limits and dates may be made.
- The test cycle is the 8-mode steady state, C1, cycle from ISO 8178-4.
- The limit values are the same as those of the ECE tractor regulation (see **Table 3.4.9**).
- It is proposed that Stage 2 limit values will be introduced 5 years after Stage 1, and that they will be equivalent to the EU on-highway truck levels of 7.0 NOx/0.15 Pm g/kWh.
  Actual limits are not yet agreed.
- The Stage 1 limit values must be met without the use of exhaust aftertreatment. This requirement may be relaxed for Stage 2.
- It is probable that the current tractor smoke Directive will be modified to make it more stringent and applicable to both the tractor and mobile off-highway equipment categories.

3.10.2 Europe - ECE Regulations

A regulation for the control of gaseous and particulate emissions from diesel powered agricultural and forestry tractors has been adopted and is awaiting publication.

Table 3.49 ECE Limit Values for Emissions from Diesel Powered Agricultural and Forestry Tractors

Effective Date	Power Band	CO g/kWh	HC g/kWh	NOx g/kWh	Pm g/kWh
*	130-560kW	5.0	1.3	9.2	0.54
*	75-130kW	5.0	1.3	9.2	0.70
*	37-75kW	6.5	1.3	9.2	0.85

* Since this is a new regulation and adoption into national legislation is optional, it is not possible for introduction dates to be included in the text.

Notes

- The test cycle is the 8-mode steady state cycle, C1, from ISO 8187-4.

- The limit values for engine >130kW (175bhp) are equivalent to the Carb '96 limits for engines >175bhp, apart from the CO limit which is more stringent.

- Stage 2 limit values have not yet been proposed.

- The limit values must be met without the use of exhaust after-treatment.

- There is currently no ECE smoke test for off-highway applications and none is under application.

3.10.3   US Federal

The EPA is developing emissions regulations for "non road" engines which will be mandatory in all of the states.

**Table 3.50 Proposed US Federal Emissions Limits for
Heavy Duty Non-Road Engines**

Effective Date	Power Band	NOx	Smoke (Opacity)		
		g/bhp.h	Accl.	Peak	Lug
1 Jan 96	175-750bhp	6.9	20%	50%	15%
1 Jan 97	100-750bhp	6.9	20%	50%	15%
1 Jan 98	50-100bhp	6.9	20%	50%	15%
1 Jan 2000	>750bhp	6.9	20%	50%	15%

Notes

- Status: proposed.

- The test cycle is the 8-mode steady state cycle, C1, from ISO 8178-4.

- Certification, compliance labelling, auditing and in-service testing is proposed.

- Banking, averaging and trading is proposed.

- Limits for HC, CO and Pm have not been proposed, but are under consideration.

- The US on-highway transient smoke test has been proposed, but is under review.

- The scope includes constant speed engines, but the test cycles must be revised to accommodate them.

- Stage 2 limits have not been proposed as a transient test cycle is under consideration.

- The final rule is expected by mid 1994.

Non-Road Engines <50bhp.

An EPA study has indicated that these engines contribute to HC and CO non-attainment and emissions proposals are expected during 1994.

3.10.4 US California

The US Clean Air Act amendments pre-empted CARB from setting emissions regulations for construction and agricultural equipment of less than 175 bhp. CARB have therefore either set, or are in the process of setting, regulations for the applications that are not covered by the Act.

Table 3.51 CARB - Heavy Duty Off-Road Diesel Cycle Emissions Limits

Effective Date	Power Band	CO g/bhp.h	HC g/bhp.h	NOx g/bhp.h	Pm g/bhp.h
1 Jan 1996	175-750bhp	8.5	1.0	6.9	0.4
1 Jan 2000	175-750bhp	8.5	1.0	5.8	0.16
1 Jan 2000	>750bhp	8.5	1.0	6.9	0.4

Notes

Status: adopted
- The test cycle is the 8-mode steady state cycle, C1, from ISO 8187-4.
- The US transient smoke test cycle and limits for heavy duty on-highway trucks is applied to diesel engines. The limit values will be revised for stage 2.
- Certification, compliance labelling, and auditing will be required.

Table 3.52 CARB - Utility, Lawn and Garden Equipment Engines Emission Limits

Effective Date	Category <25bhp and:	CO g/bhp.h	HC g/bhp.h	NOx g/bhp.h	HC+NOx g/bhp.h	Pm g/bhp.h
1 Jan 1995	<225cc	300	-	-	12.0	0.9
1 Jan 1995	>225cc	250	-	-	10.0	0.9
1 Jan 1995	<50cc Hand Held	600	180	4.0	-	-
1 Jan 1995	>50cc Hand Held	300	120	4.0	-	-
1 Jan 2000	Non hand Held	100	-	-	3.2	0.25
1 Jan 2000	Hand Held	130	5.0	4.0	-	0.25

Notes

- Status: adopted
- Test cycles are according to SAE J1088/ISO 8178-4 'G'.
- Certification, compliance labelling, and auditing will be required.

Table 3.53 CARB - Proposed Recreational and Speciality
Vehicle Engine Emissions Limits

Effective Date	Category	HC+NOx g/bhp.h	CO g/bhp.h	Pm g/bhp.h
1995	<25bhp	10.0	300	0.9
1997	>25bhp	3.2	100	0.25
1999	<25bhp	3.2	100	0.25

Notes

- Status: proposed
- 'Speciality vehicles' are used for load and passenger carrying primarily around airports, hotels, resorts and grounds. Typical payload is less than 2000 lb.
- Test cycles are according to SAE J1088/ISO 8178-4 'G'.

Appendix 4

Emissions and Fuel Economy Test Cycles and Procedures

Reproduced from Report 4/94, Section 4, by kind permission of:

CONCAWE, Brussels, Belgium

4 Emissions and Fuel Economy Test Cycles and Procedures

The current laws, designed to limit the emission of exhaust gases by motor vehicles, prescribe throughout the world maximum emission standards for the following exhaust gas components:

- carbon monoxide (CO);
- hydrocarbons (HC);
- oxides of nitrogen (NOx);
- particulates (Pm).

It is also common practice to use part or all of an emissions cycle to measure fuel economy, calculated by mass balance in conjunction with carbon dioxide measurements.

The method of gauging emission rates is determined by statutory test procedures, the objective being to establish the mass of each exhaust component emitted during the test. The mass is computed from the measured concentrations of each pollutant in the known exhaust gas volume. Exhaust species are generated when the vehicle is operated on a chassis dynamometer according to certain standard driving cycles, which are designed to simulate driving conditions in urban traffic. More recently a number of these urban cycles have been augmented with higher speed sections (e.g. the extra urban driving cycle - see **Figure 4.2**).

The following represents an overview of the major exhaust test procedures for Europe, the United States and for Japan. It shows that a large variety of test procedures exist in various countries and there are substantial differences between them with respect to driving cycle (speed and distance), vehicle preconditioning, and analytical equipment.

Evaporative emissions from gasoline powered vehicles are also controlled in many countries. These emissions are determined either by collection in activated carbon traps, or by putting the vehicle in an airtight housing (SHED) and measuring the hydrocarbon concentration. The major test procedures for evaporative emissions are also reviewed.

4.1 European Exhaust Emission and Fuel Economy Test Procedures

ECE 15 defines an urban test cycle to be used for emission measurements, as shown in **Figure 4.1**. The cycle was devised to be representative of city-centre driving (e.g. Rome) and thus has a maximum speed of only 50 km/h. The complete first ECE 15 emissions procedure consists of three tests, Type I, II and III as follows:

Type I Test Emission test cycle:

Prior to testing the vehicle must be preconditioned by driving at least 3500 km (1800 miles). The vehicle is allowed to soak for at least 6 hours at a test temperature of between 20 and 30°C. It is then started and allowed to idle for 40 seconds. The 15 mode driving cycle (**Figure 4.1**) is then repeated four times without interruption. This gives a total test cycle time of 780 sec., total distance of 4.052 km (2.5 miles) and thus an average speed of 19 km/h (11.8 miles/h). It now seems likely that the EU will abandon the 40 second idle period and commence emissions testing with an engine start at the beginning of the first ECE 15 mode cycle.

Up to Amendment 15/04, total emissions were collected in one bag. CO and HC emissions were determined by NDIR (Non-Dispersive Infra-Red) analyzers and NOx by chemiluminescent technique. From 15/04 however, emissions are measured by the *"Constant Volume Sampling"* (CVS) technique as used in the US procedure (**Figures 4.3 - 4.5**). Hydrocarbon emissions are also determined by use of a FID (Flame Ionization Detector) analyzer.

The ECE 15 cycle is very low duty (maximum speed 50 km/h) and is thus not representative of many modes of driving. Specifically it is felt to give unrealistically low figures for NOx emissions. After much discussion, an additional *"high speed"* test cycle, the Extra Urban Driving Cycle (EUDC) was agreed (see **Figure 4.2**), with maximum speed of 120 km/h. This test is carried out after the standard ECE 15 test. This combination of the ECE 15 and EUDC test cycles is required in the EU "Consolidated Emissions Directive".

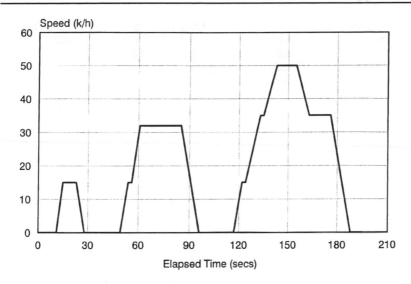

Fig. 4.1. ECE 15 Test Cycle (repeated four times)

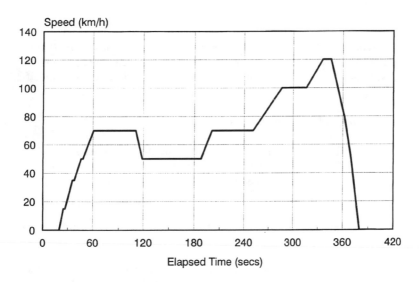

Fig. 4.2. Extra Urban Driving Cycle

Characteristics		ECE 15 Cycle	EUDC Cycle
Distance	km	4 x 1.013 = 4.052	6.955
Time	sec	4 x 195 = 780	400
Average Speed	km/h	19	62.6
Maximum Speed	km/h	50	120
Acceleration	% Time	21.6	-
Acceleration	m/s^2 (max)	-	0.833
Deceleration	% Time	13.8	
Deceleration	m/s^2(max)	-	-1.389
Idle	% Time	35.4	-
Steady Speed	% Time	29.3	-

Fig. 4.2. Extra Urban Driving Cycle (continued)

Type II Test

Warmed-up idle CO test conducted immediately after fourth cycle of Type I test; tailpipe sampling probe.

Type III Test

Chassis dynamometer procedure for crankcase emissions (idle and 50 km/h constant speed modes). The system is certified if crankcase operates at partial vacuum (as in PCV systems), or if crankcase emissions meet specified standards.

Fuel Economy and Carbon Dioxide Emissions

Fuel economy (by mass balance) and carbon dioxide emissions are also measured over the ECE 15 and EUDC cycles. One figure for the combined cycle is reported for carbon dioxide, whereas fuel economy is reported separately for the two cycles and in combination.

4.2   US Federal Light Duty Exhaust Emission and Fuel Economy Test Procedures

Since 1972 the US exhaust emission test procedure has been based on a transient cycle representative of driving patterns in Los Angeles (the LA-4-cycle). The original 1972 test procedure (US-72) is a two-phase test (Phases I and II in **Figure 4.3**) covering 7.5 miles in almost 23 minutes. From 1975 the procedure was modified such that Phase I is now repeated after a 10 minute hot soak, to form a third phase. The total test length was thus extended to 11.09 miles and the time to 31 minutes, plus 10 minutes for the hot soak.

Emission Test Cycle FTP-75

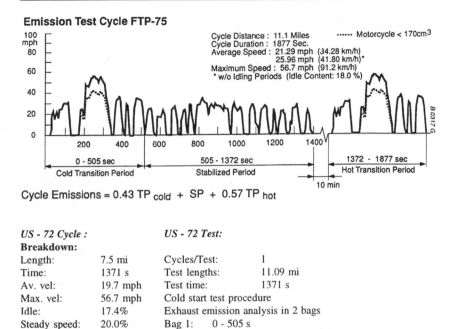

Cycle Distance : 11.1 Miles
Cycle Duration : 1877 Sec.
Average Speed : 21.29 mph (34.28 km/h)*
25.96 mph (41.80 km/h)*
Maximum Speed : 56.7 mph (91.2 km/h)
* w/o Idling Periods (Idle Content: 18.0 %)

······ Motorcycle < 170cm^3

0 - 505 sec
Cold Transition Period

505 - 1372 sec
Stabilized Period

1372 - 1877 sec
Hot Transition Period

10 min

Cycle Emissions = 0.43 TP $_{cold}$ + SP + 0.57 TP $_{hot}$

US - 72 Cycle :		*US - 72 Test:*
Breakdown:		
Length:	7.5 mi	Cycles/Test: 1
Time:	1371 s	Test lengths: 11.09 mi
Av. vel:	19.7 mph	Test time: 1371 s
Max. vel:	56.7 mph	Cold start test procedure
Idle:	17.4%	Exhaust emission analysis in 2 bags
Steady speed:	20.0%	Bag 1: 0 - 505 s
Acceleration:	34.0%	Bag 2: 56-1371
Deceleration:	28.8%	Calculation of fuel consumption is derived from the emissions

US - 75 Cycle:		*US - 75 Test:*
Breakdown:		
Length:	11.09 mi	Cycles/Test: 1
Time:	1877 + 600	Test lengths: 11.09 mi
Av. vel:	21.3 mph	Test time: 1371 s + (600 s stop) + 505 s
Max. vel:	56.7 mph	The US - 75 test is an expanded US - 72 test.
Idle:	17.3%	The first 505 s of the cycle will be repeated after a ten min.
Steady speed:	20.5%	stop at 1371 s.
Acceleration:	33.7%	Calculation of fuel consumption is derived from the
emissions.		
Deceleration:	26.5%	The emission sampling is in 3 bags:

Bag 1: 0 - 505 s (43%)
Bag 2: 506 - 1371 s (100%)
Bag 3: 0 - 505 s (57%)
after the 10 mins. stop
(Source: Ford Motor Company and AVL List GmbH)

Fig. 4.3. US Federal City Cycles (Exhaust Emission and Fuel Consumption)

Highway Cycle

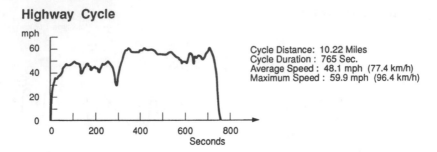

Fig. 4.4. *US Federal Highway Cycle (Fuel Consumption, NOx Emissions)*

The intention of this modified test is to produce a weighted average emission from cold start and hot start tests. It is assumed that after the first 505 seconds the engine will be stabilized, and so the stabilized portion is not repeated after the hot start, but assumed to be the same as for the cold start test. The calculation procedure weights the results from the three bags accordingly to give the required result.

The Highway Fuel Economy Test (HWFET) (**Figure 4.4**) cycle is used to measure fuel economy for the CAFE standards (see **Section 6.3**), but is also used to measure NOx emissions. In California, a standard is imposed equal to 1.33 times the city NOx limit. In the other 49 states, there is no standard for NOx over the HWFET, but data can be used to demonstrate to EPA that the vehicle is not equipped with *"cycle-beating"* devices.

Emissions are measured using a Constant Volume Sampling (CVS) system, as shown in **Figure 4.5**, and collected in three bags, for each phase of the test, i.e.:

- Bag 1   -   0 - 505 seconds;
- Bag 2   -   505 - 1370 seconds;
- Bag 3   -   0 - 505 seconds after 10 minute hot soak.

The test begins with a cold start (at 20 - 30°C) after a minimum 12 hour soak. The diluted exhaust gas volume is determined for each bag and used to calculate the mass emissions. Weighting factors for each phase are then applied to give an overall emission figure.

The US Clean Air Act Amendments require EPA to re-evaluate typical driving patterns and determine if the current FTP test cycle adequately reflects current conditions. It has done this in a number of cities and estab-

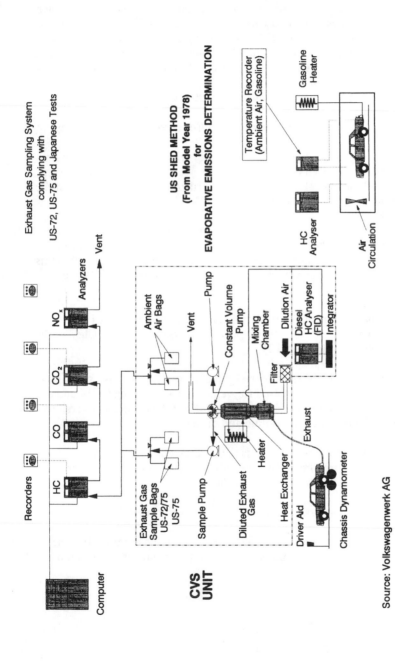

Source: Volkswagenwerk AG

Fig. 4.5. Exhaust Gas Constant Volume Sampling (CVS) Procedure (complies with US 72, US 75, Japanese and current European test requirements)

lished that some 15% of current driving is not covered, being higher in speed and more severe in acceleration. Average trip lengths are also much shorter, being about 4.9 miles compared with the test average of 7.5 miles. Also the FTP assumes that 57 per cent of all in-use starts occur with a hot catalyst compared with 30 per cent in practice. A number of revised cycles have been developed and were evaluated in 1993. It is believed that the revised procedures will most likely take the form of an extra cycle to be added to the current procedure. The EPA now has until 1995 to finalise its revision.

4.3 Japanese Exhaust Emission and Fuel Economy Test Procedures

The exhaust emission test procedures in Japan are complex, but four main test procedures have been used in the past. However, new procedures were introduced in March 1991 and these are described following a review of the tests employed prior to that date.

(1)   *10 Mode test* - a hot start urban driving cycle including accelerations up to 40 km/h, the principal cycle used for passenger cars and light vehicles powered by gasoline or LPG fuelled engines. From 1986 this procedure also applies to diesel vehicles up to 1700 kg GVW (see **Figure 4.6.1**).

(2)   *11 Mode test* - a cold start driving cycle test introduced in 1975 to supplement the 10 mode test. Speeds up to 60 km/h are reached during the cycle (see **Figure 4.6.2**).

(3)   *6 Mode test* - a test based on weighted average emissions over steady state modes. This test is used for vehicles over 2500 kg GVW and also for diesel-powered vehicles above 1700 kg GVW, different versions of the test being used for gasoline and diesel vehicles.

(4)   *Evaporative emission test* - this is based on charcoal canisters to trap emitted vapours, similar to the original US test procedure. There are no plans to introduce a SHED procedure.

The exhaust emission test equipment is shown in **Figure 4.5**.

The test procedures detailed above were replaced in the early 1990s. The new test cycles are illustrated in **Figure 4.5**. These procedures have been modified

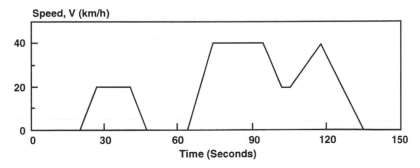

1. Stabilized hot start cycle, repeated 6 times (including one pre-cycle).
2. Distance: 0.564 km/cycle. Test duration: 14 minutes.
3. Continuous sampling of diluted exhaust gas into bag.
4. Integral analysis (comparable to US 1972 Test).
5. Modal breakdown of results not possible.
6. Determination of exhaust gas volume analogous to US 1972 Test.
7. Theoreticaly developed driving schedule.
8. Determination of evaporative emissions.

Fig. 4.6.1. Japanese 10-mode Cycle (Hot Test - from 1 April, 1973)

to include a high speed cycle and measurements will be made in terms of g/km or g/kWh, instead of ppm. Details of the changes in cycle type, measurement mode, vehicles affected and implementation dates are given below:

10-Mode Replaced by 10·15-Mode (**Figure 4.6.3**)
- Gasoline/LPG passenger cars, light and medium duty commercial vehicles, mini trucks and diesel passenger cars/light duty commercial vehicles.
- Implementation Date: 1991

The 10.15-mode test is also used to measure fuel economy.

Gasoline/LPG 6-Mode Replaced by Gasoline/LPG 13-Mode (**Figure 4.6.4**)
- Measurement Mode: ppm changed to g/kWh
- Gasoline/LPG heavy duty commercial vehicles.
- Implementation Date: 1992

Diesel 6-Mode Replaced by 10·15 Mode (**Figure 4.6.3**)
- Measurement Mode: ppm changed to g/km

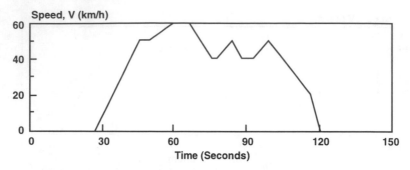

1. Cold start cycle, repeated 4 times.

2. Distance: 1.02 km/cycle. Test duration: 8 minutes.

3. Continuous sampling of diluted exhaust gas into bag.

4. Integral analysis (comparable to US 1972 Test).

5. Modal breakdown of results not possible.

6. Determination of exhaust gas volume analogous to US 1972 Test.

7. Similar to 10 Mode test, apart from cold start and increased power demand.

8. Determination of evaporative emissions.

Fig. 4.6.2. Japanese 11-mode Cycle (Cold Test - from 1 April, 1976)

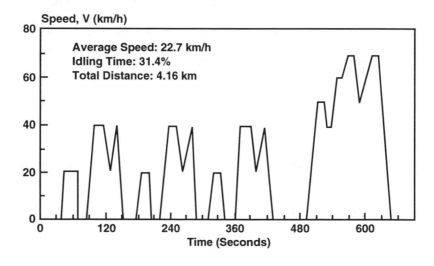

Fig. 4.6.3. Japanese Current 10.15 Light Duty Test Cycle

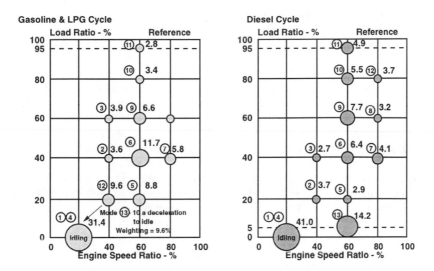

Figs. 4.6.4 & 4.6.5. Japanese Current 13-mode Test Cycles for Heavy Duty Engines

- Diesel medium duty commercial vehicles.
- Implementation Date: 1993

Diesel 6-Mode Replaced by Diesel 13-Mode (**Figure 4.6.5**)
- Measurement Mode: ppm changed to g/kWh
- Diesel heavy duty commercial vehicles.
- Implementation Date: 1994

Because of the differences in test procedures, a direct comparison of Japanese emission standards with those applied in the US and Europe cannot be made.

4.4 Evaporative Emissions Tests

Introduction

Two types of test methodology exist, the carbon canister (trap) method and the SHED (Sealed Housing for Evaporative Determination). The canister method employs weighed activated carbon traps which are connected to the fuel system at locations where fuel vapours may escape to the atmosphere (air cleaner, fuel tank vent, etc.) and is only applied in Japan. The carbon canister method is described in more detail at the end of this section.

"*The SHED*" method collects all evaporative emissions in a sealed enclosure which contains the test vehicle. The hydrocarbon concentration in the SHED atmosphere is measured and used to calculate total emissions. This technique has been used for many years in the US, and has recently been adopted for the European test procedure.

Evaporative emissions can be divided into three areas:

- **Diurnal losses:** these occur when the vehicle is stationary with the engine off and are due to emission of vapour from the fuel tank as a result of normal temperature changes which occur over a 24 hour period.

- **Hot soak losses:** these occur when a fully warmed up vehicle is left to stand, as engine heat is transferred to the fuel tank and/or carburettor.

- **Running losses:** these occur while the vehicle is being driven normally.

EU and US Federal test procedures

A summary of the current EPA and EU test procedures is given in **Table 4.1** and an outline of the new draft Federal test sequence is shown in **Table 4.2 and Figures 4.7 and 4.8**. The major differences between the current EPA and EU tests are summarized below.

- **Preconditioning:** The current US test does not precondition the canister apart from driving one LA-4 cycle, however, the new procedure will load the canister to breakthrough before driving the cold start test. The EU procedure uses a complex purge/load technique.

708

Table 4.1 Evaporative Emission Test Procedures

VEHICLE PREPARATION AND LEAK TEST	CURRENT US EPA PROCEDURE	EU PROCEDURE
CANISTER CONDITION	CANISTER CONDITION UNSPECIFIED	CANISTER CONDITION NOT SPECIFIED.
VEHICLE PRECONDITIONING	1 X LA4 DRIVE CYCLE	PURGE CANISTER BY DRIVING AT 60 KM/H FOR 30 MINS. OR EQUIVALENT AIR PURGE. 2 X DIURNAL HEAT BUILD 16 - 30°C. DRIVE ECE 15 + 2 EUDC CYCLES.
SOAK PARKING	11-35 HRS. AT 20-30°C	10-36 HRS. AT 20-30°C
FUEL DRAIN AND 40% FILL	40 ±2% OF TANK CAPACITY	40 ±2% OF TANK CAPACITY. TEMP. 10-14°C
DIURNAL TEST	15-29°C IN 1 HR, LOSSES MEASURED IN SHED	16-30°C (±1°C) OVER 60 ±2 MINS. EMISSIONS MEASURED IN SHED
DYNAMOMETER TEST	1.5 X LA4 DRIVE CYCLES COLD AND HOT EXHAUST TESTS. RUNNING LOSSES MEASURED WITH CARBON TRAPS IF IT IS DEEMED NECESSARY	ECE 15 + EUDC DRIVING CYCLES. RUNNING LOSSES NOT MEASURED. EXHAUST EMISSIONS MAY BE MEASURED
HOT SOAK IN ENCLOSURE	20-30°C IN 1 HR.	1 HR. 23-31°C
FUEL DRAIN AND 40% FILL	ABILITY TO CONTROL DIURNAL LOSSES AFTER HOT SOAK NOT TESTED	N/A
END	CALCULATION OF TOTAL EVAPORATIVE EMISSIONS (DIURNAL + RUNNING LOSSES + HOT SOAK)	CALCULATION OF TOTAL EMISSIONS = DIURNAL + HOT SOAK

- *Diurnal test:*   The current US and EU procedures are similar, but the new EPA procedure has three diurnal cycles over a higher temperature range.

- *Running losses:* These may be measured over the current EPA cycle, but in practice this is not usually carried out. They are not measured in the EU test

**Table 4.2 New US Federal Evaporative Emissions Procedure
(Implementation begins with 1996 model year)**

THREE DIURNAL TEST

FULL DRAIN/FILL	Drain tank, fill to 40% volume with test fuel (RVP 9 psi/62 kPa or 7.8 psi; 53.8 kPa for altitude testing)
SOAK PERIOD	Soak at 68-86°F (20-30°C) for 12-36h
PRECONDITIONING	Drive over the UDDS (FTP) cycle.
REFUEL/SOAK CANISTER LOADING	Fuel tank is drained and refilled. Soak for 12h (min) during which the canister is loaded with butane/nitrogen mixture (mass butane 1.5 times bed capacity)
EXHAUST EMISSIONS TEST	Full FTP Cold and Hot Start procedures
SOAK PERIOD	Vehicle stabilized at 95°F (35°C)
RUNNING LOSS	Consists of driving sequence of UDDS + NYCC + NYCC + UDDS cycles (Fuel temperature controlled according to a profile, predetermined by a drive under representative summer conditions)
HOT SOAK EMISSIONS	Measured for one hour at 95°F
SOAK PERIOD	Vehicle stabilized at 72°F (22.2°C)
THREE DAY DIURNAL	Emissions measured after three 24h ambient temperature cycles 72 to 96°F (22.2 to 35.6°C). Air circulation and temperatures may be adjusted according to correct any major discrepancy compared with fuel temperatures under outdoor summer conditions.

SUPPLEMENTAL TWO-DIURNAL TEST

CANISTER LOADING	Loaded with butane/nitrogen mixture until two-gram breakthrough
EXHAUST EMISSIONS TEST	Full FTP Cold and Hot Start procedures.
VEHICLE SOAK	6 hours at 72°F (22.2°C)
TWO DAY DIURNAL	Emissions measured after two 24h ambient temperatures cycles 72 to 96°F (22.2 to 35.6°C)

- *Hot soak:*   There is a small difference in temperature ranges between EPA and EU methods, but this is not significant in view of the high fuel temperature after the diurnal test.

The EPA has issued regulations, effective from 23 April 1993, specifying revised procedures and limits for evaporative emissions, with implementation phased-in over the 1996 to 1999 model years. The regulations apply to light- and heavy-duty vehicles and heavy-duty engines fuelled with gasoline, methanol or gasoline/methanol mixtures.

The EPA has also specified that from the 1 January 1996, the dispensing rates from gasoline and methanol pumps may not exceed 10 US gallons (37.9 litres) per minute. Facilities with throughputs below 10,000 gallons per month have been given a further two years to comply. This requirement is consistent with the dispensing rates specified in the new test measuring spillage during refuelling.

The current test procedure, which has changed little since its introduction, measures emissions from fuel evaporation during parking (diurnal emissions) and immediately following a drive (hot soak emissions).

The new procedures consist of vehicle preconditioning (including an initial loading of the carbon canister with fuel vapour), exhaust emission testing, a running loss test and three diurnal emissions cycles. The full test procedure takes 5 days to complete. Fuel spillage during refuelling (spitback) is also measured. A supplemental procedure, omitting the running loss test, but involving two diurnal cycles following the emissions cycles is included. This procedure ensures that all the emissions resulting from purging the evaporative canister are measured during the emission and diurnal cycles and do not escape during the running loss test. Because of its increased severity, the limits specified for this test are more relaxed than those for the three-diurnal sequence. The supplemental procedure can also be used in conjunction with the test procedures devised by the CARB, which are not yet in effect.

The procedures for heavy duty vehicles are similar except that the driving sequence for the running loss test consists of three consecutive UDDS cycles, which reflect the different driving pattern experienced in service. The testing of heavy-duty engines, without the vehicle chassis or body, requires that the test engine be equipped with a loaded evaporative canister and will be expected to demonstrate a sufficient level of purge during engine testing.

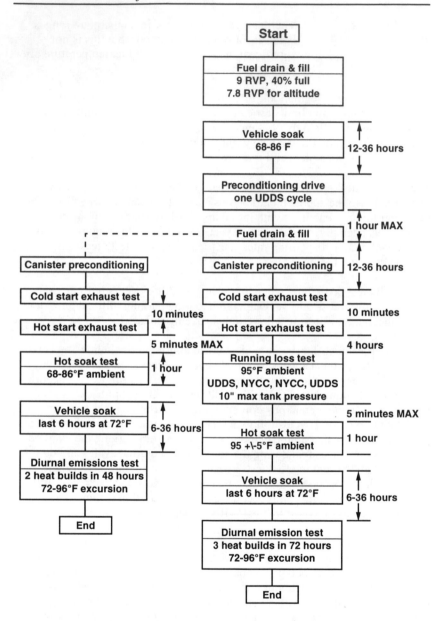

Fig. 4.7. New Federal US Emissions Test

Refuelling - Fuel Spitback Procedure.

The vehicle is refuelled at a rate of 10 USG/minute (37.9 l/minute) to test for fuel spitback emissions. All liquid fuel spitback which occurs during the test is collected in a bag. The bag (impermeable to hydrocarbons or methanol) should be designed and used so that liquid fuel does not spit back onto the vehicle body etc. The bag must not impede the free flow of displaced gasoline vapour from the filler neck and must be designed to allow the dispensing nozzle to pass through the bag. The dispensing nozzle should be a commercial model, **not** fitted with vapour recovery equipment.

The sequence for the proposed spitback test is shown in **Figure 4.8.**

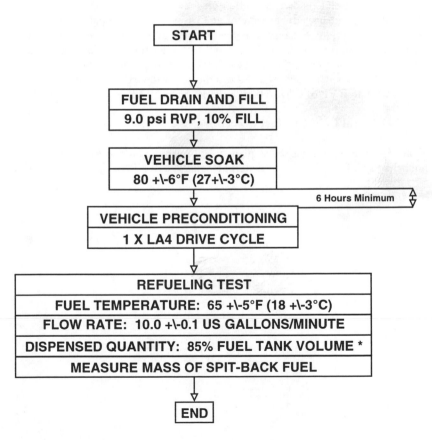

Figure 4.8 Proposed US Federal Spit-back Test Procedure

California

In California, CARB has proposed a rigorous evaporative emissions test procedure which will take almost five days to carry out. The test is similar, but not identical to, the new US Federal procedure. This procedure is summarized in **Table 4.3** and includes measurement of running losses and a three day "*real time*" diurnal emission test. It is not yet clear whether the running loss measurement must be made in a SHED mounted over a dynamometer or if a point source sampling technique can be used.

Japan - The Carbon Canister (Trap) Method

As stated earlier, the canister method employs weighed activated carbon traps which are connected to the fuel system at locations where fuel vapours may escape to the atmosphere (air cleaner, fuel tank vent, etc.) and is only applied in Japan. The vehicle is driven at 40 km/h ± 2 km/h under road conditions for 40 minutes on a chassis dynamometer at 25°C ± 5°C. Immediately after the engine is stopped, the exhaust is sealed and preweighed active carbon traps are connected to fuel tank, air cleaner and any other fuel system vents. After 1 hour the traps are reweighed and the total increase in weight must be less than 2 grams. The test fuel used is the same as for the exhaust emission testing with an RVP of 0.57 to 0.60 kg/cm^2 (56 to 59 kPa).

4.5 US Federal Cold CO Test Procedure

Table 4.4, below outlines the sequence for the above procedure. This was published as a final rule in the Federal Register, Vol 57 No. 138, dated Friday 17 July 1992.

Table 4.3 Proposed CARB Evaporative Emission Test (from 1995)

Test Sequence	Details	Time (h)	Total time (h)
FUEL DRAIN/FILL	Drain tank, fill to 40% vol with test fuel	0.3	0.3
SOAK PERIOD	Soak at 20-30°C for 12-36 h	12.0	12.3
PRECONDITIONING	Drive LA-4 cycle or 20-50 mi/day "typical" on-road driving	0.6	12.9
CANISTER PURGE	Purge with air at 25-75 grains/lb humidity at 48 ft³/h for 300 bed volumes	0.5	13.4
CANISTER LOAD	Load canister to 1.5 times nominal capacity with 50/50 butane/nitrogen at 15 g butane/h	10.0	23.4
FUEL DRAIN/FILL	Drain tank, fill to 40% vol with test fuel	0.3	23.7
SOAK PERIOD	Soak at 20-30°C for 12-36 h	12.0	35.7
EXHAUST TEST	Conduct full 3 phase FTP Cold start exhaust emission test	1.1	36.8
STABILIZE FUEL	Stabilize fuel temp to 105°F(40°C) within 1 hour	1.2	38.0
RUNNING LOSS	Drive 3 LA-4 cycles at 105°F(40°C) using point source method or SHED on dyno. Fuel temp. profile must be matched to road	1.5	39.5
HOT SOAK	Within 5 min seal in SHED at 105°F(40°C). Soak for 1 hour	1.1	40.6
STABILIZE FUEL	Stabilize fuel temp at 65°F (18°C) by artificial cooling	4.0	44.6
3 DAY DIURNAL	Park in SHED for 72 hours with SHED air temp cycles between 65°C(180°F) and 105°F(40°C) every 24 h. Measure HC emissions every 24 h.	72.0	116.6

Table 4.4 US Federal Cold CO Test Procedure

Step	Notes
1.	Winter grade fuel (optional use of FTP fuel)
2.*	Full US FTP 75 Cycle (optional use of higher temperature)
3.	No time specifications; uniform vehicle cooling; Oil temperature 20°F±3°F
4.*	12-36 hours
5.*	1 hour minimum
6.	If vehicle leaves 20°F soak area to transfer to test area and passes through a warm area it must be restabilized for 6 times the period of exposure to warmer conditions
7.*	Full US FTP 75 Cycle
8.	On dynamometer
9.*	Phase 1 (505 secs) of US FTP 75 Cycle

* Temperature Specifications (°F)	
Average	20 ± 5
Maximum Excursions	10(min), 30(max)
Three-minute Excursions	15(min), 35(max)

4.6 European Heavy Duty Exhaust Emission Test Procedure (ECE 49)

This procedure was developed as a test for medium and heavy duty diesel engines operating in Europe. Accordingly it is an engine rather than a vehicle test and basically follows the format of the obsolete US 13-mode procedure (see **Section 4.7**). Differences are that the minimum load condition uses a figure of 10% (US 2%) and the weighting factors are changed to take into account the different European driving patterns. The modes employed are as follows:

Table 4.5 ECE R49 Test Modes

Mode	Speed	Load %	Weighting Factor
1	Idle	0	0.083
2	Intermediate	10	0.080
3	Intermediate	25	0.080
4	Intermediate	50	0.080
5	Intermediate	75	0.080
6	Intermediate	100	0.250
7	Idle	10	0.083
8	Rated	100	0.100
9	Rated	75	0.020
10	Rated	50	0.020
11	Rated	25	0.020
12	Rated	10	0.020
13	Idle	0	0.083

The measuring analysers are as for the US test whilst the calculation method follows the early US procedure based on measuring exhaust flow.

The major differences between the European and US 13 mode procedures therefore lie in the minimum load level employed and the different weighting factors. These have the effect of reducing the measured hydrocarbons but increasing the measured NOx (and CO). It is important that this should be taken into account when considering US legislation levels and translating them into a European context.

As the EU moves towards more stringent heavy duty diesel engine emissions limits, a number of Member States have suggested that some form of transient test should replace the current 13-mode procedure. Some countries favour the straightforward adoption of the US transient test whilst others argue that this is not representative of European conditions. Another school of thought suggests that the 13-mode procedure be retained, albeit with the addition of some form of transient element.

4.7 US Federal Heavy Duty Exhaust Emission Test Procedures

United States 13-Mode Test

From 1973 to 1984 the United States used a steady state test for the measurement of gaseous emissions from heavy duty engines. The main details of the test procedure are as follows:

Applicability: All heavy duty gasoline and diesel engines. That is engines fitted to vehicles with a gross vehicle weight greater than 8500 lb, or a kerb weight greater than 6000 lb, or a frontal area of more than 45 square feet.

Type of Test: 13-mode steady state test bed cycle.

Table 4.6 Details of US 13-Mode Test

Mode	Speed	Load (%)	Weighting Factor
1	Idle	0	0.06667
2	Intermediate	2	0.08
3	Intermediate	25	0.08
4	Intermediate	50	0.08
5	Intermediate	75	0.08
6	Intermediate	100	0.08
7	Idle	0	0.06667
8	Rated	100	0.08
9	Rated	75	0.08
10	Rated	50	0.08
11	Rated	25	0.08
12	Rated	2	0.08
13	Idle	0	0.06667

Each mode is held for a minimum of 4.5 minutes and a maximum of 6.0 minutes;

- Intermediate speed is the peak torque speed or 60% of the rated speed, whichever is the higher;
- During each mode the specified speed shall be held to within 50 rev/min and torque at each mode must be held at the specified value ± 2% of the observed maximum torque value;
- All data, including continuous emissions traces, are to be recorded during the last two minutes of each mode.

Measuring Instruments

CO, CO_2	- NDIR
HC	- Heated FID
NOx	- Chemiluminescence analyzer

Test Parameters

Fuel flow rate	- volumetric or gravimetric determination
Air flow rate	- gravimetric determination

Allowed Temperature Ranges

Air supply to engine	- 5 to +25°C
Fuel pump inlet	- 5 to +37.8°C

United States - Transient Test

The US transient test was introduced as an option for certification of heavy duty vehicles in 1984. In 1985, it became mandatory, replacing the steady state test described above. The reasons for moving to a transient cycle were to make the test more representative of on-highway conditions and to improve repeatability.

It is a popular misconception that the US transient cycle is based on actual driving patterns assessed in the following scenario:

New York non-freeway
Los Angeles non-freeway
New York freeway
Los Angeles non-freeway (repeated)

Driving patterns were assessed from operational data measured in New York and Los Angeles, but the US transient cycle, as developed from these data, contains the following characteristics:

Cold Start	Hot Start
Congested urban	Congested urban
Uncongested urban	Uncongested urban
Expressway	Expressway
Congested urban	Congested urban

The US transient cycle tests the engine over a full range of load and speed conditions, with equal weighting within the cycle of all operational points. The cycle is run twice, first from a cold start and then, after a 20 minute soak, repeated with a hot start. These two cycles are weighted, 1/7 to 6/7 for the cold and hot cycles respectively.

The 178 seconds of motoring contained in the US transient cycle used to be considered very important. Motoring, or driving the engine by the dynamometer, tests the engine in conditions where, on the road, fuel delivery would not occur although the engine would still be rotating (travelling downhill with the accelerator closed, for example). In these cases it is important to test for fuel dribble from the injector nozzles. However, modern technology nozzles having low sac volume or valve covering the orifice (VCO) devices suffer from this to a lesser extent and the need for motoring may become open to question. Nevertheless, it seems unlikely that the EPA will alter the motoring requirement.

The emphasis is on urban driving since any US air quality problems are most acute in major cities. The cycle is therefore not typical of average driving in either the US or Europe. Speed and load traces for a heavy duty engine over the US Federal Transient Test Cycle are shown in **Figure 4.9**.

EPA has conducted tests to assess the inter-laboratory repeatability of emissions measurement using the US transient procedure. Laboratories at Detroit Diesel, Ford, EPA and Navistar took part in the programme.

Table 4.7, below, indicates average emission measurements for each laboratory in g/bhp.hr.

Table 4.7 Reproducibility of the US HD Transient Cycle

	Detroit Diesel	Ford	EPA	Navistar	Standard Deviation	Variance
CO	1.162	1.114	1.66	1.20	0.219	0.048
HC	0.11	0.113	0.103	0.14	0.014	197×10^{-6}
NOx	5.277	5.647	5.29	5.09	0.202	0.041
Pm	0.169	0.17	0.169	0.166	0.0015	2.25×10^{-6}

The improvement in repeatability and reproducibility may be due to the transient cycle but is more likely to reflect the increased accuracy of control both of the engine and general test conditions. The vastly increased number of data points must also play a large part. The test consists of three 20

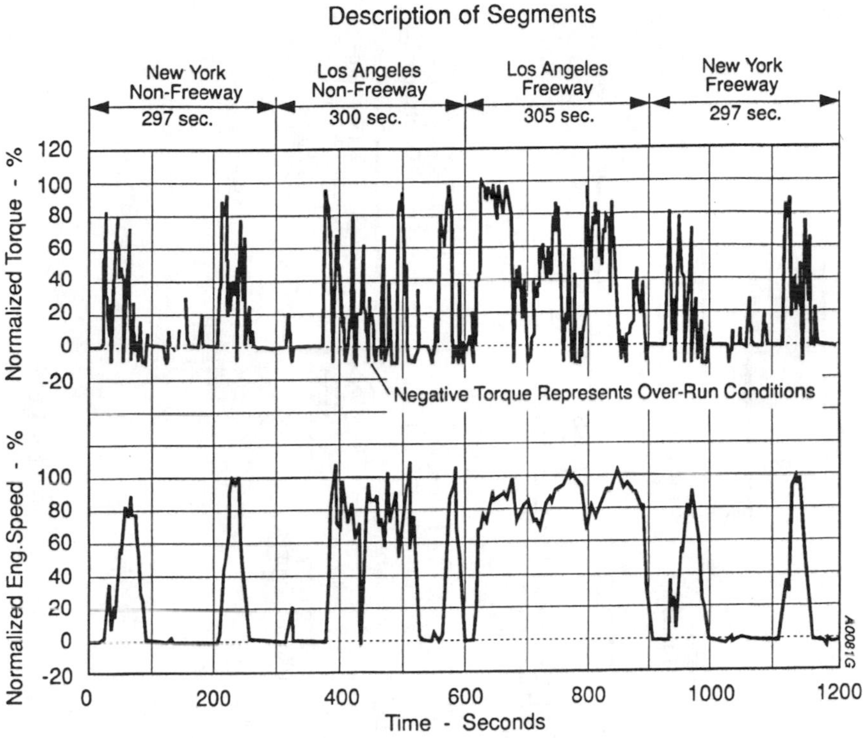

Description of Segments

Source: AVL List GmbH

Fig. 4.9. Speed and Load Traces for a Heavy Duty Engine over the US Federal Transient Test Cycle

minutes cycles (one cold start, one hot soak and one hot start cycle) and data relating to engine speed and torque, and the levels of gaseous emissions are monitored once per second. Engine control and data acquisition is by computer and the emissions results are integrated over the test cycle. Speed, torque and power data are subjected to regression analysis and must validate to within closely specified limits.

Gaseous emissions sampling is done from the primary dilution tunnel. A secondary dilution tunnel is used for the measurement of particulates. A general arrangement of a typical transient test facility is shown in **Figure 4.10**.

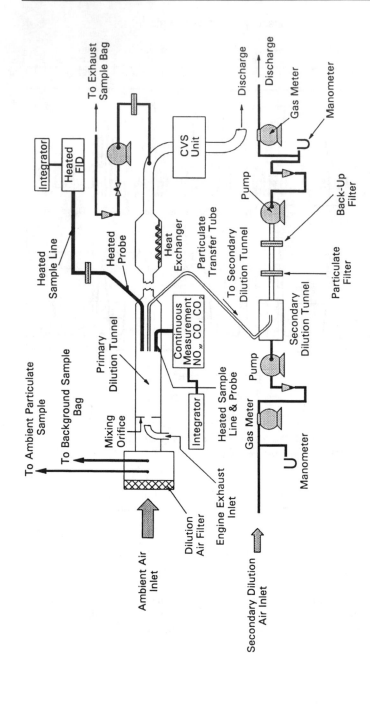

Fig. 4.10. Double Dilution Particulate Measurement System

Source: Ricardo Consulting Engineers Ltd.

The emissions results are processed in a *"wet"* form and a correction for humidity applied to NOx. The results are then weighted 1/7 and 6/7 for the cold/hot cycles respectively, and added to form an integrated result.

Particulates measurements are achieved by passing all the flow from the secondary tunnel through a 70 mm diameter Teflon coated glass fibre filter. After stabilizing in a controlled environment this is weighed. A reference paper allows correction for ambient particulates.

Full details of transient test requirements can be found in the US Federal Register 40 CFR 86 (1985).

4.8 US Enhanced Inspection and Maintenance Procedure (IM 240)

The new US Clean Air Act requires the introduction of enhanced inspection and maintenance programmes from 15 November 1992 in the most polluted areas (see **section 5.4.2**). The EPA published its proposals on 5 November 1992 and held discussions with automobile manufacturers of I&M equipment prior to publishing its final rule at the end of April 1993.

Test Procedures

Preconditioning

At the programme Administrator's discretion vehicles may be preconditioned using any of the following methods:

a) non-loading preconditioning: increase engine speed to approximately 2500 rpm in neutral for up to 4 minutes
b) drive the vehicle on the dynamometer at 30 miles/h for up to 240 seconds at road load
c) drive a preliminary transient cycle

Measurement of Exhaust Emissions

The IM 240 procedure for exhaust emissions is a transient test based on 240 seconds of Federal test certification procedures (**Section 4.2**). A trace of the vehicle emission test sequence is given in **Figure 4.11**.

Emissions of CO, HC and NOx are measured second-by-second over the cycle. The results are summed over four modes and multiplied by a factor. These figures are averaged to give a composite result. If the vehicle fails the appropriate limit, a repeat calculation can be made where the results are calculated over two phases.

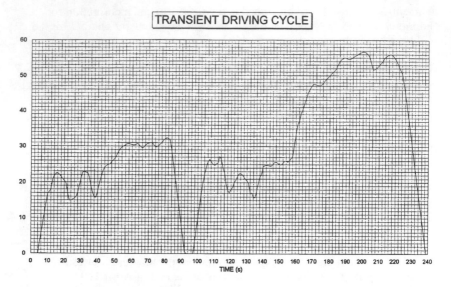

Fig. 4.11. US IM 240 Vehicle Emission Test Sequence

Composite Test	Mode	Cycle Portion
	1	0-60 secs
	2	61-119 secs
	3	120-174 secs
	4	175-239 secs

Repeat Calculation	Phase	Cycle Portion
	1	0-93 secs
	2	94-239 secs

Examination of Evaporative Emissions Control System

The integrity of the evaporative emissions control system is checked by inspection. The system is also tested for flow rate by connecting measuring equipment in series between the canister and the engine, preferably at the canister end of the hose.

724

Test Criteria and Calculation of Results

The vehicle passes the emissions test if the value for each pollutant is below the standard in the complete test. If for any pollutant the composite emission rate exceeds its standard then an additional analysis of the test results may be carried out which divides the driving cycle into two phases. The vehicle shall have deemed to have passed if the emission rate is below the standard limit in the composite result or in Phase 2 of the Repeat Calculation.

4.9 Off-Highway Test Cycle (ISO 8178)

ISO 8178, Part 4 specifies test cycles for the measurement and evaluation of gaseous and particulate emissions from reciprocating internal combustion (RIC) engines under test conditions and, with certain restrictions, under site conditions. The test is steady state, using cycles according to the given application. For the sake of brevity, only the C1 cycle applicable to larger engines is shown in the following table:

Table 4.8 ISO 8178-4 Test Cycle Type C1 for Off-road Vehicles, Industrial & Medium/High Load.

Measuring Point	1	2	3	4	5	6	7	8
Speed	Rated Speed				Intermediate Speed			Low No-load Speed
Torque	100	75	50	10	100	75	50	0
Weighting Factor	0.15	0.15	0.15	0.1	0.1	0.1	0.1	0.15

Appendix 5

In Service Emissions Testing

Reproduced from Report 4/94, Section 5, by kind permission of:

CONCAWE, Brussels, Belgium

5 In Service Emissions Testing - Vehicle Inspection and Maintenance Programmes and Legislation

This section deals with programmes and legislation for in-service emissions testing and maintenance programmes. **Section 5.1** gives EU legislation and the special requirements of EU member states. **Section 5.2** contains details of the test protocols for non-EU European countries. **Section 5.3** summarizes US legislation and **Section 5.4** provides information on legislation in the Far East.

5.1 EU Exhaust Emissions Roadworthiness Tests

Legislation with respect to in-service emissions testing for road vehicles within the EU was promulgated by Directive 77/143/EEC and subsequently amended by Directive 88/449/EEC. These directives covered buses, coaches, light and heavy goods vehicles, trailers and semi-trailers over 3.5 tonnes, taxis and ambulances. Although the directives stipulated that noise and exhaust emissions should be incorporated in the roadworthiness tests, no limits were specified. Furthermore, passenger cars were not included in that legislation. Directive 92/55/EEC, dated 22 June 1992, sets out limit values and test procedures. These are described in this section.

In the intervening period, a number of EU countries have introduced their own test regimes and limit values. These are also detailed in this section.

EU In-service Emissions Testing

The following paragraphs describe the test procedure and limit values to be applied.

Spark Ignition Engined Vehicles

The test procedure consists of two elements:

(i) A visual inspection to ensure that:

- There are no leaks from the exhaust system
- If applicable, any emissions control system is present

(ii) An idle CO test, details as follows:

No load idle test, carried out after the manufacturer's recommended engine pre-conditioning period. The limit values are shown in **Table 5.1**. An additional *"increased idle speed"* test is required for vehicles fitted with 3-way catalysts and lambda control.

Diesel Engined Vehicles

The test procedure consists of a free acceleration smoke test. That is, the engine is accelerated with the transmission in neutral (no load) from idle up

Table 5.1   EU In Service Emission Limit Values - Gasoline Vehicles

Vehicle Description	Idle CO (% vol)
All models not fitted with 3-way catalysts and lambda control	Initial type approval limit + 0.5
Where these data are not available or Member States decide not to use these reference values the following alternative will apply:	
Manufactured prior to October 1986[*]	4.5
Manufactured after 1 October 1986	3.5
All models fitted with 3-way catalysts and lambda control[#]	*Either* initial type approval limit *or* 0.5 maximum

[*]   Or the date on which Member States required the vehicles on first registration to comply with the Type Approval Directive 70/220/EEC, as amended.

[#]   An additional no load test is to be conducted at a minimum idle speed of 2000 rpm. The following limit values apply:
CO: 0.3% vol maximum
Lambda: 1 ± 0.03 or in accordance with the manufacturers specifications

to maximum (governor cut-off) speed and the smoke opacity is measured. The following maximum coefficient of light absorption is allowed:

Either:

Initial type approval limit + a tolerance of 0.5 m^{-1}

Or:

Where these data is not available or Member States decide not to use these reference values the following alternative maxima will apply:

Naturally aspirated diesel engines: 2.5 m^{-1}
Turbo-charged diesel engines: 3.0 m^{-1}

Periodicity of Testing

Table 5.2 summarizes the ages of vehicles at which testing should commence and the frequency of that testing:

5.2 Current EU Member State In-Service Emissions Legislation

Belgium

All gasoline powered cars more than four years old have to be checked annually or upon change of ownership at one of 66 state operated test centres. The requirement of 4.5%v CO max at idle must be met by all these cars, including those fitted with catalysts.

Although an opacity limit exists for diesel powered vehicles, no checks are made at the test centres on exhaust emissions from such vehicles.

Table 5.2   EU Periodicity of In-Service Emissions Testing

Vehicle Description	Age (Years)	Test Frequency (Years)
Spark Ignition Engined Passenger Cars	3	1
Heavy Commercial Vehicles, Taxis and Ambulances	1	1
Commercial Diesel Vehicles (Less than 3.5 tonnes)	4	2

Denmark

Regulation: In-service emissions testing is part of the roadworthiness test.

Scope: Cars - Inspection on change of vehicle ownership. Vans, Trucks & Buses - Periodic inspection

Frequency: Cars - As above, once vehicle is five years old.
Vans - Every other year, once vehicle is four years old.
Trucks & Buses - Annually after one year.

Gasoline Engine Test/Type/Limits:

Max 5.5% CO at idle for vehicles registered before 1984.
Max 4.5% CO at idle for vehicles registered after 1984.
Max 0.5% CO at idle for vehicles equipped with 3-way catalysts.

Diesel Engine Test/Type/Limits:

Max 3.8 Bosch Smoke Units (trucks and buses) under free acceleration.

Inspections are carried out by state operated test centres. Road side spot checks are also included in the legislation.

France

No requirements.

Germany

The following table gives details of German in-service emissions test requirements, according to regulation Abgasuntersuchung AU (Article 47a StVZO) which became effective on 1 December 1993.

Greece

An "*Automotive Emissions*" card is to be introduced, which will be issued annually after a vehicle passes an emissions test. Failure to possess a card renders the vehicle owner liable to a fine.

Italy

Regulation: Law No. 615 of 1966 (diesel engines only); no legal requirements for gasoline vehicles. Some local authorities during air quality non-attainment days allow only vehicles complying with emissions limits to operate in urban areas.

Table 5.3   Germany - In-Service Emissions Test Requirements

Vehicle Type	Gasoline	Gasoline	Diesel	Diesel
	Oxidation catalysts or no catalysts	3-way catalysts	<3.5t	>3.5t
Frequency (months)	12	24	24	12
After (months)	12	36	36	12
Type Of Testing				
Visual Check	+	+	+	+
Idle Speed, low	+	+	+	+
Idle Speed, high	-	-	+	+
Spark Timing	+	+	-	-
Dwell Angle	+	-	-	-
EGR	+	-	-	-
Secondary Air System	+	-	-	-
Catalyst CO Low Idle	(+)	+	-	-
Catalyst CO High Idle	-	+	-	-
Lambda Sensor Circuit	-	+	-	-
Full Load Stop	-	-	+	+
Opacity Free Acceleration	-	-	+	-
Bosch Number Full Load (road or chassis dynamometer)	-	-	-	+

+   Limits as stipulated by the manufacturer and conformity of production limits.

Scope: Diesel and gasoline vehicles.

Frequency: No nation-wide scheduled tests. Only road-side spot checks (diesel) or checks for local authorities annual permit (diesel and gasoline vehicles).

Test Type: Smoke opacity under free acceleration (diesel vehicles). CO at idle (gasoline vehicles)

Limits: 65% opacity for urban buses.
70% opacity for all other diesel vehicles.
Manufacturer's manual for CO of catalysts-equipped cars.
4.5%v CO and 3.5% CO at idle for non-catalyst cars registered before 1986 and after 1996 respectively.

Penalty for non-compliance: Fine and withdrawal of vehicle license until the owner resubmits the vehicle and it is found to comply.

Note: From 1989 the Italian Department of the Environment has invited all vehicle owners to have their vehicles checked on a voluntary basis. A network of test centres has been set up with the cooperation of the motor manufacturers and oil industry. After testing, the owner is advised that his vehicle requires tuning if it has exceeded the following limits:

Gasoline - 4.5% vol CO at idle.
Diesel - Legal smoke opacity limits (as above).

Netherlands

Regulation: APK/"*Milieu*" (Environmental) Inspection

Scope: Passenger cars and light spark ignition engined vans (3500 kg max.) aged three years or older have to be checked annually. Currently there are no legal limits for diesel cars or buses and heavy duty trucks. A diesel smoke opacity requirement is under consideration. Test work is also carried out to check that vehicles meet type approval limits and manufacturers are notified of the results.

Test Limits: Since 01.09.91 some cars have to meet the more stringent "*Milieu*" inspection requirements:

Gasoline and LPG vehicles registered since 1974:
Overall idle CO: 4.5% vol max.

Cars with three-way catalysts, registered since 1986:
Idle CO: 0.5% vol max

Other cars (with and without uncontrolled catalysts) and registered since 1980:
An idle CO limit per car model and in line with quoted type approval limits.

Diesel cars and vans:
No limits

Vehicles failing to meet these requirements have to be adjusted and retested. Vehicles which can not achieve the limits are not allowed on the road.

Portugal

No requirements

Spain

No requirements

United-Kingdom

Emissions testing has been introduced as part of the existing annual Department of Transport (DOT) roadworthiness test, applied to all four wheeled gasoline powered passenger cars aged three years or older. Standards and testing methods for light duty diesels and two-stroke engines are to be developed.

The test will be conducted with the engine warmed-up and at idle and the following limits will apply:

Vehicles first used on or after 01.08.83:

CO: 4.5% vol max.
HC: 0.12% vol max

Vehicles first used between 01.08.75 and 31.07.83:

CO: 6.0% vol max.

Vehicles will fail the complete DOT test even if emissions performance is the only item of failure. A free re-test will be allowed (if the vehicle has failed on emissions only) and provided the vehicle is returned to the same testing station *"within 14 days or so"*. Emissions standards will also be the subject of road-side testing. Vehicles which fail to comply will be required to be rectified *"within 14 days or so"*.

The test procedure was implemented from 1 November 1991. The UK Department of Transport is also considering the inclusion of an in-service HC standard and has stated that a proposal will be published in the near future.

Heavy Duty Vehicles

From 01.07.92, a free acceleration smoke test (see **Section 3.1.3, Table 3.6**) has been added to the heavy duty vehicle DOT test requirements.

5.3   Europe: Non-EU States In-Service Emissions Legislation

Austria

Regulation: KFG Article 57 a/32. Amendment to KDV from 18.02.91. Valid from 01.10 91

Frequency: Annual

Otto Engine/Type/Test/Limits:

Without catalysts - CO at idle speed in accordance with manufacturer's manual, max. 3.5% vol
With 3-way catalysts:
- CO at idle speed < 0.3% vol
- Lambda 0.97 to 1.03 measured at 3000 rpm idle speed
- Check on lambda sensor circuit

Diesel Engine/Type/Test/Limits:

Passenger/heavy duty - Soot measurement at free acceleration Bacharach number 4.5 (± 1.0 tolerance)

Finland

A regulation for the in-service testing of gasoline cars has been introduced effective from the 1 January 1993. Tests can be carried out along with the normal roadworthiness test by an authorized institution, repair shop or service station. The first test will be done at 4th year after registration, the second after six years and annually thereafter according to the following schedule.

Table 5.4 In-Service Testing - Gasoline Cars

Vehicle Description	Test type	CO %v	HC ppm	Lambda high idle
Registered before 1.1.78	none	-	-	-
Registered 1978-1985	idle	4.5	1000	-
Registered after 1.1.86	idle	3.5	600	-
Low Emission Vehicle	idle and>2000 rpm idle	0.3	100	0.97-1.03

Diesel powered vehicles are only inspected visually in the normal roadworthiness test. The legislated limit is 7.0 Bosch for vehicles registered before 1.1.90 and 3.5 Bosch for vehicles registered after 1.1.90.

According to a new proposal, after 1.1.95 all diesel vehicles registered after 1.1.80 are measured by an opacity method using free acceleration with limits of 2.5 l/m for naturally aspirated vehicles and 3.0 l/m for turbocharged vehicles.

Sweden

Passenger cars aged two years or older (see below) have to be inspected annually by the Swedish Motor Vehicle Inspection Company. The following limits apply:

Table 5.5   Limits for 3-way catalyst equipped cars

	CO %	HC ppm
L1	0.50	100
L2	1.0	200

* First introduced 01.10.90

Limits for 1976-1988 car models -

EGR valves are not compulsory but must function if fitted. The annual test CO limit is about 4.5% vol but the actual limit varies with model and engine. If the car fails there are certain threshold CO levels which will be reported to the vehicle owner but no further remedial action is required. These values vary between vehicle make, model and engine.

Failed vehicles have to be rectified and re-tested within one month. Vehicles which are not re-tested within that month are not allowed to be used on the road. Vehicles must be submitted for the annual test until they are five years old or have reached 80 000 km, whichever occurs first. It is expected that this limit will be extended to 160 000 km, equivalent to Californian regulations.

Switzerland

Gasoline Powered Passenger Cars/Other Light Duty Vehicles not Exceeding 3 500 kg GVW

Tested annually at authorized garages. The test comprises:

- Idle speed
- Ignition timing (with and without vacuum advance)
- CO, HC and CO_2 - Limits vary from model to model, reflecting the engine calibration employed for emissions homologation.

The emissions test certificate must be carried on the vehicle. The police conduct random spot checks and if the certificate is not displayed or is out of date the vehicle owner is liable to a fine. Vehicles failing to pass the test are considered as "*not meeting the legal requirements*" and must be withdrawn from service.

All Vehicles with Diesel Engines

Smoke opacity under full rack acceleration in neutral is measured when vehicle undergoes a regular check at an official Cantonal test station. Test interval:

- 1 year for trucks transporting dangerous goods.
- 3 years for all other diesel vehicles.

Vehicles failing the test have to be re-submitted within two to three weeks.

5.4   US Inspection and Maintenance Programmes

Introduction

There are two aspects to US in-service emissions testing. The first involves surveillance testing to ensure compliance with certification durability requirements (i.e. conformity with the 50 000 or 100 000 mile limits). The second extends the rigour of existing inspection and maintenance programmes.

5.4.1   In-Use Surveillance Testing

Non-Routine Testing

Random FTP-75 checks will be carried out by the administrator of in-use vehicles at mileages below 50/75 000 miles. Additionally, in California, an evaporative emissions test will be conducted, depending on the manufacturers certification procedure. (EPA Surveillance Test Programmes/Title 13 CCR, Section 2136 - 2140)

Continuous Vehicle Surveillance

49 States: From model year 1992, defects on certain emission-related components/systems of in-use vehicles have to be reported for a period of five years after the end of that model year. A report has to be filed once a minimum of 25 defects have occurred on an individual part. The report includes a specification of the defective component(s), a description of the failure and details of the corrective action taken. Based upon this report, the EPA may decide upon a model recall.

California: From model year 1990, defects on specified emissions-related components and systems of in-use vehicles have to be reported for a period of 5/10 years or 50 000/100 000 miles, depending on the warranty period (Title 13, CCR). The reporting procedure consists of three steps, where each step requires more and more detailed information. The reporting requirements also ask for information with regard to the number of defects, analysis of those defects and the effects on exhaust emissions. CARB will decide upon a recall, based upon the second and third report stages.

There are phase-in standards for the calendar years 1995 and 1996 applying to 1984 model year and newer vehicles. Then follows two sets of final standards, the first applying to vehicles of the 1984 model year and newer from calendar year 1997 and the second applying to vehicles certified to Tier I standards (including those certified in model years 1994 and 1995) from calendar year 1996. The limits are given in **Table 5.6**. The limits are the lowest standards that may applied. Higher limits may be adopted if approved by the Administrator.

Evaporative Emissions Control System Purge Test Standard

The Vehicle shall pass the purge test when the total volume of flow exceeds one standard litre. If total volume of flow is less than 1.0 standard litre at the conclusion of the transient driving cycle the vehicle shall fail.

Evaporative Emissions Control System Integrity

Unless inaccessible, the canister and evaporative system shall be inspected for damage or incorrect connections. The evaporative emissions control system shall be subjected to a pressure test to ensure the absence of leaks.

Table 5.6   IM 240 Short Test Emissions Standards

Standard	Effective Date	Vehicle	Age yrs	CO g/mile	HC g/mile	NOx g/mile
Phase-in	1995	LD Vehicle	<5	25.0	1.2	3.0
	1996	1984 or younger	>4	30.0	2.0	3.5
		LD Truck	<5	25.0	1.2	3.5
		1984 or younger	<4	30.0	2.0	4.0
Final	1997	LD Vehicle 1984-1995 but pre Tier I		15.0	0.8	2.0
		LD Truck 1984-1995 but pre Tier I		15.0	0.8	3.0
Final	1997	Tier I LD Vehicle (including 1994/5/6)		15.0	0.7	1.4
		Tier I LD Truck (including 1994/5/6)				
		<6000 lb GVW		15.0	0.7	2.0
		>6000 lb GVW		15.0	0.8	2.5

5.4.2   Enhanced Inspection and Maintenance Testing - The New Clean Air Act Amendments

The new US CAAA calls for the introduction of enhanced inspection and maintenance programmes from 15 November 1992 in the most polluted areas. These areas include certain *"serious"*, *"severe"* and *"extreme"* ozone non-attainment areas with populations in excess of 200 000. The Act also mandates enhanced I/M in urban areas located in an *"ozone transport region"* with 100 000 or more residents. This represents an increase in the number of areas requiring I/M programmes from 122 to 181. After long debate, and in response to a court order, the EPA issued its proposals for an I&M Programme on 13 July 1992.

These proposals were signed as a final rule during June 1993. States were required to submit their detailed inspection programme by November 1993 and promulgate the necessary State legislation.

The proposal requires 181 areas of the country to establish emissions testing programmes. The 95 non-attainment areas with "moderate" air pollution problems must introduce a basic I&M programme, 82 more polluted areas must implement an enhanced "HI-TEC" programme. EPA estimates the programme will reduce VOC emissions by 28% at a cost of US $500/tonne. The rule "should save $1.6 billion annually in clean air costs and 15 million barrels of oil a year as a result of improved fuel economy". The EPA and the Department of Transport also announced that states will be eligible to federal funds to help defray costs of the programme.

The performance standard proposed for Basic I&M is modelled after the original New Jersey programme (e.g. use of a simple idle test). States required to have only basic I&M programmes can opt for a more stringent programme and can build credits which can be used to offset other pollution control efforts.

The EPA proposed performance standards for Enhanced I&M is based on annual, centralized testing of all 1968 and later model year passenger cars and light duty trucks. A steady-state test can be used for 1968-1985 models but the High-Tech IM 240 test is required for 1986 and later model vehicles. Besides emissions tests, the I&M procedures also require a visual inspection for the presence and proper connection of the catalyst, fuel inlet restrictor and evaporative emissions control system. Pressure and purge checks are also required on the carbon canister.

The IM 240 test is a transient test based on 240 seconds of the Federal certification test procedures and is claimed to be about three times more accurate in identifying vehicles exceeding emission standards than current tests. EPA estimates that the test equipment for IM 240 will be about $150 000 and will probably cost the consumer $17 per vehicle to administer. Based on the effectiveness of the IM 240 test, EPA has concluded that it would be acceptable to require the test only once every two years. Also EPA has proposed that states design the programme such that facilities are located within five miles of where 80% of the vehicle owners work or live and that adequate test lanes are present to ensure that there is no more than a 15 minute wait in line for testing.

Congress countered that States could adopt "Alternative Enhanced I&M" programmes in lieu of the model proposed by EPA, as long as emission reduction benefits of such alternatives were comparable to EPA's programme. New Jersey has already adopted an alternative Enhanced I&M Test procedure (the 5015 loaded-mode test) which is claimed to be between half and a third cheaper to operate than the EPA procedure. EPA, as part of a political compromise with the President's Competitiveness Council, has made it somewhat easier for States to adopt decentralized programmes.

However, EPA's own studies suggest that such programmes, which allow both the test and repair to occur at the same facility, are only half as effective as programmes which require centralized, test-only facilities. States which choose the low-tech option run the risk of being required at a later date to convert to high-tech I&M if their programme is unsuccessful.

States must implement decentralized basic I&M programmes by July 1, 1993 and centralized basic programmes by January 1, 1994. The enhanced I&M programmes will need to be in place by July 1994, but EPA has proposed phasing-in the number of vehicles covered (e.g. 30% in 1994 and 100% by 1996). Also, EPA is proposing that the cut point for failing the I&M test be relaxed initially, with the full, more stringent cut-point applied by January 1, 1998.

In a March 1994 court ruling, the EPA has been ordered to approve or disapprove by July 15 1994, all State basic and enhanced automobile inspection and maintenance (I/M) programmes it has received. States that have not submitted adequate I/M programmes by that date could be immediately sanctioned. Federal sanctions include the loss of federal highway funds or a two-for-one emission offset requirement for new stationary air pollution sources.

This court decision may have widespread ramifications for the States as the EPA loses the flexibility to extend mandatory Clean Air Act deadlines, thus forcing air regulators nationwide to reassess their schedules for programme submittals.

Following this action, the EPA was sued by the Natural Resources Defense Council (NRDC), which challenged the agency's right to authorize a delay for the States in fileing SIP revisions. NRDC also sued the agency over its use of the so-called "committal SIPs" for other programmes required by the Clean Air Act, including some new source review revisions.

As a consequence, a wide range of local programmes, which vary in both complexity and stringency remain in existence.

5.4.3   Summary of Pre CAAA US I/M Programmes

Start Date

The majority of the programmes began in the years between 1982 and 1986.

Programme Type

Programme type, enforcement mechanism, frequency of inspection, and other features were generally left to the discretion of the responsible state or local officials, subject to EPA's approval that the performance criteria would be achieved.

The programme types can be classified into three categories: centralized, decentralized-computer, and decentralized-manual. Decentralized programmes generally involve having I/M tests performed at local private garages or automotive dealerships. In most cases, the establishment performing the test must have applied for and received an inspection license from the state. A vehicle which fails the inspection text may be repaired at the same facility and then re-inspected. Approximately 65% of the programmes are decentralized.

Centralized programmes involve the testing of vehicles at specialized test facilities capable of handling a high volume of vehicles. Centralized systems may be contractor-run, state-run, or locally-run. The centralized programme, however, is never run by an organization which is involved in automotive repair. Therefore, vehicles which fail have to be repaired at another location and are usually returned to the centralized site for re-inspection.

Enforcement

The method of enforcement differs significantly. The most popular one (used by approximately 50% of the programmes) is to require compliance with the I/M programme requirements as a prerequisite for vehicle registration renewal. Approximately 39% of the programmes rely on the *"sticker"* method, which requires that all vehicles included in the programme display a current sticker, usually on the windscreen. If a vehicle is required to be tested and does not display a current sticker, the vehicle owner may be liable to a fine. The third method is called *"computer matching."* It identifies vehicles which have not undergone I/M testing.

EPA audits have identified some I/M programmes which appear to be less effective than other. One major problem is related to low failure rates in decentralized I/M programmes with manual analyzers. The two major causes of the low failure rates are poor instrument quality control and in some cases, cheating.

Frequency

Virtually all of the programmes require annual testing. California, Indiana, Maryland, New Mexico, and Oregon, however, require that their tests be performed once every two years.

Type of test

Inspection and maintenance tests can take many forms. In the simplest one, vehicle exhaust gases are sampled while the engine is idling, and those vehicles with excessive CO or HC emissions are required to undergo remedial maintenance and pass a retest. Some programmes include a tampering check which can range from a quick visual inspection to see that emissions controls are in place, to a more thorough check which can include efforts to identify misfuelling with leaded gasoline.

Exhaust Emissions Test Mode

There are a number of different test modes that measure vehicle emissions, including:
- idle test
- two-speed test (idle and high speed idle in neutral)
- loaded mode test

An idle test, which is used by over half of the programmes, gathers emissions data at a single engine idle speed. A few programmes use a two-speed idle test which checks emissions at both high (2500 rpm in neutral) and low (idle) engine speeds. Only one state uses a loaded mode test for pass/fail decisions. There are five *"tamper only"* I/M programmes, which do not employ an emissions measurement test mode. In addition, several programmes use combinations of the test modes listed above.

Tamper Test

Although nearly all programmes have some type of tamper test, they are not necessarily required for all vehicles. For example, many programmes require a tamper test only when vehicles are subject to waivers to the emissions cutpoint requirements (see overleaf). Also 1974 and older models may not require inspection as they are not catalyst equipped.

Safety Test

A number of I/M programmes are conducted in conjunction with a safety inspection.

Waivers

Some 75% of the I/M programmes limit the cost a vehicle owner incurs in order to repair vehicles that do not pass the I/M test. The remainder of the programmes do not allow waivers. Where waivers exist, the limit varies from a low of $15 to a high of $300. In most of the programmes that have waivers, the amount excludes any repair costs associated with tampering. A number of programmes have cost ceilings that are based on the stated repairs necessary to pass testing. Several combine the tamper exclusion and stated repairs concepts, while some use neither. Finally, a few programmes have *"parts"* or *"parts plus labour"* methods of determining cost ceilings, as well as complicated *"percentage repair cost reduction"* guidelines.

Eligible Vehicles

Each programme determines which model year vehicles are required to undergo I/M testing. Many of the programmes base this decision on the model year method. For example, a programme (in 1994) with a ten-model-years-method indicates that 1984 and newer vehicles must be tested. The number of model years required varies, from a low of nine years in Detroit up to 21 in Nevada.

Another way to specify the age under which vehicles are required to be tested is by simply specifying the oldest model year. The most popular first model year appears to be 1968, which coincides with the introduction of the first nationwide US exhaust emissions controls used on vehicles. This initial model year can be modified periodically as long as the overall emissions reduction provided for in the State Implementation Plan is achieved.

Vehicles of all ages are required to be tested in some states.

Exemptions

All I/M programmes provide for at least some exemption. These exemptions can be in one or more of the following categories:

Weight
Type of vehicle
Fuel types
Cooling system.

For the most part, vehicles over a certain weight are excluded from I/M tests. This exemption category has an average around 8 500 lb.

Motorcycles are the only vehicles that appear to be exempted by most programmes.

Programmes offer exemptions for two major fuel types: diesel fuel and alternative fuels. All the programmes, except Kentucky, currently exempt diesel-powered vehicles. Several also offer exemptions for vehicles that operate on other fuels, such as methanol, ethanol, compressed natural gas (CNG), or LPG. Vehicles which can be operated on alcohol/gasoline blends are treated as gasoline fuelled vehicles and are, therefore, not exempt.

While, in some cases, the cooling system of a vehicle may affect its emissions as well as the design of its emissions control system, only one state, Virginia, currently has an exemption for air-cooled vehicles.

Emissions Standards (Cutpoints)

CO and HC emissions cutpoints have a significant impact on vehicle failure rates. Most of the I/M programmes have CO and HC emissions standards that usually vary according to the model year of the vehicle being tested.

Failure rates bear a strong relationship to cutpoints. Erring too far in one direction (high stringency) will fail many vehicles, at probable excessive overall consumer costs, for a modest incremental gain in emissions benefits. Erring too far in the other direction will pass many high emitters and achieve only a modest overall reduction in emissions.

The majority of programmes have adopted as cutpoints the Federal Short Test Limits, 1.2% CO and 220 ppm HC as n-hexane, for 1981 and newer vehicles. The Short Test limits are also known as the *"Section 207 (b) limits"* of the Clean Air Act where the standards were authorized. It is interesting to note that these are the limits adopted by EPA for triggering the Section 207 (b) Emissions Warranty provision of the US emissions compliance programme. These cutpoints do not necessarily reflect vehicle manufacturers' specifications for idle CO monitoring.

These programmes do not set generalized (model year) cutpoints, basing standards instead on various vehicle characteristics. Limits are established through consultation with vehicle manufacturers and EPA and a combination of engineering judgement and some testing. California's programme has maximum CO/HC levels based on each vehicle's make, model year, engine and emissions control system. In most of the cases where specialized standards are set, maximum levels for 1981 and newer vehicles are generally in keeping with a range near the EPA recommended 1.2% CO and 220 ppm HC levels.

California - Diesel Smoke Law

In April 1991 CARB introduced a new smoke law for road-side checks. The inspection consists of a "*snap-idle*" test, employing a full flow end-of-line opacity meter (SAE J1243, May 1988). Three accelerations to full governed speed are to be conducted - each acceleration should take approximately 5 to 7 seconds.

The limit is set at 40% opacity as the average of three accelerations. However, some vehicles, for reason of age or "*special exemption*" will be allowed up to 55% opacity.

For the period April 1991 to April 1992:

1. Pre 1991 engines with opacity in the range 40-55% will be given a warning.

2. 1991 engines or later in excess of 40% opacity will be cited.

3. All trucks will be cited if their opacity exceeds 55%

A series of penalties and fines, repair and recertification is in place and, in extreme circumstances, the truck may be impounded. At the end of the first year CARB was planning to evaluate the results to decide whether to extend the 55% limit or lower the opacity limit to 40% for all vehicles.

5.5   Central and South American Countries

5.5.1   Mexico

Mexico City - Inspection and Maintenance

An I&M programme was introduced by the Mexico City Metropolitan Area (MCMA) in 1989. By June 1990 about 750 repair shops and 32 government stations were licensed by the government of the City (Departamento Del Distrito Federal or DDF) or the state of Mexico to carry out the inspections covering about 2.7 million vehicles.

Inspections are currently performed twice a year and consist of a brief visual inspection of emission control components, a visual check of exhaust smoke and an analysis of exhaust gas emissions for gasoline-powered vehicles. CO and HC emissions are measured at idle and at 2500 rpm in neutral. The analysers are manually operated units built to the specifications issued by the California Bureau of Automotive Repair (BAR) in 1974 and 1980 (BAR74 or BAR80). An inspector is responsible for issuing pass/fail certificates.

Enforcement is by a visual sticker system. A dated window sticker is issued by the inspection station at the completion of the test. Display of an expired sticker results in a fine and impoundment of the vehicle for 24 hours.

The authorities are dissatisfied with the degree of enforcement and are currently studying ways to upgrade the programme. Under consideration are:

- new automated equipment to BAR90 which record data on a pass/fail basis and restrict tampering

- centralized testing of high-use vehicles (including some 90,000 taxis and minibuses) and diesel-powered vehicles to more stringent test procedures.

- improved monitoring including the possibility of requiring certification for annual registration

5.6   Far East

5.6.1   Japan

Emissions testing forms an integral part of the Japanese roadworthiness test ("Shaken"). Vehicles must be submitted for testing once they are three years old and thereafter every two years. The limits applied are:

CO: 4.5% vol max.
HC: 1200 ppm max. (4-stroke engines)
        7800 ppm max. (2-stroke engines)

The measurements are made by NDIR at idle. If the emissions exceed the prescribed limits then some re-tuning is permitted.

5.6.2 South Korea

Table 5.7 South Korea - In-Service Emissions Limits: Gasoline & LPG Engines

CO Limits (% vol)

Model Year	Action	Limits	
		Until 09.90	**From 10.90**
1987/88	Rectification	1.3 - 9.0	1.3 - 4.4
1987/88	Penalty	9.1 and above	4.5 and above

HC Limits (ppm)

Model Year	Action	Limits	
		Until 09.90	**From 10.90**
1987	Rectification	1200	1200-4800
1987	Penalty	—	Above 4800
1988	Rectification	220	220-880
1988	Penalty	—	Above 880

Table 5.8 South Korea - In-Service Emissions Limits: Diesel Engines

Smoke Limits (% Opacity)

Until 09.90	**From 10.90**
50	40

Appendix 6

Fuel Consumption and CO_2 Regulations

Reproduced from Report 4/94, Section 6, by kind permission of:

CONCAWE, Brussels, Belgium

6 Fuel Consumption and CO_2 Regulations

The degree of control of fuel consumption exercised by legislators varies throughout the world. In the US and Japan, improvements in fuel consumption have been enforced. In Europe, government control is limited to mandatory publication of vehicle fuel consumption data (in France and the United Kingdom), and voluntary commitments have been made by motor manufacturers in several countries for improvements in fuel consumption. Other measures which have been taken by some countries to reduce fuel consumption include lower speed limits, and higher taxes on vehicles with high fuel consumption.

6.1 Europe - EU Fuel Consumption & CO_2 Regulations

There are currently no formal fuel consumption or CO_2 limits for motor vehicles within the EU. However, the European Union adopted on 13 May 1992 a proposal for an energy tax to be levied at $3 on a barrel of oil (or equivalent fuels) and rising gradually to reach $10 a barrel by the year 2000. The EU member states have already pledged to stabilize CO_2 emissions at 1990 levels by the end of the century. However, it appears that a number of member states have misgivings about the tax and future action seems conditional upon the EU's major trading partners taking similar steps.

CO_2 emissions and fuel consumption limits formed a major part of the Terms of Reference for the 1992/93 Work Programme of the MVEG (Motor

Vehicle Emissions Group of the Commission of the European Communities). The following topics were on the agenda:

- Agreement on a method of measurement.

- Establishment of a base-line to evaluate future improvements.

- Evaluation of the potential of technologies (either available or under development) for reducing CO_2 emissions/fuel consumption.

- Establishment of reference values to be used in a rating system for future car models, or categories of vehicles.

- Possible sanctions for non-attainment of reference values.

- Establishment of the possible contribution of road traffic to the overall objective of controlling CO_2 in the EU.

Subsequently, The European Commission issued on 25 March 1993 a draft amendment to Council Directive 80/1268/EC, relating to the measurement of fuel consumption of motor vehicles, to include the measurement and reporting of carbon dioxide emissions in its scope. Tests are carried out over the Urban and Extra Urban test cycles as described in the 91/441/EEC Directive for exhaust emissions. One figure for the combined cycle is reported for carbon dioxide, whereas fuel consumption is reported for the two cycles separately and in combination. The draft was approved on the 1 January 1994.

6.2 Europe - Non-Eu Fuel Consumption & CO_2 Regulation

Switzerland

The Ministry of Transport and Energy has declared its intention to introduce regulations to reduce fuel consumption. It is considering several options for implementation:

- Assurance of, or voluntary conformance to, a limit jointly defined by the Ministry and the respective manufacturers or importers.
- Conditions imposed on each manufacturer/importer to reduce its fleet average fuel consumption in a series of steps.
- Credit system to achieve a national fleet average fuel economy level.

6.3 US Fuel Consumption Regulations

The Energy Policy and Conservation Act, passed in December 1975, amended the Motor Vehicle Information and Cost Saving Act to require improvements in passenger car fuel consumption up to a figure of 27.5 miles per US gallon (8.55 l/100 km) by 1985, as shown in **Table 6.1**. Vehicle manufacturers are required to test sufficient vehicles to allow a fuel consumption figure to be assigned to each product line produced by them. From these figures a sales weighted average fuel consumption figure is calculated for all the passenger cars produced by the manufacturer concerned. This figure must be below the level specified for the appropriate model year. A manufacturer whose fleet average fuel consumption does not meet the standard is subject to a fine of $5 for each vehicle produced, for every 0.1 miles/US gal that the standard is exceeded. However, these fines may be offset by credits accrued in other model years.

The standards are based on the combined city/highway fuel consumption figures, and are known as the CAFE standards (Corporate Average Fuel Economy). Since 1979 model year the programme has been expanded to cover light-duty trucks as well as passenger cars. Since 1986 however, because of low oil prices and a return to larger cars the US motor industry has found it difficult to meet these limits. The US Department of Transportation therefore "*rolled back*" the limits for 1986-88 to 26 miles/USG (9.05 l/100 km) for passenger cars. Thus the US manufacturers have not been forced to pay the fines described above. For 1989 the limit was set at 26.5 miles/USG (8.88 l/100 km), and in 1990 it returned to 27.5 miles/USG (8.55 l/100 km).

Each vehicle must carry a label specifying the vehicle fuel consumption as determined by EPA, an estimate of the annual fuel cost based on 15 000 miles of operation, and the range of fuel economy achieved by similar sized vehicles of other makes. The figures quoted are not expected to give an accurate estimate of the fuel consumption that the owner will achieve under normal driving conditions but are to allow a comparison between different models of vehicle.

In calculating the CAFE fuel economy figure, a manufacturer may include in the fleet average figure any electric or hybrid vehicles produced. The procedures defined by EPA for calculating the equivalent fuel economy of electric vehicles takes account only of the gasoline energy equivalent of the electricity which powers the vehicle disregarding losses during electricity production and distribution. The calculated fuel economy figures are therefore high (around 185 miles/US gal), giving an incentive for development of such vehicles.

Table 6.1 US Fuel Consumption Standards in litres/100 km
(miles/US gal)

Model year	Passenger cars	Light trucks[a] Combined	Light trucks[ab] (2WD)	Light trucks[ab] (4WD)
1978	13.07 (18.0)	-	-	-
1979	12.38 (19.0)	13.68 (17.2)	13.68 (17.2)	14.89 (15.8)
1980	11.76 (20.0)	16.8 (14.0)[c]	14.70 (16.0)	16.80 (14.0)[b]
1981	10.69 (22.0)	16.22 (14.5)[c]	14.08 (16.7)[d]	15.68 (15.0)[b]
1982	9.80 (24.0)	13.45 (17.5)	13.07 (18.0)[b]	14.70 (16.0)
1983	9.05 (26.0)	12.38 (19.0)	12.07 (19.5)	13.45 (17.5)
1984	8.71 (27.0)	11.76 (20.0)	11.59 (20.3)	12.72 (18.5)
1985	8.55 (27.5)	12.07 (19.5)[e]	11.94 (19.7)[e]	12.45 (18.9)[e]
1986	9.05 (26.0)	11.76 (20.0)	11.47 (20.5)	12.06 (19.5)
1987	9.05 (26.0)	11.47 (20.5)	11.20 (21.0)	12.06 (19.5)
1988	9.05 (26.0)	11.47 (20.5)	11.20 (21.0)	12.06 (19.5)
1989	8.88 (26.5)	11.47 (20.5)	10.94 (21.5)	12.38 (19.0)
1990	8.55 (27.5)	11.76 (20.0)	11.47 (20.5)	12.38 (19.0)
1991	8.55 (27.5)	11.65 (20.2)	11.37 (20.7)	12.32 (19.1)

Notes:

a. Light trucks defined as less than 6000 lbs GVW in 1979, less than 8500 lbs 1980 - 91.
b. Separate 2WD/4WD standards or combined light truck standard may be used 1982-1991.
c. Relaxation granted for 1980-81 trucks with engines not based on passenger cars.
d. Revised mid-year to 18.0 mpg (13.02 l/100 km).
e. Revised in October 1984 to 21.6 mpg for 2WD, 19.0 mpg for 4WD and 21.0 mpg combined.

These provisions have been extended to include 'flexible-fuel' vehicles which can operate on gasoline, ethanol, methanol or any mixture of these fuels. A CAFE credit of up to 1.2 miles/US gal is allowed for these vehicles. The exact figure is calculated from a complex formula based on the difference in fuel economy between the vehicle running on alcohol and gasoline. This ruling will be in effect from 1993-2005, but may be extended beyond that date. The credit of 1.2 mi/USG applies to a CAFE standard of 27.5 mi/USG, and is scaled down for lower standards as currently apply, but no lower than 0.7 mi/USG.

In addition to the average fuel economy figure required of each manufacturer, taxes are levied on vehicles which do not achieve certain minimum fuel economy figures. These standards are:

1984	19.5 miles/US gal (12.1 l/100 km)
1985	21.0 miles/US gal (11.2 l/100 km)
1986	22.5 miles/US gal (10.5 l/100 km)

Proposals for a further drastic tightening of Corporate Average Fuel Economy (CAFE) limits were debated during 1990 but were blocked in September that year. The proposals set out to increase CAFE standards by 40% by the year 2001, improving average fuel economy for new cars from 27.5 miles/USG to 40.1 miles/USG (8.55 l/100km to 5.87 l/100km).

These proposals were re-introduced in 1991 and again ran into difficulties. A ruling in February 1992, in the U.S Appeal Court, held that CAFE restricted the production of larger cars, which are deemed safer. Conversely, a draft US Department of Energy report suggests that more fuel efficient cars offer the potential for large reductions in tailpipe emissions, moderate reductions in evaporative running loss emissions and small reductions in diurnal emissions.

6.4 Japanese Fuel Consumption Guidelines

In January 1993 fuel economy targets for passenger cars in the year 2000 were officially published. The targets were drawn up by MITI and MOT, based on recommendations of a committee set up in 1990. These are the first such guidelines since 1979, when fuel economy limits were set for 1985. Since then there has been a steady decline in fuel efficiency, especially over the last few years. This is due to the trend to larger engines and more widespread use of automatic transmissions etc. The current targets apply only to gasoline passenger cars but the government are believed to be considering similar regulations for trucks.

Formally the targets are expressed in terms of the weighted average fuel economy for vehicles within the specified weight ranges. They are not mandatory, will not become a regulation, and there is no suggestion of penalties such as a "gas guzzler" tax. However, for domestic manufacturers there is strong incentive to comply in order to maintain the goodwill of MITI. Fuel economy targets are specified over the new 10.15 mode cycle for three vehicle weight categories as shown in **Table 6.2**. The estimated figure for the total car population is for information only and is not a target for individual manufacturers.

So as not to unfairly advantage or disadvantage OEM's whose car production is all at the top or bottom of one of the three main weight categories, these have been further subdivided into 6 subranges as shown in **Table 6.2**.

Table 6.2 Japanese Passenger Car Fuel Economy Targets for 2000

Classification	Gross Vehicle Weight kg	Fuel Economy Target km/l	Improvement relative to 1990
Light cars	<827.5	19.0	7.3%
Small cars	827.5 - 1515.5	13.0	8.3%
Normal cars	>1515.5	9.1	11.0%
Estimate for total population		13.5	8.5%
Vehicle Weight sub-ranges			
Sub-Class 1	<702.5	19.2	6.5%
Sub-Class 2	702.5 - 827.5	18.2	7.0%
Sub-Class 3	827.5 - 1015.5	16.3	7.2%
Sub-Class 4	1015.5 - 1515.5	12.1	7.9%
Sub-Class 5	1515.5 - 2015.5	9.1	9.5%
Sub-Class 6	>2015.5	5.8	13.6%

6.5 Other Far Eastern Requirements

Australia

The FCAI have proposed reductions of the national average fuel consumption of the new car fleet with a target of 5% from 1989 levels by 1995 and further reductions thereafter as follows:

	1/100km
1995	8.7
2000	8.2
2005	8.0

South Korean Fuel Efficiency Requirements

The Korean Ministry of Energy and Resources issued a notice based on the Rationalisation of Energy Consumptions Act on 17 August 1992. It requires manufacturers (excluding importers) to meet new fuel efficiency standards from 1 January 1996. More restrictive standards will be introduced from 1 January 2000. From the 1 September 1992 all manufacturers, including importers, have been required to display the level of fuel efficiency on their cars according to five classifications for each engine displacement class. Non-compliance results in a fine of 500 million Won.

Table 6.3 South Korean Fuel Efficiency Requirements

Vehicle Class (Engine Displacement, cc)	From 1.1.1996	From 1.1.2000
<800	23.4	24.6
800-1100	20.3	21.3
1100-1400	17.3	18.1
1400-1700	15.4	16.1
1700-2000	11.4	12.0
2000-2500	9.9	10.4
2500-3000	8.5	8.9

Appendix 7

Fuel Composition and Quality Regulations

Reproduced from Report 4/94, Section 7, by kind permission of:

CONCAWE, Brussels, Belgium

7 Fuel Composition and Quality Regulations

7.1 Lead in Gasoline

7.1.1 Europe

Within the EU the maximum lead content of leaded gasolines is required to be within the range of 0.15 to 0.4 g/l by Directive 78/611/EEC. In practice all countries are at 0.15 g/l maximum, except Portugal which is still at 0.4 g/l. In 1985 another Directive (85/210/EEC) allowed unleaded gasoline (0.013 g Pb/l max) to be marketed and in addition required the introduction of a premium unleaded grade of 95 RON/85 MON from 1 October 1989. This Directive also encouraged Member States to provide incentives (e.g. through taxation) to promote the use of unleaded grades. Subsequently several Member States banned leaded regular completely, as allowed by Directive 87/416/EEC.

Unleaded gasoline (95 RON Europremium) is widely available in all West European countries and sales are growing. In 1992 over 50% of gasoline sales in Europe were unleaded. Leaded regular has almost completely disappeared. Austria has banned manufacture of all leaded gasoline from February 1993 and sales from October 1993; the leaded 98 RON grade has been replaced by an unleaded 98 RON grade which must contain valve seat recession (VSR) protection additives. In Sweden a similar ban may be introduced (together with the introduction of "Reformulated" gasoline) from January/July 1995 (manufacture /sales). In Eastern Europe unleaded gasoline is available in all countries on a limited basis, generally as 95 RON Europremium, although there is some variation in octane levels.

7.1.2 North America

In the US the EPA imposed a drastic reduction in the permitted lead level in gasoline, from 1.1 g/USgal to 0.5 g/USgal (0.13g/l) in 1985 and then to 0.1 g/USgal (0.026 g/l) from January 1986. This is considered to be the lowest lead level which will allow continued operation of older engines. However, sales of leaded regular have declined to around 1.5% in 1993, and leaded gasoline will be banned in the USA from January 1995. This is the same date from which Reformulated Gasoline must be sold in specified areas which do not meet ambient ozone targets (see **Section 2.5.4**). The reformulated gasoline requirements also include a ban on the use of other heavy metals without a specific waiver. A waiver has been applied for a manganese additive, but to date has not been approved.

In California sales of leaded gasoline were banned from January 1992.

In Canada sales of leaded gasoline has been banned since December 1990. A manganese additive is however permitted and used in all grades.

7.1.3 South America

Brasil has been a totally unleaded market for several years, due largely to the use of ethanol in their gasoline to boost octane quality (**Section 7.4.4**).

Chile and Argentina launched unleaded gasoline in 1992, but penetration so far is low (about 1%). **Guatemala** took the unusual step of going completely unleaded in 1991. **Mexico** has also had ULG for several years, but data on sales are not readily available. Finally, ULG is widely available in most **Caribbean** Island markets and also **Bermuda**, which is totally unleaded.

7.1.4 Far East

Australia introduced ULG in 1986 and penetration has grown to around 40%. The maximum lead content in leaded gasoline varies between states from 0.3 g/l in Victoria to 0.84 g/l in Northern Territories. New South Wales has recently proposed to reduce lead content from their current level of 0.4 g/l to 0.15 g/l in 1995, but this is still under discussion.

New Zealand has had ULG since 1987 and penetration is now 37%, with a total ban on leaded gasoline under consideration for 1996.

Japan introduced unleaded regular in 1975 and has been totally unleaded since the early 1980s, when a premium ULG was introduced whose sales have now grown to around 18%.

South Korea and Taiwan have both had ULG since 1986/7, and **Korea** has now banned sales of leaded gasoline as from January 1993. Many other countries have introduced ULG at various octane levels since 1991-2 as shown in **Table 7.3**, the latest being Brunei in late 1992. **Thailand** introduced an unleaded premium grade in 1993 and will replace leaded regular by unleaded regular from January 1995.

7.1.5 Middle East

Turkey introduced ULG 95 during 1992. Elsewhere only the **United Arab Emirates (UEA)** is known to have launched ULG in 1992, but data for other countries is difficult to find.

Table 7.1 Maximum Lead Content and ULG Sales in Europe - 1992

Region (1)	Total Sales 1992	Lead content	1992 Gasoline Market Share %ULG of Total				Tax Incentive
			Superplus 97-99RON	Premium 94-96RON	Regular 90-93RON	Total ULG	US cents/l
Country	'000 m³	g/l					(inc. VAT)
West Europe	**154,809**	**0.15-0.4**	**8.9**	**30.0**	**11.8**	**50.7**	
Austria	3,581	0.15	-	33.7	33.5	67.2	9/9.6
Belgium	3,042	0.15	14.8	30.5	-	45.3	10.6
Denmark	2,340	0.15	20.1	39.0	11.5	70.6	11.6/12.4/12.7
Finland	2,654	0.15	7.9	62.7	-	70.6	12.4
France	23,217	0.15	31.3	2.6	-	33.9	7.3/7.7
Germany	41,788	0.15	7.2	37.6	40.2	85.0	7.1
Greece	3,463	0.15	-	16.7	-	16.7	7.9
Iceland	180	0.15	-	24.4	51.1	75.6	3.7/7.1
Ireland	1,156	0.15	22.1	10.8	-	32.9	4.4/4.8
Italy	21,009	0.15	-	13.5	-	13.5	6.6
Luxembourg	679	0.15	22.2	35.5	-	57.7	7.1/6.9
Netherlands	4,790	0.15					8.8
Norway	2,274	0.15	18.8	50.0	-	68.8	11.3
Portugal	2,085	0.40	-	12.9	-	12.9	9.2
Spain	n/a	0.15	yes	yes	-	n/a	4.6
Sweden	5,704	0.15	5(est)	63.0	-	68.0	8.4
Switzerland	5,376	0.15	2.3	62.5	-	64.8	5.8
UK	31,453	0.15	5.8	40.7	-	46.5	6.9/7.7
East Europe			**-**	**7.1**	**-**	**7.1**	
Bulgaria	1,700		-	4.7	-	4.7	0.3
Czechia/Slovakia	n/a		yes	yes	-	n/a	3.0
Hungary	1,928		-	9.1	?	9.1	6.6/7.1
Poland	5,471		?	7.1	-	7.1	2.4
Romania	n/a		-	yes	-	n/a	-

See notes following **Table 7.3**.

Table 7.2 Lead Content and ULG Availability in North and South America - 1992

Region (1)	Total Sales 1992	Lead content	1992 Gasoline Market Share %ULG of Total				Tax Incentive
Country	'000 m³	g/l	Superplus 97-99RON	Premium 94-96RON	Regular 90-93RON	Total ULG	US cents/l (inc. VAT)
North America	**475,985**						
Bahamas	183	0.84	59.0	-		59.0	-
Canada	33,299	Banned	15.8	6.5	77.7	100	-
USA	437,690	0.026 (6)	19.2	11.5	67.8	98.5	-
Mexico	n/a		-	-	yes	n/a	-
Guatemala	476	Banned	-	41.2	58.8	100	-
El Salvador	288		-	1.0	-	1.0	-
Dominican	686		5.6	-	-	5.6	-
Puerto Rico	3,363	not sold	-	59.9	40.1	100	-
South America	**30,620**						
Argentina	6,329	?	-	0.8	-	0.8	-
Brasil (4)	22,165	Banned	42.8	-	57.2	100	-
Chile	2,126	?	-	-	1.2	1.2	-

Table 7.3 Lead Content and ULG Availability in Far East and Middle East - 1992

Region (1)	Total Sales 1992	Lead content	1992 Gasoline Market Share %ULG of Total				Tax Incentive
Country	'000 m³	g/l	Superplus 97-99RON	Premium 94-96RON	Regular 90-93RON	Total ULG	US cents/l (inc. VAT)
Far East	**88,358**						
Australia	16,838	0.3-0.84	-	0.8	39.4	40.2	-
Brunei	222	0.15	4.1	-	-	4.1	-
Guam	171	0	25(est)	-	75(est)	100	-
Hong Kong	394	0.15	57.6	-	-	57.6	6.5
Japan	47,516	0	18.4	-	81.6	100	-
Malaysia	4,333	0.15	27.4	-	-	27.4	1.6
New Zealand	2,671	0.45	-	-	37.1	37.1	2.1
Singapore	701	0.15	40.9	-	8.8	49.7	9/10.9
South Korea	5,200	0	-	-	90(est)	90(est)	-
Taiwan	5,977	0.34	-	yes	yes	46	-
Thailand	4,335	0.15(7)	yes	11.9	-	11.9	3.0
Middle East	**4,700**						
Turkey	3,940	0.84	-	0.5	-	0.5	-
UAE	760		5.3	-	-	5.3	

Notes

1. Regional totals are based only on the countries listed which market ULG
2. Sales of leaded gasoline banned from October 1993
3. Includes former East Germany
4. Brazilian premium is 95% ethanol, regular is blended with 23% ethanol
5. Sales of leaded gasoline banned from January 1993
6. Unleaded only from January 1995
7. Unleaded regular only from 1.1.95

n/a = data not available   yes = grade is available but share uncertain
? = uncertain whether grade is available   (est)= estimated.

7.1.6 Africa

ULG is known to be available in several North African countries (**Tunisia, Morocco** etc.) but no data are available. In the rest of Africa ULG is not generally available. **South Africa** has reduced its gasoline lead content from 0.6 to 0.4 g/l from 1988 and plans to introduce unleaded gasoline in 1995.

7.2 Reformulated Gasoline

7.2.1 Europe

Sweden

An 'Environmental Classification' system for diesel fuel was introduced in Sweden in 1991 (see **Section 7.7.1**). During 1993 a similar classification was developed for gasolines, comprising 4 different classes. Class 4 is equivalent to CEN standard, and Class 3 to the current Swedish standard. Specifications for Class 2 gasolines (leaded and unleaded) have been drawn up and will be introduced from 1 December 1994. The specifications for Class 1 should be developed during 1995 following a review of the results of the US Auto/Oil Programme. The specifications are optional and use will be encouraged by tax relief, 4 ore/l for Class 3 and 10 to 12 ore/l for Class 2.

Table 7.4 Swedish Environmental Gasolines

| Property | Class 2 | | Class 3 |
	Catalyst	Non-cat	
Sulphur %m (max)	0.01	0.03	0.1
RVP (S/W) kPa (max)	70/95	70/95	75/95
RVP (S/.W) kPa (min)	45/65	45/65	45/65
E100 %v (min)	47/50	47/50	43/45
FBP °C (max)	205	200	215
Benzene %v (max)	3	3	5
see Note A	5.5	6	-
Oxygen %m (max)	2	2	2
Lead mg/l (max)	5	5	13
Phosphorus mg/l (max)	nil	2	nil
Additives	*	*	*
Inlet Valve Cleanliness	**	**	**
Fuel System Cleanliness	***	***	***

Note A   Empirical Limit [$\dfrac{\text{Aromatics } \% \text{ v}}{13}$ + Benzene %v] (max).

*   Additives according to the Law on Chemical Products 1985.426, must not contain ash-forming constituents.

**   Minimum 9 demerits according to Mercedes M102E (CEC F-05-T-92).

***   Maximum 4% according to Peugeot 205 GTI (GFC TAE I-87).

Class 1 limits will be developed during 1995.

Finland

In Finland an interim reformulated gasoline "Citygasoline" was introduced commercially from January 1993, which was modestly supported with tax incentives, see **Table 2.21**. From March 1994 a more severely reformulated fuel with 1% benzene and 100 ppm sulphur has been introduced commercially. Marketing companies sell only unleaded gasoline because of the extra tax on leaded grades.

Table 7.5 Finnish Reformulated Gasoline

Property	Reformulated Gasoline	Citygasoline	Standard Gasoline
Oxygen %m	2.0 - 2.7	2.0 - 2.7	-
Benzene %v (max)	1.0	3.0	5.0
Sulphur ppm m (max)	100	400	1000
RVP (S/W) kPa (max)	70/90	70/90	80/100
Fuel Tax FIM/l	2.424	2.424	2.474
Tax Incentive FIM/l	0.05	0.05	0
Extra Tax (leaded) FIM/l	0.45	0.45	0.45

7.2.2 United States

Federal States

The US Clean Air Act includes a requirement for *"Reformulated Gasoline"* to be sold in major cities which fail to meet ambient ozone standards. Other areas with similar problems can also opt into the programme. Following discussions with the oil industry (the so called "Reg-Neg" process) a NPRM was published in the Federal Register on 31 March 1992. The EPA announced the final rule for the reformulated gasoline programme on December 15 1993.

The programme will be implemented in two phases. Phase I of the programme begins on 1 January 1995 and Phase II begins on 1 January 2000. The EPA expects the Phase I programme to achieve a 15 to 17 per cent reduction in both volatile organic compounds (VOC) and in toxic emissions from motor vehicles compared with 1990. The Phase II programme will achieve a 25 to 29 per cent reduction in VOC, a 20 to 22 per cent reduction in toxics emissions and a 5 to 7 per cent NOx reduction. All reductions are relative to the average 1990 US baseline quality.

Reformulated gasoline is required to be sold in the nine worst ozone non-attainment areas (with populations over 250,000) which are: Baltimore, Chicago, Hartford, Houston, Los Angeles, Milwaukee, New York, Philadelphia and San Diego. So far thirteen other states have opted to join the programme: Connecticut, Delaware, Kentucky, Maine, Maryland, Massachusetts, New Hampshire, New Jersey, New York, Pennsylvania, Rhode Island, Texas and Virginia. The cities and states listed above represent about 30 per cent of all gasoline sold in the United States. If all ozone non-attainment areas decide to opt into the programme over 50 per cent of the gasoline sold in the US would have to be reformulated. Distribution patterns from terminals may result in a further 5 per cent being supplied to areas not requiring reformulated gasoline.

All reformulated gasoline must contain a minimum of 2.0 per cent oxygen by weight, a maximum of 1.0 per cent benzene by volume and must not contain heavy metals. Sulphur, T90E and olefins content are not reduced, but may not be higher than a refiners' 1990 average. A summary of the regulation is given in **Table 7.6**. In addition they must meet certain VOC, air toxics and NOx emissions performance requirements, judged against qualities produced in 1990. Emissions performance will be calculated on the basis of empirical "models".

For the first three years (1995-1997) refiners will be allowed to use a "simple model" to certify their reformulated gasolines. Besides the compositional constraints with respect to oxygen, benzene and heavy metals, the simple model is designed to reduce VOC emissions (by limiting maximum RVP), and total air toxics.

From 1 January 1998, refiners will be required to use a "complex model" for certification. The complex model is a set of equations correlating a gasolines properties to its emissions characteristics. Refiners can also use this complex model for the first three years if they wish, which should give them more flexibility in meeting the requirements. The complex model uses additional fuel parameters including oxygenate type, sulphur content, olefins content and fuel distillation characteristics. Refiners can comply with the standards either on a batch (per gallon) basis or on a quarterly average basis. Average limits are more severe overall, but have more latitude on a per gallon basis.

Supplies of conventional gasolines will also be regulated to prevent any increase in emissions (the so-called "anti-dumping" rule). For the first three years olefins, sulphur, and T9OE will not be allowed to exceed their 1990 values by more than 25 per cent and aromatics and benzene will be controlled by means of a formula (BEE, see footnote **Table 7.6**). From 1997, emissions of benzene, toxics and NOx will not be allowed to exceed 1990 values and VOC emissions will be controlled by regional RVP limits.

Table 7.6 US Reformulated Gasoline Requirements
Fixed Specification Requirements

Parameter	Batch Basis	Average Basis
Benzene %v/v max	1.0	0.95 1.3 allowed on a batch
Oxygen %m		2.0 - 2.7
Heavy Metals		None without an EPA waiver
T90E Sulphur } Olefins		Average no greater than refiners 1990 average
Detergent Additives		Compulsory (see **Section 7.2.3**)

Phase I : Emission Targets (1995-1997 Simple and Complex Models)
(All emission reductions are relative to 1990 baseline quality)

	Simple model (1995-1997)		Simple + Complex	Complex Model (1995-1999)		
Parameter	RVP (psi, min)		Toxics (% redn)	VOC (% reduction) (min)		NOx (% reduction) (min)
Region	1 South*	2 North*	all	1 South	2 North	all
Batches	7.2	8.1	15.0	35.1*	15.6	0.0
Average	7.2	8.1	16.5	36.6*	17.1	1.5

*Calculated relative to the Clean Air Act baseline of 8.7 psi.
 Reduction relative to 7.8 psi is similar to Region 2

Phase II (2000 onwards) - Complex Model
(All emission reductions are relative to 1990 baseline quality)

	VOC (% reduction) (min)		Toxics (% redn)	NOx (% reduction) (min)
Parameters				
Region	1 South*	2 North*	all	all
Batches	27.5	25.9	20.0	5.5
Average	29.0	27.4	21.5	6.8

* See map **Figure 7.1**

Statutory Baseline Parameters - 1990 Average Quantity

Gravity	59.1	Distillation	
Benzene	1.6 %v	T50	207°F
Aromatics	28.6 %v	T90	332°F
Olefins	10.8 %v	E200F	46 %v
Sulphur	338 ppm	RVP	8.7 lb/m^2

Benzene Exhaust Emissions, BEE g/mile
= 1.884 + (0.949 x % benzene) + (0.113 x [% aromatics - % benzene])

All refiners, blenders and importers, even those not supplying reformulated gasoline, will be required to provide data of the characteristics of their gasoline production to the EPA for the baseline year 1990, in terms of batches of finished gasoline or components used in blending. Similar data will be required for subsequent years. This data will be used by the EPA to judge a suppliers performance against the requirements of the Act and to prevent dumping of unsuitable components into the conventional gasoline pool. Incremental volumes will be judged against the parameters of the statutory baseline. (see **Table 7.6**).

The EPA also announced a NPRM which would require refiners of reformulated gasolines to use oxygenates of which at least 15 per cent were derived from renewable sources in 1995, rising to 30% hereafter. In practical terms this amounts to the use of alcohol and ethers derived from alcohol such as ETBE. The proposal would require that renewable ethers be used in the summer months to avoid increases in evaporative emissions, but alcohols or ethers could be used during the rest of the year. A final rule has been passed but is being challenged in the courts.

7.2.3 Clean Air Act Amendments Requirement for Deposit Control Additives - Proposed Rule

The 1990 CAAA requires that "effective January 1 1995, all gasolines in the US must contain additives to prevent the accumulation of deposits in engines and fuel supply systems". The Act provides no definition of additives or deposits and no guidance as to which parts of the fuel system are to be considered. The final rule was to be promulgated by November 15 1992 and adopted before October 15 1994. In practice a NPRM proposing requirements was not issued by EPA until December 1993.

The EPA had some experience to work on because, since January 1 1992, all gasolines sold or supplied in California have been required to contain deposit control additives which are certified by the CARB for effectiveness.

As with the CARB requirements, the EPA proposal defines additive performance by requiring certification of additives in a port fuel injector (PFI) keep clean test and the BMW intake valve deposit (IVD) test, using fuels with certain minimum specification requirements. Note that in addition CARB requires a PFI clean up test.

The PFI keep clean test uses the 2.2-litre Chrysler turbocharged engine. Under the EPA's proposal 10,000 miles must be accumulated on a vehicle using repetitions of a standard PFI test cycle comprising 15 minutes at 55 mph, followed by a 45-minute shutdown soak period. As the method is not yet fully standardized the EPA is proposing to use a draft ASTM procedure.

The EPA will require a performance equivalent to the CARB standard; i.e., no injector may experience a flow restriction greater than 5%.

The BMW IVD Test uses a 1985 BMW 318i four-cylinder eight-valve naturally aspirated engine. Deposits are accumulated in a road driving cycle involving 70 per cent highway, 20 per cent suburban and 10 per cent city driving. The CARB pass requirement is a maximum average deposit on the four intake valves of not more than 100 mg after 10,000 miles. The EPA is considering alternative limits of 100 mg maximum on any single valve after 10,000 miles or 25 mg on any single valve after 5,000 miles. Either requirement is significantly more severe than the CARB requirement. The CARB, however, also stipulates a PFI clean-up requirement. In this test deposits must be built up on at least one injector to give a minimum of 10 per cent flow restriction. Then the cleanliness of all injectors must be restored to less than 5 per cent flow restriction within a further 10,000 miles of test cycles using the test gasoline.

Recognizing that it would be impractical to fully certify all deposit control additives before the 1 January 1995 deadline, the EPA have instituted an interim certification procedure. In this the EPA proposes that all gasolines sold must contain a detergent additive, which has been given an interim certification number at a specified treatment level. Gasolines proved under the CARB programme will automatically qualify. Interim detergency certification numbers will be allocated to only four classes of deposit control additive with molecular weights of at least 900:

- polyalkylamines
- polyetheramines
- polyalkylsuccinimides
- polyalkylamino-phenols

To qualify, the additive manufacturers must recommend the minimum concentration of additive to meet the performance standards and provide a test method to identify the additive in its pure form.

Beginning January 1 1996, all gasolines will have to contain detergent additives certified by the EPA for effectiveness. The certification establishes the minimum amount of additive that must be used and limits for some properties of the base gasoline (T9OE, olefins, aromatics, sulphur and oxygenates by type). The EPA is proposing four major generic categories for certification depending on the definition of the base fuel quality i.e.:

- National
- Specific to the Petroleum Administration for Defense Districts (PADD)
- Fuel specific
- CARB (PADD V)

Gasolines blended with detergents certified under the generic national certification can be marketed anywhere in the US. However, the base gasoline cannot exceed the limits given in **Table 7.7**. Such additive certification requires passing BMW and PFI tests in four different test fuels taken from four different refineries or distribution systems and at least two different PADDs. **Table 7.8** lists the minimum properties for each of these fuels. If the properties of the base gasoline exceed the limitations given in **Table 7.7**, then an additive must be used meeting a more severe national certification level using fuels with higher values than those given in **Table 7.8**.

Gasolines sold only within a given PADD must still have base gasoline qualities meeting the requirements given in **Table 7.7** but the test fuel qualities can be restricted to those specified for that district. Like the national requirements, base gasolines having properties exceeding those in **Table 7.7** must use deposit control additives certified in test fuels with properties representing the more severe base gasoline. Gasolines using detergents with fuel-specific certification must remain segregated from production through distribution to the end user. The test fuels in a segregated system must meet or exceed the segregated gasoline's 65th percentile level for critical fuel properties. Gasolines using detergents certified for deposit control effectiveness by CARB will automatically qualify for national certification in 1995 and PADD V certification in 1996. For each detergent blending facility, weekly gasoline samples must be taken and analyzed for the four critical properties. This data base serves to demonstrate that base gasolines continue to have properties within those detailed in **Table 7.7**. If any single measurement exceeds specification, a statistical distribution curve for each property must be prepared (using the past year's data). The 65th percentile then becomes the minimum level for certification of the test fuel. Such an occurrence may require new testing and detergent certification.

New statistical distribution curves must be prepared twice a year, employing a full year's data. If the new 65th percentile for any property exceeds the previous 75th percentile, then the detergent must be certified with the test fuel at the new 65th percentile. The detergent blender has 90 days to comply with the new requirement.

Because it will be very difficult to enforce compliance by means of analytical testing for additive addition, the EPA propose to institute a system in which fuel marketers must maintain records of additive inventories and usage. For blending installations using automated additive injection equipment a weekly mass balance will be required. For non-automated systems an additive balance record must be obtained by recording every addition made (even to batches as small as road tankers). These records must be kept for at least five

Table 7.7 Maximum Allowed Compositional Limits for Base Gasoline

Generic Certification	Sulphur %m/m	T90E ° F	Olefins %v	Aromatics %v
National	0.085	356	18.7	41.2
PADD I	0.071	358	22.2	42.5
PADD II	0.089	352	14.4	38.3
PADD III	0.075	358	18.2	39.5
PADD IV	0.106	344	19.4	31.15
PADD V	0.04	352	11.5	44.2

Table 7.8 Required Minimal Fuel Parameter Values for Certification

Test Fuel		Sulphur %m/m	T90E ° F	Olefins %v	Aromatics %v	MTBE %v	ETOH %v
National	TF 1	0.033	340	-	-	None	10
	TF 2	-	340	10.7	-	15	None
	TF 3	0.033	-	10.7	-	None	None
	TF 4	-	336	-	29.2	None	None
PADD I	TF 1	0.036	344	-	-	None	10
	TF 2	-	344	13.3	-	15	None
	TF 3	0.036	-	13.3	-	None	None
	TF 4	-	338	-	29.2	None	None
PADD II	TF 1	0.035	340	-	-	None	10
	TF 2	-	340	9.5	-	15	None
	TF 3	0.033	-	8.9	-	None	None
	TF 4	-	338	-	28.6	None	None
PADD III	TF 1	0.030	344	-	-	None	10
	TF 2	-	344	12.7	-	15	None
	TF 3	0.030	-	12.7	-	None	None
	TF 4	-	340	-	29.1	None	None
PADD IV	TF 1	0.052	329	-	-	None	10
	TF 2	-	329	11.2	-	15	None
	TF 3	0.045	-	10.5	-	None	None
	TF 4	-	331	-	24.6	None	None
PADD V	TF 1	0.015	335	-	-	None	10
	TF 2	-	336	7.0	-	15	None
	TF 3	0.016	-	7.0	-	None	None
	TF 4	-	332	-	32.3	None	None

years and no tolerance below the amount stipulated for certification will be allowed. Weekly analyses of the critical fuel parameters must also be made, initially for six months to characterize the gasoline pool and then indefinitely to ensure compliance.

7.2.4 Far East

Thailand

Thailand introduced new specifications on the 1 January 1993 incorporating a number of changes, including two types of premium leaded and unleaded grades, Type 2 for Bangkok and other cities and Type 1 for rural areas. The Type 1 specifications contain a number of compositional constraints including maximum limits for benzene, aromatics and oxygenates and a requirement for additives to control injector and inlet valve deposits (see **Table 7.9**). The Type 2 specifications in addition require a minimum concentration of MTBE. The specification for the regular grade will also require detergency additives from 1 January 1995 at the time when it becomes unleaded.

Table 7.9 Thai Gasolines - Compositional Constraints

		Premium Leaded		Premium Unleaded	
Property	Regular	Type 1	Type 2	Type 1	Type 2
Lead content g/l	0.15*	0.15	0.15	0.013	0.013
Sulphur %m min	0.15	0.15	0.15	0.10	0.10
Benzene %v max	3.5	3.5	3.5	3.5	3.5
Aromatics %v max					
from 1.1.1994	-	50	50	50	50
from 1.1.2000	-	35	35	35	35
MTBE %v					
min	-	-	5.5	-	5.5
max	10.0	10.0	10.0	10.0	10.0
PFI/IVDC Additive**					
before 1.1.1995	-	+	+	+	+
after 1.1.1995	+	+	+	+	+

*   Reduces to 0.013 from January 1995
** Required to meet the Californian BMW 318i Test - see **Section 7.2.3**

7.3 Gasoline Volatility

7.3.1 Europe

In Europe, volatility limits are incorporated in the new European specification for unleaded gasoline, CEN EN-228 (see **Section 2.5.1 and 8.1**). There are eight volatility classes from which each country shall specify one for each defined period of the year and for a defined region to suit national requirements. Normally limits for three periods are chosen - summer, intermediate and winter - according to climatic conditions.

In Sweden three "Environmental Classes" of gasoline are under consideration which include relatively low volatility limits (see **Section 7.2.1**). Classes 2 and 3 have been proposed to Parliament and will probably be implemented from 1 December 1994. The limits for Class 1 will be published during 1995 following a review of the US Auto/Oil Programme.

Finland have also introduced from 1 January 1993 alternative gasoline specifications with RVPs reduced by 10 kPa compared with standard gasolines. The use of these gasolines is encouraged by tax incentives (see **Section 2.5.2**).

7.3.2 USA

Since 1989 the EPA has imposed maximum limits on gasoline vapour pressure during the summer months (May-September) varying between 9.5 and 10.5 lb RVP for different states. Since May 1992 these limits have been replaced by the more stringent Phase 2 limits given in **Figure 7.1**. These are simplified into North and South zones with maximum RVPs of 9.0 psi (62.1 kPa) (May-Sept.) and 7.8 psi (53.8 kPa) (June-Sept.) respectively. A permanent waiver of 1 psi (7kPa) is allowed for ethanol blends.

From 1 January 1995, RVP will be further restricted in those areas where legislation requires "reformulated gasolines" to 8.1 psi (55.8 kPa) and 7.2 psi (49.6 kPa) respectively (see **Section 2.5.4**). From 1996 an even lower limit of 7.0 psi (48.2 kPa) will be required in California for "Phase 2 reformulated gasolines".

7.3.3 Canada

Canadian provincial ministers agreed to limit gasoline volatility to 10.5 psi (72.5 kPa) starting summer 1990, and to institute vapour controls throughout the gasoline distribution and marketing system, also starting in 1990. Since then British Columbia has put a maximum summertime limit on RVP of 9 psi in the Lower Fraser Valley region, but proposals are being considered for waivers for the use of ethanol.

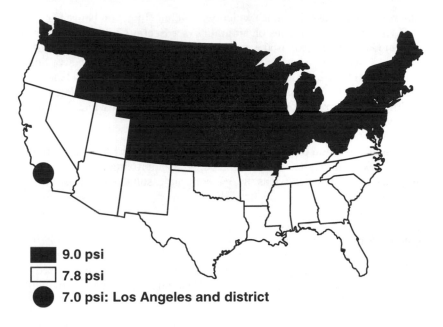

9.0 psi
7.8 psi
7.0 psi: Los Angeles and district

Fig. 7.1. EPA Phase II RVP Regulations

7.4 Benzene

Europe

The EU Directive (85/210/EEC), which required the introduction of unleaded gasoline, specifies a benzene level of 5%v maximum. From 1.10.89 this also applied to leaded grades marketed in EU countries. This limit is also generally accepted in other countries in Europe, including Finland, Norway, Switzerland and Sweden.

However in some countries, notably Austria, Italy, Finland, Sweden and Germany, lower benzene and aromatics limits have been introduced or are under consideration. Austria has reduced benzene content to 3%v maximum from 1 September 1990. (Regulation No. 239. Minister of Environment, Youth and Family, 5 March 1990. Subsequently replaced by Regulation No. 123. Minister of Environment, Youth and Family, 1 March 1992). In Italy the benzene content of the total gasoline pool is limited to 3.0%v from 1 January 1993. This is on the basis of a three month sales weighted production average for each refinery by a voluntary agreement between the Environment Ministry and the oil companies. Sweden is intending to introduce optional

771

"Environmental Classifications" which will include limits on benzene content of 3%v, as well as restrictions on aromatics (see **Section 7.2**). In 1990 the German Government proposed a reduction of 1%v and although this was rejected by the EU as a barrier to trade within the EU, more restrictive limits (to 1%v max) have been again proposed at the request of Land Baden-Wurttemburg. Both Finland and Sweden have introduced reformulated gasolines with benzene contents limited to 3%v, the sale of which is encouraged by tax incentives.

A new grade was introduced commercially in Finland on 1.5.94, containing less than 1% benzene. Its current market share is approximately 95% and it enjoys the same tax incentive as the 3% benzene gasoline (see **Table 7.5**).

USA

In the USA benzene content will be limited to 1%v maximum (or 0.95%v period average, with a 1.3%v absolute maximum) from 1 January 1995 by implementation of the regulations requiring *"reformulated"* gasoline (see **Section 7.2.2**). Total aromatics are indirectly controlled by the requirement to demonstrate a 15% reduction in the emissions of "air toxics".

In California, the *"Phase 2 Reformulated Gasoline"* required from March 1996 also has a 1%v benzene limit (or 0.8%v average, with an absolute maximum of 1.2%v). Total aromatics are also limited to 25%v (or 22%v average, absolute maximum of 30%v)

Other countries

Benzene contents are limited to 5%v maximum in Australia. New Zealand will reduce its benzene limit to 3%v after 1995. Singapore is likely to reduce the benzene content of gasolines to below 3%v by 2000. South Korea currently has limits of 6%v benzene and 55%v aromatics but the benzene limit will be reduced to 5%v from 1 January 1996. Thailand has had a reformulated gasoline standard in place since 1992 with a benzene limit of 3.5%v.

7.5 Oxygenates in Gasoline

This section gives information on oxygenates in gasoline other than that described in **Section 7.2** on "Reformulated Gasolines".

7.5.1 European Union Regulations

The EU adopted in December 1985, Directive 85/536/EEC on oxygenates in gasoline. The Directive specifies that:

- by 1988 Member States must ensure there are no legal and administrative obstacles to the sale of gasoline blends containing oxygenates and suitable for use in spark ignition engines designed to operate on gasoline.

- the components and concentrations which are deemed to meet this requirement are outlined in a Technical Annex (see **Table 7.10**).

The following are acceptable for use as substitute fuel components:

- mono-alcohols with an atmospheric boiling point lower than the final atmospheric boiling point laid down in the national gasoline standards;

- ethers, with molecules containing 5 or more carbon atoms, and with atmospheric boiling point lower than the final atmospheric boiling point laid down in the national gasoline standards.

Member States must permit fuel blends containing levels of oxygenates not exceeding the level set out in column A of the table in the technical Annex to the Directive. If they so desire, they may authorize proportions of oxygenates above these levels.

However, if the levels so permitted exceed the limits set out in column B of the Annex table, the pumps which dispense the fuel blend must be very clearly marked accordingly, in particular to take account of variations in the calorific value of such fuels. The main points of the Annex table are set out in **Table 7.10**.

7.5.2 USA

The use of new components in unleaded gasoline must be approved by the EPA, who must ensure that they will not adversely affect emission control systems.

In the case of oxygenates, the EPA has ruled that aliphatic alcohols and glycols, ethers and polyethers may be added to the fuel provided that the amount of oxygen in the finished fuel does not exceed 2.7% mass. However, note should be taken of new limits set out in the Clean Air Act Amendments. Methanol is excluded from this approval.

Table 7.10 Oxygenates limits set out in Directive 85/536/EEC

	A (% vol)	B (% vol)
Methanol, suitable stabilizing agents must be added[a]	3%	3%
Ethanol, stabilizing agents may be necessary[a]	5%	5%
Iso-propyl alcohol	5%	10%
Tertiary butyl alcohol	7%	7%
Iso-butyl alcohol	7%	10%
Ethers containing 5 or more carbon atoms per molecule[a]	10%	15%
Other organic oxygenates defined in Annex Section I	7%	10%
Mixture of any organic oxygenates defined in Annex Section I[b]	2.5% weight oxygen, not exceeding the individual limits fixed above for each component	3.7% weight oxygen, not exceeding the individual limits fixed above for each component

Notes:
[a] In accordance with national specifications or, where these do not exist, industry specifications.
[b] Acetone is authorized up to 0.8% by volume when it is present as by-product of the manufacture of certain organic oxygenate compounds.
Not all countries permit levels exceeding those in column (A) even if the pump is labelled.

This is known as the *"substantially* similar" ruling, as these components are considered to be substantially similar to fuels in widespread use before the requirement for EPA approval. This EPA ruling originates from July 1981, when the oxygen content limit was set at 2.0% mass max. In February 1991, in response to a request from the Oxygenated Fuel Association (OFA), EPA revised the ruling to increase the maximum oxygen limit to 2.7% mass.

A number of specific proposals have been granted waivers allowing their use as follows:

- *"Gasohol"* consisting of gasoline with 10% vol ethanol is permitted. This contains 3.5% mass oxygen. (April 1979).

- A mixture of TBA and methanol up to a maximum concentration of 3.7% mass oxygen provided that methanol does not form more than 50% of the mixture. (November 1981).

- Methanol up to 5.0% volume plus at least 2.5% volume co-solvent (ethanol, propanols or butanols) plus corrosion inhibitor, with maximum oxygen content 3.7% mass. This is known as the *"Dupont waiver"*. (January 1985).

In addition to the foregoing, one section of the 1990 Clean Air Act calls for cities which do not meet ambient air CO level targets in winter to use gasoline containing oxygenates to give 2.7% mass oxygen.

This has applied during four winter months (Nov-Feb) since 1992. In a few cities this requirement may be applied for a longer period. The area covered by the legislation includes 41 cities which account for some 31% of total winter US gasoline sales. The programme will be closely monitored by the EPA Administrator who will have the power to delay implementation for up to 2 years in individual areas if he perceives specific oxygenate supply problems. Trading of oxygen credits within the area is to be permitted.

California has applied for a waiver to allow the state to set the maximum oxygen level at 2%m. The state of Alaska suspended its oxygenates programme in Fairbanks and Anchorage after numerous complaints. The EPA and API have separately announced new studies on the effects of MBTE on health.

7.5.3 Brazil

Since 1979 the use of ethanol has been mandatory in Brazil through their government *"Proalcool"* programme. Two grades of fuel are available:

- *"Alcool"* is neat hydrated ethanol, as produced by conventional distillation containing 93% ethanol and 7% water, often referred to as E93.

- *"Gasolina"* is a blend of anhydrous ethanol and hydrocarbon streams containing 22% (± 1%) ethanol and with octane quality specified as 80 MON minimum.

Specially designed vehicles are produced in Brazil to operate on one or the other of these fuels. The market for alcool (E93) vehicles has varied enormously over the last few years, depending upon tax incentives.

Problems with crude supplies, imbalances in petroleum fuel demand and uncertainties over sugar subsidies leading to sugar shortages resulted in the proportion of alcohol in 'gasolina' being reduced from 22 to 12 per cent. However this led to an increase in emissions and the recent law setting revised emissions standards for Brazil also restores the composition of Brazil's gasohol mixture to 22 per cent alcohol and 78 per cent gasoline.

7.5.4 Other Countries

In South Africa, 8-12% alcohol has been incorporated into the 93 RON grade manufactured by the SASOL oil-from-coal process for a number of years. The alcohol must contain a minimum of 85%m ethanol, the balance being mainly iso-propanol and n-propanol. The new 95 RON and 91 RON unleaded gasoline specifications require a maximum of 2.8 or 3.7%m oxygen (7.5 or 9.5%v alcohol) respectively.

South Korea has an oxygen content limit of 0.5%m minimum, effective from 1 January 1993. Thailand has introduced optional specifications for gasolines containing 5.5 to 10%v MTBE from January 1993. These are supported by tax incentives (see **Section 7.2.4**).

7.6 Diesel Fuel Sulphur Content

7.6.1 European Union Legislation

The EU's Council of Environment Ministers, at a meeting in March 1987, agreed to a Directive reducing the maximum sulphur content of all gas oils, except those used by shipping or for further processing, to 0.3%m/m and allowing Member States to set a stricter limit of 0.2%m/m in heavily polluted areas. Member States were required to implement the Directive (EEC/85/716) by 1.1.89.

Additionally, the Commission was requested to come forward with a proposal before 1.12.91 to set a single limit to be applied throughout the Union. As a result of that request a Directive (93/12/EEC) dated 23.04.93 was adopted requiring:

- A maximum limit of 0.2%m/m will be applied for all gas oils, middle distillates and diesel fuel by 1 October 1994.

- A maximum limit of 0.05%m/m for diesel fuel to be implemented by 1 October 1996. Member States should ensure the *"progressive availability"* of a diesel fuel with a sulphur content of 0.05%m/m from 1 October 1995. This is to allow the implementation of the *"Clean Lorry"* Directive (see **Section 2.1.1**).

- For all gas oils and middle distillates (excluding aviation kerosine) further reductions in sulphur content beyond 0.2%m/m are under consideration. These will be based upon a Commission proposal which will then be implemented by 1 October 1999 at the latest.

- For aviation kerosine, new sulphur limits are also being reviewed.

Denmark

Denmark introduced a tax incentive for low sulphur (0.05% mass) diesel fuels with effect from 01.07.92. Two grades are involved (CEN and Bus diesel fuels) (see **Table 7.11**).

Table 7.11 Tax Incentives for Public Bus Service Diesel Fuel in Denmark

	A CEN Quality	B CEN 0.05% S Quality	C Public Bus Service Ultra Light Diesel
Tax (DKr/litre)	2.04	1.94	1.74
Tax Incentive	0	0.10	0.30

Italy

Italy conforms to the EU 1994 limit of 0.2%m/m, enforced by Decree in the major cities and elsewhere by voluntary agreement between the government and the oil companies (from 1.1.93 in northern and southern Italy and from 1.10.93 in central Italy).

7.6.2 Other European Countries

Austria

Austria has a sulphur limit of 0.15%m/m for diesel, but a lower limit of 0.1%m/m is applied for HGO (heating oil). The sulphur content of automotive diesel fuel will further reduce to 0.05%m/m, with effect from 01.10.95. Regulation No. 123. Minister of Environment, Youth and Family, 1 March 1992.

Finland

The sulphur content of diesel fuel in Finland has been limited in the national specification to 0.2%m/m since 1 January 1989. Also qualities are produced with market specifications of 0.05%m/m maximum. On 01.07.93 a new tax law (number 1561) was applied, giving incentives for "sulphur free" diesel fuel.

**Table 7.12 Finland - Characteristics and Tax Structures
for Diesel Fuel**

Property	Reformulated Diesel	Standard Diesel
Sulphur % m/m (max)	0,005	0,2
Aromatics % v (max)	20	-
Cetane Index (min)	47	46
or Cetane number (min)	47	
Fuel tax FIM/l (1.6.94)	1.601	1.751
Tax Incentive FIM/l	0.15	0

Norway

In Norway the maximum sulphur content is 0.5%m/m for both automotive and industrial gas oils, although typical sulphur contents are around 0.2%m/m, due to the processing of North Sea crude oils. There is, however, a combined tax/duty levy which penalises high sulphur gas oils, as follows:

Table 7.13 Norwegian Diesel Sulphur Content Tax Levies

Sulphur Content (%m/m)	Tax (NoK/Tonne)	US$/Tonne
<0.05	0	0
0.05-0.25	70	10.8
0.25-0.50	140	21.6

Fuels with higher sulphur contents may be used in industrial installations with flue gas treatment, but a further 70 NoK/tonne tax applies for each additional 0.25%m/m sulphur.

Sweden

In Sweden a maximum sulphur content of 0.2%m/m is applied to gas oils (Swedish Standard SS EN590). However, with effect from 1.1.91 a tax levy of $9/m^3 has been imposed on gas oils with a sulphur content between 0.1 and 0.2%m/m. In addition, from the same date, there are environmental classifications for diesel fuel, with tax relief for low sulphur and low aromatic grades. These are described later in this section.

Switzerland

In Switzerland, the sulphur content of diesel fuel was reduced from 0.2%m/m to 0.05m/m from 1 January 1994.

7.6.3 North America

In the US a sulphur limit of 0.05%m/m has been adopted by the EPA and made effective from 1 October 1993. In California 0.05%m/m was introduced as part of a new specification (see **Section 7.2.2**).

In Canada a Memorandum of Understanding has been signed between the Government (Environment and Transport Canada) and the majority of petroleum marketers for retail and fleet supplies of diesel fuel to conform to a sulphur limit of 0.05%m/m beginning October 1994.

Table 7.14 Maximum Permitted Sulphur Content of Automotive Diesel Fuel

Country	Maximum sulphur content % m/m	Comments
EU Countries	0.2	Effective 1.10.94
	0.05	All EU countries by 1.10.96
Belgium	0.2	Effective 1.10.89
Denmark	0.2	Effective 1.11.88 [1]
Ireland	0.3	
Italy	0.2	[2]
France	0.3	
Germany	0.2	
Greece	0.3	
Luxembourg	0.2	Effective 1.10.89
Netherlands	0.2	Effective 1.10.88
Portugal	0.3	
Spain	0.3	
United Kingdom	0.2	BS EN 590:1993
Other Countries		
Austria	0.15	From 1986
"	0.05	Effective 01.10.95
Canada	0.4	
	0.05	By agreement from 1.10.94
China	0.2	
Finland	0.2	Effective 1.1.89 [3]
India	0.5	
Japan	0.2	By agreement from October 1992
	0.05	From May 1997
Mexico	0.05	Mexico City from Oct. 1993, elsewhere from December 1994
Norway	0.5	Average 0.25% from January 1990 [4]
South Africa	0.55	[5]
South Korea	1.0	
Singapore	0.3	From 1.7.96
Sweden	0.2	From 1.1.93 [6]
Switzerland	0.05	From 1.1.94
Taiwan	0.3	From July 1993
	0.05	Planned possibly by 1997
Thailand towns	0.5	Current
	0.25	From 1.1.96
	0.05	From 1.1.2000
rural	1.0	Current
USA	0.05	From 1 October 1993

notes:
(1)   See **Table 7.11**.
(2)   Italy conforms to the EU limit of 0.2%m, enforced by Decree in the major cities and elsewhere by voluntary agreement between the government and the oil companies (from 1.1.93 in northern and southern Italy and from 1.10.93 in central Italy).
(3)   See **Table 7.12**.
(4)   See **Table 7.13**.
(5)   Fuel produced in the SASOL oil-from-coal process is essentially sulphur free.
(6)   See **Table 7.17**.

The Mexican Social Development Secretariat (SEDESOL) requires the Mexican petroleum industry (PEMEX) to supply diesel fuel to a sulphur content of 0.05%m/m from October 1993 in Mexico City and throughout Mexico from December 1994.

7.6.4 Far East

The Japanese oil industry agreed to reduce sulphur content to 0.2%m/m from October 1992 and to 0.05%m/m from May 1997.

Taiwan reduced the sulphur content of diesel fuel from 0.5%m/m to 0.3% in July 1993 and is planning to reduce it further to 0.05%m/m, possibly by 1997.

The Ministry of Commerce of Thailand issued a specification for diesel fuels on the 28 August 1993 giving a phased reduction of sulphur for diesel fuel used in Bangkok and other cities (Type 2) from its current level of 0.5%m/m max to 0.25%m/m from 01.01.96 and 0.05%m/m from 01.01.2000. The sulphur content of diesel fuel used in country areas (Type 1) remains at 1.0%m/m.

7.7 Diesel Fuel Compositional Specifications

7.7.1 Europe

Denmark

A study was made following the introduction on July 1 1992 of an interim diesel fuel quality for public transportation buses. Based on the study, tax rebates were approved for low sulphur fuels, CEN quality with 0.05%m/m sulphur and an "ultra light" diesel. At the same time the diesel tax was increased in two steps - 1 July and 1 October, 1992 - to the EU minimum level. Also the right of VAT-registered business to deduct part of the tax was withdrawn. Details of the qualities are shown in **Table 7.15**.

Table 7.15 Danish Diesel Fuel Characteristics

Characteristics	A CEN Quality	B CEN 0.05% S Quality	C Public Bus Service Ultra Light Diesel
Sulphur (%m/m)	0.20	0.05	0.05
95% Distillation (°C)	370	370	325
Density (kg/m^3)	820-860	820-860	820-855
Cetane Number	49	49	50
Cetane Index	46	46	47
Tax (DKr/litre)	2.04	1.94	1.74
Tax Incentive	0	0.10	0.30

Sweden

From 1.1.91 *"Environmental Classifications"* were introduced for diesel fuel with tax relief for both sulphur content and composition. These were further revised in 1992. **Table 7.16** summarizes the position:

Table 7.16 Swedish 1991 Diesel Fuel Classification

Fuel Characteristic	Class 1	Class 2	Class 3
Sulphur Content %m/m (max)	0.001	0.02	Standard fuel to SS 155435
Aromatics %v/v (max)	5	20	As Above
Distillation Range (IBP-FBP °C)*	180-300	180-300	As Above
Tax Rate ($/m³)**	152	185	210

Notes:

* The distillation range for Class 1 and 2 grades is significantly lighter than conventional European diesel fuels. The oil industry proposed that, for Class 2 Fuel, the limits should be relaxed, as follows, to improve availability:
Sulphur Content %m/m (max)   0.03
Distillation Range (IBP-FBP °C)   180-340

** These taxes exclude the penalties for gas oils containing more than 0.1%m/m sulphur.

Class 1: City public transport buses.
Class 2: Urban area transport.
Class 3: Grade for long haul traffic and off-highway applications. However, the tax penalties for exceeding 0.1%m/m sulphur suggest that most Class 3 product was at this sulphur level.

In January 1992, the specifications were again revised, as shown in the following table:

Table 7.17 Sweden - 1993 Diesel Fuel Classifications

Fuel Characteristic	Urban Diesel 1	Urban Diesel 2	Standard Grades* SS.EN.590
Sulphur Content %m/m (max)	0.001	0.005	0.20
Aromatics %v/v (max.)	5	20	-
PAH % vol (max)	0.02	0.1	-
Distillation:			
IBP (min) °C	180	180	-
10% (min) °C	-	-	180
95% (max) °C	285	295	**
Density (kg/m^3)	800-820	800-820	#
Cetane Number	50	47	##
Tax Rate ($/m^3)$^{(1)}$	126	165	199

Notes:

*   In addition to the urban grades, one summer and three winter standard
     grades are specified.
** 95% distillation varies with grade:
       Summer: 370°C max
       Winter: 340°C max
\#   Density varies with grade:
       Summer: 820-860 kg/m^3.
       Winter: 800-845 kg/m^3 (-26°C CFPP grade).
       Winter: 800-840 kg/m^3 (-32 and -38°C CFPP grades).
\#\# See **Section 8**.
(1) 1994 tax rates exclude added value tax.

7.7.2 North America

California

In California, the California Air Resources Board (CARB) has adopted a diesel fuel specification of 0.05% mass sulphur and 10% vol aromatics from 1 October 1993. The intention is to provide fuel quality that will ensure low emissions, and other fuels are allowed, provided the supplier can demonstrate equivalent emissions to a reference fuel from engine test data. The current specification for the reference fuel is shown in **Table 7.18**:

Table 7.18 Californian Diesel Reference Fuel Specification

Property	Limit
Sulphur % mass (max)	0.05
Aromatics % vol (max)	10.0
Polycyclic aromatics, % mass (max)	1.4
Nitrogen, ppm mass (max)	10.0
Natural cetane number, (min)	48.0
API gravity	33-39
Viscosity at 40°C, mm^2/s	2.0-4.1
Flash Point °F (min)	130
Distillation °F (°C):	
IBP	340-420 (170-215)
10%	400-490 (205-255)
50%	470-560 (245-295)
90%	550-610 (290-320)
FBP	580-660 (305-350)

The reference fuel is produced from straight-run California fuel by hydro-dearomatization and testing is carried out over the US Federal heavy-duty transient procedure (see **Section 4.7**). in a 1991 US heavy duty emissions standard engine.

The EPA proposes to replace the Fluorescent Indicator Absorption Method (ASTM D1319) by the Supercritical Fluid Chromatograph Method (ASTM D1586) in any references as the means of measuring aromatics content of diesels fuels.

7.8 Alternative Fuels

This Section covers the specification requirements for alternatives to conventional hydrocarbon fuels such as methanol, ethanol, compressed natural gas (CNG), hydrogen, vegetable oils and esterified vegetable oils.

7.8.1 Europe

Vegetable Oil Methyl Esters (Biodiesel)

The European Commission has put forward a draft Council Directive for a specification for vegetable oil methyl esters (biodiesels). The proposal is presented in the framework of EU's ALTENER Programme for the promotion of alternative fuels. Within this programme the EU has the objective of securing a five per cent market share of total motor fuel consumption for biofuels, of which it is expected that biodiesel will form the major share. Some countries, notably Austria and Italy, have already produced their own specifications (see **Tables 7.19** to **7.21**).

Biodiesel production units are in operation or under construction in Austria, Belgium, Germany, France and Italy and others are planned. Total production in Europe could reach 200,000 tons by 1995.

Rapeseed methyl ester diesel fuels are already sold in Italy but can only be marketed outside retail outlets. A Government Decree fixes a maximum of 125,000 tons per year to be exempted from gas oil excise tax. When claiming tax exemption producers have to show that at least 80% of the raw vegetable oil used derives from "set-aside" crops. The Italian specification for rapeseed methyl esters, CUNA NC 635-01, is given in **Table 7.20**.

Table 7.19 EU Draft Specification for Vegetable Oil Methylester Diesel Fuel (Biodiesel)

Properties		Limit	Analytical Method
A. Fuel Specific Properties	**Units**		
Density at 15°C	g/cm³	0.86-0.90	ISO 3675
Kinematic viscosity at 40°C	mm²/s	3.5-5.0	ISO 3104
Flash point	°C	min. 100	ISO 2719
Cold filter plugging point CFPP	°C	summer max. 0 winter max. < -15	D EN 116
Sulphur content	%wt	max. 0.01	ISO 8754/ EN 41
Distillation:			
5% vol. evaporated at	°C	to be indicated	ASTM-1160/ISO 3405
95% vol. evaporated at	°C	to be indicated	
Carbon residue			
Conradson			
(10% by vol. residue on distillation at reduced pressure)	%wt	max. 0.30	ISO 10370
Cetane number	-	min. 49	ISO 5165/DIN 51773
Ash content		max. 0.01	EN 26245
Water content (Karl Fischer)	mg/kg	max. 500	ISO 6296/ASTM D 1744
Particulate Matter	g/m³	max. 20	DIN 51419
Copper corrosion (3h/50°C)	corrosion -rating	max. 1	ISO 2160
Oxidation stability	g/m³	max. 25	ASTM D 2274
B. Methyl Ester Specific Properties	**Units**		
Acid value	mg KOH/g	max. 0.5	ISO 660
Methanol content	%wt	max. 0.3	DIN 51413,1
Monoglycerides	%wt	max. 0.8	GLC
Diglycerides	%wt		GLC
Triglycerides	%wt		GLC
Bound glycerine	%wt	max. 0.2	calculate
Free glycerine	%wt	max. 0.03	GLC
Total glycerine	%wt	max. 0.25	calculate
Iodine number	-	max. 115	DIN 53241/IP 84/81
Phosphorous content	mg/kg	max. 10	DGF C-VI 4

Note : Many of the test methods have yet to be finalized.

Table 7.20 Italian Specification for Vegetable Oil Methylester Diesel Fuel

Property	Units	Limits	Test Method
Appearance	Visual	Clear & Bright	
Density	@ 15°C kg/m^3	0.86-0.90	ASTM D1298/ISO 3675
Flash point	PM °C, min.	100	ASTM D93
Cloud point	°C max.	0	ASTM D97
Kinematic viscosity at 40°C	mm^2/s	3.5-5.0	ASTM D189/ISO 3104
Distillation	°C		ASTM D86
IBP min.		300	
95%v max		360	
Sulphur content	%m max.	0.01	ASTM D1552/ISO 8754
Carbon Residue Conradson (CCR)	%m max.	0.5	ASTM D189/ISO 10370
Water content	ppm	700	ASTM D1744
Saponification number	mg/KOH/g min.	170	NGD G33-1976
Total acidity	mg/KOH/g max.	0.05	ASTM D664
Methanol content	%m max.	0.2	GLC
Methyl ester	%m min.	98	GLC
Monoglycerides	%m max.	0.8	GLC
Diglycerides	%m max.	0.2	GLC
Triglycerides	%m max.	0.1	GLC
Free glycerine	%m max.	0.05	GLC
Phosphorus	ppm max.	10	DGF GIII 16A-89

Table 7.21 Austrian Specification for Vegetable Oil Methylester Diesel Fuel (Önorm Vornorm C1190)

Property	Units	Limits	Test Method
Density	@ 15°C kg/m^3	0.86-0.90	DIN 51 757
Flash point	PM °C, min.	55	ONORM C 1122/ISO 2719
CFPP	°C max.	-8	ONORM EN 116
Kinematic viscosity at 20°C	mm^2/s	6.5-9.0	ISO 3104/ISO 3105
Sulphur content	%m max.	0.02	ONORM C 1134
Carbon Residue Conradson (CCR)	%m max.	0.1	DIN 51 551
Cetane number	min.	48	ISO 5165
Neutralisation value	mgKOH/g, max.	1	ONORM C 1146
Methanol content	%m max.	0.2	GLC
Free glycerine	%m max.	0.03	GLC/enzymatic
Total glycerine	%m max.	0.25	GLC/enzymatic

7.8.2 United States

As described in **Section 2.5.5** the Clean Air Act Amendments include legislation on fuel composition and emissions performance, as well as vehicle emission limits. President Bush's original proposal called for a major shift to the use of *"clean alternative fuels"*, i.e. methanol, ethanol, CNG, LPG and hydrogen. However, as the debate progressed the emphasis shifted from alternatives to reformulated fuels, i.e. conventional gasoline whose composition has been modified to reduce exhaust emissions.

Contrary to President Bush's original proposals, the final version of the bill contains no mandate for the introduction of alternative fuels. Instead it describes performance criteria for *"Clean alternative fuels"* which may include:

Methanol and ethanol (and mixtures thereof)
Reformulated gasoline
Natural gas and LPG
Electricity and any other fuel which permits vehicles to attain legislated emission standards.

Since the emission standards set in the CAAA appear likely to be achievable by future conventional vehicles it is likely that *"conventional"* gasoline and diesel will qualify as clean fuels under certain specific circumstances.

The Act does make provision for a Clean Fuels programme which will apply from 1998 to fleets of 10 or more vehicles that are capable of being centrally refuelled (but NOT including vehicles that are garaged at personal residences under normal circumstances) which operate in areas which have problems achieving air quality standards. This programme mandates emission standards for these vehicles which are the same as those specified in California's Low Emission Vehicle (LEV) programme (see **Table 2.7 and Section 2.1.3**)

This part of the CAAA also specifies a pilot programme for the introduction of lower emitting vehicles in California, beginning in 1996. Under this programme 150,000 clean fuel vehicles must be produced for sale in California in 1996 and this figure will rise to 300 000 in 1999. These vehicles will initially be required to meet Transitional Low Emission Vehicle (TLEV) standards. These limits remain in force until 2000 when the LEV standards outlined above come into operation.

Appendix 8

Fuel Quality Specifications— National Standards

Reproduced from Report 4/94, Section 8, by kind permission of:

CONCAWE, Brussels, Belgium

8 Fuel Quality Specifications

8.1 Europe - CEN Standards for Gasoline, Diesel Fuel and Automotive LPG

In 1988 the EU mandated the European Standards Organization (CEN) to develop comprehensive specifications for unleaded gasolines (premium and regular grades), diesel fuel and automotive LPG. The standards were circulated to the national bodies on August 13 1992 for a formal vote before 13 October 1992. They were then officially ratified by CEN on March 16 1993. Member states were required to adopt them as national standards by September 1993 and withdraw conflicting national standards by the same date.

Unleaded Gasoline - EN 228:1993

The EU mandate required the gasoline specifications to cover all major items and eliminate the three category classifications A, B and C in the existing EN 228:1987 standard. In this specification the A category contained the mandatory limits, the B category limits to be specified on the national level and the optional C category any other items which were allowed to be specified by national bodies.

Apart from the octane requirements of the regular grade, all relevant characteristics and test methods are now specified in this European Standard. Provisions are included for national bodies to select seasonal grades from the eight volatility classes during a defined period of the year for a defined region of its country. These have to be specified in the national annex to the

EN 228:1993 specification. Regular grade (if required) octane levels must also be included in the national annex. Details are given in **Table 8.1**. Of the 18 CEN member states, 14 voted in favour and 2 voted against the standard which, according to CEN/CENELEC rules, means that it was accepted.

Diesel Fuel - EN 590:1993

Of the 18 CEN member states, 16 voted in favour of the standard. It was also agreed that the CEN Technical Committee 19 Diesel Working Group WG24 should work on a test method for aromatics and an alternative to the CFPP method, to better predict low temperature performance. The methods for ash content, cetane index, oxidation stability and particulate matter are also under review.

The EN 590:1993 standard specifies six CFPP grades for temperate climates and five different classes for arctic climates. Each country shall detail requirements for summer and winter grades and may include intermediate and/or regional grades, which can be justified by national meteorological data. Details are given in **Table 8.2**.

Automotive LPG - EN 589:1993

Four grades are specified based on seasonal limits for minimum vapour pressure during the winter months. Each country must specify which winter grade it adopts in an annex to its national standard. No minimum is set for the summer period. Details are given in **Table 8.3**.

8.2 France - The Cahier Des Charges/UTAC Labelling Systems

The French motor manufacturers have developed an unofficial performance related fuel quality labelling system called the "Cahier des Charges". In addition to meeting national specifications (**Section 8.3**), oil companies may choose to submit a dossier of information to the motor industry and have their fuels approved. They can then claim that products are "approved by the French motor industry". The scheme was introduced in 1989 and the specifications are updated annually.

Until 1994 the scheme was jointly administered by Renault and the PSA Group. However, on 11 March 1994 it was agreed that the French Transport Ministry Technical Advisory Committee (UTAC) could issue certificates of quality for gasoline and diesel fuel. This certification will replace the "Cahier des Charges de Qualité". The new labelling system was announced in June 1994 and is expected to be launched later in the summer.

Table 8.1 CEN Unleaded Gasoline Specification (EN228:1993)

Property	Premium	Regular	Test Method
RON (min)	95.0	(1)	ISO 5164
MON (min)	85.0	(1)	ISO 5163
Lead g/l (max)	0.013	0.013	EN 237
Benzene %v (max)	5.0	5.0	EN 238
Sulphur %m (max)	$0.10^{(2)}$	$0.10^{(2)}$	EN 24260 ISO 8754
Gum mg/100 ml (max)	5	5	EN 5
Copper Corrosion	1	1	ISO 2160
Appearance	Clear and Bright		Visual
Oxidation Stability : Minutes (min)	360		ISO 7536
Density : kg/m^3	725-780		ISO 3675
Oxygenates	as per directive 85/536/EEC		
Water Tolerance	no water segregation		
Acidity	(4)		ISO 1388

Volatility				Class				
(Notes 3,5)	1	2	3	4	5	6	7	8
RVP hPa	350-700	350-700	450-800	450-800	550-900	550-900	600-950	650-1000
E70 % vol	15-45	15-45	15-45	15-45	15-47	15-47	15-47	20-50
VLI max (RVP +7E70)	900	950	1000	1050	1100	1150	1200	1250
E100 % vol	40-65	40-65	40-65	40-65	43-70	43-70	43-70	43-70
E180 % vol min	85	85	85	85	85	85	85	85
FBP°C max	215	215	215	215	215	215	215	215
Residue % max	2	2	2	2	2	2	2	2

Notes:
(1) Must be specified in National standard
(2) Sulphur reduced to 0.05%m max. from 01.01.95.
(3) Test methods, with the exception of RVP, to ISO 3405. RVP is tested according to EN 12, which is suitable for oxygenates contents meeting column A of EU Directive 85/536/EEC.
(4) The acidity of fuel ethanol used as blendstock shall not exceed 0.007%m (as acetic acid).
(5) See also **Figure 8.1.**
(6) The use of dyes, markers and performance additives is allowed, but no phosphorous containing compounds.

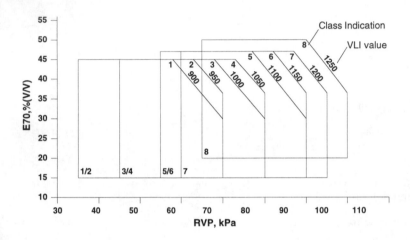

Fig. 8.1. Relationship between RVP, E70 and VLI for the Eight Volatility Classes

Table 8.2 CEN Diesel Fuel Specification EN 590 : 1993

Properties applying to all grades		Test Method
Flash Point PMCC °C (min)	55	ISO 2719
Ash %m (max)	0.01	EN 26245
Water mg/kg (max)	200[(1)]	ASTM D1744
Particulates mg/l (max)	24	DIN 51419
Copper Corrosion 3h at 50°C (max)	1	ISO 2160
Oxidation Stability g/m^3 (max)	25	ASTM D2274
Sulphur %m (max)	0.20	EN 24260/ISO 8754
Carbon Residue (10% btms) %m (max)	0.30[(3)]	ISO 10370

Temperate Climate Grades (Grades A to F)

		Test Method
CFPP (max)	Note [(4)]	EN 116
Density at 15°C kg/m^3	820-860	ISO 3675/ASTM D4052
Viscosity at 40°C mm^2/s	2.00-4.50	ISO 3104
Cetane Number (min)	49	ISO 5165
Cetane Index (min)	46	ISO 4264
Distillation °C:		
10%v rec at	report	ISO 3405
50%v rec at	report	
65%v rec at (min)	250	
85%v rec at (max)	350	
95%v rec at (max)	370	

Arctic Grades (Grades 0 to 4)

Grade	0	1	2	3	4	
CFPP (max)	-20	-26	-32	-38	-44	EN 116
Cloud Point °C (max)	-10	-16	-22	-28	-34	ISO 3015
Density at 15°C kg/m^3 (min)	800	800	800	800	800	ISO 3675/ASTM D4052
Density at 15°C kg/m^3 (max)	845	845	840	840	840	
Viscosity at 40°C mm^2/s (min)	1.50	1.50	1.50	1.40[(5)]	1.20[(5)]	ISO 3104
Viscosity at 40°C mm^2/s (max)	4.00	4.00	4.00	4.00	4.00	
Cetane Number (min)	47	47	46	45	45	ISO 5165
Cetane Index (min)	46	46	46	43	43	ISO 4264
Distillation °C:						
10%v rec at (max)	180	180	180	180	180	ISO 3405
50%v rec at	report	report	report	report	report	
95%v rec at	340	340	340	340	340	

Notes:
(1) A limit of 500 mg/kg may be specified by countries with inherently wet distribution systems until December 1995.
(2) Sulphur limit will be reduced towards 0.05% m. maximum, in line with EU directives or national standards.
(3) Based on fuel without ignition improver additives. If a higher value is found, fuel should be tested by ASTM D 4046 for presence of nitrates. If present the limit does not apply.
(4) Six grades (A, B, C, D, E and F), with CFPP limits from plus 5°C to minus 20°C, in 5°C intervals.
(5) Arctic classes may exhibit poor lubricity characteristics and corrective measures (lubricity additives) may have to be used.

Table 8.3 CEN Automotive LPG Specification EN 589:1993

Property		Test Method
Odour	Distinctive and unpleasant down to 20% LEL	Annex A EN 5589
Density	Optional	ISO 8973
Hydrogen Sulphide	Pass	ISO 8819
MON min	89.0	Annex B EN 589
Vapour Pressure Absolute		ISO 4256
kPa min	$250^{(1)}$	
kPa max	1550	
Dienes molar % max	0.5	ISO 7941
(as 1,3 Butadiene)		
Sulphur mg/kg max	200	EN 24260
(after stenching)		
Evaporative Residue mg/kg max	100	NF-M 41-015
Copper Corrosion 1h at 40°C max	1	ISO 6251 [2]
Water	none	by visual inspection
	pass	ASTM D2713

Notes:
(1)   Four Winter Grades A, B, C and D with minimum absolute vapour pressures of 250 kPa at -10, -5, 0 and +10°C respectively. Grade and date range to be specified in national annexes to standard. No minimum vapour pressure for summer period.
(2)   Addition of corrosion inhibitors for the sole purpose of biasing the test method is prohibited.

A "Certification Committee" will be in charge of the general administration of the scheme, and delivers certificates upon the recommendation of technical committees. These committees are also responsible for controlling conformity of production in service and for recommending changes to the technical content of the UTAC label. Representatives of UTAC, the French administration, the oil industry, motor manufacturers and consumer associations will be able to attend meetings of these committees.

The norms and specifications of the "Cahier des Charges" which have to be met by gasoline and diesel fuel are summarized in **Tables 8.4** and **8.5**. The new system will use these specifications, apart from the sulphur content of unleaded 95 RON gasoline, which is relaxed to 300 ppm.

The scheme incorporates a quality assurance system for final inspection, testing, storage and distribution of automotive fuels, equivalent to ISO 9003.

Table 8.4 Gasoline Specification for UTAC Labelling

Gasoline must also meet administrative specifications and additives must be approved by DHYCA

	Leaded Gasoline	Unleaded 98	Unleaded 95
RON min	97	98	95
MON min	86.5	88	85
Distillation : E70 (**% evaporated at 70°C**)	≤ 40% from 1/6 to 30/9 ≤ 45% from 1/12 to 28/2 ≤ 43% for the other months		
VLI **K = RVP + 7x E70**	VLI < 850 from 1/6 to 30/9 <1150 from 1/12 to 28/2 < 1000 for the other months		
Equivalent Vapour Pressure - **Grabner at 37.8°C (hPa)**	350-700 from 1/6 to 30/9 550-900 from 1/12 to 28/2 450-800 for the other months		
Oxygenates		No alcohols and no ketones Ethers ≤ 15% vol[1]	
Lead Content	0.08 - 0.15 g/l		5 mg/l max
Silicon Content		≤ 2g/ml	
Sulphur		≤ 200 ppm	≤ 300 ppm
Carburettor Cleanliness **Renault R5 (CEC F-03-T-81)**		Merit >8	
Injector cleanliness **- 205 GTI (GFC-TAE-1-87)**		Flow rate losses < 4%	
Intake valve cleanliness **- Opel Kadett** **(CEC F-04-A-87)**		Merit ≥ 9	
- Mercedes M 102 E **(CEC F-05-A-94)**		Merit ≥ 9	
Octane requirement **increase** **(Renault 22700, 22710 or BTC CEC PF 28)**	$KLSA^{(2)} \leq 12°$		KLSA ≤ 8°
Black sludge **(DKA/M 102E, RL 140)**			Merit ≥ 9
Camshaft Bearing Wear- **Petter W1** **(CFC-L-02 A 78, 36 h)**		Bearing Weight loss ≤ 25 mg°	

(1) MTBE ≤ 15% vol
     ETBE ≤ 15% vol, with residual alcohols (ethanol, TBA etc.) ≤ 1% vol
(2) KLSA = "*Knock Limited Spark Advance*"

**Table 8.5 Automotive Diesel Fuel and "Grand Froid"
Specifications for Labelling**

Diesel fuels must also meet administrative specifications and additives must
be approved by DHYCA

Density at 15°C:	820-850 kg/m³ (820-855, if cetane index is ≥ 50 and cetane number is ≥ 51)
Viscosity at 40°C mm²/s	2-4.5
Distillation:	
%v **Evaporated at 250°C**	<60
%v **Evaporated at 350°C**	>85
%v **Evaporated at 370°C**	>95
Cetane Number	>50
Cetane Index (ASTM D 976)	>49
Aromatics [1],[2] %m	<30%
Naphthenes [1],[2] %m	<40%
Oxidation Stability (Insolubles) (ASTM D2274) mg/100 ml.	1.5
Anti-Corrosion: (ASTM D 655-83 procedure A)	<5% (*"moderate rusting"*)
Sulphur %m	<0.15
Foaming Tendency	[4]
Biological Property (M 07070 method)	"light contaminations"
Total Acid Number (NF T 60 112) mg KOH/g	0.08 max
Detergents	[5]
Cold Flow Properties[3]	1 Nov. to 31 March
Cloud (max)	5°C
Point (max)	15°C
CFPP	

(1)   Mass spectrometry; Method to be defined.
(2)   To be defined - Possible introduction of polyaromatic limits.
(3)   *"Grand Froid"* quality available in at least 30% of filling stations, from
       mid-December to the end of February (Cloud Point-8°C (max), CFPP-
       20°C (max)).
(4)   Presence of additive to be confirmed by demonstration of efficiency.
(5)   85% max flow rate loss at 0.1mm needle lift.

Unlike the "Cahier des Charges de Qualité", companies producing fuels of the required quality will be able to claim that their products are "UTAC approved". As with the previous system, approval will be available to companies outside France.

8.3 National Standards

Gasoline lead limits, compositional constraints, volatility, benzene contents and oxygenate contents are discussed in **Sections 7.1 to 7.5**. Sulphur limits and any special compositional constraints on diesel fuel quality are given in **Sections 7.6 and 7.7**. National Standards governing other properties of automotive gasoline and diesel fuels are given in **Tables 8.6 to 8.16** as follows:

Table 8.6	European National Specifications for Leaded Motor Gasoline
Table 8.7	European National Standards for Unleaded Motor Gasoline
Table 8.8	European Standards for Automotive Diesel Fuel
Table 8.9	US National Specifications for Automotive Spark Ignition Engine Fuel (ASTM D4814-91)
Table 8.10	US National Specifications for Automotive Diesel Fuels (ASTM D975-91)
Table 8.11	Japanese National Specification for Gasoline (JIS K2202-1988).
Table 8.12	Japanese National Specification for Automotive Diesel Fuel
Table 8.13	South Korean National Specifications for Automotive Motor Gasoline
Table 8.14	South Korean National Specifications for Automotive Diesel Fuel
Table 8.15	Thailand — Motor Gasoline Specifications
Table 8.16	Thailand — Diesel Fuel Specifications
Table 8.17 and 8.18	South African National Specification for Gasoline (SABS 299- 1972, revised 1988 ; SABS 1598:1993, issued 6.12.93)
Table 8.19	South African National Specification for Automotive Diesel Fuel

Table 8.6 European National Specifications for Leaded Motor Gasoline

Marketing Area	RON (1) min	MON (1) min	RVP (bar)(3)	VLI (2)	E70 % v	E100 % v	E180 %v min	FBP °C max	Residue % vol max
Belgium	P 97.5	-	0.45-0.95	-	15-45	40-70	90	215	2.0
Denmark	P 98.0	P 88.0	S 0.45-0.80 I 0.58-0.90 W 0.70-0.95	700-1100 800-1200 900-1300	15 20 25	44-68 45-70 45-72	90	210	2.0
Finland (5)	P 99.0	P 87.4	S 0.6-0.8 I 0.7-0.9 W 0.85-1.0	<1050 <1150 <1250	20-40 23-43 25-45	40-63 43-66 45-68	90	210	2.0
France	P 97.0-99.0	P 86.0	S 0.45-0.79 I 0.50-0.86 W 0.55-0.99	≤ 900 ≤ 1000 ≤ 1150	10-47	40-70	85	215	2.0
Germany	P 98.0	P 88.0	S 0.45-0.70 W 0.60-0.90	-	S 15-40 W 20-45	S 42-65 W 45-70	90	215	2.0
Greece	P 96-98	-	S 0.62 max W 0.80 max		10 min	30-65	85	215	2.0
	R 90		S 0.65 max W 0.80 max		10 min	30-65	85	215	2.0
Italy	P 97.0	P 87.0	S 0.40-0.74		10-45	30-70	85	220	2.0
Norway	P 98.0	P 87.0	S 0.45-0.90 W 0.70-1.05	>1100 >1300	S 15 min W 20 min	36-68 45-72	85	220	2.0
Portugal	P 98.0 R 90.0	P 87.0	S 0.35-0.70 I 0.45-0.80 W 0.60-0.95		15-45 15-47	40-65 43-70	85	215	2.0
Spain	P 97.0 R 92.0	P 87.0 R 82.0	S 0.48-0.64 W 0.55-0.78		10-45	30-70	80	210	2.0
Sweden (5)	P 98.0	P 87.0	S 0.45-0.70 W 0.65-0.95	<1050 <1250	S 18-45 W 20-47	S 43-68 W 45-70	90	215	2.0
Switzerland	P 98.0	P 88.0	S 0.45-0.70 W 0.60-0.90	<990 <1230	S 15-42 W 20-47	S 40 min W 42 min	85	215	-
UK and Ireland			(4)	(4)					
4 star	97.0	86.0	S 0.45-0.80	<1050	15-45	40-65	90	220	2.0
3 star	93.0	82.0	I 1.03 max	<1240					
2 star	90.0	80.0	W 0.6-1.15	<1360					

(1) P = Premium, R = Regular.
(2) VLI is calculated as RVP (kPa) x 10 + (% at 70°C) x 7.
(3) S = Summer, W = Winter, I = Intermediate.
(4) Recommended seasonal volatility characteristics, values equivalent to **figure 2** of BS4040.
(5) Leaded gasoline no longer available.

798

Table 8.6 European National Specifications for
Leaded Motor Gasoline (Continued)

Marketing Area	Oxidation Stability min	Existent gum mg/100 ml max	Copper Corrosion 3h50°C max	Sulphur % m max	Density at 15°C kg/m3	Reference
Belgium	-	5	1	0.10	P 720-770	NBN T52- 705 1990
Denmark	480	4	1	0.10	P 730-770	Danish Petroleum Institute 01/10/90
Finland[5]	530	5	1	0.1	725-770	
France		10	1 b	P 0.15	P 720-770	(6)
Germany	-	5	1	0.10	P 730-780	DIN 51600 Jan. 1988
Greece (P)	360	4	1	0.10	720-770	Greek Government Gazette (P) 556/93 (R) 581/91
(R)	360	4	1	0.15	700-750	
Italy	-	8	1	0.20	P 725-770	NC 623-01 Nov. '89
Norway	-	5	1	-	725-780	
Portugal	360	5	1	0.10	P 720-770 R 710-760	Ministério da Economia Portaria Feb. '94
Spain	240	5	1b	0.13	P 720-780 R 710-760	Royal Decree 1485/1987
Sweden [5]	360	5	1	0.10	P 725-775	SS 15 54 21 13/03/91
Switzerland	240	5	1	0.10	-	SN 181161/1 Jan. '86
UK and Ireland	240	5	1	0.20	-	BS 4040 1988 (Amended 01/10/94)

(6)   Premium: CSR04-N (Aug. 1991).

Table 8.7 European National Specifications for Unleaded Motor Gasoline

Marketing Area	RON (1)	MON (1)	RVP (bar)	VLI (2)	E70 % v	E100 % v	E180 % v min	FBP °C max	Residue % vol max	Density (kg/m³)	Sulphur % m/m*	Reference
Austria	S 98.0	87.0	S 0.35-0.70	<950	15-45	40-65	>85	215	2	735-780	0.05	Ö-Norm EN 228
	P 95.0	85.0	I mixtures									
	R 91.0	82.5	W 0.55-0.90	<1150	15-47	43-70	>85			720-770		
Belgium	S 98.0	88.0	S 0.45-0.80	<1050	15-45	40-65	>85		2		0.1	NBN T-52-705 1990
	P 95.0	85.0	I mixtures							730-780		
	R 90.0	80.0	W 0.60-0.95	<1200	15-47	43-70	>85			725-775		
										720-770		
Denmark	S 98.0	88.0	S 0.45-0.80	<1050	15-45	40-65	>85		2		0.1	Danish Pet. Institute 01/10/90
	P 95.0	85.0	I 0.58-0.90	<1150	15-47	43-70	>85			730-780		
	R 92.0	83.0	W 0.70-0.95	<1200	15-47	43-70	>85			730-770		
										710-750		
Finland	S 98.0	88.0	S 0.6-0.8	<1050		40-63					0.1	EN228: 1993
	P 95.0	85.0	I 0.7-0.9	<1150		43-66						
			W 0.85-1.0	<1250		45-68						
	S 99.0	88.0	S 0.6-0.7	<1000	20-40	43-63	91				0.01	(3)
	P 95.0	85.0	I 0.7-0.8	<1100	23-43	45-66						
			W 0.8-0.9	<1200	25-45	48-70						
France	P 95.0		S 0.35-0.70	<900	15-45	40-65	>85		2	730-780		
		85.0	I 0.45-0.80	<1000	15-45	40-65	>85				0.1	NF EN 228: 1993
			W 0.55-0.90	<1150	15-47	43-70	>85					
Germany	S 98.0		S 0.35-0.70	<950	15-45	40-65	>85					EN 228: 1993
	P 95.0	88.0	I mixtures					215	2	725-780	0.1	
	R 91.0		W 0.55-0.90	<1150	15-47	43-70	>85					
		85.0										
		82.5										
Greece	P 95	850	S 0.35-0.70	900	15-45	40-65	85	215	2	725-780	0.1	
			W 0.45-0.80	1000			85	215				EN 228: 1993
Italy	95.0	85.0	S 0.35-0.70	<900	15-45	40-65	>85			725-780		
			I 0.45-0.80	<1000	15-45	40-65	>85				0.1	UNI-CUNA EN228: 1993 1.10.93
			W 0.55-0.90	<1100	15-47	43-70	>85					
Netherlands	P 95	85	S 0.40-0.80	<1050	15-45	40-65	>85	215	2		0.1	
			I mixtures									NEN-EN 228 1993
			W 0.60-0.95	<1200	15-47	43-70	>85					
Portugal	95.0	85.0	S 0.35-0.70	<900	15-45	40-65	>85	215	2	735-785	0.1	
			I 0.45-0.80	<1050	15-45	40-65	>85					Porteria
			W 0.60-0.95	<1200	15-47	43-70	>85					125/89
Spain	P 95.0	85.0	S 0.48-0.64		15-45	40-65	>85	215	2			
			W 0.55-0.78		15-45	40-65	>85			735-785	0.1	Royal Decree 1485/1987 (6)
Sweden	P 95.0	85.0	S 0.45-0.80	<1050	15-45	40-65	85	215	2		0.1	SIS SS-EN 228 (March 1994)
			W 0.60-0.60	<1200	15-47	43-85				725-775		
Switzerland	P 95.0	85.0	S 0.35-0.70	<950	15-45	40-65	>85	215	2	725-780		SN EN 278: 1993
			I mixtures								0.1	
			W 0.60-0.95	<1200	15-47	43-70	>85					
UK	S 98.0	87.0	S0.45-0.80	<1050	15-45	40-65	>85	215	2	725-780	0.1	BS EN 228: 1993 (01.10.94)
	P 95.0	85.0	W0.65-1.00	<1250	20-50	43-70	>85					

(1) S=ULG98, P=Premium, R=regular
(2) VLI=RVP (kPa)x10+7E70, S=Summer, I=Intermediate, W=Winter
(3) See also **Table 7.5**, "City gasoline"
(4) VLI=850 for ULG98 only
(5) Stability specifications as for leaded grades. Lead, benzene and oxygenate limits are not included (see **Section 7.1, 7.4** and **7.5** respectively).
(6) EN 228: 1993 is expected to be implemented from winter 1994; Class 2 (S), Class 4 (W).
* Reducing to 0.05% m/m by 31.12.95.

Table 8.8 European National Specifications for Automotive Diesel Fuel

Market Area	Nat. Stan. reference (date)	Grade	Centane number min	Centane index min	Cloud point °C max	Pour point °C max	CFPP °C max	Density at 15°C kg/m³	Flash point PM°C min	KV @ 20°C mm²/sec	KV @ 40°C mm²/sec
Austria	Ö-Norm EN 590-93 (01.02.94)	Winter Intermediate Summer	49	46	-	-	-20 -15 +5	820-860	55	-	2.0-4.5
Belgium & Luxembourg	NBN	Winter Intermediate Summer	49	46	-	-	0 -5 -15	820-870	55	-	2.0-4.5
Denmark		CEN Diesel	49	46	-	-		820-860	55	-	2.0-4.5
		Bus Diesel	50	47	-	-		820-855	-	-	
Finland	EN 590	C 1 3 4	49 47 45 45	46 46 43 43	-16 -28 -34		-5 -26 -38 -44	820-860 800-845 800-840 800-840	55 55 55 55		2.0-4.5 1.5-4.0 1.4-4.0 1.2-4.0
	Market Specification	Sulphur free/S Sulphur free/W Arctic	49 47 45	45 47 43	-5 -29 -40		-15 -34 -44	820-850 800-830 800-840	56 56 56		2.0-3.5 1.4-2.6 1.2-2.5
France	EN 590: 1993	Summer Winter Grand Froid	49	46			0 -15 -20	820-860	55		20-4.5
Germany	DIN EN 590: 1993	Summer Winter	49	46			0 -10 -20	820-860	55 (see remarks)		2.0-4.5
Greece	EN 590: 1993	Summer Winter	49	46			+5 -5	820-860	55	-	2.0-4.5
Ireland	IS EN 590: 1993	Summer Winter	50	50	0	-	- -12	820-860	55	-	2.0-4.5
Italy	UNI-CUNA EN 590-93 1/10/93	Summer Winter	49	46	-	-	0 -10	820-860	55		2.0-4.5

Table 8.8 European National Specifications for Automotive Diesel Fuel (Continued)

Market Area	Nat. Stan. reference (date)	Grade	Centane number min	Centane index min	Cloud point °C max	Pour point °C max	CFPP °C max	Density at 15°C kg/m³	Flash point PM°C min	KV @ 20°C mm²/sec	KV @ 40°C mm²/sec
Netherlands		Winter					-15				
		Intermediate					-5				
		Summer					0				
Norway	No national standard	Summer	-	47	0	-	-	820 min	55	-	
		Winter					-11				
Portugal	EN 590: 1993	Summer	49	46	-	-	0	820-860	55	-	2.0-4.5
		Winter	50	45			-6				
Spain	(2)	Summer	50		+4	-	0	825-860	55	-	5.2 max.
		Winter			-1		-8				4.3 max.
Sweden	SIS 15 54 35 (13/03/91)	Urban Diesel 1	50	50	*0	-	-10	800-820	56	-	1.2-4.0
		Summer (TD1 Grade)-Winter			-16		-26				
		Urban Diesel 2	47	47	*0		-10	800-820	56		1.2-4.0
		Summer (TD2 Grade)-Winter			-16		-26				
		Normal Summer Diesel D 10	49	46	*0		-10	820-860	56		2.0-4.5
		Winter Diesel 1 D 26	47	46	-16		-26	800-845	56		1.5-4.0
		Winter Diesel 2 D 32	48	46	-22		-32	800-840	56		1.5-4.0
		Winter Diesel 3 D 36	45	43	-38		-28	800-840	56		1.4-4.0
Switzerland	SN EN 590	Summer (01.05-30.09)	49	46	-	-	-10	820-860	55	-	2.0-4.5
		All year	47	46	-	-	-20	800-845	55	-	1.5-4.0
UK	BS EN 590: 1993		49	46	-	-	-15	820-860	55	-	2.0-4.5

Table 8.8 European National Specifications for Automotive Diesel Fuel (Continued)

Market Area	Distillation (°C) IBP min	10% min	50% max	65% min	85% max	90% max	95% max	Remarks
Austria				250	360	-	370	
Belgium & Luxembourg				250	350		370	
Denmark						-		
				250	350		370	
							325	see **7.7.1**
Finland C				250	350		370	
1		180					340	
3		180					340	
4		180					340	
Sulphur free/S							350	Further details in
Sulphur free/W							310	**Table 7.12**
Arctic		180					340	
France				250	350		370	AFNOR T60103
				"	"		"	(equivalent to
				"	"		"	>52°C PM)
Germany					350		370	
				250				
Greece						-	370	Greek Government
				250	350			Gazette 336/94
Ireland				250	350		370	2.0-4.5
Italy				250			370	2% Max.
					350			64.5% Max
Netherlands					350		370	
				250				
Norway			280			355		Government regulation
Portugal				250	360		370	
Spain				250	350		370	
Sweden	180						285	Further details of the urban diesel grades.
	180						295	TD1 and TD2 will be found in **Section 7.7.1.**
				250	350		370	
		180					340	
		180					340	
		180					340	
Switzerland	180			250	350		370	
	180(max)						340	
UK				250	350		370	(1) Alternative Standards Class A2: Off-Highway use.

Note: (1) For Sulphur limits, see **Table 7.11**
(2) EN 590: 1993 is expected to be implemented from winter 1994; Class B (S), Class D (W).

Table 8.9 US National Specifications for Automotive Spark Ignition Engine Fuel (ASTM D4814-93a)

A new specification was issued in 1988 to cover gasoline and its blends with oxygenates such as alcohols and ethers. This specification, which was further revised in 1993, is not a legal requirement except in a few states which have adopted it as such. It can consequently be overruled by U.S. Federal legislation on volatility, as described in **Section 7.3.2.**

Table 8.9.1 Volatility

Six vapour pressure/distillation classes and five vapour lock protection classes are specified. The appropriate fuel volatility is specified by a designation that uses a letter from each of the two tabulations according to the region and season of sale.

| Vapour pressure/ distillation class | Reid vapour pressure kPa max (1,2) | Distillation temperature, °C at % evaporated[1] | | | | End point max | Dist. residue % mass | Vapour/liquid ratio[3] | | Vapour lock protection class |
		10 max	50 min	50 max	90 max			Test tempera- ture	V/L max	
AA	54	70	77	121	190	225	2	-	-	-
A	62	70	77	121	190	225	2	60	20	1
B	69	65	77	118	190	225	2	56	20	2
C	79	60	77	116	185	225	2	51	20	3
D	93	55	77	113	185	225	2	47	20	4
E	103	50	77	110	185	225	2	41	20	5

1) If Federal legislation restricts RVP to a level lower than specified in the standard, distillation limits shall be consistent with corresponding RVP limits as above.

2) Dry test methods must be used for gasoline/alcohol blends

3) Version of test D 2533 using mercury must be used for oxygenate blends

Table 8.9.2 Octane Quality

Octane quality is not specifically controlled by the ASTM specification, being left to *"commercial practice"*. However, EPA regulations do require a grade with a minimum antiknock index [(RON + MON)/2] of 87 to be sold. The ASTM specification lists current antiknock Indices in Current Practice (i.e. grades) as follows:

Antiknock index [(RON + MON)/2]	Application
88	*Leaded Fuel*
	For most vehicles that were designed to run on leaded fuel
	Unleaded Fuel
87	Designed to meet antiknock requirements of most 1971 and later model vehicles
89	Satisfies vehicles with somewhat higher antiknock requirements
91 and above	Satisfies vehicles with high antiknock requirements

Notes:

As required by EPA (Reg. 40 CFR part 80), reductions in octane for altitude are allowed

Unleaded gasoline with an antiknock index of 87 should also have a minimum MON of 82

Permissible reductions in antiknock index for altitude and seasonal variation are given in tables in the specification

Table 8.9.3 Other ASTM D 4814-93 Gasoline Specifications

Copper strip corrosion max	Existent gum mg/100 ml max	Sulphur, % mass max		Oxidation stability min	Water Tolerance
		Unleaded	Leaded		
No. 1	5	0.10	0.15	240	(1)

(1) Maximum phase separation temperatures are specified by region and by month.

Lead Content

Grades	Lead Content g/US gal (max)	Lead Content g/l (max)
Leaded	4.2	1.1
Unleaded	0.05	0.013

In addition phosphorous is limited in unleaded gasoline to 0.005 g/US gal (0.0013 g/l)

Table 8.10 US National Specifications for Automotive Diesel Fuels (ASTM D975-91)

Grade of diesel fuel oil	Flash point °C min	Cloud point °C max	Water & sediment % vol max	Carbon residue On 10% residue m % mass max	Ash % mass max	Distillation temperature 90% °C min	max	Viscosity Kinematic at 40°C mm²/s min	max	Saybolt at 100°C(sus) min	max	Sulphur content max	Copper strip corrosion max	Cetane number min
No1 1-DA volatile distillate fuel oil for engines requiring frequent speed and load change.	38	(b)	0.05	0.15	0.01	-	238	1.3	2.4	-	34.4	0.05	No. 3	40(d)
No. 2 2-DA distillate fuel oil of lower volatility for engines in industrial and heavy mobile service	52	(b)	0.05	0.35	0.01	282 (c)	338	1.9	4.1	32.6	40.1	0.05	No. 3	40(d)
No 4-DA heavy distillate fuel or blend of distillate and residual fuel for low/medium speed engines in non-automotive applications	55	(b)	0.05	-	0.10	-	-	5.5	24.0	45.0	1.25	2.00	-	30(d)

Table 8.10 US National Specifications for Automotive Diesel Fuels (ASTM D975-91) (Continued)

Notes:

(a) To meet special operating conditions, modifications of individual limiting requirements may be agreed upon between purchaser, seller and manufacturer.

(b) It is unrealistic to specify low-temperature properties that will ensure satisfactory operation on a broad basis. Satisfactory operation should be achieved in most cases if the cloud point (or wax appearance point) is specified at 6°C above the tenth percentile minimum ambient temperature for the area in which the fuel will be used. Appropriate low temperature operability properties should be agreed upon between the fuel supplier and purchaser for the intended use and expected ambient temperatures.

(c) When cloud point less than -12°C is specified, the minimum viscosity shall be 1.7 cSt (or mm2/s) and the 90% point shall be waived.

(d) Low atmospheric temperatures as well as engine operation at high attitudes may require use of fuels with higher cetane ratings.

Table 8.11 Japanese National Specification for Gasoline
(JIS Specification K2202:1988).

The standard defines two unleaded grades.

	Premium (No.1 grade)	Regular (No.2 grade)
Octane number		
Research	96 min	89 min
Motor	-	-
Distillation		
IBP, °C	-	-
10%, °C	70 max	70 max
50%, °C	125 max	125 max
90%, °C	180 max	180 max
FBP, °C	220	220
Residue %	2.0 max	2.0 max
RVP, kPa	45-80*	45-80*
Density	0.783 max	0.783 max
Cu corrosion	1 max	1 max
Exist. Gum (Unwashed)**	20 max	20 max
(mg/100 ml) (Washed)	5 max	5 max
Oxidation stability	240 min	240 min

*   0.95 bar max for cold climate use
**   Unwashed gum specification is not mandatory, and test method has
      been deleted, but limit remains in the National Standard.

Table 8.12 Japanese National Specification for Automotive Diesel Fuel
JIS KK2204 - 1988

Class	Reaction	Flash Point °C min	Distillation temperature 90% point °C max	Pour Point °C max	CFPP °C max	Carbon Residue on 10 % Residuum % mass	Cetane[a] Index min	Kin. Visc at 38°C (mm²/s) min	Sulphur content % m/m
Special No.1		50	360	+ 5			50	2.7	
No.1		50	360	- 2.5	- 1		50	2.7	
No.2	Neutral	50	350	7.5	5	0.1 max	45	2.0	0.20
No.3		45	330[b]	- 20	- 12		45	2.0	max
Special No.3		45	330	- 30			45	1.8	(c)

Notes:
(a) Cetane number may be used in place of cetane index (the same numerical standards apply). The cetane index formula used
     is specified in the standard.
(b) If viscosity is $\geq$ 4.7 mm²/sec, 90% point should be < 350°C
(c) 0.05m/m max from 1997.

Table 8.13 South Korean National Specifications for Automotive Motor Gasoline

Classification	No.1 Premium	No.2 Medium	No.3 Regular	No.4 Unleaded
Research Octane Number (min)[1]	95	91	88	91
Motor Octane Number (min)	87	83	80	83
Distillation				
10%°C		70 (max)		
50%°C		125 (max)		
90%°C		190 (max)		
Residue (% vol)		2 (max)		
Water & Deposits (% vol)		0.01 (max)		
Copper corrosion (50°C, 3 hours)		1 (max)		
Vapour pressure (37.8°C, kg/cm^2)		0.45-0.85[2]		
Oxidation stability (minutes)		480 (min)		
Existent washed gum (mg/100 ml)		5 (max)		
Sulphur (% mass)		0.1 (max)		
Colour		Yellow (See *Note 3*)		
Pb (g/l)		0.3 (max)		0.013 (max)
P (g/l)	-	-	-	0.0013 (max)

Notes:

1. Octane number must be checked by the Research or Motor method.
2. For cold weather operation a maximum vapour pressure of 0.92 kg/cm^2 is permitted.
3. Leaded grades (No.1, No.2 and No.3) are coloured for easy identification.
4. For the leaded grades, additive varieties and tolerance limits must comply with the environmental protection law.
5. Premium No.1:   High grade or premium gasoline
   Medium No.2:   Military gasoline, not available at filling stations
   Regular No.3:   Low grade or regular gasoline
   Unleaded No.4:   Unleaded gasoline

Table 8.14 South Korean National Specifications for Automotive Diesel Fuel

Classification	Number 1		Number 2	
	Summer	Winter	Summer	Winter
Cetane index (min)	50	50	45	45
Pour point (°C max)	-10	-25	- 5	-18
Flash point (PM°C min)	40	40	50	50
KV at 37.8°C (mm^2/s)	1.4-2.5	1.4-2.5	2.0-5.8	2.0-5.8
90% distillation temp. (°C max)	330	330	360	360
Carbon residue on 10% residium (max)	0.15	0.15	0.2	0.2
Ash (% wt max)	0.01	0.01	0.01	0.01

Note: Number 1 grade is for military use only and not available at filling stations

Table 8.15 Thailand - Motor Gasoline Specifications

Properties		Regular	Premium Leaded	Premium Unleaded
RON	min	87.0	95.0	95.0
MON	min	76.0 [1]	84.0	84.0
Lead Content	g/l max	0.15 [1]	0.15	0.013
Sulphur Content	%m max	0.15	0.15	0.10
Phosphorus Content	g/l max	-	-	0.0013
Copper Corrosion	3h @ 50°C max	1	1	1
Oxidation Stability	min	360	360	360
Existent Gum	g/100ml max	0.004	0.004	0.004
Distillation:	°C			
10 % evap	max	70	70	70
50% evap	min	70	70	70
	max	110	110	110
90% evap	max	170	170	170
end point	max	200	200	200
RVP	@ 37.8°C kPa max	62	62	62
Benzene Content	%v max	3.5	3.5	3.5
Aromatic Content:	%v max			
from 1.1.1994	-	-	50	50
from 1.1.2000	-	-	35	35
Water:	%m max			
non-alcohol blends		none	none	none
alcohol blends		0.7	0.7	0.7
Oxygenated	MTBE %v			
Compounds:				
minimum		-	5.5 [2]	5.5 [2]
maximum		10.0 [4]	10.0 [3]	10.0 [3]
PFI/IVDC additive			yes	yes

Notes

    (1)   Regular will be unleaded from 1.1.1995 (0.013 gPb/l max)
    (2)   For Type 2 Reformulated Gasoline only.
    (3)   Methanol blend must be at 3.0 %v max.
    (4)   To be added from 1.1.1995. Additive must meet Californian BMW test.

Table 8.16 Thailand - Diesel Fuel Specification

Properties	Units	Limit	Type 1	Type 2
Specific Gravity @ 15.6/15.6°C		min	0.81	0.81
		max	0.87	0.87
Cetane Number		min	47	47
Calculated Cetane Index		min	47	47
Viscosity @ 40°C	mm^2/s	min	1.8	1.8
		max	4.1	4.1
Pour Point	°C	max	10	10
Sulphur Content	%wt			
-before 1 September, 1993		max	1.0	0.5
-from 1 September, 1993		max	-	0.5
-from 1 January, 1996		max	-	0.25
-from 1 January, 2000		max	-	0.05
Copper Strip Corrosion	No.	max	1	1
Carbon Residue	%w/w	max	0.05	0.05
Water & Sediment	%v/v	max	0.05	0.05
Ash	%w/w	max	0.01	0.01
Flash Point	°C	min	52	52
Distillation				
-90% recovered	°C	max	357	357
Colour		max	4.0	4.0
Detergent Additive			yes	yes

Type 1 - for the whole country except Bangkok, Nonburi, Patumthani, Samuthprakarn, Samuthsakorn, Samuthsongkram, Phuket, Cholburi, Nakornpathom and Chachoengsao

Type 2 - for exceptions to Type 1 areas.

Table 8.17 South African National Specification for Gasoline (SABS 299- 1972, revised 1988)

Classification[1] (By Minimum Research Octane Number)	87	93[2]	97
Distillation:			
10%°C (max)		65	
50%°C		77-115	
90%°C (max)		185	
FBP°C (max)		215	
Residue (vol)		2.0	
RVP, kPa, max		75	
Vapour/Liquid Ratio at 325 kPa (max)[+]			
at 50°C*		20	
at 55°C		20	
Induction Period, Minutes, (min)		240	
Lead Content (as Pb), g/l (max)		0.4	
Existent Gum, mg/100 ml (max)		4.0	
Potential Gum, 2½ h at 100°C,			
mg/100 ml (max)		4.0	
Sulphur Content, % mass (max)		0.15	
Copper Strip Corrosion (3 h at 50°C)			
Classification (max)		1	
Water Tolerance**		Pass	
Total Acidity***, mg KOH/g		0.03	

*   Applicable only to petrol supplied in coastal regions between 1 April and 30 September (inclusive).

+   The V/L ratio calculated as described in ASTM D439 may be used as an approximation, but ASTM Method D2533 shall be the referee procedure. However, when blends containing alcohol(s) are tested in accordance with ASTM D2533, the method must be modified by substituting mercury for glycerine in the pressure control system.

**   Applicable only to blends containing alcohol(s) or other oxygenated compounds or both.

*** Applicable only to fuels derived from coal and to blends containing alcohol(s) or other oxygenated compounds or both.

(1) In any particular area only two of the three grades will be available. For the *"Coastal areas"* (nominally areas at altitudes below 1200 metres), 97 and 93 RON grades are usually supplied. Above this altitude, only the 87 and 93 RON fuels are generally marketed.

(2) The 93 RON grade manufactured in the SASOL oil-from-coal process contains 8-12% alcohols (mainly ethanol plus some higher alcohols)

Table 8.18 South African Unleaded Gasoline Specifications
(SABS 1598:1993 issued 6 December 1993)

Property		Coastal	Inland
		Grade 95	Grade 91
RON	min	95	91
MON	min	85	81
MON (blends with >0.2%v/v alcohol)	min	87	83
Density at 20°C (kg/m^3)		710-785	710-785
Distillation °C:			
E10	max	65	
E50		77-115	
E90	max	185	
FBP		215	
residue %v/v	max	2.0	
VLI summer max		950	890
winter max		1000	940
Lead Content gPb/l max		0.013	
Induction Period	min/min	360	
Existent Gum mg/100ml	max	4	
Potential Gum mg/100ml	max	4	
Sulphur Content %m/m	max	0.10	
Copper Strip Corrosion			
3h at 50°C	max	1	
Total Acidity [1] **mgKOH/g**	max	0.03	0.03
Oxygen Content [2] **%m/m**	max	2.8	3.7

Notes
(1) Applicable to gasolines containing alcohol only
(2) Any alcohol blend into the fuel shall contain a minimum of 85%m/m ethanol with the balance mainly iso- and n-propanol. Ethers containing 5 or more carbon atoms may be incorporated

Table 8.19 South African National Specification for Automotive Diesel Fuel

Characteristic	Requirement
Cetane Number* (min)	45
CFPP,°C (max)	-4**
Density at 15°C, kg/l	-
Flash Point, PM °C (min)	55
KV at 40°C, mm²/sec	1.6-5.3
Distillation, 90% recovery °C (max)	362

* In the case of a fuel that does not contain an ignition improver the calculated cetane index (Method IP 218 - ASTM D976) may be used as an approximation but the cetane number shall be used in cases of dispute.

** Unless otherwise acceptable - see note (1), below

(1) A product with a maximum CFPP of:
    (a)   0°C and supplied between 15 April and 14 May (inclusive) or between 1 September and 30 September (inclusive)
    (b)   3°C and supplied between 1 October and 14 April (inclusive) may be considered to be acceptable.

Appendix 9

Legislative and Reference Fuels

Further information on the fuels listed may be obtained from:

CEC Reference Fuels: Coordinating European Council
                         61 New Cavendish Street
                         London W1M 8AR, England

Phillips Petroleum: Phillips Petroleum
                         Borger, Texas

Howell Reference Fuels: Howell Hydrocarbons Inc.
                         7811 South Presa Street
                         San Antonio, Texas 78223-3596

CEC reference and legislative fuels available in Europe:

Fuel type and application	CEC Reference
1. Premium gasoline, leaded and unleaded (legislative)	
1.1 Leaded: European Exhaust Emissions Test and Fuel Consumption	RF-01-A-80
1.2 Unleaded: European Exhaust Emissions Test and Fuel Consumption	RF-08-T-85
2. Diesel fuel (legislative)	
2.1 European Exhaust Emissions Test and Fuel Consumption	RF-03-A-80 RF-03-A-84

Fuel type and application	CEC Reference
3. **Regular gasoline, unleaded (legislative)**	
3.1 Australian, Californian, Canadian, Japanese, and U.S. Exhaust Emissions	RF-05-A-83
3.2 Gasoline, unleaded, for development testing of gasoline engines	RF-10-A-90
5. **Gasoline, leaded, for antiknock tests**	
5.5 High delta R100°C, low octane, low lead	RF-24-A-83
5.6 High delta R100°C, high octane, low lead	RF-25-A-83
5.7 High sensitivity, low octane, low lead	RF-26-A-83
5.8 High sensitivity, high octane, low lead	RF-27-A-83
6. **Gasoline, unleaded, for antiknock tests**	
6.1 North America, Japan, Australia:	
6.1.1 Moderate sensitivity, low octane	RF-28-A-84
6.1.2 High sensitivity, high octane	RF-29-A-84
6.2 Europe, conventional hydrocarbon components only:	
6.2.1 High sensitivity, high octane	RF-30-A-85
6.3 Europe, containing 10% oxygenates (3% vol. methanol, 2% vol. TBA, 5% vol. MTBE):	
6.3.1 RF-30-A-85 gasoline oxygenated as above	RF-30-OXY-A-85
6.3.2 RF-31-85 gasoline oxygenated as above	RF-31-OXY-A-85
6.4 Moderate sensitivity, low octane	RF-31-A-84
6.5 High sensitivity, high octane	RF-32-A-84

Fuel type and application	CEC Reference

7. <u>Gasoline, leaded, for volatility tests</u>

7.1 Europe:

7.1.1	Hot weather, about 35°C	RF-40-A-84
7.1.2	Moderate cold weather, -15°C to +5°C	RF-41-A-84
7.1.3	Severe cold weather, about -30°C	RF-42-A-84

7.2 Europe, containing 10% oxygenates (3% vol. methanol, 2% vol. TBA, 5% vol. MTBE):

7.2.1	RF-40-A-84 gasoline oxygenated as above	RF-40-OXY-A-84
7.2.2	RF-41-A-84 gasoline oxygenated as above	RF-41-OXY-A-84
7.2.3	RF-42-A-84 gasoline oxygenated as above	RF-42-OXY-A-84

8. <u>Gasoline, unleaded, for volatility tests</u>

8.1 North America:

8.1.1	Hot weather, about 35°C	RF-43-A-83
8.1.2	Moderate cold weather, -15°C to +5°C	RF-45-A-83
8.1.3	Severe cold weather, -30°C	RF-47-A-83

9. <u>Gasoline, intake system icing and automatic-choke calibration tests</u>

9.1 Europe, leaded:

9.1.1	Automatic-choke calibration, moderate cold weather, -15°C to +5°C	RF-41-A-84
9.1.2	Intake-system icing and automatic-choke calibration, about 3°C	RF-44-A-83

Fuel type and application	CEC Reference

9.2 North America, unleaded:

9.2.1 Automatic-choke calibration, moderate
cold weather, -15°C to +5°C   RF-45-A-83

9.2.2 Intake-system icing and automatic-
choke calibration, about 3°C   RF-48-A-83

10. Hydrocarbon fluid and oxygenated fluids
for corrosion, elastomer, plastics and paint tests

10.1 Hydrocarbon fluid:

10.1.1 Toluene/isooctane/diisobutylene blend

10.2 Hydrocarbon fluid with oxygenates:
Toluene/iso octane/di-isobutylene blend   RF-50-A-84

10.2.1 RF-50-A-84 +
5% ethanol/1 ppm acetic acid blend   RF-51-A-84

10.2.2 RF-51-A-84 +
15% methanol/3 ppm formic acid blend   RF-52-A-84

10.2.3 RF-50-A-84 +
53% methanol/30 ppm formic acid blend   RF-53-A-84

11. Diesel fuel for evaluation of engines/vehicles

11.1 Europe, Japan, Australia,
New Zealand   RF-70-A-86

11.2 South America   RF-72-A-87

11.3 Europe   RF-73-A-93

12. Gasoline: Lubricating oil tests

12.1 CEC Engine tests - Petter W1, Ford
Cortina, Motobecane AV7L,
Peugeot 204, Ford Kent, Johnson
outboard, Crescent outboard,
OMC 4 Euro outboard   RF-80-A-87

Fuel type and application	CEC Reference
12.2 Super premium gasoline, leaded:	
12.2.1 CEC engine tests - Fiat 132, Fiat 600D, Piaggio Vespa 18055	RF-82-A-81
12.3 Premium unleaded gasoline CEC valve train and 2-T tests	RF-83-A-91
13. Automotive diesel fuels: Lubricating oil tests	
13.1 Diesel fuel, low sulfur:	
13.1.1 CEC engine tests - Mercedes-Benz OM616, OM602A, OM364A, VW PV1431 test	RF-90-A-92
13.2 Diesel fuel, high sulfur:	
13.2.1 CEC engine tests - Petter AV1, Petter AVB, MWM KD 12E (methods A and B), Ford Tornado	RF-91-A-81

Phillips Petroleum Reference Fuels

RUBBER SWELLING TEST FLUIDS

Fuel A (Isooctane)
Fuel B (70 percent Isooctane; 30 percent Toluene)
Fuel C (50 percent Isooctane; 50 percent Toluene)
Hydrocarbon Fluids Type I, II, III and VII

These test fluids are used to determine the change in properties of rubber resulting from immersion in liquids as proposed in various military and ASTM Test Methods.

ASTM NAPHTHA

ASTM Precipitation Naphtha

Prepared specifically to meet the requirements of high quality solvents for use in ASTM methods.

PETROLEUM ETHER

Petroleum Ether (30-60C)

Petroleum ether (30-60C) meets the Federal Specification O-E-1751a (with Amendment) for ether, petroleum, technical grades; ACS specification for petroleum ether.

FLASH POINT CHECK FLUID

Para-xylene

Specifically prepared for use as a standard in determining flash point of volatile solvents according to ASTM D 56 and D 1310.

CALIBRATING FLUID

MIL-F-27351, High Flash

For use in calibration of fuel metering and controls of aircraft engines.

ASTM KNOCK TEST REFERENCE FUELS

ASTM Isooctane

Certified by ASTM to meet all ASTM requirements for use in determining the knock characteristics of fuels. Octane number is 100 as defined by ASTM.

ASTM n-heptane

Certified by ASTM, octane number is 0 as defined by ASTM.

ASTM 80 Octane Number Blend

A blend of n-heptane and isooctane; certified by ASTM.

BLENDS OF ASTM REFERENCE FUELS

Fuels are available with Octane Number from 0 to 100

Blended with ASTM isooctane and ASTM n-heptane, the most common fuels have octane numbers of 60 and 82 through 99.

TOLUENE STANDARDIZATION FUELS

Reference Fuel Toluene

An ASTM specification toluene used in making toluene standardization fuels.

Toluene Standardization
Fuel 652
   (Motor Octane Number
   57.8; Research Octane
   Number 65.2)

Toluene Standardization
Fuel 934
   (Motor Octane Number
   81.1; Research Octane
   Number 93.4)

Toluene Standardization   Other octane numbers are
Fuel 1137   available on request.
   (Motor Octane Number
   100.8; Research Octane
   Number 113.7)

AVIATION CHECK FUELS

ASTM Aviation Check Fuels   Reference Fuels for use in
Check Fuel 100/130   checking knock characteristics of
Check Fuel 115/145   aviation gasolines by ASTM D 909 or
                                 D 2700.

SECONDARY DIESEL REFERENCE FUELS

ASTM Secondary Cetane Fuels   These fuels are used in place of the
T-17 and U-10   more expensive primary
                                 reference fuels — n-cetane and
                                 heptamethylnonane — in routine
                                 ratings of diesel fuels.

ASTM Diesel Check Fuels   Diesel check fuels to be used in
40 and 50   checking the rating characteristics
                                 of diesel engines as defined in ASTM
                                 D 613. These two fuels are not for
                                 blending except at the 50/50 level to
                                 obtain a cetane standard of 45.

DIESEL EMISSION TEST FUEL

D-2 Diesel Control Fuel

This fuel is produced from typical refinery streams without cetane improvers. D-2 diesel control fuel is certified to meet the specifications as described in chapter one, Environmental Protection Agency, Part 86.177-6.

GMR 995 FUEL

GMR 995 Fuel

Fuel certified by General Motors research for use in sequence IId and sequence IIId engine testing. Leaded fuel.

J-FUEL

J-Fuel

Fuel certified for use in sequence VD engine testing. Unleaded fuel.

The following technical fuels are available from Howell Hydrocarbons, Inc.

Gasolines
EEE Clear - Unleaded Emission Reference Fuel
EEE Leaded - Reference Fuel
Injector Durability Fuels
Volatility Test Fuels
NO_x Sensor Fuel
Oxygenated Gasoline Blends
Oxidation Stability Fuels

Diesel Fuels
I-H Cat Diesel
I-D Cat Diesel
I-G Cat Diesel
1% Natural Sulfur Diesel
EPA Exhaust Emission Diesels
Service Accumulation Diesels
Detroit Diesel 5Y13
Detroit Diesel 5Y14
VVF-800 (All Grades)
MIL-F-46162

Turbine Fuels
MIL-T-25524 - Thermally Stable Turbine Fuel
MIL-T-5624 - Turbine Jet Fuel (With or Without Additives)(JP-4 and
    JP-5)
JP-8 (With or Without Additives)
Jet A
Jet A-1

Performance Products
001 Racing Fuel (100 Octane Leaded Racing Fuel)
002 Racing Fuel (104.5 Octane Leaded Racing Fuel)
007 Racing Fuel (112 Octane Leaded Racing Fuel)
SB #6 (95 Octane Super Premium No Lead)

Appendix 10

Octane Properties of Hydrocarbons

Conversion of Millilitres of Tetraethyllead per U.S. Gallon in Isooctane to Octane Numbers Above 100

TEL, mL U. S. gal	0.00	0.01	0.02	0.03	0.04	0.05	0.06	0.07	0.08	0 .09
0.0	100.00	100.14	100.28	100.42	100.55	100.68	100.81	100.94	101.07	101.19
0.1	101.32	101.44	101.56	101.68	101.80	101.92	102.03	102.14	102.26	102.37
0.2	102.48	102.58	102.69	102.80	102.90	103.01	103.11	103.21	103.31	103.41
0.3	103.51	103.60	103.70	103.79	103.89	103.98	104.07	104.16	104.25	104.34
0.4	104.43	104.52	104.61	104.69	104.78	104.86	104.94	105.03	105.11	105.19
0.5	105.27	105.35	105.43	105.51	105.59	105.66	105.74	105.82	105.89	105.97
0.6	106.04	106.11	106.19	106.26	106.33	106.40	106.47	106.54	106.61	106.68
0.7	106.75	106.81	106.88	106.95	107.01	107.08	107.15	107.21	107.27	107.34
0.8	107.40	107.46	107.53	107.59	107.65	107.71	107.77	107.83	107.89	107.95
0.9	108.01	108.07	108.13	108.18	108.24	108.30	108.35	108.41	108.47	108.52
1.0	108.58	108.63	108.69	108.74	108.79	108.85	108.90	108.85	109.01	109.06
1.1	109.11	109.16	109.21	109.26	109.31	109.36	109.41	109.46	109.51	109.56
1.2	109.61	109.66	109.71	109.76	109.80	109.85	109.90	109.94	109.99	110.04
1.3	110.08	110.13	110.17	110.22	110.26	110.31	110.35	110.40	110.44	110.49
1.4	110.53	110.57	110.62	110.66	110.70	110.74	110.79	110.83	110.87	110.91
1.5	110.95	110.99	111.04	111.08	111.12	111.16	111.20	111.24	111.28	111.32
1.6	111.36	111.40	111.43	111.47	111.51	111.55	111.59	111.63	111.66	111.70
1.7	111.74	111.78	111.81	111.85	111.89	111.92	111.96	112.00	112.03	112.07
1.8	112.11	112.14	112.18	112.21	112.25	112.28	112.32	112.35	112.39	112.42
1.9	112.46	112.49	112.52	112.56	112.59	112.62	112.66	112.69	112.72	112.76
2.0	112.79	112.82	112.86	112.89	112.92	112.95	112.98	113.02	113.05	113.08
2.1	113.11	113.14	113.17	113.21	113.24	113.27	113.30	113.33	113.36	113.39
2.2	113.42	113.45	113.48	113.51	113.54	113.57	113.60	113.63	113.66	113.69
2.3	113.72	113.75	113.77	113.80	113.83	113.86	113.89	113.92	113.95	113.97
2.4	114.00	114.03	114.06	114.09	114.11	114.14	114.17	114.20	114.22	114.25
2.5	114.28	114.30	114.33	114.36	114.38	114.41	114.44	114.46	114.49	114.52
2.6	114.54	114.57	114.60	114.62	114.65	114.67	114.70	114.72	114.75	114.77
2.7	114.80	114.83	114.85	114.88	114.90	114.93	114.95	114.97	115.00	115.02
2.8	115.05	115.07	115.10	115.12	115.15	115.17	115.19	115.22	115.24	115.26
2.9	115.29	115.31	115.34	115.36	115.38	115.41	115.43	115.45	115.48	115.50
3.0	115.52	115.54	115.57	115.59	115.61	115.63	115.66	115.68	115.70	115.72
3.1	115.75	115.77	115.79	115.81	115.84	115.86	115.88	115.90	115.92	115.94
3.2	115.97	115.99	116.01	116.03	116.05	116.07	116.09	116.12	116.14	116.16
3.3	116.18	116.20	116.22	116.24	116.26	116.28	116.30	116.32	116.34	116.36
3.4	116.38	116.41	116.43	116.45	116.47	116.49	116.51	116.53	116.55	116.57
3.5	116.59	116.61	116.63	116.64	116.66	116.68	116.70	116.72	116.74	116.76
3.6	116.78	116.80	116.82	116.84	116.86	116.88	116.90	116.91	116.93	116.95
3.7	116.97	116.99	117.01	117.03	117.05	117.06	117.08	117.10	117.12	117.14
3.8	117.16	117.18	117.19	117.21	117.23	117.25	117.27	117.28	117.30	117.32
3.9	117.34	117.36	117.37	117.39	117.41	117.43	117.44	117.46	117.48	117.50
4.0	117.51	117.53	117.55	117.57	117.58	117.60	117.62	117.63	117.65	117.67
4.1	117.69	117.70	117.72	117.74	117.75	117.77	117.79	117.80	117.82	117.84
4.2	117.85	117.87	117.89	117.90	117.92	117.94	117.95	117.97	117.99	118.00
4.3	118.02	118.03	118.05	118.07	118.08	118.10	118.11	118.13	118.15	118.16
4.4	118.18	118.19	118.21	118.23	118.24	118.26	118.27	118.29	118.30	118.32
4.5	118.33	118.35	118.37	118.38	118.40	118.41	118.43	118.44	118.46	118.47
4.6	118.49	118.50	118.52	118.53	118.55	118.56	118.58	118.59	118.61	118.62
4.7	118.64	118.65	118.67	118.68	118.70	118.71	118.73	118.74	118.76	118.77
4.8	118.79	118.80	118.81	118.83	118.84	118.86	118.87	118.89	118.90	118.91
4.9	118.93	118.94	118.96	118.97	118.99	119.00	119.01	119.03	119.04	119.06
5.0	119.07	119.08	119.10	119.11	119.13	119.14	119.15	119.17	119.18	119.19
5.1	119.21	119.22	119.24	119.25	119.26	119.28	119.29	119.30	119.32	119.33
5.2	119.34	119.36	119.37	119.38	119.40	119.41	119.42	119.44	119.45	119.46
5.3	119.48	119.49	119.50	119.52	119.53	119.54	119.55	119.57	119.58	119.59
5.4	119.61	119.62	119.63	119.64	119.66	119.67	119.68	119.70	119.71	119.72
5.5	119.73	119.75	119.76	119.77	119.78	119.80	119.81	119.82	119.83	119.85
5.6	119.86	119.87	119.88	119.90	119.91	119.92	119.93	119.95	119.96	119.97
5.7	119.98	119.99	120.01	120.02	120.03	120.04	120.06	120.07	120.08	120.09
5.8	120.10	120.12	120.13	120.14	120.15	120.16	120.18	120.19	120.20	120.21
5.9	120.22	120.23	120.25	120.26	120.27	120.28	120.29	120.30	120.32	120.33
6.0	120.34	...	...	...	...	...	...	...	...	...

Source: 1992 ASTM Book of Standards, Vol. 05.04, p. 62, 1992.

SUMMARY OF DATA ON THE KNOCK RATINGS OF HYDROCARBONS
(A.P.I. Research Project 45)

	Research octane number (CRC-F_1)				Motor octane number (CRC-F_2)			
	ml TEL/US gal			Blending octane number	ml TEL/US gal			Blending octane number
Hydrocarbon	0.0	1.0	3.0		0.0	1.0	3.0	
Alkanes								
n-Pentane	6.71	74.9	88.7	62	61.9	77.1	83.6	67
2-Methylbutane	92.3	+0.37	+1.0	99	90.3	100.0+		104
2,2-Dimethylpropane	85.5	97.4	+0.1	100	80.2	93.0	99.9	90
n-Hexane	24.8	43.4	65.3	19	26.0	51.1	65.2	22
2-Methylpentane	73.4	84.6	93.1	83	73.5	87.3	91.1	79
3-Methylpentane	74.5	85.0	93.4	86	74.3	87.5	91.3	81
2,2-Dimethylbutane	91.8	+0.1	+0.6	89	93.4	+0.6	+2.10	97
2,3-Dimethylbutane	+0.3	+2.1			94.3	+0.4	+1.79	107
n-Heptane	0.0	10.0	43.5	0	0.0	25.4	46.9	0
2-Methylhexane	42.4		73.2	41	46.4		74.5	42
3-Methylhexane	52.0		74.7	56	55.0		81.0	57
2,2-Dimethylpentane	92.8		+0.4	89	95.6		+2.43	93
2,3-Dimethylpentane	91.1	98.6	+0.3	87	88.5	99.0	+0.29	90
2,4-Dimethylpentane	83.1	93.7	96.6	77	83.8	93.0	99.1	78
2,2,3-Trimethylbutane (triptane)	+1.8			113	+0.1		+3.07	113
3-Ethylpentane	65.0	75.2	85.0	64	69.3	81.2	88.0	73
n-Octane			24.8	-19		0.7	28.1	-15
2-Methylheptane	21.7	34.4	57.6	13	23.8	45.0	61.4	24
3-Methylheptane	26.8	37.5	59.6	30	35.0	53.5	68.0	30
4-Methylheptane	26.7	38.7	61.1	31	39.0	55.4	70.1	48
3-Ethylhexane	33.5	46.3	61.1	49	52.4	65.9	80.0	49
2,2-Dimethylhexane	72.5	85.4	93.3	67	77.4	90.0	95.2	76
2,3-Dimethylhexane	71.3	82.5	91.7	71	78.9	88.4	93.7	76
2,4-Dimethylhexane	65.2	77.6	87.3	65	69.9	83.8	89.0	70
2,5-Dimethylhexane	55.5	68.0	81.6		55.7	71.6	82.9	
3,3-Dimethylhexane	75.5	86.2	94.6	73	83.4	95.4	+0.0	81
3,4-Dimethylhexane	76.3	88.4	94.7	67	81.7	92.5	7.1	80
2,2,3-Trimethylpentane	+1.2	+3.7		105	99.9	+0.7	+2.0	112
2,2,4-Trimethylpentane (*iso*-octane)	100.0	+1.0	+3.0	100	100.0	+1.0	+3.0	100
2,3,3-Trimethylpentane	+0.61	+2.7		100	99.4	+0.6	+1.9	110
2,3,4-Trimethylpentane	+0.22	+1.2		97	95.9	+0.2	+0.7	102
n-Nonane				-17				-20
2,2,5-Trimethylhexane				91				88
2,3,5-Trimethylhexane				81				78
2,2-Dimethyl-3-ethylpentane	+1.8			108	99.5		+0.8	112
2,4-Dimethyl-3-ethylpentane	+0.5			88	96.6		+0.4	93
2,2,3,3-Tetramethylpentane	+3.6			123	95.0		99.4	113
n-Decane				-41				-38
2,7-Dimethyloctane				20				20
4,5-Dimethyloctane				48				61
2,5,5-Trimethylheptane				31				41
3,3,5-Trimethylheptane				77	88.7		+0.2	87
2,2,3,3-Tetramethylhexane				126	92.4		96.7	112
Alkenes								
1-Butene				144				126
2-Butene				153				130
3-Methylpropene				170				139
1-Pentene	90.9		98.6	119	77.1	81.3	82.9	109
cis-2-Pentene				154				137
trans-2-Pentene				150				134
2-Methyl-1-butene	+0.2		+0.3	146	81.9		84.2	133
3-Methyl-1-butene				129				125
2-Methyl-2-butene	97.3	99.0	99.2	176	84.7	85.5	85.8	141

Notes: Figures prefixed by a + sign in the above table represent ml TEL/US gal of *iso*-octane to give an equivalent rating.

1.0 ml/US gal = 1.2 ml/UK gal = 0.264 ml/l
3.0 ml/US gal = 3.6 ml/UK gal = 0.793 ml/l

Hydrocarbon	Research octane number (CRC–F$_1$) ml TEL/US gal 0.0	1.0	3.0	Blending octane number	Motor octane number (CRC–F$_2$) ml TEL/US gal 0.0	1.0	3.0	Blending octane number
Alkenes (contd.)								
1-Hexene	76.4		91.7	97	63.4		76.3	94
trans-2-Hexene	92.7		98.4	134	80.8		83.2	129
trans-3-Hexene	94.0			137	80.1		82.3	120
2-Methyl-1-pentene	95.1		99.3	126	78.9		86.7	114
3-Methyl-1-pentene	96.0		+0.05	113	81.2		82.2	114
4-Methyl-1-pentene	95.7		+0.5	112	80.9		84.5	108
2-Methyl-2-pentene	97.8	99.3	99.5	159	83.0	84.7	85.0	148
cis-3-Methyl-2-pentene				125				113
trans-3-Methyl-2-pentene	97.2		100	130	81.0		84.0	118
cis-4-Methyl-2-pentene	99.3			130	84.3			128
trans-4-Methyl-2-pentene	99.3			130	84.3			128
2-Ethyl-1-butene	98.3		+0.3	143	79.4		82.0	129
2,3-Dimethyl-1-butene				148				129
3,3-Dimethyl-1-butene	+1.7			137	93.5			121
2,3-Dimethyl-2-butene	97.4	98.1	98.5	185	80.5		84.0	144
1-Heptene				68				46
5-Methyl-1-hexene				96				80
2-Methyl-2-hexene	90.4			129	78.9			119
2,4-Dimethyl-1-pentene	99.2		+0.36	142	84.6		87.3	124
4,4-Dimethyl-1-pentene	+0.4		+0.8	144	85.4		87.7	136
2,3-Dimethyl-2-pentene	97.5		99.5	165	80.0		83.3	145
2,4-Dimethyl-2-pentene	100			135	86.0			123
2,3,3-Trimethyl-1-butene	+0.5	+0.7	+1.2	145	90.5	92.3	93.7	130
1-Octene	28.7	43.8	63.5		34.7	46.6	57.7	
2-Octene	56.3	69.9	78.7	75	56.5	67.9	73.0	68
trans-3-Octene	72.5	84.6	89.4	95	68.1	77.7	81.2	85
trans-4-Octene	73.3	85.4	91.8	99	74.3	82.8	84.2	101
2-Methyl-1-heptene	70.2	79.6	87.9	77	66.3	73.3	79.6	
6-Methyl-1-heptene	63.8	74.8	87.2	74	62.6	69.9	76.6	69
2-Methyl-2-heptene	75.9			91	71.0			102
6-Methyl-2-heptene	71.3	84.6	90.2	75	65.5	77.0	80.5	
2,3-Dimethyl-1-hexene	96.3			118	83.6	86.7	88.1	109
2,3-Dimethyl-2-hexene				144				
Di-*iso*-butylene	+0.5	+0.9	+1.1	168	88.6	89.1	90.1	151
2,3,3-Trimethyl-1-pentene	+0.6		+0.9	138	85.7		87.2	129
2,4,4-Trimethyl-1-pentene	+0.6		+1.0	164	86.5		88.8	153
2,3,4-Trimethyl-2-pentene	96.9		+0.02	142	80.9		84.4	130
2,4,4-Trimethyl-2-pentene	+0.3	+0.4	+0.6	148	86.2	87.3	88.0	139
3,4,4-Trimethyl-2-pentene	+0.3		+0.7	151	86.4		87.7	144
1-Nonene				35				22
Aromatics								
Benzene				99	+2.7			91
Methylbenzene (toluene)	+5.8			124	+0.3	+1.00	+1.7	112
Ethylbenzene	+0.8	+0.8	+0.8	124	97.9	100.0	+0.2	107
1,2-Dimethylbenzene (*o*-xylene)				120	100.0		+0.0	103
1,3-Dimethylbenzene (*m*-xylene)				145				124
1,4-Dimethylbenzene (*p*-xylene)				146				127
n-Propylbenzene	+1.5	+3.8	+4.3	127	98.7	+0.1	+0.2	129
iso-Propylbenzene (cumene)	+2.1	+3.4		132	99.3	+0.2	+0.5	124
1-Methyl-2-ethylbenzene				125				111
1-Methyl-3-ethylbenzene	+1.8			162	100			138
1-Methyl-4-ethylbenzene				155	97.0			115
1,2,3-Trimethylbenzene				118				105
1,2,4-Trimethylbenzene (*pseudo*-cumene)				148	+0.9			124
1,3,5-Trimethylbenzene (mesitylene)	>+6.0			171	+6.0			137

Notes: Figures prefixed by a + sign in the above table represent ml TEL/US gal of *iso*-octane to give an equivalent rating.

1.0 ml/US gal = 1.2 ml/UK gal = 0.264 ml/l
3.0 ml/US gal = 3.6 ml/UK gal = 0.793 ml/l

Hydrocarbon	Research octane number (CRC–F₁)				Motor octane number (CRC–F₂)			
	ml TEL/US gal			Blending octane number	ml TEL/US gal			Blending octane number
	0.0	1.0	3.0		0.0	1.0	3.0	
Aromatics (contd.)								
n-Butylbenzene				114	95.3			117
iso-Butylbenzene				122				118
sec-Butylbenzene				116				117
tert-Butylbenzene	>+3.0			138	+0.8			127
1-Methyl-4-*n*-propylbenzene				152				139
1-Methyl-3-*iso*-propylbenzene				154				136
1-Methyl-4-*iso*-propylbenzene (cymene)	+1.4			150	97.7			133
1,3-Diethylbenzene	>+3.0		>+3.0	155	97.0		+0.2	144
1,4-Diethylbenzene				151	96.4			138
1,2,3,5-Tetramethylbenzene				154	+0.2			128
Cycloalkanes								
1,1-Dimethyl*cyclo*propane				116				108
1,1,2-Trimethyl*cyclo*propane	+1.5		>+3.0	133	87.8		93.0	123
1,1-Diethyl*cyclo*propane				114				108
Ethyl*cyclo*butane	41.1			30	63.9			62
*Cyclo*pentane				141	85.0	91.4	95.2	141
Methyl*cyclo*pentane	91.3	99.5	+0.5	107	80.0	89.4	93.0	99
Ethyl*cyclo*pentane	67.2	72.3	79.5	75	61.2	72.7	80.7	67
cis-1,3-Dimethyl*cyclo*pentane	79.2		91.2	98	73.1		86.8	85
trans-1,3-Dimethyl*cyclo*pentane	80.6		93.2	91	72.6		87.1	75
Dimethyl*cyclo*pentane	84.2		95.9	96	76.9		88.7	86
n-Propyl*cyclo*pentane	31.2	43.1	59.8	27	28.1	43.3	60.5	27
iso-Propyl*cyclo*pentane	81.1	89.6	94.3	83	76.2	85.7	89.4	81
n-Butyl*cyclo*pentane	–3		29.6	–4	–2		36.7	5
iso-Butyl*cyclo*pentane	33.4	47.9	59.2	34	28.2	40.6	58.1	37
tert-Butyl*cyclo*pentane				112				112
*Cyclo*hexane	83.0	92.9	97.4	110	77.2	85.4	87.3	97
Methyl*cyclo*hexane	74.8	83.5	88.2	104	71.1	82.0	86.2	84
Ethyl*cyclo*hexane	46.5	54.0	65.1	43	40.8	52.3	65.4	39
cis-1,2-Dimethyl*cyclo*hexane	80.9	89.2	94.3	85	78.6	87.2	90.7	82
trans-1,2-Dimethyl*cyclo*hexane	80.9	89.8	94.5	85	78.7	87.3	90.8	84
cis 1,3 Dimethyl*cyclo*hexane	71.7				71.0			
trans-1,3-Dimethyl*cyclo*hexane	66.9	75.7	83.5	67	64.2	78.3	83.8	65
cis-1,4-Dimethyl*cyclo*hexane	67.2	78.0	84.7	68	68.2	80.0	85.0	66
trans-1,4-Dimethyl*cyclo*hexane	68.3	75.1	82.8	64	62.2	77.4	83.4	59
n-Propyl*cyclo*hexane	17.8	25.6	42.8		14.0	29.4	47.7	
iso-Propyl*cyclo*hexane	62.8	70.1	79.6	62	61.1	74.3	81.4	62
1,1,3-Trimethyl*cyclo*hexane	81.3	89.5	94.8	85	82.6	91.2	95.8	92
n-Butyl*cyclo*hexane			22.5	–8		4.4	25.3	–4
iso-Butyl*cyclo*hexane	33.7	42.4	56.4	39	28.9	40.3	58.3	39
sec-Butyl*cyclo*hexane	51.0	59.5	71.2	44	55.2	64.2	74.6	58
tert-Butyl*cyclo*hexane	98.5	+0.2	+0.8	116	89.2	92.3	96.3	101

Notes: Figures prefixed by a + sign in the above table represent ml TEL/US gal of *iso*-octane to give an equivalent rating.

1.0 ml/US gal = 1.2 ml/UK gal = 0.264 ml/l
3.0 ml/US gal = 3.6 ml/UK gal = 0.793 ml/l

Sources: **API Research Project 45**
Technical Data on Fuel by J.W. Rose and J.R. Cooper, 7th Edition, The British National Committee, World Energy Conference, London, 1977.

Appendix 11

CEC Code of Practice
Cold Weather Performance Test
Procedure for Diesel Vehicles
CEC M-11-T-89

Reprinted with permission of CEC, Coordinating European Council for the development of performance tests for lubricants and engine fuels.

Introduction

This Code of Practice contains test procedures to determine aspects of the performance of a vehicle/fuel combination at low temperature.

Startability: The ability to start the vehicle using normal starting procedures.

Operability: Operability and vehicle driveability under fuel waxing conditions.

Driveability: Vehicle driveability, including assessment of noise and smoke emissions, during the warm-up phase following a cold start. This assessment is conducted at temperatures above that at which waxing will influence vehicle performance.

The procedures may be performed independently or in combination as illustrated below.

NOTE: A procedure for Cold Driveability is still under development. It will be added to this Code of Practice at a later date.

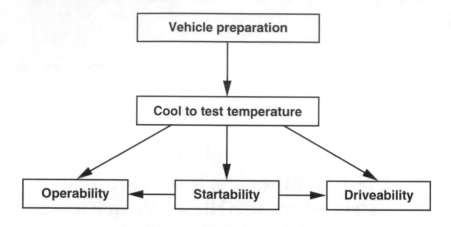

A. Preparation of Test Vehicle

A.1 Measurements Required

The following measurements are required on the test vehicle or engine:

Thermocouple No.	Point of Measurement
1	Cooling liquid (except where air cooled) in the upper hose and before the thermostat.
2	Oil in the sump, for example, at the end of the dipstick or, if practicable, in the drain plug or in the oil gallery.
3	Inlet air at the entry of the engine (under bonnet/hood).
4	Air under the bonnet/hood, in the longitudinal axis of the vehicle at the rear of the engine compartment (optional).
5	Fuel at the inlet of the main filter.
6	Fuel at the outlet of the main filter.
7	Fuel in the fuel tank (representing the bulk temperature).

Thermocouple No.	Point of Measurement
8	Fuel in the recycle line close to the fuel tank.
9	Fuel in the pre-filter (optional).
10	Electrolyte in the battery (optional).
11	Ambient air.

(b) Pressures:

Instrument No.	Point of Measurement
1	Fuel at the inlet of the pre-filter (if fitted).
2	Fuel at main filter inlet.
3	Fuel at main filter outlet.
4	Fuel transfer pressure (rotary pumps) (optional).

(c) For field testing a screened instrument, as used by the meteorological department, to measure outside ambient air temperature.

(d) A revolution counter to measure engine speed (not essential for operability tests).

(e) A recorder to measure the battery voltage of the vehicle during cold starting tests. A paper speed of 25 mm/s is well suited to this measurement.

(f) An instrument to record the equivalent road speed of the vehicle during the test. For tests on a chassis dynamometer, the signal corresponds to the speed of the rollers, or, if not, to the speed of the engine (rpm) for manual transmission vehicles only. For tests on the road or on the track, the signal is obtained by means of a fifth wheel. The engine speed must not be used as an indication of the speed of the vehicle when the latter is equipped with automatic transmission because of the losses in the torque converter. A recorder speed of 120 mm/min is well suited to this type of measurement. Where it is not possible to use such an instrument, the vehicle speed is noted from the speedometer on the dashboard, once this has been calibrated.

A.2 Test Vehicle Preparation and Performance Checks

The following operations will be carried out if possible but may be modified if necessary:

(a) Lubricants:

(i) Engine: Drain the engine oil before the test and refill with the quantity and the grade recommended by the constructor for the minimum test temperature.

(ii) Transmission and auxiliaries: For the parts which will operate during the test, check that the appropriate grades are present and, if not, drain and refill with the quantity and the grade recommended by the constructor for the minimum test temperature.

(b) Filters: Fit new filters for:

(i) Fuel.

(ii) Oil.

(iii) Air.

(c) Cooling system: Check the level and suitability of the antifreeze, for the minimum test temperature. Check the thermostat and the start point of the cooling fan. Check the tension of the drive belts.

(d) Battery: Use a fully charged battery.

Guidance given in paragraphs (i) and (ii) may be of help in ensuring the battery is in good condition. (Record data in Table 2.)

(i) Check the nominal capacity (C25) of the vehicle battery. Charge completely, then discharge to the intensity I = C25/20, note the time t when the voltage is equal to 10.5 V (for a battery of 12 V), calculate the nominal capacity by the product I.t. If the calculated capacity is below the manufacturer's specification, use a new battery and recharge until the voltage does not vary by more than 2% during two hours consecutively.

(ii)   Where a new battery is used, stabilize its capacity by means of three consecutive cycles of discharge and charge made at the intensity I = C25/20. Discharge until the voltage is equal to 10.5 V, and recharge until the voltage does not vary by more than 2% over two consecutive hours.

(iii)   If necessary refer to the battery manufacturer.

(e)   Electrical circuit: Check the conformity of the starter, of the alternator and of the regulator.

(f)   Preheater plugs: Check that all the preheater plugs are working effectively.

(g)   Starting accessories: Check that type, specification and function are correct.

(h)   (i)   The fuel tank of the vehicle is to be used for cold operability testing. Auxiliary fuel tanks may be used, for cold starting or cold driveability tests. In this case, a fuel line which must also be as short as possible, must be joined at the earliest point to the line of the original vehicle. The diameter of the auxiliary fuel line should be equivalent to that of the original vehicle. The auxiliary fuel tank must be placed so as to give a pressure in the feed line of the vehicle, similar to that given by the original tank, when the latter is 50% full.

(ii)   Check for cleanliness and absence of leaks.

(i)   Injection pump: Check and adjust if necessary:

(i)   Initial timing.

(ii)   Advance, if necessary.

(iii)   Hot idle speed.

(iv)   Maximum engine speed no load.

(v)   Check "residual fuel" setting of the fuel pump (as defined in manufacturing service manual).

(j) <u>Injectors:</u> For cold startability tests the injectors should be between 1000 and 30,000 km of service use.

   (i) Check opening pressure before tests and adjust if outside tolerance band.

   (ii) Fit new heat-shield washers.

   (iii) If necessary, injectors should be pre-conditioned after refitting to the engine.

(k) <u>Engine</u>: Should have completed at least 3000 km or have completed a running in procedure indicated by the manufacturer.

For cold startability tests:

   (i) Check the compressions on a hot engine using manufacturer's recommended procedure.

   (ii) Check and adjust the valve clearances.

(l) For cold startability tests or cold operability tests:

   (i) Check that the air circulation around the engine and around the fuel supply system is not restricted by any non-standard equipment.

(m) <u>Tires</u>: Inspect the tires to ensure suitability for test conditions. Check pressures.

(n) <u>Road Test</u>: For vehicles, a road test is recommended to check correct operation of the vehicle and produce data for setting the dynamometer (see Section C).

B. Test Fuel

B.1 Characteristics

The following key characteristics of test fuel should be measured. For cold startability:

(a) Density at 15°C

(b) Distillation

(c) Viscosity at 20°C

(d) Viscosity at 40°C

(e) Sulfur content

(f) Cetane number

For cold operability:

(a) Cloud point

(b) Cold filter plugging point

(c) Pour point

Other tests may be made on the fuel if necessary.

The results of measurements will be noted in the column "Measurement" on Table 3. Under "Observations," include all information concerning the use of additives.

B.2 Storage and Handling of Test Fuels

(a) The test fuels must be kept in closed containers to avoid contamination. Before use, warm up the fuel sufficiently to ensure that there is no wax out of solution and roll or shake the container to ensure homogeneity.

(b) A test fuel may not be reused.

(c) The quantity of fuel which must be in the vehicle is:

   (i) At the discretion of the tester for cold start and driveability tests.

   (ii) For operability tests, equal to 50% of the vehicle fuel tank capacity. For trucks with tank capacity above 200 L, a volume of 100 L of fuel may be used especially if the available quantity of test fuel is restricted.

C. Test Facilities

C.1 Engine Tests for Cold Startability

The cold room must be capable of operating over the temperature range required for the tests. The temperature of the cold room must be maintained within ±1°C of the required temperature.

C.2 Vehicle Tests on Chassis Dynamometer

The chassis dynamometer must be capable of operating over the temperature range required for the tests. The temperature of the chassis dynamometer chamber must be maintained within ±1°C of the required temperature.

When the vehicle is being driven on the chassis dynamometer, the air speed must be equal to vehicle speed or not be different by more than 15 km/h.

Note: The vehicle speed indicator must be calibrated if it is used during the tests.

C.3 Setting of Chassis Dynamometer Load for Cold Operability Tests

To obtain the best correlation with road conditions, the test bench should be adjusted to simulate as closely as possible the vehicle load on a horizontal road over the speed range achieved during the test. Where possible it is advised that the chassis rolls are "motored" for a period prior to commencing testing. This will eliminate cold friction effects on the vehicle loading.

(a) Passenger cars: The dynamometer load should be adjusted to simulate as accurately as possible the road load of the vehicle over the speed range of the test, paying particular attention to matching the road load at the test speed of 110 km/h. Note the time taken to accelerate from a steady speed of 60 km/h to 110 km/h with a warmed-up engine.

Data for setting the dynamometer may be obtained from manufacturer's information or from road tests of vehicle performance, e.g., coast down measurements.

(b) Trucks: The dynamometer should be adjusted at the test speed of 80 km/h to equal 75% of the maximum power available at the wheels of the vehicle at this speed. Dynamometer coefficients should be adjusted to give a realistic ratio between rolling resistance and wind resistance effects.

D. Test Procedure for Cold Startability

D.1 Test Preparation

The following operations must be carried out before each test:

(a) Warm up the vehicle to a temperature above 20°C or otherwise ensure elimination of any traces of wax.

(b) Remove the battery. Store and recharge, ideally at 20°C, according to paragraph A.2(d). In any case this operation should not be carried out below 10°C to avoid serious loss of battery performance. Use an auxiliary battery during the recharge and for completion of the preparation.

(c) Drain engine oil after five tests or at end of each working week.

(d) Drain the fuel tank:

    (i) by the drain plug if fitted.

    (ii) if not, by suction through the pump feed line or if possible by using a flexible pipe through the filler.

(e) Fit a new fuel filter.

(f) Fill the fuel tank with test fuel, as described in B.2(c). Ensure the temperature is at least 5°C above the fuel cloud point.

(g) Disconnect the return line from the fuel tank, start the engine and recover the test fuel returned.

(h) When 1-2 liters of fuel have been recovered, reconnect the return pipe to the fuel tank.

(i) Drive at high load or at high engine speed to ensure correct operation and that all air has been purged from the system.

(j) Check idle speed and stability.

(k) Reconnect original recharged battery.

D.2 Cooldown of the Test Cell

(a) Regulate wind speed as close as possible to 15 km/h during the cooldown.

(b) Reduce the temperature of the test cell to 5°C above the cloud point of the fuel. Maintain the temperature at this level for not less than 2 hrs. Then cool down so that the test temperature is reached after 12 hours. The temperature reduction rate must be linear.

> **Note:** At the start of the controlled cooldown the fuel temperature should be only about 8°C above that of the air. The pre-soak period at 5°C above fuel cloud point is intended to achieve this. Other means to achieve this objective could be accepted, such as pre-cooling of the fuel to the appropriate temperature before charging to the vehicle tank. If fuel is more than 8°C above the air temperature at the start of cooldown this will have to be considered when interpreting the results.

(c) Maintain test temperature for four hours.

D.3 Startability Test

The test method may be applied either to an engine on a stand, or to one installed in a vehicle.

When the cooldown procedure outlined in D.2 has been carried out, proceed as follows:

(a) Use the starting procedure recommended by the vehicle or engine constructor. Use all recommended starting aids (e.g., block heaters at very low temperatures, glo plugs, ether sprays, etc.). If a recommendation does not exist, the following procedure is suggested:

    (i) Place the gear-lever in neutral. (In position "N" for vehicles equipped with automatic transmission.)

    (ii) Operate the starting aid(s).

    (iii) Declutch.

    (iv) Fully depress the acceleration pedal.

(b)   Start a timing device and turn the key to operate the starter motor in sequences of 30 seconds cranking, unless otherwise stated by the manufacturer. If possible record cranking speed as this may explain a poor result if too low.

(c)   In case of non-start, release the key, stop the stopwatch and wait 1 minute between each sequence. (If the vehicle manufacturer advises longer than 30 seconds cranking the waiting period between successive attempts should be extended, e.g., 1 minute cranking, 2 minutes wait, etc.)

(d)   Repeat the above action (a) to (c) until a start has been achieved with a maximum of 3 attempts.

(e)   Keep the starter motor operating after the initial firings until the engine can auto rotate.

Release the accelerator pedal when the engine speed reaches a suitable level, for example:

(i)   2000 rpm for passenger cars.

(ii)   1000 rpm for trucks.

For vehicles equipped with automatic fast idle, release the accelerator completely and note idle speed. For other vehicles, regulate the idle speed at 1000 revs/min, by means of the hand accelerator, if fitted, or by the foot pedal, after noting the idling speed.

If the engine stalls during the first ten seconds of idling speed in auto-rotation, restart the engine, as from paragraph D3(a). If the engine again stalls, terminate the cold start test. Report the result as "Stall at start" and commence a new test as from paragraph D.1.

D.4 Method of Evaluation

The starting performance is evaluated by means of the parameters noted below as measured during the test.

(a)   Time to start:

Defined as being the total time of action on the starter to obtain engine auto-rotation for at least ten seconds without starter.

(b) Time for engine speed rise:

Defined as being the total time of operation of the starter motor and of the time of auto-rotation of the engine just to the point where the engine accelerates, to the speed defined in para. D.3(e), followed by an operation time for at least ten seconds without stalling.

(c) Number of attempts to start.

(d) Number of stalls.

(e) Time of stalls after first auto-rotation.

(f) Idle speed.

D.5 Reporting of Results

Results should be reported in the form shown in the attached Table 4.

E. Operability Test Method

Operability problems during a test, or road driving, are usually due to blockage of the main fuel filter or other parts of the fuel system. Blockage is generally caused by wax, although water may be the cause in some cases on the road. As a blockage develops the pressure drop across the fuel filter increases. The first phenomenon observed is slight fluctuation in speed, "surge," which may become more severe as the test continues. If the pressure drop continues to increase, significant pedal adjustment will be required to maintain the vehicle speed. Eventually the speed of the car will start to reduce, even with the pedal fully depressed, and the vehicle may finally stall.

The operability test method is carried out on a complete vehicle, either on a chassis dynamometer, a level road or track. The detailed procedure is written for chassis dynamometer testing. This should be followed as closely as practical for tests on the road or track. Section E.1 applies to vehicles carrying up to 9 people (including the driver) or with a maximum gross weight of 3.5 tons. All other vehicles are tested according to Section E.2. This is consistent with regulation 70/156/EEC.

E.1 Test Sequence for Passenger Cars

After following the preparation, cooldown and engine starting procedures outlined in D.1-D.3, the following test method should be used:

(a) Switch on all electric accessories at their maximum power (head-lights, lights, de-icing, vent, ...), and keep, or leave, the engine at fast idling for 30 seconds after cranking.

(b) Start a stopwatch, for the beginning of the operability test, and accelerate through the gears so as to reach 60 km/h on the fourth gear within approximately 35 seconds. For each vehicle determine a suitable engine speed for gear change. Thereafter always change at this engine speed in order to ensure consistency from test to test. For vehicles with automatic transmission, allow the transmission to dictate the shifts with the gear selector in "D."

If a stall occurs, restart the engine immediately and perform the acceleration again. Note the occurrence of the stall and record the length of time when the engine was not operating.

(c) Drive at 60 km/h in fourth gear.

(d) At 3 min 35 s, accelerate at full load up to 110 km/h, within approxi-mately 25 sec. Calculation of demerits commences at 4 mins (see Section E.3).

(e) Drive at 110 km/h in the highest gear and maintain, if possible, for 30 mins.

(f) The total test time is 34 mins including the acceleration phase.

(g) Data shall be recorded as described in Section E.3.

Note: If the engine fails to start using the above starting procedure, use all other available means (external power supply, roller driving, ...) to continue the test.

E.2 Test Sequence for Trucks and Agricultural Tractors

After following the preparation cooldown and engine starting procedures in D.1-D.3, the following test method should be used:

(a) Switch on all electric accessories at their maximum power (head-lights, lights, de-icing, vent, ...). Run the engine at 1000 rpm for 2 minutes, or follow manufacturer's specific idling recommendations.

For vehicles fitted with air brake systems the idling period should be extended to a minimum of 6 minutes with a maximum of 9 minutes. The idling time should be held constant for all tests on the same vehicle.

(b) For trucks, start a stopwatch to mark the beginning of the operability test, accelerate at full load through the gears. For each vehicle determine a suitable engine speed for gear change. Thereafter always change at this engine speed in order to ensure consistency from test to test.

If a stall occurs restart the engine immediately and perform the acceleration again. Note the occurrence of the stall and record the length of time when the engine is not operating.

Drive for 30 minutes at 80 km/h, the bench load being set so that the engine is at 75% maximum power.

(c) For agricultural tractors, start a stopwatch for the beginning of the operability test, drive for 5 minutes at 15 km/h followed by 25 minutes at 20 km/h. Choose the gears so that the engine is running at approximately 80% of rated speed.

(d) Data shall be recorded as described in Section E.3.

Note: If the engine fails to start using the above starting procedure, use all other available means (external power supply, roller driving, ...) to continue the test.

E.3 Method of Evaluation

Depending upon the severity of wax plugging in the fuel system, the performance of a diesel vehicle will be affected in different ways. These are, in order of severity:

1) No observable affect upon performance.

2) Slight fluctuation of speed (surge), engine misfire, or the need for significant pedal adjustment to maintain speed.

3) Inability to maintain speed even with pedal fully depressed.

4) Stalling of the engine.

The <u>most severe</u> of these occurrences is used to calculate an overall demerit rating for the tests, using the procedures below. The evaluation procedures are defined in a way which enables the result to be represented on a continuous scale from 0 to 100.

(a)   (i)   Occurrence of surge (i.e., an increase in speed without change of pedal position) or misfire on the level at 110 km/h for passenger cars and at 80 km/h for trucks. Occurrences shall be recorded at the end of each minute during the 30-minute test period. Calculate a demerit rating as shown below:

$$\text{Demerit a(i)} = \left(\begin{array}{c} \text{No. of minutes during} \\ \text{which incidents are recorded} \end{array} \right) \times \frac{10}{30}$$

The maximum demerits possible under this heading are 10.

(ii)   The need for significant pedal adjustment to maintain test speed. Occurrences shall be recorded at the end of each minute during the 30-minute test period. A demerit rating is calculated as follows:

	No. of Minutes	Weighting Factor	Rating
Slight (or nil) pedal adjustment	w	0	-
Significant pedal adjustment	y	3	3 x y
Need to use full throttle	z	6	6 x z

The maximum rating is 180 (when z = 30 minutes). The demerit is then calculated as:

$$\text{Demerit a(ii)} = \text{Rating} - 18$$

so that a maximum demerit of 10 is achievable.

(iii)   Calculate a total demerit for this section as:

$$A = \text{demerit a(i)} + \text{demerit a(ii)}$$

The maximum possible demerit is 20.

(b)   Severity of speed loss and vehicle failure will be assessed on a scale from 20 to 100. (Speed loss will be assessed during the 30-minute period of constant high-speed driving (80 km/h for trucks, 110 km/h for cars) defined in the procedure.) The assessment will be defined as follows:

(i)   The vehicle operates throughout the 30-minute period but may suffer loss of speed at some time during the test.

$$\text{Demerit b(i)} = 20 + \left(\frac{M}{30} \times 40\right)$$

where M = the total number of minutes <u>during</u> which the required speed cannot be maintained.

(ii)   The vehicle suffers a driving stall before the end of the 30-minute test.

$$\text{Demerit b(ii)} = 100 - \left(\frac{N}{30} \times 40\right)$$

where N = the number of minutes operation before occurrence of the driving stall.

(iii)   The demerit for this section is B = demerit b(i) <u>or</u> demerit b(ii), whichever is applicable.

Note: M and N are to be calculated during the 30-min driving period. Idling and acceleration times are excluded.

E.4 Reporting of Results of Operability Procedure

The result shall be reported as demerit A *or* demerit B, whichever is applicable.

Graphical Representation of Cold Operability Results

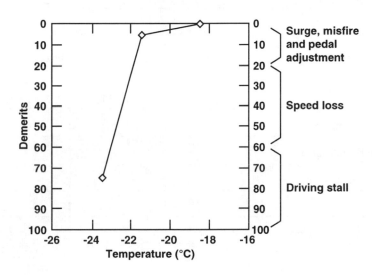

Table 1 Characteristics of Test Vehicle

Make & model _____ Year _____

Registration No._____ Chassis No._____ Laboratory _____

Operator _____ Test No. _____ Date _____

ENGINE	Specified	Measured
Type of engine		
Number of engine		
Position of engine		
Cycle 2/4 stroke		
Number of cylinders		
Capacity (cm³)		
Bore (mm)		
Stroke (mm)		
Type of injection Direct/Indirect		
Type of combustion chamber		
Compression ratio (:1)		
Compression pressure (bar)		
Maximum power (kW)		*
Valves gap Inlet (mm) Exhaust (mm)		
Heating plugs Make/Type		
Post-heating Yes/No		
Other starting aids		
Heating of inlet air Yes/No		
Supercharging system Make/Type		

Table 1 Characteristics of Test Vehicle (Cont)

ENGINE (Cont)		Specified	Measured
Cooling of supercharged air	Yes/No		
High idle speed	(rpm)		
Hot idle speed	(rpm)		
Maximum speed no load	(rpm)		
Antipollution system E.G.R. Others			
Grade of oil	Temperature/Grade Temperature/Grade		
Oil sump capacity + filter	(l)		

FUEL SYSTEM

Screen	Surface (mm²) Gage (mm)		
Prefilter	Make/Type Position		
Filter	Make/Type Position		
Reheater	Make/Type		
Fuel pump	Make/Type		
Injection pump	Make Type Initial timing (c.a.) Initial timing (mm) Extra advance (c.a.)		
Control system of extra advance			
Control system of extra delivery			
Inlet/transfer injection pump pressure (bar)			

Table 1 Characteristics of Test Vehicle (Cont)

FUEL SYSTEM (Cont)			Specified	Measured
Injectors	Make Type Calibration	(bar)		
Fuel tank	Capacity Position	(l)		

COOLING SYSTEM

Level of protection	(°C)		
Drive system of cooling fan			
Thermostat opening temperature	(°C)		

TRANSMISSION

Type gear box				
Manual/Automatic				
Number of gears				
Double ratio	Yes/No			
Front/rear wheel drive				
Speed at 1000 rpm in top gear		(km/h)		*
Grade of oil of gear box	Temperature/Grade Temperature/Grade			
Capacity of gear box		(l)		
Axle oil grade	Temperature/Grade Temperature/Grade			
Capacity of axle(s)		(l)		
Tires	Type/Size Pressure	(bar)		

Table 1 Characteristics of Test Vehicle (Cont)

AUXILIARIES (List of the auxiliaries which consume power)

PERFORMANCES

Maximum speed	(km/h)		
Weight	(kg)		

(*) Optional

OBSERVATIONS: _____

_____

_____

_____

_____

Table 2 Preparation Sheet of Test Vehicle

Make & model _____ Year _____

Registration No. _____ Chassis No. _____ Laboratory _____

Operator _____ Test No. _____ Date _____

Sequence					
S	D	O	*		
			Thermocouples	n°1 Cooling system	
				n°2 Oil in sump	
				n°3 Inlet air	
				n°4 Air under bonnet (optional)	
				n°5 Fuel at filter entry	
				n°6 Fuel at injection pump entry	
				n°7 Fuel in fuel tank	
				n°8 Fuel outlet injection pump (opt.) .	
				n°9 Fuel in pre-filter (if it exists)	
				n°10 Electrolyte in battery	
				n°11 Ambient air	
			Manometers	n°1 Fuel at pre-filter entry/outlet	
				n°2 Fuel at filter entry/outlet	
			Calibration of speedometer (optional)		
			Drainage	Engine	
				Gear box....................................	
				Axle(s)	
				Auxiliary(ies)	

Table 2 Preparation Sheet of Test Vehicle (Cont)

Sequence					
S	D	O	*		
			New filters	Fuel	
				Oil	
				Air	
			Antifreeze mixture	Top level	
				Protection	
			Drive belts		
			Thermostat...		
			Start point of cooling fan...		
			Battery	Check of nominal capacity	
				Charge	
				New battery (eventually)	
			Check of electrical circuit	Starter	
				Alternator	
				Regulator..	
			Preheating plugs	Replacement......................................	
				Operation	
				Connection	
			Starting accessories	Specification	
				Operation	
			Fuel supply circuit	Cleanliness	
				Leak	

Table 2 Preparation Sheet of Test Vehicle (Cont)

Sequence S	D	O	*		
			Injection pump	Initial timing	
				Extra advance....................................	
				Hot idle speed	
				Maximum speed no load....................	
			Injectors	Calibration	
				Replacement heat-shield washer........	
			Engine	Compressions	
				Valve gaps..	
				Check of air circulation....................	
				Check of fuel system.........................	
			Tires..		

* S =Startability Test
  D =Driveability Test
  O =Operability Test

Table 3 Characteristics of Test Fuel

Reference _____

Laboratory _____ **Date** _____

		Method	Specified	Measured
Density at 15°C	(kg/L)	ASTM D 1298		
Distillation	IBP (°C)	ASTM D 86		
	5% (°C)			
	50% (°C)			
	65% (°C)			
	85% (°C)			
	95% (°C)			
	FBP (°C)			
Viscosity at 20°C	mm²/s	ASTM D 445		
Viscosity at 40°C	mm²/s	ASTM D 445		
Sulfur Content	(mass %)	ASTM D 2622		
Cetane Number Measured		ASTM D 613		
Cloud Point	(°C)	ASTM D 2500		
Cold Filter Plugging Point	(°C)	EN 116		
Pour Point	(°C)	ASTM D 97		
Water	(% volume)	ASTM D 1744		
Ash	(% mass)	ASTM D 482		
Sediment	(% mass)			
Total Acidity	mg KOH/g	ASTM D 664		

OBSERVATIONS_____

_____

_____

_____

_____

Table 4 Cold Startability Test

Laboratory: _____ Date: _____ Test No.: _____

Make & Model of Engine/Vehicle: _____ Year: _____

Registration No.: _____ Chassis No.: _____ Engine No.: _____

Fuel: _____

Temperatures												Pressures	
Time	1	2	3	4	5	6	7	8	9	10	11	2	3
0 s													
10 s													

Time to Start (secs)	
Time for Engine Speed Rise (secs)	
Number of Attempts	
Number & Time of Stalls	
Idle Speed (rpm)	

Table 5 Cold Operability Test - Passenger Cars

Laboratory: _____ Date: _____ Test No.: _____

Make & Model of Engine/Vehicle: _____ Year: _____

Registration No.: _____ Chassis No.: _____ Engine No.: _____

Fuel: _____ Test Result: _____ (Demerits)

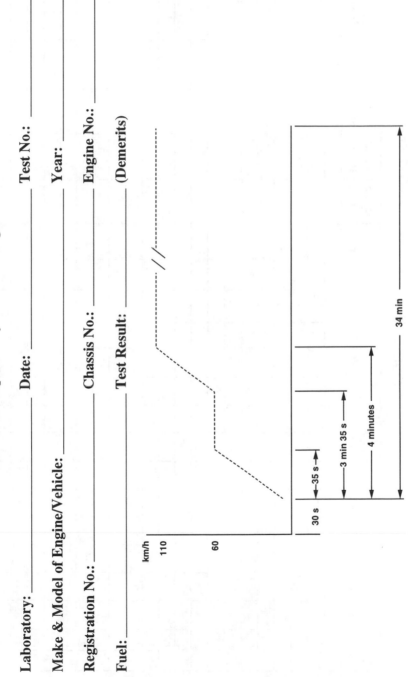

Table 5 (Cont)

Time (mins)	Temperatures											Pressures		rpm	Speed km/h	Gear Postn.	Surge	Misfire	Pedal Adjust
	1	2	3	4	5	6	7	8	9	10	11	1	2						
0																			
1																			
2																			
3																			
4																			
5																			
6																			
7																			
8																			
9																			
10																			
11																			
12																			
13																			
14																			
15																			
16																			
17																			
18																			
19																			
20																			

Table 5 (Cont)

Time (mins)	Temperatures											Pressures		rpm	Speed km/h	Gear Postn.	Surge	Misfire	Pedal Adjust
	1	2	3	4	5	6	7	8	9	10	11	1	2						
21																			
22																			
23																			
24																			
25																			
26																			
27																			
28																			
29																			
30																			
31																			
32																			
33																			
34																			

Table 6 Cold Operability Test - Trucks

Laboratory: _____ Date: _____ Test No.: _____

Make & Model of Engine/Vehicle: _____ Year: _____

Registration No.: _____ Chassis No.: _____ Engine No.: _____

Fuel: _____ Test Result: _____ (Demerits)

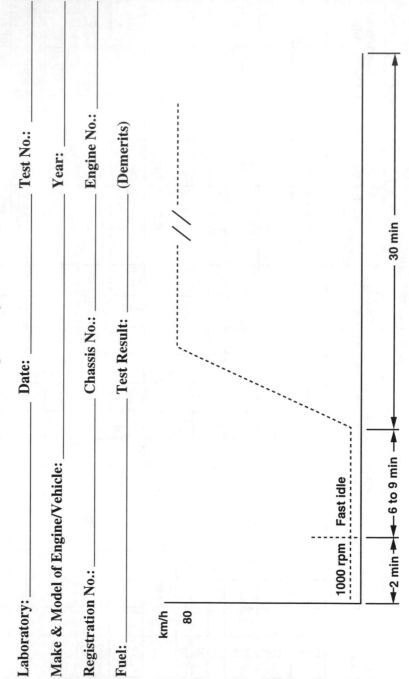

Table 6 (Cont)

Continue idle until air warning off — 1000 rpm

Maximum time → Minimum time

Time (mins)	Temperatures											Pressures			rpm	Surge	Misfire	Pedal Adjust
	1	2	3	4	5	6	7	8	9	10	11	1	2	3				
0																		
1																		
2																		
3																		
4																		
5																		
6																		
7																		
8																		
9																		
10																		
11																		
12																		
13																		
14																		
15																		
16																		
17																		
18																		
19																		
20																		

Table 6 (Cont)

Time (mins)	Temperatures											Pressures			rpm	Speed km/h	Gear Postn.	Surge	Misfire	Pedal Adjust
	1	2	3	4	5	6	7	8	9	10	11	1	2	3						
21																				
22																				
23																				
24																				
25																				
26																				
27																				
28																				
29																				
30																				
31																				
32																				
33																				
34																				
35																				
36																				
37																				
38																				
39																				
40																				
41																				

Appendix 12

Glossary of Terms

Abnormal combustion
Combustion in which knock, preignition, run-on or surface ignition occurs, i.e., combustion which does not proceed in the normal way (where the flame front is initiated by the spark and proceeds throughout the combustion chamber smoothly and without detonation).

Additive
Material added in small amounts to finished petroleum products to improve certain properties or characteristics.

Adsorption
The adhesion by weak forces of materials to the surface of solid bodies in which they are in contact often as a monomolecular film, although the layer can sometimes be two or more molecules thick.

Advance
When used in relation to the timing of a spark for initiating combustion, it is the number of degrees of crankshaft rotation that the spark fires earlier than a fixed or optimum setting. It is often expressed as degrees before TDC (Top Dead Center).

Air-fuel ratio
The proportions, by weight, of air and fuel supplied for combustion.

Alcohols
A group of colorless organic compounds, each of which contains a hydroxyl (OH) group. The simplest alcohol is methanol, CH_3OH.

Alkane
A hydrocarbon having the general formular C_nH_{2n+2}. Also called a paraffin.

Alkylation
A refinery process for producing high-octane components consisting mainly of branch chain paraffins. The process involves combining light olefins with isoparaffins in the presence of a strong acid catalyst such as sulfuric acid or hydrofluoric acid.

Alkyl group
A group of atoms, derived from an alkane (paraffin), having the general formula C_nH_{2n+1}, which forms part of a molecule. Examples are the methyl group (CH_3), the ethyl group (C_2H_5), etc.

Alternative fuel
An alternative to gasoline or diesel fuel which is not produced in a conventional way from crude oil.

Amide
A compound containing the group $CONH_2$. The hydrogen atoms on the nitrogen can also be substituted by other groups.

Amine
There are three types of amine. Primary amines are compounds containing the group NH_2 attached to an alkyl or aryl radical. Secondary amines have the group NH attached to two alkyl and/or aryl groups, and tertiary amines have three alkyl or aryl groups attached to the nitrogen atom.

Aniline Point
The minimum temperature for complete mixing of equal volumes of aniline and the test sample when evaluated by ASTM D 611. Often used to provide an estimate of the aromatic hydrocarbon content of a mixture.

Antiknock additive
An additive which, when added in small amounts to a gasoline, improves the octane quality of the fuel by suppressing knock.

Antiknock Index
The average of the RON and MON for a fuel. Used as a measure of the octane quality of a gasoline, particularly in North America.

API Gravity
An arbitrary scale representing the gravity or density of liquid petroleum products in terms of API degrees, in accordance with the formula:
API Gravity (degrees) = (141.5/Sp Gravity 60/60°F) – 131.5

The higher the API gravity, the lighter the material.

Aromatic
A hydrocarbon based on a six-membered benzenoid ring. See Appendix 1.

Aryl
A hydrocarbon group containing a benzene ring where the benzene ring is directly attached to the rest of the molecule.

Autoignition
The spontaneous ignition of a mixture of air and fuel without an ignition source. The autoignition temperature is the minimum temperature at which this takes place.

Aviation gasoline
Special grades of gasoline produced for aircraft reciprocating engines. The antiknock quality is defined by a lean mixture rating (ASTM D 2700) and a supercharge rating (ASTM D 909). The Vapor Pressure is generally somewhat lower than for motor gasoline and the distillation range can be narrower ideally from about 30°C to about 150°C.

Azeotrope
A mixture of liquids whose distillation characteristics do not conform to Raoult's law, i.e., one that boils at a higher or lower temperature than that of any of its constituents. Azeotropic distillation is used as a method of separating materials having very similar boiling points, and which form suitable azeotropes.

Benzole
A mixture of aromatic hydrocarbons containing a high proportion of benzene, obtained from the distillation of coal tar.

Biocides
Additives used for killing microbiological growths, which often occur in the water interface at the bottom of storage tanks, particularly those containing middle distillates.

Black products
A general term for any refinery stream or product containing residuum.

Black smoke
Smoke and particulates emitted from a diesel engine when under load.

Blending number
A value assigned to a compound or component that will enable it to be blended linearly with other fuel components so that the value of the

finished blend can be predicted. They can refer to a number of different properties such as octane, vapor pressure, etc.

Blowby

Fuel and gases that escape from the combustion chamber, past the pistons, into the crankcase.

Boiling range

The spread of temperature over which a fuel, or other mixture of compounds, distills.

Bosch number

A measure of diesel smoke determined by passing the exhaust gas through a white filter paper. The darkening of the paper is determined using a reflectometer, and Bosch numbers are reported on a scale from 0 (clear) to 10 (black).

Branch chain

A description of a paraffinic hydrocarbon in which the carbon atoms are arranged in a branch form and not in a straight line.

Breakdown time

Also known as the Induction Period. The time, in minutes, for a sample of gasoline when tested for oxidation stability by ASTM D 525, to show a break point, i.e., when a pressure drop of 2 psi occurs over a period of 15 minutes in the bomb containing the sample and oxygen at 100°C.

Brightstock

Heavy lube oil obtained by refining residuum.

Carbonium ion

A positively charged carbon radical in which the charge is due to the loss of one electron from the carbon atom.

Carburetor

The device in some engine fuel systems that mixes fuel with air in the correct proportions and delivers this mixture to the intake manifold.

Carburetor foaming

The formation of a foam in a carburetor caused by the rapid boiling of fuel as it enters a hot carburetor. The foam cannot support the weight of the float bowl so that more and more fuel enters the carburetor causing an increase in pressure and forcing excess fuel out through the vent and/or metering jets, so that an over-rich mixture is obtained.

Carburetor percolation
This occurs when the fuel in a carburetor bowl starts to boil, either during or after a hot soak, forcing excess fuel into the inlet manifold via the vent or metering jet, so that an over-rich mixture results.

Catalyst
A substance that influences the speed and direction of a chemical reaction without itself undergoing any significant change.

Catalytic converter
A device in the exhaust system of an engine containing a catalyst so that reactions can occur which convert undesirable compounds in the exhaust gas into harmless gases.

Catalytic cracking
A refinery process in which heavy hydrocarbon streams are broken down into lighter streams by the use of a catalyst and high temperatures.

Catalytic desulfurization
A refinery process in which sulfur is removed from a hydrocarbon stream by combining it with hydrogen in the presence of a catalyst and then stripping out the hydrogen sulfide thus formed.

Catalytic reforming
A refinery process which converts low-octane quality naphtha to a high-octane blendstock (catalytic reformate) in the presence of a catalyst, mainly by converting naphthenes and paraffins into aromatics. There are many commercially licenced versions of this process.

Caustic soda
Sodium hydroxide (NaOH), a strongly alkaline chemical.

Cetane
A paraffinic hydrocarbon, hexadecane ($C_{16}H_{34}$). The straight chain isomer, n-cetane or n-hexadecane, is a primary reference fuel on which the cetane number scale for measuring the ignition quality of diesel fuels is based. It has a cetane number of 100. The other primary reference fuel is 2,2,4,4,6,8,8 heptamethyl nonane which has a cetane number of 15.

Cetane Index
An approximation of Cetane number based on an empirical relationship with density and volatility parameters such as the mid-boiling point.

Cetane number
A measure of the ignition quality of diesel fuel based on ignition delay in an engine. The higher the cetane number, the shorter the ignition delay and the better the ignition quality.

Chassis dynamometer
Equipment used to measure the power output of a vehicle at the drive wheels.

Climate chamber
A room or chamber, usually containing a chassis dynamometer, in which various climatic conditions can be reproduced. Temperature control is most commonly used but humidity, air pressure, sunshine, and rain can also be reproduced in a repeatable manner.

Cloud Point
The temperature at which a sample of a petroleum product just shows a cloud or haze of wax crystals when it is cooled under standard test conditions, as defined in ASTM D 2500.

CNG
Compressed natural gas.

Coal tar
One of the products from the destructive distillation of coal, the other main products being gas and coke.

Coking
A refinery process that is an extreme form of thermal cracking in which fuel oil is converted to lighter boiling liquids and coke.

Cold Filter Plugging Point (CFPP)
A measure of the ability of a diesel fuel to operate satisfactorily under cold weather conditions. The test measures the highest temperature at which wax separating out of a sample can stop or seriously reduce the flow of fuel through a standard filter under standard test conditions.

Compression ignition
The form of ignition which initiates combustion in a diesel engine. The rapid compression of air within the cylinders generates the heat required to ignite the fuel as it is injected.

Compression ratio
The volume of the cylinder and combustion chamber when the piston is at BDC (Bottom Dead Center) divided by the volume when the piston is at TDC.

Conjugated olefin
An organic non-cyclic hydrocarbon with alternate double and single bonds, e.g., 1,3 butadiene $CH_2=CHCH=CH_2$.

Conversion process
A process which converts heavy products to lighter products.

CFR (Cooperative Fuel Research) engine
A single-cylinder, overhead valve, variable compression ratio engine used for measuring octane or cetane quality.

Corrosion inhibitor
An additive used in a fuel or other liquid that protects metal surfaces from corrosion.

Cracking
A type of refinery process that involves converting large, heavy molecules into lighter, lower boiling point ones.

Crude oil
Naturally occurring hydrocarbon fluid containing small amounts of nitrogen, sulfur, oxygen and other materials. Crude oils from different areas can vary enormously.

Cyclic dispersion
The cycle-to-cycle variations in cylinder pressure that occur when an engine is running under otherwise constant conditions.

Dehydrogenation
A chemical reaction that involves removing hydrogen atoms from alkanes or naphthenes to give olefins or aromatics.

Demerit rating
A numerical rating system in which increasingly high numbers represent increasingly poor performance. It is often used in evaluating vehicle driveability or the cleanliness of engine parts.

Demulsifier
An additive used for breaking oil in water emulsions.

Density
Mass of a substance per unit volume.

Delay period
See ignition delay.

Detergent
A fuel detergent is an oil-soluble surfactant additive that maintains the cleanliness of engine parts by solubilizing deposits or materials likely to deposit in the engine fuel system.

Detonation
Often used to describe the uncontrolled explosion of the last portion of a fuel in the combustion chamber. See also Knock.

Diesel Index
An obsolescent measure of ignition quality in a diesel engine, defined as:
$$\text{Diesel Index} = 0.01(\text{Aniline point})(\text{API gravity})$$

Diesel knock
An abnormal form of combustion that occurs in a compression ignition engine and which is associated with long ignition delays. Often occurs when the engine is cold.

Di-iso-propyl ether (DIPE)
An oxygenated compound having the formula $C_3H_7\text{-}O\text{-}C_3H_7$ suitable for use as a high octane gasoline blend component.

Direct injection (DI)
A type of diesel engine in which the fuel is injected directly into the cylinder.

Dispersant
A surfactant additive designed to hold particulate matter dispersed in a liquid.

Distillation
The general process of vaporizing liquids in a closed vessel, condensing the vapors, and collecting the condensed liquids. Since liquids vaporize generally in order of their boiling points, it provides a method of separating materials according to their volatility.

DON

Distribution Octane Number — a measure of the way octane quality is distributed across the boiling range of a gasoline, as measured using a modified CFR engine.

Drag coefficient

A measure of the air resistance of a vehicle as it is being driven.

Drag reducing agent

A fuel additive which can reduce resistance to flow so that the capacity of a pipeline is increased.

Driveability

The response of a vehicle to the throttle. Good driveability requires such characteristics as smoothness of idle, ease of starting when hot or cold, smoothness during acceleration without hesitations or stumbles, and absence of surging at constant throttle when cruising. Separate tests are used for hot weather and cold weather driveability, in which numerical assessments of performance are assigned to each type of malfunction.

Elastomer

Synthetic rubber-type materials frequently used in vehicle fuel systems.

Emulsification

The formation of a dispersion of one liquid in another where the liquids are not miscible with each other, such as oil in water. Emulsifying agents, which are surfactant materials, will stabilize such emulsions and prevent them from separating into two layers.

Ethers

A class of organic compounds containing an oxygen atom linked to two groups which can be alkyl and/or aryl.

Ethyl tertiary butyl ether (ETBE)

An oxygenated compound having the formula $C_2H_5\text{-}O\text{-}C_4H_9$ suitable for use as a high octane gasoline blend compound.

Evaporative loss controls

Devices used on vehicles, service station pumps, tanks, etc., to prevent losses of light hydrocarbons by evaporation. Such controls on vehicles usually involve the use of a canister filled with activated charcoal into which vents from the fuel tank, carburetor, etc., are fed so that the vapors are adsorbed onto the charcoal. The canister is regenerated by drawing intake air through the canister while the engine is running so that the hydrocarbons are desorbed and burned in the engine.

Exhaust gas recirculation (EGR)

The recycling of some exhaust gas back into the inlet manifold so as to lower combustion temperature and, hence, reduce nitrogen-oxide emissions.

FIA hydrocarbon analysis

A fluorescent indicator adsorption test procedure defined by ASTM Procedure D1319 for the determination of hydrocarbon types in terms of aromatics, olefins and saturates. The sample is adsorbed on silica gel containing a mixture of fluorescent dyes and is then desorbed down an activated silica gel column. The hydrocarbons are separated by types and their positions indicated under ultraviolet light.

Flammability limits

Mixtures of air and petroleum vapors will only burn or explode within a certain range of concentrations. The lean limit (or lower explosive limit) is where the mixture has just enough hydrocarbon to burn and the rich limit (or upper explosive limit) is where it is almost too rich to burn.

Flash Point

The lowest temperature at which vapors from a petroleum product will ignite on application of a small flame under standard test conditions.

Fluidizer

A high-boiling-point, thermally stable organic liquid used as an additive in gasoline to reduce deposits on the undersides of intake valves.

Fractionation

The separation by boiling point of mixtures of compounds by a distillation process. The degree of separation, i.e., the fractionation efficiency, will depend on the design of the fractionating tower.

Free radical

A radical is a group of atoms such as the methyl group (CH_3) that is part of a larger molecule. Such a group will not normally exist on its own since it has a free electron, but when it does it is called a free radical and it will rapidly react with other materials such as oxygen, sometimes forming further free radicals.

Freezing point

The temperature, determined under standard conditions, at which crystals of hydrocarbons formed on cooling disappear when the temperature of the fuel is allowed to rise.

Fuel injector
A device for injecting fuel into a piston engine. They are used in all diesel engines and some gasoline engines where they replace the carburetor.

Fuel oil
A term usually applied to a heavy residual fuel although it can also be applied to heavy distillates.

Fungibility
The ability to interchange or mix products from different sources or of different compositions without interactions occurring.

Gasohol
A blend of 90% gasoline and 10% ethanol used as an automotive fuel in some states in the U.S.

Glow plug
A plug-type electrical heater used as a cold starting aid in indirect injection diesel engines.

Gum
The oxidation product arising from the storage of automotive fuel such as gasoline. It is barely soluble in gasoline and so may separate out and form a sludge. In an engine it will form sticky deposits which can cause malfunctions.

Hartridge unit
A measurement of the black smoke emitted from a diesel engine in which the opacity of the smoke is determined.

Heat of combustion
Also called the thermal, calorific or heating value and refers to the heat liberated when a fuel is burned. The Upper or Gross heating value includes the latent heat of water from combustion which is condensed in the test procedure. In an engine the water is exhausted in the vapor form and so a correction is made to give the net or lower heating value by subtracting the latent heat of condensation of any water produced.

Heat of vaporization
Also called Latent Heat of Vaporization. The heat associated with the change of phase from liquid to vapor at constant temperature.

Heptane
An alkane or paraffin having seven carbon atoms with the formula C_7H_{16}. Normal heptane, in which the carbons are arranged in a straight chain, is a primary reference fuel with Research and Motor octane values of zero. There are 9 isomers of heptane altogether.

Hydrocarbon
A compound made up of hydrogen and carbon only.

Hydrocracking
A refinery process in which heavy streams are cracked in the presence of hydrogen to yield high-quality middle distillates and gasoline streams.

Hydrodesulfurization
A refinery process in which sulfur is removed from petroleum streams by treating it with hydrogen to form hydrogen sulfide which can be removed from the oil as a gas by stripping.

Hydrofining
A proprietary name for one version of the hydrodesulfurization process.

Hydrogenation
Treatment of a stream with hydrogen, usually to remove sulfur or to stabilize it by saturating double bonds.

Hydrophilic group
An organic group which has an affinity for water, such as the hydroxyl group (·OH) or an acid group.

Hydrophobic group
The opposite of Hydrophilic. Most hydrocarbon groups come into this category.

Hydroskimming refinery
A simple refinery consisting only of process units for distilling, catalytically reforming and hydrotreating.

Ignition delay
The period between the start of injection and the ignition of a fuel in a diesel engine.

Indirect injection (IDI)
A type of diesel engine in which the fuel is sprayed into a prechamber to initiate combustion, rather than directly into the cylinder, as in a Direct Injection engine.

Induction period
See breakdown time.

Injectors
See fuel injectors.

Intake system icing
The formation of ice in the carburetor or parts of some injector systems of spark ignition engines that can cause vehicle malfunctions such as stalling and loss of power. It occurs only during cool humid weather when there is enough moisture in the intake air to condense and then freeze in the carburetor due to the temperature drop caused by the evaporation of the gasoline. If the ambient temperature is too low (below about 0°C) there is not enough water in the atmosphere to give icing, and if the temperature is too high (above about 15°C) then the temperature depression is not enough to freeze the water.

Isomers
Compounds which have the same composition in terms of the elements present but which have the individual atoms arranged in different ways. Thus there are two isomers of butane (C_4H_{10}):

$$CH_3\text{-}CH_2\text{-}CH_2\text{-}CH_3 \qquad \text{and} \qquad CH_3\text{-}\underset{\underset{\displaystyle CH_3}{|}}{CH}\text{-}CH_3$$

n-butane   isobutane

Isomerization
A refinery process which converts normal or straight chain hydrocarbons that have a poor octane quality into high-octane branch chain isomers. Thus n-butane is converted into isobutane, etc.

Isooctane
The hydrocarbon 2,2,4 trimethylpentane which has 8 carbon atoms and is used as a primary reference fuel with assigned values of RON and MON of 100.

Jerk Pump
A term used to describe the cam-operated plunger-type diesel fuel injector pump.

Kerosene (or Kerosine)
A refined petroleum distillate of which different grades are used as lamp oil, as heating oil, and as fuel for aviation turbine engines.

Knock

In a spark ignition engine it is the autoignition (sometimes called detonation) of the end gas in the combustion chamber and causes the characteristic knocking or pinging sound. It can cause damage to the engine and can be overcome by increasing the octane quality of the fuel or by engine modifications. In diesel engines it is caused by excessive pressures in the combustion chamber and is avoided by the use of higher cetane number fuels.

Knock sensor

A detector, usually fixed to the cylinder head, that detects when knock is occurring in a spark ignition engine and actuates a mechanism such as one which retards the ignition to overcome the knock.

Latent heat

The heat associated with a change of phase such as going from a solid to a liquid at constant temperature (latent heat of fusion) or from a liquid to a gas at constant temperature (latent heat of vaporization or, simply, heat of vaporization).

Lead alkyl

A class of lead compounds, most commonly with methyl and/or ethyl groups attached to the lead atom, which are used as antiknock compounds in gasoline. See also TEL and TML.

Lead antagonism

Some compounds, and particularly those containing sulfur, are antagonistic to lead in that when they are present in a gasoline they reduce the antiknock benefit given by lead alkyls. The sulfur compounds themselves vary, according to their chemical composition, in the degree to which they are antagonistic.

Lead response

The extent to which the octane quality of a stream or component is improved by the addition of lead alkyls.

Lean mixture

An air-fuel mixture which has an excess of air over the amount required to completely combust all the fuel.

Low Temperature Flow Test (LTFT)

A test to predict the low-temperature performance of a diesel fuel; gives a better correlation with field performance for U.S. vehicles and fuels than the Cold Filter Plugging Point (CFPP) test used in Europe and elsewhere.

LPG
Liquefied Petroleum Gas which consists mainly of propane and/or butane, and which can be stored as a liquid under relatively low pressure for use as a fuel.

Lubricity
The ability of a fuel (or oil or grease) to lubricate. It has particular relevance to low viscosity, low sulfur and low aromatic diesel fuels, made to meet stringent exhaust emissions legislation.

Markers
Chemicals that are added to fuel and can be detected by a color reaction with another chemical. Used to detect theft, contamination, tax evasion, etc.

Mercaptans
Compounds also known as thiols, having the group -SH. They have an extremely unpleasant odor and are removed from automotive fuel components to avoid customer complaints. Very low concentrations are often added to LPG to give it a distinctive warning odor.

Merit rating
A numerical scale in which high numbers represent good performance. See also Demerit rating.

Merox treating
A proprietary refinery process for removing mercaptans from petroleum streams.

Metal deactivator
A fuel additive that deactivates the catalytic oxidizing action of dissolved metals, notably copper, on hydrocarbons during storage.

Methanol
Methyl alcohol, CH_3OH, the simplest of the alcohols. It has been used, together with some of the higher alcohols, as a high-octane gasoline component and is a useful automotive fuel in its own right.

Methyl tertiary butyl ether (MTBE)
An oxygenated compound having the formula $CH_3 \cdot O \cdot C_4H_9$, used widely as a high-octane component of gasoline.

Misfire
Failure to ignite the air-fuel mixture in one or more cylinders without stalling the engine.

MMT

Methylcyclopentadienyl Manganese Tricarbonyl, an antiknock additive sometimes used in conjunction with lead and sometimes used on its own in unleaded gasolines. Also known as Hitech 3000.

Motor Octane number

A measure of the antiknock quality of a fuel as determined by the ASTM D 2700 method. It is a guide to the antiknock performance of a fuel under relatively severe driving conditions as can occur under full throttle, i.e., when the inlet mixture temperature and the engine speed are both relatively high.

Multifunctional additive

An additive or blend of additives having more than one function.

Naphtha

Loosely defined term covering a range of light petroleum distillates, used as chemical and reformer feedstocks, gasoline blend components, solvents, etc.

Naphthenes

A group of hydrocarbons having a cyclic structure with the general formula C_nH_{2n}. Examples are cyclohexane and cyclopentane.

Natural gas

A naturally occurring gas, consisting mainly of methane.

Naturally aspirated engine

An engine in which the intake air entering the system is at atmospheric pressure.

Neutralization Number

An indication of the acidity or alkalinity of an oil or fuel.

Nozzle coking

Deposit formation in the nozzle of a metering-type fuel injector.

Nucleation (of wax crystals)

A function of wax crystal modifier additives, that provides nuclei onto which wax molecules attach themselves when they come out of solution from a diesel fuel as it cools below its cloud point.

Nucleator (of wax crystals)
Component of a cold flow improver additive which creates nuclei onto which wax molecules attach themselves as they come out of solution, when a middle distillate fuel is cooled to below its cloud point.

Octane number
A measure of the antiknock performance of a gasoline or gasoline component; the higher the octane number, the greater the fuel's resistance to knock. There are two main types of octane number, the Research Octane Number (RON) and the Motor Octane Number (MON), which are based on different engine operating conditions and, therefore, relate to different types of driving mode. Both are based on the knocking tendencies of pure hydrocarbons; n-heptane has an assigned value of zero and isooctane a value of 100. The octane number of a fuel is the percentage of isooctane in a blend with n-heptane that gives the same knock intensity as the fuel under test when evaluated under standard conditions in a standard engine.

Above a level of 100, the octane rating is based on the number of milliliters of tetraethyl lead per gallon which is added to isooctane to give the same knock intensity as the fuel under test.

Octane requirement
The octane number of a reference fuel (which can be a primary reference fuel or a full boiling range fuel) that gives a trace knock level in an engine on a test bed or a vehicle on the road when being driven under specified conditions.

Octane Requirement Increase (ORI)
The increase in octane requirement that occurs in an engine over the first several thousand miles of its life, due to buildup of carbonaceous and other deposits in the combustion chamber. It is influenced by driving mode, gasoline composition and the presence of lead and other additives.

Oil dilution
The dilution of the lubricating oil by gasoline or partially combusted gasoline, which can find its way past the piston rings into the crankcase, particularly during cold starting. Because it lowers the viscosity of the lubricant, it increases wear. However, the lighter portions of the gasoline are evaporated off as the oil heats up but any remaining material from the gasoline may reduce the effectiveness of the lubricant.

Olefin
An unsaturated hydrocarbon, that is, one containing one or more double or triple bonds.

Oleophilic group
A chemical group attached to a molecule having an affinity for oily materials.

Oleophobic group
The opposite of oleophilic.

Operability limit
The lowest ambient temperature at which a diesel fuel will just function satisfactorily without wax separation causing filter plugging problems.

Otto cycle
The 4-stroke cycle of most piston engines, i.e., intake, compression, power and exhaust.

Oxidation
Loosely it is the chemical combination of oxygen to a molecule, although strictly it has a much broader meaning. It can be part of a manufacturing process or it can represent the deterioration of organic materials such as gasoline or diesel fuel due to the slow combination of oxygen from the air.

Oxygenate, Oxygenated compound
Terms which have come to mean compounds of hydrogen, carbon and oxygen that can be added to gasoline to boost octane quality or to extend the volume of fuel available. Examples are methanol, ethanol and methyl tertiary butyl ether (MTBE).

Paraffin
A hydrocarbon having the general formula C_nH_{2n+2}. Also called an Alkane.

Particulates
Particles, as opposed to gases, emitted from the exhaust systems of vehicles.

Pintle
A type of fuel injector nozzle in which a shaped extension at the end of the injector needle controls the initial rate of fuel injection through the orifice of the nozzle. When the needle is fully lifted, there is full flow.

Pipestill
The primary distillation equipment used in a refinery in which the crude oil is heated in a furnace and passed into a fractionating tower, where it is split into different boiling range fractions.

Polymerization

The combination of two or more molecules of the same type to form a single molecule having the same elements in the same proportion as in the original molecule, but with a higher molecular weight. The product is a polymer. The product of two or more dissimilar molecules is known as copolymerization.

It is also a refinery process in which propenes and butenes from cracking processes are combined to form heavier olefins having a boiling range that makes it suitable for use as a gasoline blend component (polymer).

Port fuel injection

Fuel injectors which inject into the intake port rather than directly into the cylinders. They can be electronically or mechanically operated.

Pour point

The lowest temperature at which a petroleum product will just flow when tested under standard conditions, as defined in ASTM D 97.

Preflame reaction

The chemical reactions which take place in an air-fuel mixture in an engine prior to ignition.

Preignition

The premature ignition of the fuel-air mixture in a combustion chamber, i.e., before the spark from the plug. It can be caused by glowing deposits, hot surfaces or the autoignition of the fuel itself.

Primary Reference Fuel (PRF)

For use in spark ignition engines, it is a blend of n-heptane and isooctane used as a primary standard for knock evaluations. The octane value (both RON and MON) of a primary reference fuel is the percentage of isooctane in the blend with n-heptane.

For compression ignition engines, primary reference fuels are used to define the cetane quality of a fuel and are usually n-cetane (cetane number of 100) and heptamethyl nonane (cetane number of 15).

Pyrolysis gasoline (pygas)

Terms frequently used to describe naphtha produced by steam cracking and used as a gasoline component or a gasoline.

Quench

The removal of heat during combustion from the end gas or outside layers of the air-fuel mixture by the cooler walls of the combustion chamber.

R100°C

The Research Octane number of the part of a gasoline distilling up to 100°C using a standard distillation apparatus (ASTM D 86). The difference between the RON of the whole fuel and the R100°C is the ΔR100°C.

Ra

The universally recognized and most-used international parameter of roughness. It is the arithmetic mean of departures of the roughness profile from the mean line.

Reflux ratio

A term used in connection with distillation and refers to the ratio of condensed sidestream returned to a distillation column to the amount taken off as a sidestream.

Reforming

Sometimes used as a short form of Catalytic Reforming which is a process for making high-octane components from naphtha. Also, it is a mild thermal cracking process for naphtha.

Reformulated Gasoline

A gasoline blended to minimize undesirable exhaust and evaporative emissions.

Refutas Chart

A temperature-viscosity chart devised to show a linear relationship for Newtonian fluids.

Reid Vapor Pressure (RVP)

A measure of the vapor pressure of a liquid as measured by the ASTM D 323 procedure; usually applied to gasoline or gasoline components.

Repeatability

The maximum difference between duplicate test results carried out on the same sample by the same operator using the same test equipment, above which the test is considered suspect.

Reproducibility

The maximum difference between test results carried out on the same sample by different operators in different laboratories, above which the test is considered suspect.

Research Octane number (RON)

A measure of the antiknock quality of a gasoline as determined by the ASTM D 2699 method. It is a guide to the antiknock performance of a fuel when vehicles are operated under mild conditions such as at low speeds and low loads.

Residue, Residuum
The non-volatile portion of a crude oil resulting from distillation.

Response
The way a fuel responds to treatment with additives such as lead alkyls, cold flow improvers, cetane improvers, etc., which are used to improve particular properties of the fuel.

Reynolds Number
Proportional to the ratio of inertial force to viscous force in a flow system. The **critical** Reynolds number corresponds to the transition from turbulent flow to laminar flow as the velocity is reduced.

Rich mixture
A fuel-air mixture that has more fuel than the stoichiometric ratio.

Road octane number
Usually the octane number of a Primary Reference Fuel (PRF) that just gives trace knock in a vehicle on the road or chassis dynamometer when tested under specified conditions.

Run-on
A condition in which a spark ignition engine continues to run after the ignition has been switched off. Also known as "after-running" or "diesel-ing."

Saturated compound
A paraffinic hydrocarbon (alkane), i.e., a hydrocarbon with only single bonds and no double or triple bonds.

Scavenger
Term applied to the halogen compounds (usually dibromoethane and/or dichloroethane) present in a lead antiknock compound to prevent lead compounds such as lead oxides and sulfates from building up in the combustion chamber.

Sensitivity
The difference between the Research Octane number and the Motor Octane number of a gasoline. It is a measure of the sensitivity of the fuel to changes in the severity of operation of the engine.

Shale oil

A largely hydrocarbon mixture derived from oil shale (a naturally occurring deposit) by distillation. It is a potential future alternative to crude oil.

SHED

An acronym for "Sealed Housing for Evaporative Determination," a sealed chamber in which a vehicle is placed in order to determine the amount of evaporative losses that occur from its fuel system.

Silicone

Organic compounds containing silicon, often used as antifoaming agents.

Solvent oil

An alternative term for fluidizer.

Spark knock

The most common form of knock, so-called because it is influenced by the spark timing.

Squish

The squeezing of part of the air-fuel mixture out of the end gas region in certain types of cylinder head as the piston reaches the end of the compression stroke. It promotes turbulence and further mixing of the air and fuel and minimizes the tendency to knock.

Steam cracking

A petrochemical process for the production of ethylene in which naphtha is cracked in the presence of steam.

Stoichiometric air-fuel ratio

The exact air-fuel ratio required to completely combust a fuel to water and carbon dioxide.

Stoke

A unit of kinematic viscosity (the quotient of the dynamic viscosity and the density). One stoke is 1 cm^2/s. A more convenient unit is the centistoke where 1 cSt = 0.01 St or 1 mm^2/s.

Storage stability

The ability of a fuel to resist deterioration on storage due to oxidation.

Straight chain

A descriptive term applied to a hydrocarbon in which all the carbon atoms are arranged consecutively in a straight line.

Supercharger
A mechanical device that pressurizes the intake air or air-fuel mixture to an engine and so increases the amount delivered to the cylinders and, hence, the power output.

Surface ignition
The ignition of the air-fuel mixture in a combustion chamber by a hot surface rather than by the spark.

Surfactant additive
When applied to fuels it is an organic compound with oleophobic and oleophilic groups that will form a coating on metal and other surfaces with the oleophilic group sticking into the hydrocarbon and the oleophobic group attaching itself to the surface. In this way it can protect surfaces and act as a detergent or dispersant by partially solubilizing deposits.

Susceptibility
The extent to which the octane quality of a gasoline stream is improved by the addition of lead alkyls.

Sweetening process
A refinery process for converting mercaptans (thiols) into nonodorous compounds.

Synthesis gas
A mixture of carbon monoxide and hydrogen obtained from coke or natural gas by partial oxidation or steam reforming.

TEL
See tetraethyl lead

Terne plate
Steel sheet coated with a lead/tin alloy and often used for the construction of vehicle fuel tanks.

Tertiary amyl methyl ether (TAME)
An oxygenate used as a gasoline blend component.

Tetraethyl lead (TEL)
A volatile lead compound, $Pb(C_2H_5)_4$, widely used as an antiknock additive in gasoline.

Tetramethyl Lead (TML)
A volatile lead compound, $Pb(CH_3)_4$, widely used as an antiknock additive in gasoline and having a higher volatility than TEL. It is particularly useful for improving the octane quality of the front end of a gasoline.

Thermal cracking
A refinery process for converting heavy streams into lighter ones by heat treatment.

Thermostat
A device used for the automatic regulation of temperature.

Toluene
A relatively volatile aromatic compound, $CH_3 \cdot C_6H_5$, present in catalytic reformate and widely used as a solvent. It has excellent octane qualities and can be used as a gasoline blend component.

Turbocharger
A device, driven by engine exhaust gas pressure, for pressurizing the intake air or air-fuel charge of an engine, so as to increase the mixture delivered to the cylinders and, hence, increase power output.

Unsaturated compounds
Hydrocarbons having one or more double or triple bonds. Such compounds are reactive and will combine with other elements such as oxygen or hydrogen. Olefins are unsaturates.

Valve seat recesssion
The wearing away of a valve seat in the cylinder head of an engine. Not usually a problem when lead is present in gasoline because the lead combustion products act as a lubricant. Engines designed for unleaded gasoline have hardened valve seats.

Vapor-liquid ratio
The ratio, at a specified temperature and pressure, of the volume of vapor in equilibrium with liquid to the volume of liquid charged, at a temperature of 0°C. It can be measured using test procedure ASTM D 2533 and is used to define the tendency for a gasoline to vaporize in the fuel system of a vehicle.

Vapor lock
A gasoline supply failure to the engine of a vehicle due to vaporization of the fuel preventing the pump from delivering an adequate supply of fuel. Factors favoring vapor lock are high ambient temperatures, low ambient pressure, volatile gasoline and vehicle designs where heat from the engine can give high fuel line temperatures.

Vapor Lock Index (VLI)
An index that combines the Reid Vapor Pressure and the Percentage Evaporated at 70°C of a gasoline, and is a measure of the likelihood of a gasoline to cause vapor lock in vehicles on the road.

Vapor pressure

The pressure exerted by the vapors derived from a liquid at a given temperature and pressure. The Reid Vapor Pressure, as determined using the test procedure ASTM D 323, is used to define the vapor pressure of a gasoline.

Venturi

In a carburetor it is the narrowing of the air passageway. This increases the velocity of the air moving through it and induces a vacuum which is responsible for the discharge of fuel through the jets into the air stream.

Visbreaking

A refinery process for thermally cracking residual fuel oil, originally to reduce its viscosity but now to produce cracked streams.

Viscosity

A measure of the resistance to flow of a liquid.

Volatility

The property of a liquid that defines its evaporation characteristics. Highly volatile liquids boil at low temperatures and evaporate rapidly.

Wankel engine

A rotary engine in which a three-lobe rotor containing combustion chambers turns eccentrically in an oval chamber.

Wax

High-molecular-weight, generally straight chain paraffins having limited solubility in diesel fuel.

Wax antisettling additive (WASA)

An additive that reduces the tendency for wax crystals to settle out on storage of diesel fuel.

Wax antisettling flow improver (WAFI)

An additive that improves the cold flow characteristics of a diesel fuel and also reduces the tendency for wax crystals to settle out during storage.

Wax appearance point

A measure of the likelihood of wax coming out of solution as temperature is reduced. It is the temperature at which separated wax just becomes visible when tested under standard conditions, as defined by ASTM D 3117, and gives a similar result to cloud point.

Wax Precipitation Index (WPI)
A function of Cloud Point (CP) and the spread between cloud point and pour point (CP-PP), used to correlate with the average estimated minimum operating temperature of vehicles in a CRC field test.

White product
Products which do not contain any residue from distillation.

White smoke
The smoke emitted during a cold start from a diesel engine, consisting largely of unburnt fuel and particulate matter.

Wide cut
A fraction from a distillation column which has a wide boiling range. Wide cut fuels are of interest for both diesel and spark ignition engines, particularly for military use, when a normal fuel might not be available.

Wobbe Index
The Wobbe Index of a gas is proportional to the heating value of the quantity of gas that will flow subsonically through an orifice in response to a given pressure drop. It is calculated from the formula:

$$W = H/d^{1/2}$$

where W is the Wobbe Index, H is the volumetric heating value of the gas, and d is the specific gravity.

Xylene
An aromatic hydrocarbon, $(CH_3)_2C_6H_4$, present in catalytic reformate and widely used as a solvent.

Abbreviations

AAMA	American Automobile Manufacturers Association
ACEA	Association des Constructeurs Européen d'Automobiles (formerly CCMC)
ACORC	Australian Cooperative Octane Requirement Committee
AECD	Auxiliary Emission Control Device
A/FR	Air/fuel ratio
API	American Petroleum Institute
AQIRP	US Auto/Oil Air Quality Improvement Research Program
ASTM	American Society for Testing and Materials
ATDC	After Top Dead Center
BDC	Bottom Dead Center
BS	British Standards

BS&W	Bottoms Sediment and Water
BSI	British Standards Institution
BTC	British Technical Council
BTDC	Before Top Dead Center
BThU	British Thermal Unit
CAFE	Corporate Average Fuel Economy
CARB	California Air Resources Board
CEFIC	Conseil European des Federations de l'Industrie Chimique (European Chemical Industry Council)
CBI	Cloud Point Blending Index
CCMC	Committee of Common Market Constructors of Automobiles (now ACEA)
CEC	Coordinating European Council (for the Development of Performance Tests for Lubricants and Engine Fuels)
CEN	Comité Européen de Normalisation (European Standards Committee)
CFPP	Cold Filter Plugging Point
CFR	Cooperative Fuel Research (Committee)
CI	Compression ignition or Cetane Index
CNG	Compressed Natural Gas
CO	Carbon Monoxide
CO_2	Carbon Dioxide
CONCAWE	The oil companies' European organization for environment, health and safety
CORC	Cooperative Octane Requirement Committee (Europe)
CP	Cloud Point
CPD	Cloud Point Depressant
CRC	Coordinating Research Council (U.S.)
CR-50	Anti-knock containing a mixture of methyl and ethyl lead compounds
CU	Conductivity Unit
CVS	Constant Volume Sampling System (FTP)
Delta R (ΔR)	Difference between the RON of a gasoline and the RON of the part boiling up to 100°C
DI	Direct Injection
DIN	Deutsches Institut fur Normung (German Standards Institute)
DIPI	Deposit Induced Preignition
DON	Distribution Octane Number
E70	Percent Evaporated at 70°C
E100	Percent Evaporated at 100°C
EC	European Community (now called EU)
ECE	Economic Commission for Europe (UNO)
EEC	European Economic Community (the original six member states—1958 Rome Treaty)
EFEG	Engine Fuels Emissions Group

EGR	Exhaust Gas Recirculation
EN	European Norme (Standard)
EP	End Point
EPA	Environmental Protection Agency (U.S.)
ETBE	Ethyl tertiary butyl ether
EtOH	Ethanol, Ethyl alcohol
EU	European Union (formerly EC)
EUDC	Extra-Urban Driving Cycle
EUROPIA	European Petroleum Industries' Association
FBP	Final Boiling Point
FEVI	Front-end Volatility Index
FFV	Flexible Fueled Vehicle
FI	Fuel Injection
FIA	Fluorescent Indicator Adsorption
FIE	Fuel Injection Equipment (f.i.e.)
FISITA	Federation Internationale des Societes d'Ingenieurs des Techniques de L'Automobile
FTP	Federal Test Procedure
FVLI	Flexi-Vapor Lock Index
GC	Gas Chromatography
GLC	Gas/liquid Chromatography
GTBA	Gasoline grade tertiary butyl alcohol
HC	Hydrocarbons
HSU	Hartridge Smoke Units
HTA	Hydrogenated Tallow Amine
HUCR	Highest Useful Compression Ratio
HWFET	Highway Fuel Economy Test (part U.S. FTP)
IBP	Initial Boiling Point
IDI	Indirect Injection
IFP	Institut Française du Petrole
IGT	Institute of Gas Technology (U.S.)
ILEV	Inherently Low Emission Vehicle (EPA definition)
IMechE	Institution of Mechanical Engineers (UK)
IP	Institute of Petroleum (UK)
IPA	Isopropyl alcohol
ISO	International Organization for Standardization
JPI	Japanese Petroleum Institute
KLSA	Knock Limited Spark Advance
LEV	Low Emission Vehicle (CARB emission standard)
LNG	Liquefied Natural Gas
LPG	Liquefied Petroleum Gas
LTFT	Low Temperature Flow Test
MBN	Motor Octane Blending Number
MDFI	Middle Distillate Flow Improver
MeOH	Methanol, Methyl alcohol
MMT	Methylcyclopentadienyl Manganese Tricarbonyl

Mn	Manganese
MON	Motor Octane number
MTBE	Methyl tertiary butyl ether
MVMA	Motor Vehicle Manufacturers' Association (U.S. now AAMA)
NACE	National Association of Corrosion Engineers
NGV	Natural Gas Vehicle
NO_x	Oxides of Nitrogen
NMA	N-methyl aniline
NMHC	Non-Methane Hydrocarbons
NMOG	Non-Methane Organic Gases
NPAH	Nitrated Polycyclic Aromatic Hydrocarbons
NPRA	National Petroleum Refiners Association
OEM	Original Equipment Manufacturer
OFP	Ozone Forming Potential
OTA	Office of Technology Assessment (U.S.)
ORI	Octane Requirement Increase
PAH	Polycyclic Aromatic Hydrocarbon
PNA	Polynuclear Aromatics
POM	Polycyclic Organic Matter
Pb	Lead
PCV	Positive Crankcase Ventilation
PFI	Port Fuel Injector
PPD	Pour Point Depressant
PPM	Parts per million (usually by weight) - sometimes ppm
PRF	Primary Reference Fuel
PTB	Pounds per thousand barrels
RAF	Reactivity Adjustment Factor
R100	RON of the part of a gasoline boiling up to 100°C
RBN	Research Blending Number
RON	Research Octane Number
RSI	Runaway Surface Ignition
RUFIT	Rational Utilization of Fuels in Transport
RVP	Reid Vapor Pressure
SAE	Society of Automotive Engineers (U.S.)
SCF	Standard Cubic Feet (of gas at normal temperature and pressure)
SE	Specific Energy
SHED	Sealed Housing for Evaporative Determination
SI	Système Internationale (international system of metric units)
T10	Temperature (°C) at which 10% of a gasoline distills
T50	Temperature (°C) at which 50% of a gasoline distills
TAME	Tertiary amyl methyl ether
TBA	Tertiary butyl alcohol
TBI	Throttle Body Injector

TCF	Trillion Cubic Feet
TDC	Top dead center
TEL	Tetraethyl lead
TLEV	Transitional Low Emission Vehicle (CARB Emission Standard)
TML	Tetramethyl lead
UDDS	Urban Dynamometer Driving Schedule
ULEV	Ultra Low Emission Vehicle (CARB Emission Standard)
UTAC	French Transport Ministry Technical Advisory Committee
VBN	Viscosity Blending Number
VI	Viscosity Index
V/L	Vapor/Liquid Ratio
VLI	Vapor Lock Index
VOC	Volatile Organic Compounds
VOLFE	Volumetric Fuel Economy
WAFI	Wax Antisettling Flow Improver
WAP	Wax Appearance Point
WASA	Wax Anti-Settling Additive
WCM	Wax Crystal Modifier
WOT	Wide Open Throttle
WPI	Wax Precipitation Index
ZEV	Zero Emissions Vehicle

Index

A

A

A

A

B

B

B

C

C

C

C

C

C

D

D

D

D

D

D

D

D

D

E

E

E

E

E

E

F

F

F

G

G

G

G

G

G

G

G

Gasoline manufacture (continued)
    product demand pattern
        crude prices, influence of, 30-32, 529-531
        refinery design, influence on, 32-34
        yield *vs.* market demand, middle distillate, 387
    refineries
        complex (conversion) refineries, 32-34
        heat integration in, 32
        hydroskimming (simple) refineries, 30-32
        petrochemical processing in, 32
        yield *vs.* refinery schemes, middle distillate, 388, 394
    steam cracking, 45-46
    sweetening (copper chloride, inhibitor), 50-51
    thermal cracking
        chemistry of, 38-39
        delayed coking, 40
        description and disadvantages of, 40-41
        flexicoking, 42
        fluid coking, 41-42
        middle distillate/gasoline yields from, 388, 394
        thermal *vs.* catalytic naphthas, 39
        visbreaking, 41
    Ultraforming (Amoco), 47
    see also Future trends; Reformulated gasoline
Gasoline properties
    appearance, 247-248
    composition, major limits on
        benzene, 248, 771-772
        hydrocarbons, 248-249
        lead (U.S.), 806. *see also* Lead
        manganese, 249
        mercaptans, 251
        oxygenates. *See* Oxygenated blend components
        sulfur. *See* Sulfur
        water/particulates. *See under* Contamination, fuel
        see also specific countries; European Community/European Union (EC/EU);
            Legislative fuels; Reference fuels
    conductivity, electrical
        discussion of, 76, 252-253
        tests for (ASTM D 3114, 4308), 253
    corrosivity
        discussion of, 254
        of copper: copper strip corrosivity test (ASTM), 254
        corrosion inhibitors. *See under* Gasoline additives
        in fuel tanks, 85-86
        of gasoline/oxygenate blends, 285-286
        reference fuels for, 820
        of steel: ASTM, NACE tests, 254

G

G

G

H

H

H

I

I

J

J

K

L

L

L

M

M

N

N

N

O

O

O

O

O

P

P

P

Powerforming (Exxon), 47
Preignition. *See under* Gasoline combustion
Product differentiation, 538-539
Puerto Rico
    lead content/ULG availability, 760

Q

Quality, fuel. *See* Fuel quality

R

Racing fuels
    antiknock additives for
        lead compounds, 323-324
        methanol and ethanol, 324-325
        MMT, toxicity of, 324
        MTBE, TAME, 325, 326
        nitrogen compounds, 326
        water, 325-326
    ethanol
        as antiknock additive, 324
        properties of, 323
    gasoline blends
        avgas as a blend component, 321-322
        benzene, unacceptability of, 320
        gasoline, limitations of, 320
        Hector Fuel, 320
        isooctane, triptane, cyclohexane in, 320
        methane/nitromethane, prohibition against, 322
        pure hydrocarbons, octane qualities of (tabulation), 320
        RON/MON requirements (typical racing engines), 320, 322
        TEL, unavailability of, 321, 324
    hydrazine, 328
    hydrocarbons
        octane qualities of (tabulation), 320
    methanol
        discussion of, 322
        as antiknock additive, 324
        as blend component, advantages/disadvantages of, 325
        blends with nitromethane, 322, 328
        methanol/nitro-methane, prohibition against (Formula I), 321
        properties of (tabulation), 323
    MMT, toxicity of, 324
    nitromethane
        combustion reactions of, 327

R

R

S

S

S

S

S

S

S

S

S

T

T

U

U

U

V

V

W

W

W

Y

Z